AF774732

Agronomy and Field Crops

Agronomy and Field Crops

Editor: Jamie Hanks

Callisto Reference
www.callistoreference.com

Callisto Reference,
118-35 Queens Blvd., Suite 400,
Forest Hills, NY 11375, USA

Visit us on the World Wide Web at:
www.callistoreference.com

© Callisto Reference, 2018

This book contains information obtained from authentic and highly regarded sources. Copyright for all individual chapters remain with the respective authors as indicated. All chapters are published with permission under the Creative Commons Attribution License or equivalent. A wide variety of references are listed. Permission and sources are indicated; for detailed attributions, please refer to the permissions page and list of contributors. Reasonable efforts have been made to publish reliable data and information, but the authors, editors and publisher cannot assume any responsibility for the validity of all materials or the consequences of their use.

ISBN: 978-1-63239-911-3 (Hardback)

The publisher's policy is to use permanent paper from mills that operate a sustainable forestry policy. Furthermore, the publisher ensures that the text paper and cover boards used have met acceptable environmental accreditation standards.

Trademark Notice: Registered trademark of products or corporate names are used only for explanation and identification without intent to infringe.

Printed in China.

Cataloging-in-Publication Data

Agronomy and field crops / edited by Jamie Hanks.
p. cm.
Includes bibliographical references and index.
ISBN 978-1-63239-911-3
1. Agronomy. 2. Field crops. 3. Agriculture. 4. Crops. I. Hanks, Jamie.
SB91 .A37 2018
630--dc23

Table of Contents

Preface

In my initial years as a student, I used to run to the library at every possible instance to grab a book and learn something new. Books were my primary source of knowledge and I would not have come such a long way without all that I learnt from them. Thus, when I was approached to edit this book; I became understandably nostalgic. It was an absolute honor to be considered worthy of guiding the current generation as well as those to come. I put all my knowledge and hard work into making this book most beneficial for its readers.

Agronomy is the application of science and technology for the use of plants for food, land reclamation, fiber etc. It overlaps with subjects such as plant biotechnology, soil science and plant physiology. This book contains some path-breaking studies in the field of agronomy. It is a vital tool for all researching or studying agronomy and field crops, as it gives incredible insights into emerging trends and concepts. The extensive content of this book provides the readers with a thorough understanding of the subject.

I wish to thank my publisher for supporting me at every step. I would also like to thank all the authors who have contributed their researches in this book. I hope this book will be a valuable contribution to the progress of the field.

Editor

1

New Insights into How Increases in Fertility Improve the Growth of Rice at the Seedling Stage in Red Soil Regions of Subtropical China

Yilin Li[1], Weiming Shi[1], Xingxiang Wang[1,2]*

1 State Key Laboratory of Soil and Sustainable Agriculture, Institute of Soil Science, Chinese Academy of Sciences, Nanjing, China, 2 Jiangxi Key Laboratory of Ecological Research of Red Soil, Yingtan, Jiangxi, China

Abstract

The differences in rhizosphere nitrification activities between high- and low- fertility soils appear to be related to differences in dissolved oxygen concentrations in the soil, implying a relationship to differences in the radial oxygen loss (ROL) of rice roots in these soils. A miniaturised Clark-type oxygen microelectrode system was used to determine rice root ROL and the rhizosphere oxygen profile, and rhizosphere nitrification activity was studied using a short-term nitrification activity assay. Rice planting significantly altered the oxygen cycling in the water-soil system due to rice root ROL. Although the oxygen content in control high-fertility soil (without rice plants) was lower than that in control low-fertility soil, high rice root ROL significantly improved the rhizosphere oxygen concentration in the high-fertility soil. High soil fertility improved the rice root growth and root porosity as well as rice root ROL, resulting in enhanced rhizosphere nitrification. High fertility also increased the content of nitrification-induced nitrate in the rhizosphere, resulting in enhanced ammonium uptake and assimilation in the rice. Although high ammonium pools in the high-fertility soil increased rhizosphere nitrification, rice root ROL might also contribute to rhizosphere nitrification improvement. This study provides new insights into the reasons that an increase in soil fertility may enhance the growth of rice. Our results suggest that an amendment of the fertiliser used in nutrient- and nitrification-poor paddy soils in the red soil regions of China may significantly promote rice growth and rice N nutrition.

Editor: Wagner L. Araujo, Universidade Federal de Vicosa, Brazil

Funding: This research was financially supported by the National Natural Science Foundation of China (No. 31471948), the Special Fund for Agro-scientific Research in the Public Interest (No. 201203050-3) and the GanPo 555 Talents Program of Jiangxi Province, China. The funders had no role in study design, data collection and analysis, decision to publish, or preparation of the manuscript.

Competing Interests: The authors have declared that no competing interests exist.

* Email: xxwang@issas.ac.cn

Introduction

Nitrification, an important component of the nitrogen cycle, involves two aerobic microbial processes. In nitrification, ammonia (NH_3) is first converted to nitrite (NO_2^-); then, NO_2^- is further converted to nitrate (NO_3^-) [1,2]. Nitrification occurs in a variety of environments, such as sediments and soils. However, in waterlogged environments, such as paddy soils and flooding waters, nitrification is limited by the low availability of oxygen, which results in low NO_3^- availability. An increased number of hydroponic experiments have demonstrated that rice absorbs and assimilates NO_3^- in preference to ammonium (NH_4^+) [3,4]. Rice growth, yield, net N acquisition and N use efficiency (NUE) have been shown to be superior using mixed NH_4^+ and NO_3^- nutrition compared with pure NH_4^+ cultivation [3,5,6]. In paddy fields, NO_3^- concentrations often fluctuate and are difficult to control compared with hydroponic culturing, primarily because of denitrification in the bulk soil and nitrification in the rhizosphere. In our previous soil culture study [7], different rice cultivars (Indica and Japonica) showed a significant difference in the nitrification activity in their rhizospheres, and rhizosphere nitrification was closely related to rice growth and N nutrition as well as rice N accumulation and NUE.

The oxidation of NH_3 is a microbial process and is thought to be the rate-limiting step for nitrification [8]. Recently, according to metagenomic [9,10] and cultivation [11] methods, another type of NH_3 oxidiser, the ammonia-oxidising archaea (AOA) group, has been found to be abundant in the natural environment [12] and has been hypothesised to be a dominant group among the ammonia-oxidising prokaryotes in the soil [13]. Because ammonia-oxidising bacteria (AOB) and AOA are both strictly aerobic chemolithoautotrophic microorganisms, nitrification can only occur in the presence of oxygen and is restricted in the oxic upper few millimetres of flooded soils [14–16].

However, rice plants form numerous adventitious roots containing aerenchyma. These roots promote radial oxygen loss (ROL) from the roots to the rhizosphere, supporting aerobic microbial nitrification [17,18]. Furthermore, ROL alleviates the phytotoxicity of reduced compounds (e.g., Fe^{2+}, Mn^{2+}, H_2S) in rice [19]. ROL is of biological significance for submerged plants and has been investigated extensively using various methods [20–23]. A Clark-type oxygen microelectrode has been used to measure the rice ROL and the detailed oxygen vertical profiles in the field, indicating that rice plants at 21 d had 20% air saturation at the root surface. Diffusion of O_2 away from the root was able to create oxic conditions up to a distance of 0.4 mm away from the root [23]. The diameter of this oxygen microelectrode was only 25 μm, which should have caused minimal damage to the soil and rice plants.

Two paddy soils (representing low and high fertilities) in red soil regions with low nitrification activity were examined by oxygen microelectrode techniques to study the relationship between rice ROL and rhizosphere nitrification and the contribution of root ROL to rice N nutrition. Red soils, which can be classified as acrisols and ferralsols according to the Food and Agriculture Organization (FAO) nomenclature [24], are widespread in tropical and subtropical China, covering 11.8% of the nation's land area. The red soil region contains abundant water resources and heat resources and is one of the principal rice production areas in China [25]. Due to intensive leaching and weathering, long-term inappropriate utilisation and management, and the inherent fragility of hill ecosystems, the red soils are characterised by low pH and fertility, as well as low nitrification activity [26,27]. When natural factors, such as parent materials, topography, and climate, are the same in the paddy soil-forming process, time and human factors, such as long-term fertilisation, are the primary influencing factors in improving soil fertility. Two paddy soils collected from paddy fields with different fertility levels were defined as low fertility (i.e., reclamation of approximately 15 years, long-term inorganic and organic fertilisation in the rice season) and high fertility (i.e., reclamation great than 50 years, long-term inorganic and organic fertilisation in the rice season and manure application in the fallow winter). Although increasing soil fertility may improve rice root growth, little is known about whether it could change the rice root aerenchyma development and ROL. Our previous study showed that oxygen concentrations in paddy soil (without rice plants) with high fertility were much lower than those measured in low-fertility paddy soil collected from red soil regions (unpublished data). In this study, we hypothesised that rice plants grown in high-fertility paddy soil would improve the root aerenchyma development and ROL, which would lead to increased oxygen in the rhizosphere soil, and, thus, enhance rhizosphere nitrification activity.

Materials and Methods

Paddy soils and rice plant cultivars

Paddy soils derived from Quaternary red clay (hydragric anthrosols according to the FAO nomenclature [24]) were collected from the Chinese Academy of Sciences Red Soil Ecological Experimental Station, located in Yingtan (28°15′20″ N, 116°55′30″ E), Jiangxi Province, which is one of the major rice production areas in China. The cultivation practice in the area is two crops per year. The Yingtan region has a typical subtropical monsoon climate with an annual precipitation of 1785 mm and a mean annual temperature of 17.8°C. Two paddy soils (0–20 cm) collected from paddy fields with different fertility levels were defined as low fertility and high fertility, and their properties are described in Table 1. The soil samples were air-dried, ground, sieved through 0.71-mm mesh, and stored for further incubation experiments. The plant residues were carefully removed by hand. One of the most popular Indica rice cultivars in Jiangxi Province, Hesheng 10, was used in the experiment.

Soil incubation and sampling

A rhizobox, as described by Li *et al.* [28], was used to collect soil samples at various distances from the rice root surface. Two pieces of nylon nets (with a 30-μm mesh size) were used to divide the rhizobox into three compartments to facilitate the isolation of the rice root from the rhizosphere soil. Three rice plants were planted in the middle compartment of each rhizobox. The mesh of the nylon net was fine enough to allow the water and nutrient elements to pass through. A 600-g sample of the paddy soil (air dried and sieved) was mixed thoroughly with urea (120 mg N kg^{-1}) and KH_2PO_4 (93 mg kg^{-1}) to fill each rhizobox (the fertiliser applications were performed according to the conventional amounts used by the local farmers to support rice growth adequately). A rhizobox without rice plants was used as the control (CK). One centimetre of surface water was maintained after rice germination, using deionised water every morning and evening throughout the incubation. The rice plants were incubated at 25°C with a 16-h photoperiod during the experiment. Three replicate rhizoboxes for each treatment were used, and soil samples were collected at 40, 50 and 60 d after rice germination; an additional six replicate rhizoboxes were used for in situ measurements of soil oxygen concentration and rice root ROL 50 d after germination.

A previous experiment in this laboratory showed that between 40 and 60 days after rice germination, the rice roots were fully developed and filled the middle compartment of the rhizobox so that the nylon nets would not influence the rice root growth. Therefore, the soil adhering to the rice roots in the inner rhizobox was defined as root surface soil, and the soil in the outer rhizobox within a 4-mm distance from the nylon nets was defined as rhizosphere soil. To prevent the water layer from interfering with the soil mineral N and nitrification activity determination, no watering occurred on the evening of the day prior to sampling to maintain a thin (1–2 mm) water layer. The sampling was conducted at 10:00 am, and each treatment (three replicates) was first frozen at −20°C for 2 h to harden the paddy soil for subsequent slicing. First, the rhizobox was carefully split along the two nylon nets into three compartments: middle, right, and left parts. Next, a piece of organic glass board (2 mm thick, 8 cm×8 cm) was inserted into the inner side of the right or left part of the rhizobox. Then, a soil sample of 2-mm thickness was pushed aside, and the soil was sliced to obtain a soil sample located 2 mm away from the root surface. Next, another piece of organic glass board (2 mm thick, 8 cm×8 cm) was inserted to obtain a layer located 4 mm away from the root surface. Using the same procedure, soil samples were obtained at distances of 6, 8, 10, 20, 30 and 40 mm from the root surface. The soil samples collected in the right and left compartments at the same distance were mixed together for the assessments. The root surface soil was separated from the rice roots using sterilised tweezers. All soil samples were collected at the same time as the plant sampling to assay for mineral N and nitrification activity.

Mineral N assays

Fresh soil samples were extracted with 2 mol L^{-1} of KCl (soil:solution ratio: 1:10), and the extracts were measured colourimetrically for NH_4^+ and NO_3^- [26] using a continuous flow autoanalyser (model Autoanalyzer 3, Bran + Luebbe, Hamburg, Germany).

Short-term nitrification activity assays

Short-term nitrification estimates are typically used for the nitrification activity assay. The principle underlying this method is based on the determination of NO_2^- after the incubation of soil

Table 1. Properties of the paddy soils used in the experiment.

Property	Low fertility	High fertility
pH (water:soil, 2.5:1)	4.75	5.02
Organic matter (g kg^{-1})	17.5	39.1
Total N (g kg^{-1})	0.85	1.74
Total P (g kg^{-1})	0.32	0.67
NH_4^+-N (mg kg^{-1})	0.72	2.27
NO_3^--N (mg kg^{-1})	0.17	0.21
Clay (%)	36.3	20.9
Average yield of double-cropping rice per year (kg hm^{-1})	7500	10000
Reclamation years	15	>50

samples with $NaClO_3$ in the absence of NH_4^+ for 24 h at 25°C [29,30]. Briefly, each soil sample (5 g) was shaken with 2.5 mL of $NaClO_3$ (75 mmol L^{-1}, which prevents oxidation of NO_2^-) at 170 rpm on a rotary shaker (model HZ-9611K, Hualida Laboratorial Equipment Co., Ltd., Taicang, China) for 30 min, and then incubated aerobically in an incubator for 24 h at 25°C. After incubation, NO_2^- was extracted from the soil samples into a total volume of 15 mL using two solutions (5 mL of deionised H_2O followed by 10 mL of 2 mol L^{-1} KCl) by shaking at 170-rpm for 30 min on a rotary shaker, as described previously. The contents were mixed thoroughly and immediately filtered. Five millilitres of the clear filtrate was pipetted into glass test tubes, followed by 3 mL of the buffer (0.19 mol L^{-1} NH_4Cl, pH 8.5) and 2 mL of the reagent (1% sulphanilamide, 0.05% N-(1-naphthyl) ethylene-diamine dihydrochloride and 10% phosphoric acid mixed solution), for NO_2^- determination. The contents were again vigorously shaken and allowed to stand for 15 min at room temperature. For a control measurement, soil samples were extracted as previously described after incubation with $NaClO_3$ at −20°C. The amount of NO_2^- produced was measured based on the absorbance at 520 nm in a spectrophotometer. Pre-experiments using the paddy soil without planted rice (waterlogged for 40 d) to observe the kinetics of ammonia oxidation in the first 24 hours were performed and revealed that nitrite was produced linearly over time ($p<0.05$).

In situ measurements of soil oxygen concentrations

An additional three replicate rhizoboxes were used for the in situ measurement of soil oxygen concentrations 50 d after germination. The incubation conditions were the same as described above except that a 1-cm water layer was maintained in the rhizoboxes for the in situ measurement.

The oxygen microelectrode was a miniaturised Clark-type oxygen electrode with a guard cathode (OXY25, ø = 20–30 µm, Unisense, Aarhus, Denmark [31]). The soil oxygen profile measurements were performed in the right and left compartments of each rhizobox and in the centre of the range at a distance of 2 mm from the nylon nets for all treatments. All measurements were performed with a 50-µm depth interval using a microelectrode with a diameter of approximately 25 µm, and the periods for "wait before measure" and "measure" were both set as 3 s.

In vitro measurements of rice root oxygen loss in an agar microcosm

Three rhizosphere boxes were split into three compartments each, and the rice plants with soil were placed into a plastic bucket filled with deionised water. The soil adhering to the rice roots was removed by repeated immersion in the deionised water. After the soil was removed, the rice roots were separated according to the individual plants and were then individually placed into beakers with International Rice Research Institute (IRRI) rice nutrient solution (pH 5.5) [32] for 2 h. The roots of each individual rice plant were then inserted into a container filled with IRRI rice nutrient solution (pH 5.5) containing 0.9% agar [23] before the agar was allowed to solidify at 38°C. The container was combined with one small culture dish (6 cm in diameter, 2.5 cm in depth) and one large culture dish (15 cm in diameter, 2.5 cm in depth). The rice root (to be measured) was inserted into the small culture dish, whereas the other roots were evenly distributed in the large culture dish. Next, paraffin oil was poured onto the surface of the solidified agar in the small culture dish to prevent air penetration. To obtain comparable results, the sample root was placed horizontally in the agar. Because the solidified agar was a clear medium, it allowed observation of the rice roots and provided fixation support for the roots. To measure the oxygen levels of the outer and inner rice roots, the microelectrode had to be sufficiently thin to longitudinally penetrate the root. Therefore, a much slimmer microelectrode (OXY10, ø = 8–10 µm, Unisense, Aarhus, Denmark [31]) was used for the measurement. Newly formed roots with lengths of approximately 45 mm were selected and measured at 25 mm from the root tip. The measurements were conducted after the rice roots had been fixed in the agar for approximately 2 h; these measurements were repeated for different rice plants with different newly formed roots with the same length as described above. All measurements were conducted with a 50-µm depth interval, and the periods for "wait before measure" and "measure" were both set to 3 s.

Because oxygen concentration is very sensitive to temperature, soil and agar temperatures were monitored and recorded during the oxygen measurements. All of the oxygen measurements, including soil oxygen concentration and root ROL, were performed in a specialised laboratory without electrical noise interference at a temperature of 20°C.

Plant dry matter and N accumulations assays

The rice tissues were separated into shoots and roots and heated at 105°C for 30 min to terminate enzyme activity. The samples were then oven-dried at 70°C for 48 h to a constant weight. The dried plant material was ground and digested for total N determination using the Kjeldahl method. A 5-mL aliquot from the 100-mL digested solution was then analysed for N using a continuous flow auto-analyser (model Autoanalyser 3, Bran + Luebbe, Hamburg, Germany).

Plant NRA and GSA assays

Fresh leaves and roots of both treatments were immediately frozen with liquid N and then stored in a refrigerator at −70°C for the later measurement of glutamine synthetase activity (GSA) and NO_3^- reductase activity (NRA).

The level of GSA was assayed using an in vitro method. Frozen plant material (0.5 g) was ground in a precooled mortar with sand and a pestle after adding liquid N and then homogenised in an extraction buffer (pH 7.2, 10 mL g^{-1} fresh weight) containing 0.5 mmol L^{-1} of EDTA and 50 mmol L^{-1} of K_2SO_4. The homogenates were centrifuged at 20,000 *g* for 20 min. Next, 1.2 mL of the clear filtrate was pipetted into a centrifuge tube, to which was then added 0.6 mL of imidazole-HCl (pH 7.0, 0.25 mol L^{-1}), 0.4 mL of sodium glutamate (pH 7.0, 0.3 mol L^{-1}), 0.4 mL of ATP-Na (pH 7.0, 15 mmol L^{-1}), 0.2 mL of $MgSO_4$ (0.5 mol L^{-1}) and 0.2 mL of hydroxylamine (1 mol L^{-1}). After the mixture was incubated at 25°C for 20 min, the reaction was terminated by adding 0.8 mL of acidic $FeCl_3$ [24% (W/V) trichloroacetic acid and 10% (W/V) $FeCl_3$ in 18% HCl]. The amount of γ-glutamyl hydroxamate was measured with a spectrophotometer at 540 nm. One unit of GS activity was defined as the amount of enzyme that catalysed the formation of 1 μmol γ-glutamyl hydroxamate per min at 25°C [33].

The level of NRA was assayed by another in vitro method. Frozen plant material (0.5 g) was ground in a precooled mortar with sand and a pestle after adding liquid N and then homogenised in an extraction buffer (10 mL g^{-1} fresh weight) containing 5 mmol L^{-1} of EDTA, 5 mmol L^{-1} of cysteine and 25 mmol L^{-1} of potassium phosphate buffer (pH 8.7). The homogenates were centrifuged at 20,000 *g* for 20 min. Next, 0.4 mL of the clear filtrate was pipetted into a centrifuge tube, and the filtrate was added to 1.0 mL of KNO_3 (0.1 mol L^{-1}) and 0.6 mL of NADH (2 mg mL^{-1}). After the mixture was incubated at 25°C for 30 min, the excess NADH was oxidised by the addition of 0.5 mL of sulphanilamide (1%). The mixture was centrifuged at 20,000 *g* for 15 min. The amount of NO_2^- produced was measured after the addition of N-(1-naphtyl) ethylene-diamine dihydrochloride (0.02%) using an absorbance of 540 nm in a spectrophotometer [34].

Plant POR assay

For this assay, 0.4–0.6 g of fresh adventitious roots was sampled according to published methods [13,35,36]. POR was determined as , where POR is the root porosity (%), P_{gr} is the mass of the pycnometer with water and ground roots (g), P_r is the mass of the pycnometer with water and roots (g), r is the mass of roots (g), and P is the mass of the pycnometer with water (g).

Data analysis

All statistical analyses were performed using SPSS version 13.0. A one-way ANOVA with a homogeneity of variance test was used, followed by a least significant difference (LSD) test to check for quantitative differences between treatments. The statistically significance level was set at $p<0.05$.

Results

Distribution of soil nitrification activity

The nitrification activities of the root surface and rhizosphere soils increased over time, whereas those in the bulk soil did not change significantly (Fig. 1A, B and C). The maximal nitrification activities were found 2 mm away from the root surface in the rhizosphere soil, and their values were 2.76 (0.33) μg kg^{-1} h^{-1} for the low-fertility treatment and 4.32 (0.92) μg kg^{-1} h^{-1} for the high-fertility treatment for the last sampling period (Fig. 1C). Minimal nitrification activities appeared in the root surface and bulk soils (20–40 mm away from the root surface). Throughout the sampling period, the nitrification activities in the paddy soil with high fertility were always significantly higher than those in the soil with low fertility, except for the root surface soil 40 d after germination (Fig. 1A). Compared with the control, rice planting generally increased the rhizosphere soil nitrification activity significantly, but no increased effect of rice planting on nitrification occurring in the root surface or bulk soil was observed in either treatment. For example, the nitrification activity 2 mm away from the root surface was 1.8, 2.8 and 5.2 times higher than that in the control in the low-fertility soils at 40, 50 and 60 d after germination, respectively, and the corresponding activity in the high-fertility soil was 1.7, 2.9 and 3.9 times higher than the control, respectively (Fig. 1A, B, C and D).

Distribution of soil nitrate concentration

The NO_3^- concentration did not change over time. Although there was a strong absorption of NO_3^- by the rice roots, the horizontal distribution of soil NO_3^- at different distances from the root surface showed no concentration gradient (Fig. 2A, B and C). This result might have been due to the strong mobility of NO_3^- in the soil. The NO_3^- concentrations in the high-fertility soil were always higher than those in the low-fertility soil. This difference which might have contributed to the higher nitrification activity of the high-fertility soil described above (Fig. 1A, B, and C). The NO_3^- concentration in the soil planted with rice did not differ significantly from that in the control (Fig. 2A, B, C and D).

Distribution of the soil ammonium concentration

The NH_4^+ concentration in the soil decreased over time (Fig. 3), suggesting that a large quantity of NH_4^+ was absorbed by the rice roots. This conclusion is partially supported by the distinct depletion of NH_4^+ in the zone near the root surface (Fig. 3A, B, and C), where the NH_4^+ concentration increased with increasing distance from the rice roots. In the period 40–60 d after germination, the NH_4^+ concentrations of the high-fertility treatment in the bulk soil were always higher than those of the low-fertility treatment, except for the measurement taken 40 d after germination (Fig. 3A). This difference increased over time (Fig. 3A, B, and C). As the NH_4^+ concentrations in the control increased slightly with time, the NH_4^+ concentrations in the control for the high-fertility treatment were significantly higher than those in the control for the low-fertility treatment (Fig. 3D). Due to the NH_4^+ uptake by rice roots, the content of NH_4^+ in the treatments with rice was significantly lower than that in the control (without rice), especially on the last sampling day. For example, the NH_4^+ concentrations 40 mm away from the rice root (bulk soil) were 19.9% and 26.9% of those in the control at 60 d after germination in the low- and high-fertility soils, respectively, and those in the root surface soil were only 9.75% and 8.86% of the control.

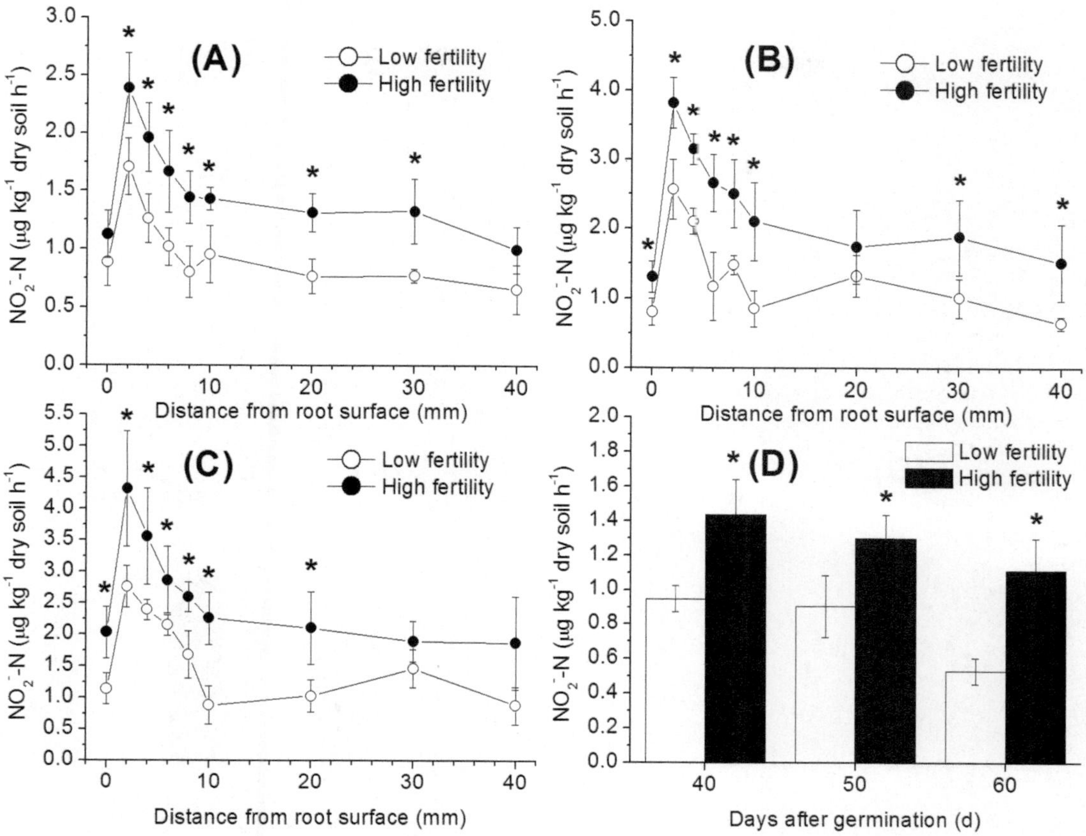

Figure 1. Nitrification activities measured at various distances from the rice root surface in paddy soils with different fertility levels (A) 40 d after germination, (B) 50 d after germination, (C) 60 d after germination, relative to the (D) control (without rice plants). The bars indicate ±SD. Dots and bars with * indicate a significant difference ($p<0.05$) between the two fertility levels.

Rice plant nitrogen assimilation, growth, and nitrogen accumulation

The NRA and GSA in the rice leaves and roots did not change over time, and the levels of NRA and GSA in the rice leaves were significantly higher than those in the roots (Table 2). In addition, high soil fertility ameliorated the NRA and GSA levels significantly; the leaf and root NRAs increased by 48.2% and 49.7% in the high-fertility soil compared to those in the low-fertility soil, whereas the leaf and root GSA levels in the high-fertility soil were, respectively, 37.6% and 103.1%, greater than those in the low-fertility soil (Table 2).

The rice biomass and N accumulation measurements differed significantly between the two fertility levels. The high-fertility soil increased the rice plant biomass by 90.4%, 65.2%, and 45.6% compared with the low-fertility soil when measured 40, 50, and 60 d after germination, respectively. The corresponding N accumulations increased by 47.9%, 48.6%, and 29.9% at the measurements taken 40, 50, and 60 d after germination, respectively (Table 2). High soil fertility markedly improved the growth and N uptake of the rice.

Rice root morphological structure

We observed large changes in the morphological structure of the rice roots and found that the major characteristics of the developmental level of the root system such as root biomass, adventitious root number and root diameter were much greater in the high-fertility soil than in the low-fertility soil (Table 3). Furthermore, the characteristics of the rice root, such as POR, which indicated the developmental degree of aerenchyma, were significantly higher in the high-fertility soil than in the low-fertility soil (Table 3). In summary, the high-fertility soil markedly improved the growth of the rice root and the root porosity.

Distribution of the soil oxygen concentration

The oxygen dissolved in the water layer remained stable. The average values were approximately 231 and 243 μmol L^{-1} in the low-fertility and high-fertility treatments ($n=6$; Fig. 4A), respectively, and 218 and 209 μmol L^{-1} in their respective controls ($n=6$; Fig. 4B). Rice planting significantly increased the content of dissolved oxygen in the water layer. The oxygen concentration decreased with soil depth so rapidly that it became zero at a depth of approximately 2.6–4.2 mm below the soil surface (Fig. 4A, B). Interestingly, the oxygen concentration and oxygen penetration depth in the rhizosphere soil with high fertility were significantly higher than the corresponding values in the rhizosphere soil with low fertility, whereas a reverse trend existed in the controls of these two fertility levels, although certain values at certain depths showed no significant difference due to the presence of abnormal peaks. For example, at a depth of 2 mm from the soil surface, the average dissolved oxygen concentrations in the low- and high-fertility rhizosphere soils were 33.1 and 68.9 μmol L^{-1}, respectively ($n=6$; Fig. 4A), and their corresponding controls were 24.5 and 13.7 μmol L^{-1}, respectively ($n=6$; Fig. 4B). Among the six repetitions, the abnormal peaks appeared more often in the rice planting treatments than in the control. This finding may be a result of the rice root ROL.

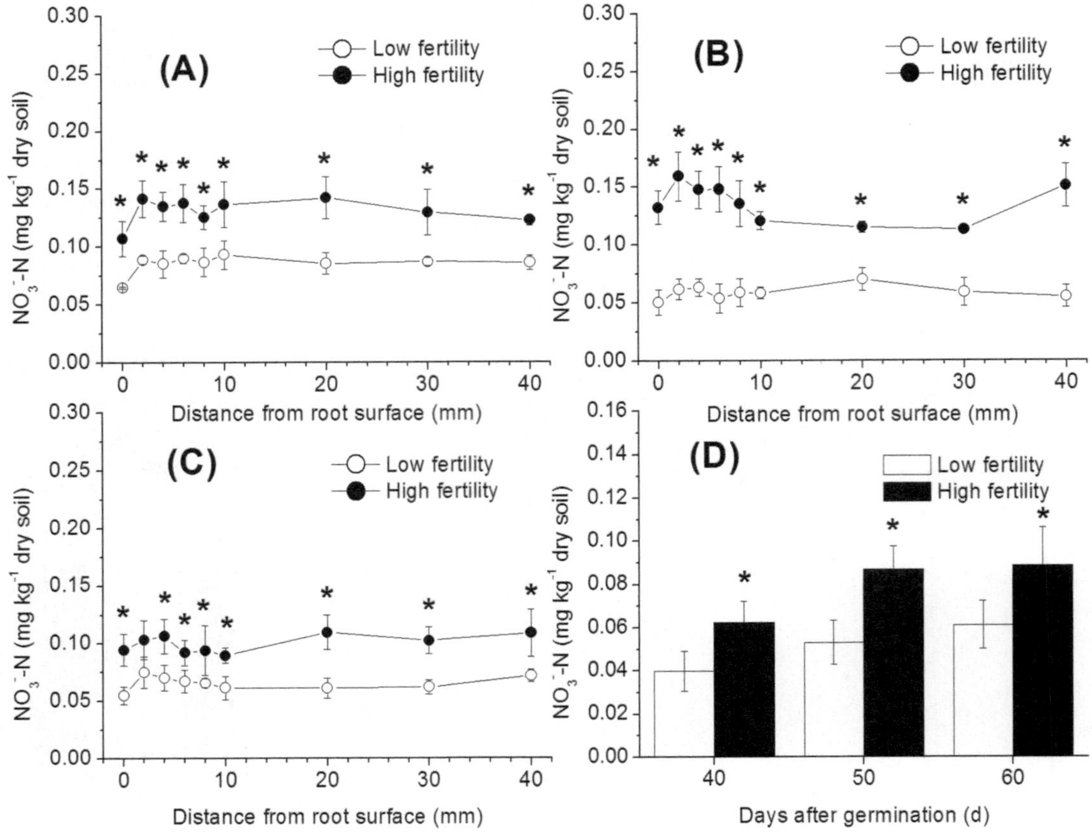

Figure 2. NO_3^--N measured at various distances from the rice root surface in paddy soils with different fertility levels (A) 40 d after germination, (B) 50 d after germination, and (C) 60 d after germination, relative to the (D) control (without rice plants). The bars indicate ±SD. Dots and bars with * indicate a significant difference ($p<0.05$) between the two fertility levels.

Root radial oxygen loss

The oxygen in the agar was detected at an average distance of 400 μm from the root surface in the low-fertility treatment (Fig. 5A), in comparison with an average distance of 600 μm from the root surface in the high-fertility treatment (Fig. 5B). This finding indicated that the root ROL level of rice in the high-fertility soil was significantly higher than that in the low-fertility soil. The oxygen concentration decreased rapidly with distance near the rice root surface. When the microelectrode was in contact with the root surface, the average oxygen concentrations were 50.2 and 97.7 μmol L^{-1} in the low- and high-fertility treatments, respectively. The oxygen concentration remained unchanged inside the rice root, and the oxygen concentrations in the roots of the low-fertility treatment were lower than those in the high-fertility treatment (Fig. 5A, B). For example, the average oxygen concentrations in the low-and high-fertility treatments were 55.4 and 103.3 μmol L^{-1}, respectively.

Discussion

The penetration depths of oxygen in the soil surface and in the rhizosphere generally depend on both rice root ROL and the oxygen consumption caused by soil organisms and chemical oxidation [37]. In fact, rice planting could significantly alter (and, most likely, improve) the cycling of oxygen in the water and soil system due to rice root ROL, and our results confirmed this hypothesis (see Fig. 4A, B). For example, the dissolved oxygen concentration in the water layer was approximately 231 and 243 μmol L^{-1} in the low- and high-fertility treatments (Fig. 4A), respectively, compared with 218 and 209 μmol L^{-1} in their corresponding controls (Fig. 4B). In planted microcosms (planting with 110-d-old rice seedlings), oxygen is still detectable to a depth of at least 40 mm [38], which appears to be the maximum depth reported to date [23]. Interestingly, our results showed that the oxygen concentration in the rhizosphere soil with high fertility was significantly higher than that in the rhizosphere soil with low fertility (Fig. 4A), whereas the soil oxygen concentration in the high-fertility soil control was significantly lower than that in the low-fertility soil control (Fig. 4B). Because the control paddy soil without rice plants had no exogenous replenishment of oxygen as occurred for the rice root ROL, except for the dissolved oxygen provided by the flooding water, the high organic matter content in the high-fertility soil (see Materials and Methods) led to more severe oxygen depletion. This result may have been due to the higher oxygen consumption required by the degradation of organic matter and reduced compounds in the high-fertility soil compared with the low-fertility soil [37]. Although the oxygen content in the control high-fertility soil was lower than that in the control low-fertility soil (Fig. 4B), the soil oxygen concentration in the high-fertility paddy soil with rice plants was higher than that in the low-fertility soil (Fig. 4A), potentially due to the high rice root ROL in the high-fertility soil (Fig. 5B).

The difference in dissolved oxygen concentrations in rhizosphere soils between the high- and low-fertility soils appeared to be related to the differences in rice root ROL in these soils. When rice plants were grown in the high-fertility soil, the rice root growth

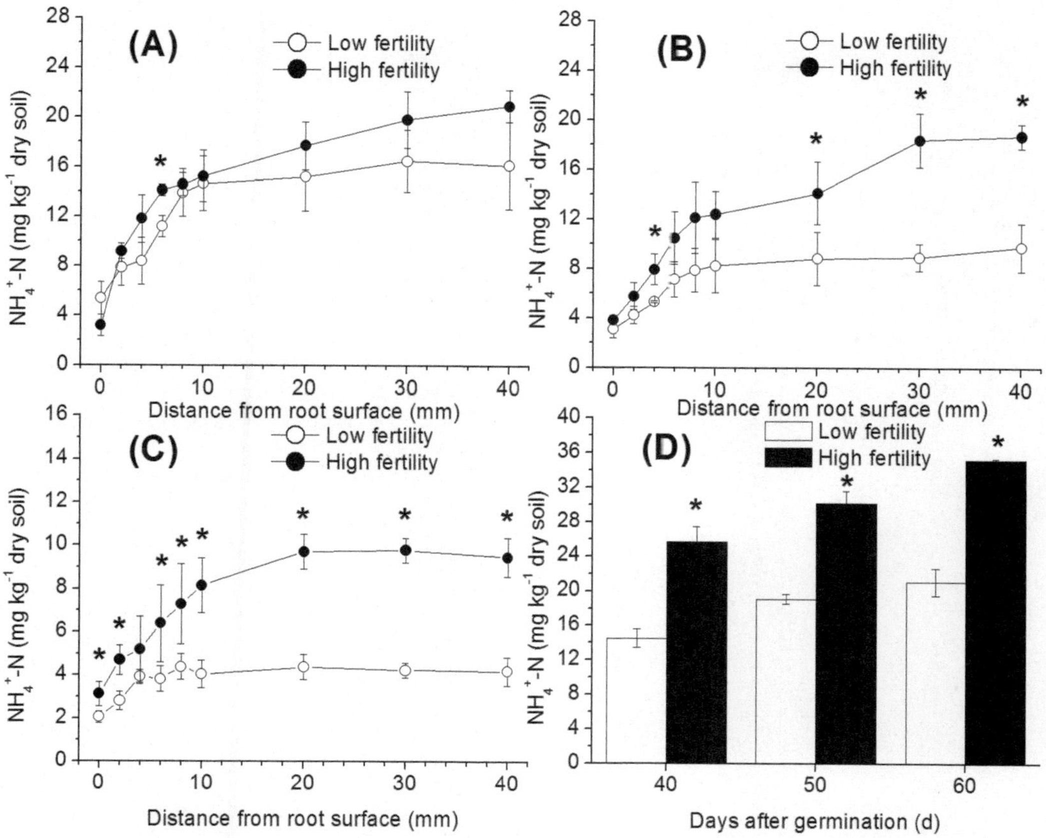

Figure 3. NH_4^+-N measured at various distances from the rice root surface in paddy soils with different fertility levels (A) 40 d after germination, (B) 50 d after germination, and (C) 60 d after germination, relative to the (D) control (without rice plants). The bars indicate ±SD. Dots and bars with * indicate a significant difference ($p<0.05$) between the two fertility levels.

was greatly improved; the root biomass, adventitious root number, and root diameter of the rice grown in the high-fertility soil were much greater than the corresponding values for the rice grown in the low-fertility soil (Table 3). A similar observation has previously been reported for wheat [39]. Not only did the rice root biomass, adventitious root number and root diameter change, but root porosity (see Table 3) was also improved by high soil fertility. This phenomenon does not appear to have been previously reported, as its positive feedback mechanism is unclear. ROL from rice roots estimated by Kludze *et al.* [19], showed a significant and positive correlation with POR. Oxygen efflux from root to soil represented 30%–40% of oxygen transportation from shoot to root [17]. Well-developed root systems and aerenchyma formation resulted in higher rice root ROL in the high-fertility soil (Fig. 5) [7] and consequently increased the oxygen content in the rhizosphere soil (Fig. 4A). Mugnai et al. [40] investigated maize roots and found that the root apex transition zone (intercalated between the apical division zone and the elongation zone [41]) plays central roles in both sensing and adapting to root hypoxia and emits the greatest amount of nitric oxide. This discovery motivated us to further investigate whether the root apex transition zone of rice plays a key role in altering the ROL and, correspondingly, the rhizosphere soil oxygen distribution.

An agar microcosm was used to further quantify the rice root ROL. The results indicated that the root ROL of rice planted in the high-fertility soil was greater than that in the low-fertility soil. The average oxygen concentrations at the root surface were 50.2 and 97.7 μmol L^{-1}, and they decreased rapidly to nearly zero at distances of approximately 400 μm and 600 μm from the root surfaces in the low- and high-fertility treatments, respectively (Fig. 5A, B). The average oxygen concentrations in the interior of the rice roots in the low- and high-fertility treatments were approximately 55.4 and 103.3 μmol L^{-1}, respectively (Fig. 5A, B). Similar results have been obtained by Revsbech *et al.* [23], who found that oxygen leakage created an oxic zone extending approximately 150 μm away from the fine lateral root (for 21-d-old rice); the inner oxygen concentration of the root (42-d-old rice) was approximately 75 μmol L^{-1}. The rice root ROL can also promote the development of aerobic niches in the rhizosphere to restrict the accumulation of phytotoxic compounds (e.g., Fe^{2+}, Mn^{2+}, H_2S) as well as denitrification [42] and maintain aerobic microbial processes, such as nitrification [14,19].

Due to long-term fertiliser changes, soil fertility increased significantly and promoted soil nitrification, as widely reported in previous studies of paddy soil [8,16] and upland soil [2,43]. A low clay content contributed to a high soil porosity, which promoted gas transport by diffusion and advection, thereby supporting soil nitrification. The high ambient NH_4^+ pool in the high-fertility soil also improved its nitrification activity (Fig. 1 and 5). The maximal nitrification activities were found in the rhizosphere soil located 2 mm away from the root surface, and decreased with distance from the root surface (Fig. 1). In our previous study, the nitrification activity gradient also revealed nitrification activity differences in different rice cultivars. It also showed that the maximum nitrification activity detected in the rhizosphere soil, i.e., the maximum nitrification activities of the Indica- and

Table 2. Rice plant N assimilation, growth, and nitrogen accumulation[a].

Fertility level	Sampling date (d)	NRA (μg g^{-1} FW h^{-1})		GSA (mmol g^{-1} FW min^{-1})		Biomass (g DW plant^{-1})	N accumulation (mg N plant^{-1})
		Leaf	Root	Leaf	Root		
Low fertility	40	7.21±0.93 ^{b}	4.06±0.33 ^{c}	1.85±0.14 ^{c}	0.08±0.02 ^{b}	0.49±0.10 ^{d}	16.8±1.77^{d}
	50	6.84±0.76 ^{b}	4.61±0.82 ^{bc}	2.09±0.05 ^{bc}	0.09±0.01 ^{b}	0.85±0.09 ^{c}	23.4±0.93 ^{c}
	60	7.38±0.30 ^{b}	4.14±1.08 ^{c}	2.35±0.15 ^{bc}	0.11±0.04 ^{b}	1.32±0.15 ^{b}	34.8±3.23 ^{b}
High fertility	40	10.6±1.38 ^{a}	5.85±0.69 ^{ab}	2.56±0.45 ^{b}	0.18±0.02 ^{a}	0.93±0.14 ^{c}	24.8±3.59 ^{c}
	50	10.7±2.41 ^{a}	6.86±0.49 ^{a}	2.58±0.25 ^{b}	0.18±0.01 ^{a}	1.41±0.23 ^{b}	34.8±3.88 ^{b}
	60	10.5±0.49 ^{a}	6.45±1.22 ^{a}	3.51±0.41 ^{a}	0.21±0.06 ^{a}	1.92±0.15 ^{a}	45.1±3.24 ^{a}

[a]Values represent means ± SDs with 3 replicates (with each replicate composed of rice plant tissue sampled from different rhizoboxes of each treatment). Different letters in the same column indicate a significant difference at the $p<0.05$ level.

Japonica- planted soils occurred 6 and 2 mm away from the rice roots, respectively [28]. The use of a rhizobox is convenient for distinguishing between the root surface and rhizosphere soil, but this design produces a dense root-zone and thus leads to a generally more oxidised environment, which may enhance nitrification. In contrast, more competition for NH_4^+ in this design may lower nitrification.

Based on our previous experience, a water layer of 1 cm that is maintained with applications of deionised water every morning and evening throughout the incubation would make it difficult to slice the soil. Additionally, mineral N, especially NO_3^-, would be dissolved in the overlying water because the oxygen concentration was much higher in the water layer than in the soil profile. If we continued to maintain a 1-cm water layer, the concentration of NO_3^- in the soil would be very low and might even remain undetected because the paddy soil that we used is derived from Quaternary red clay and has relatively low pH values (usually 4–5.5). The nitrification activity in this paddy soil is much weaker than that in neutral paddy soil [26]. The NO_3^- concentration in the paddy soil is correspondingly low. For this reason, we chose not to water on the evening of the day prior to the sampling date. The water layer was actually 1–2 mm deep when we measured the mineral N and nitrification activity. The advantage of this approach is that the soil is easy to slice, moreover, the NO_3^- concentration can be measured more effectively. However, the disadvantage of this approach is that it imperceptibly stimulates nitrification and results in overestimates of the soil NO_3^- concentration. In this experiment, the NO_3^- concentration in the soil planted with rice did not differ significantly from that in the control. Because the nitrification activity in the rice planting soil was much higher than that in the control (Fig. 1D), the majority of the NO_3^- resulting from increased nitrification might have been absorbed primarily by the rice roots. NO_3^- can be absorbed rapidly by rice roots or can diffuse rapidly to other parts of the soil. Therefore, the NO_3^- distribution in flooded paddy soil is relatively even. However, advection caused by evaporation from the leaves and following water transport to the roots increases transport processes in the soil to a greater extent than that resulting from diffusion. This pattern, was also confirmed by our previous experiment [28]. The low pH caused by acidic root exudates [44] and NH_4^+ (nitrification substrate) exhaustion in the root surface soil (Fig. 3A, B, and C) strongly inhibited root surface nitrification in both of the studied soil types (Fig. 1A, B, and C) [14]. Due to the high nitrification activity in the high-fertility soil, the NO_3^- concentrations in the high-fertility soil were higher than those in the low-fertility soil (Fig. 2). This difference, might have caused more NO_3^- uptake and assimilation by the rice plants grown in the high-fertility soil (the NRA results in Table 2 may support this hypothesis).

The NO_3^- that is converted via nitrification in the root surface and rhizosphere soils is very important for N nutrition of rice [3,4]. Extensive hydroponic experiments have demonstrated that rice growth, yield, net N acquisition, N translocation to the shoot, and NUE are greater with a mixed NH_4^+ and NO_3^- supply than with NH_4^+ alone [3,5,6]. In the present study, the NO_3^- concentration in the rice-planted soil was not significantly different from that measured in the control (Fig. 2), although the nitrification activity in the rice-planted soil was much higher than that in the control (Fig. 1), thus indicating that greater amounts of NO_3^- absorbed by the rice root might have produced the results described above. NO_3^- reductase is a key enzyme involved in NO_3^- assimilation in crops, and its activity is strongly dependent on the external NO_3^- concentration [28]. In this experiment, the rice leaf and root NRA in the high-fertility soil increased by 48.2% and 49.7%,

Table 3. Rice root growth and morphology[a].

Fertility level	Sampling date (d)	Root biomass (g DW plant^{-1})	Adventitious root number (plant^{-1})	Root diameter (mm)	POR (%)
Low fertility	40	0.11±0.02^{d}	22±3^{d}	0.64±0.05^{d}	14.1±1.46 ^{e}
	50	0.20±0.03 ^{c}	42±3 ^{c}	0.78±0.06 ^{c}	16.2±1.13^{d}
	60	0.24±0.01 ^{c}	49±3 ^{c}	0.90±0.02 ^{b}	16.9±1.38 ^{cd}
High fertility	40	0.21±0.04 ^{c}	28±3^{d}	0.75±0.04 ^{c}	19.6±1.44 ^{bc}
	50	0.35±0.02 ^{b}	61±13 ^{b}	0.90±0.03 ^{b}	22.0±2.92 ^{ab}
	60	0.42±0.06 ^{a}	82±7 ^{a}	1.06±0.09 ^{a}	23.1±2.36 ^{a}

[a]Values represent means (means ± SDs), and different letters in the same column indicate a significant difference at the $p<0.05$ level. Data were pooled for n = 3, except for adventitious root diameter (n = 12), which was measured at the middle of each root selected for the average length of the whole root.

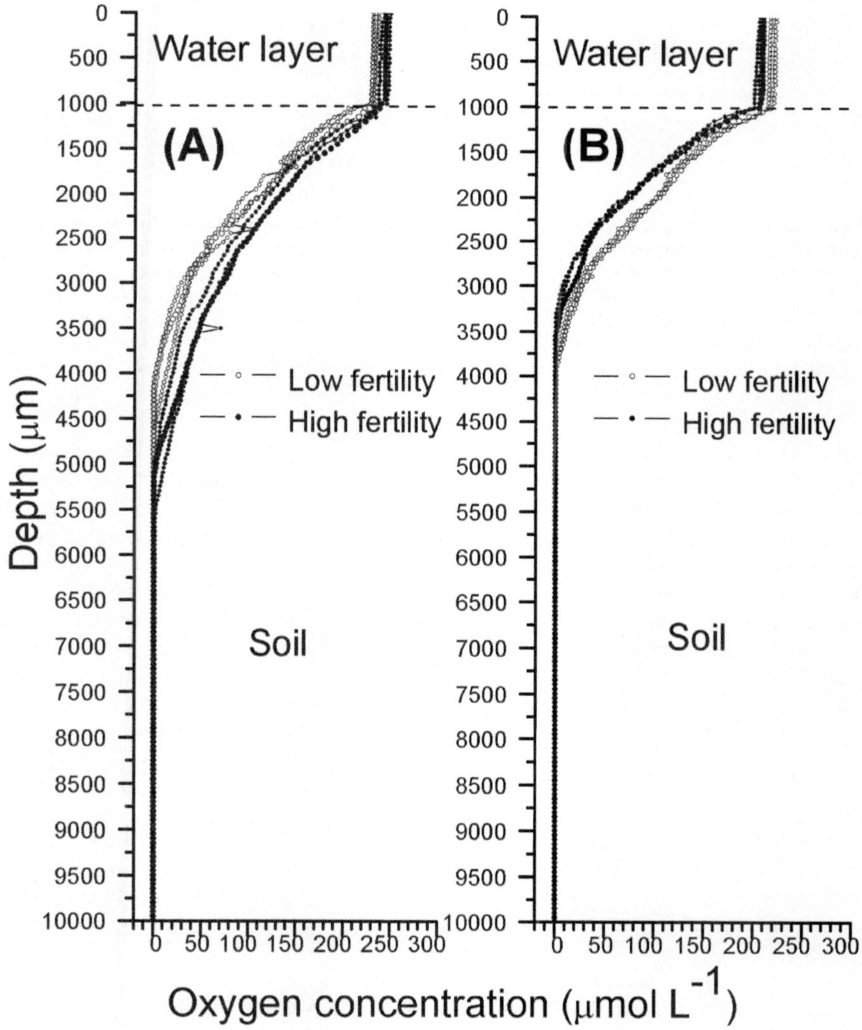

Figure 4. A representative graph showing oxygen profiles of (A) the rhizosphere soil (2 mm away from the root surface) and (B) the control (without rice plants) 50 d after germination in paddy soils with different fertility levels at 20°C. The oxygen microelectrode used was 20–30 μm in diameter (OXY25, Unisense, Aarhus, Denmark [31]). Three rhizoboxes for each treatment were used in the in situ measurements, and the oxygen concentration profile was determined in the right and left compartments of each rhizobox. All measurements were conducted in the centre of the range marked by a 2-mm distance from the nylon nets in the low- and high-fertility soils with and without the rice plant treatments. Hence, the measurements were repeated six times in each treatment. Three results (measured in the right compartment) are shown, and each curve shows one individual soil oxygen profile.

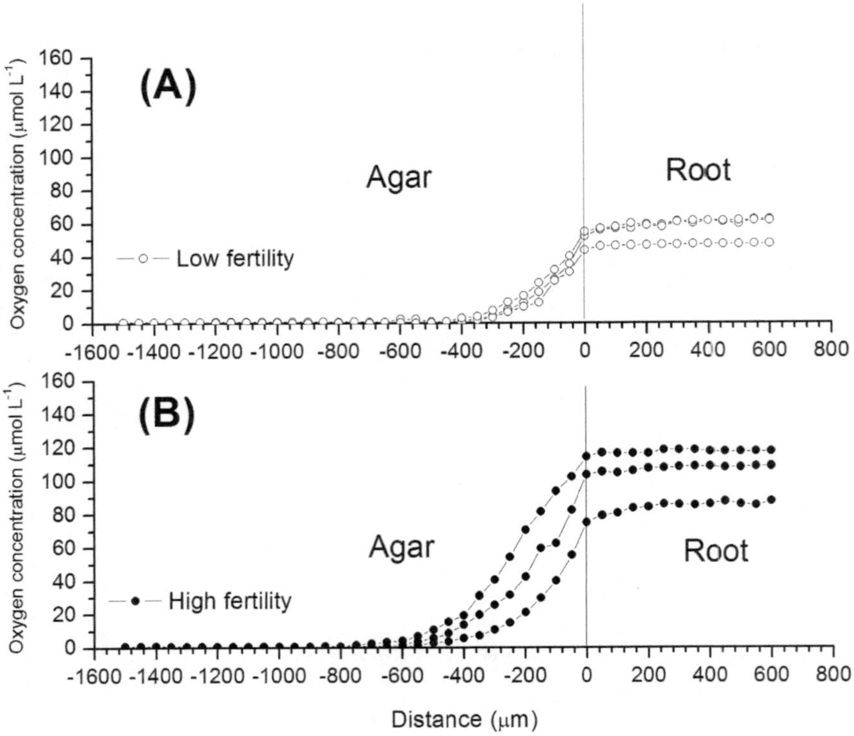

Figure 5. A representative graph showing oxygen profiles of the rhizosphere and within the rice root, 50 d after germination, growing in the (A) low- and (B) high-fertility soils at 20°C. The rice plant roots were separated from the soil and placed in agar with a nutrient solution. Newly formed roots with uniform 45-mm lengths were selected, and both were measured at a position 25 mm away from the root tips. The average diameters of the roots growing in the low- and high-fertility conditions were 0.74 and 0.93 mm at the measured points, respectively. The measurements were repeated in different rice plants and with different newly formed roots, with lengths ranging from 30 to 55 mm. An OXY10 oxygen microelectrode (Unisense, Aarhus, Denmark [31]) 8–10 μm in diameter was used to penetrate the rice root longitudinally. Three rhizoboxes were used for the individual rice root ROL measurements. Each curve shown in the figure indicates an individual rice root.

respectively, compared with those in the low-fertility soil (Table 2), which also confirmed that more NO_3^- was absorbed and assimilated by the rice planted in the high-fertility soil compared with that planted in the low-fertility soil. Greater amounts of NO_3^- absorbed and assimilated by rice plants will also accelerate NH_4^+ uptake and assimilation, perhaps through the NO_3^--specific induction of additional pathways for NH_4^+ assimilation [3], as also confirmed by the GSA results (Table 2). Hence, the rice N accumulation and rice growth of plants grown in the high-fertility soil were improved compared with those of plants grown in the low-fertility soil (Table 2).

Conclusions

The present study investigated reasons that increasing soil fertility would enhance the growth of rice. The study adopted a new perspective focusing on rice root ROL and its corresponding soil oxygen profile. The fertile soil had high porosity (because of the low clay content) and a high NH_4^+ content, which supported soil nitrification. A high rhizosphere oxygen pressure distribution caused by the rice root ROL, which was attributed to the improved rice root growth and root porosity observed in the high-fertility soil, also contributed to the promotion of rhizosphere nitrification. The mechanism accelerating the development of rice root aerenchyma in the high-fertility soil requires further exploration. Investigation of the mechanisms of N loss and the quantification of N loss were not performed in this study but will be addressed by our laboratory in future research. Long-term tillage and fertilisation (especially organic fertiliser) would significantly increase soil fertility, but a large amount of organic matter would decrease the soil oxygen content. Fortunately, rice planting altered the water layer and soil oxygen status via root ROL, and rice in the fertile soil further improved the rhizosphere soil oxidisation environment. Our results suggest that increased fertilisation could lead to increased soil aeration in red soil regions in China, which might promote rice growth and rice N nutrition.

Acknowledgments

We thank Dr. O. Pedersen at the Freshwater Biological Laboratory, Institute of Biology, University of Copenhagen, Denmark, for his patience and experienced instruction in oxygen microelectrode measurement and careful English revisions of this manuscript.

Author Contributions

Conceived and designed the experiments: YLL WMS XXW. Performed the experiments: YLL. Analyzed the data: YLL. Contributed reagents/materials/analysis tools: YLL XXW. Wrote the paper: YLL WMS XXW.

References

1. Abbasi MK, adams WA (1998) Loss of nitrogen in compacted grassland soil by simultaneous nitrification and denitrification. Plant and Soil 200: 265–277.
2. He J Z, Shen JP, Zhang LM, Zhu YG, Zheng YM, et al. (2007) Quantitative analyses of the abundance and composition of ammonia-oxidizing bacteria and

ammonia-oxidizing archaea of a Chinese upland red soil under long-term fertilization practices. Environmental Microbiology 9: 2364–2374.

3. Kronzucker HJ, Siddiqi MY, Glass ADM, Kirk GJD (1999) Nitrate-ammonium synergism in ice: a subcellular flux analysis. Plant Physiology 119: 1041–1046.
4. Kronzucker HJ, Glass ADM, Siddiqi MY, Kirk GJD (2000) Comparative kinetic analysis of ammonium and nitrate acquisition by tropical lowland rice: implications for rice cultivartion and yield potential. New Phytologist 145: 471–476.
5. Raman DR, Spanswick RM, Walker LP (1995) The kinetics of nitrate uptake from flowing nutrient solutions by rice: Influence of pretreatment and light. Bioresourse Technology 53: 125–132.
6. Duan YH, Zhang YL, Ye LT, Fan XR, Xu GH, et al. (2007) Responses of rice cultivars with different nitrogen use efficiency to partial replaced nitrate nutrition. Annals of Botany 99: 1153–1160.
7. Li YL, Fan XR, Shen QR (2008) The relationship between rhizosphere nitrification and nitrogen-use efficiency in rice plants. Plant, Cell and Environment 31: 73–85.
8. Chen XP, Zhu YG, Xia Y, Shen JP, He JZ (2008) Ammonia-oxidizing archaea: important players in paddy rhizosphere soil? Environmental Microbiology 10: 1978–1987.
9. Venter JC, Remington K, Heidelberg JF, Halpern AL, Rusch D, et al. (2004) Environmental genome shotgun sequencing of the Sargasso Sea. Science 304: 66–74.
10. Schleper C, Jurgens G, Jonuscheit M (2005) Genomic studies of uncultivated archaea. Nature Reviews Microbiology 3: 479–488.
11. Könneke M, Bernhard AE, de la Torre JR, Walker CB, Waterbury JB, et al. (2005) Isolation of an autotrophic ammonia-oxidizing marine archaeon. Nature 437: 543–546.
12. Francis CA, Roberts KJ, Beman JM, Santoro AE, Oakley BB (2005) Ubiquity and diversity of ammonia-oxidizing archaea in water columns and sediments of the ocean. Proceedings of the National Academy of Sciences of the United States of America 102: 14683–14688.
13. Leininger S, Urich T, Schloter M, Schwark L, Qi J, et al. (2006) Archaea predominate among ammonia-oxidizing prokaryotes in soils. Nature 442: 806–809.
14. Arth I, Frenzel P (2000) Nitrification and denitrification in the rhizosphere of rice: the detection of processes by a new multi-channel electrode. Biology and Fertility of Soils 31: 427–435.
15. Briones AM, Okabe S, Umemiya Y, Ramsing NB, Reichardt W, et al. (2002) Influence of different cultivars on populations of ammonia-oxidizing bacteria in the root environment of rice. Applied and Environmental Microbiology 68: 3067–3075.
16. Wang YN, Ke XB, Wu LQ, Lu YH (2009) Community composition of ammonia-oxidizing bacteria and archaea in rice field soil as affected by nitrogen fertilization. Systematic and Applied Microbiology 32: 27–36.
17. Armstrong W (1979) Aeration in higher plants. Advances in Botanical Research 7: 225–332.
18. Armstrong W, Justin SHFW, Beckett PM, Lythe S (1991) Root adaptation to soil waterlogging. Aquatic Botany 39: 57–73.
19. Kludze HK, DeLaune RD, Patrick WH (1993) Aerenchyma formation and methane and oxygen exchange in rice. Soil Science Society of America Journal 57: 386–391.
20. Armstrong W (1964) Oxygen diffusion from the roots of some British Bog plants. Nature 204: 801–802.
21. Armstrong W (1971) Radial oxygen losses from intact rice roots as affected by distance from the apex, respiration and waterlogging. Physiologia Plantarum 25: 192–197.
22. Kludze HK, DeLaune RD, Patrick WH (1994) A colorimetric method for assaying dissolved oxygen loss from container-grown rice roots. Agronomy Journal 86: 483–487.
23. Revsbech NP, Pedersen O, Reichardt W, Briones A (1999) Microsensor analysis of oxygen and pH in the rice rhizosphere under field and laboratory conditions. Biology and Fertility of soils 29: 379–385.
24. FAO (1998) FAO, ISRIC, ISSS World Reference Base for Soil Resources. World Soil Resources Reports 84. Rome, Italy.
25. Zhong WH, Cai ZC (2007) Long-term effects of inorganic fertilizers on microbial biomass and community functional diversity in a paddy soil derived from quaternary red clay. Applied Soil Ecology 36: 84–91.
26. Yuan F, Ran W, Shen QR, Wang DZ (2005) Characterization of nitrifying bacteria communities of soils from different ecological regions of China by molecular and conventional methods. Biology and Fertility of Soils 41: 22–27.
27. Zhang MK, Xu JM (2005) Restoration of surface soil fertility of an eroded red soil in southern China. Soil and Tillage Research 80: 13–21.
28. Li YL, Zhang YL, Hu J, Shen QR (2007) Contribution of Nitrification Happened in Rhizospheric Soil Growing with Different Rice Cultivars to N Nutrition. Biology and Fertility of Soils 43: 417–425.
29. Berg P, Rosswall T (1985) Ammonium oxidizer numbers, potential and actual oxidation rates in two Swedish arable soils. Biology and Fertility of Soils 1: 131–140.
30. Alef K, Nannipieri P (1995) Methods in Applied Soil Microbiology and Biochemistry. In: Berg P, Rosswall T, Editors.Assay of nitrification (short-term estimations).New York: Harcourt Brace Academic Press. pp.241–242.
31. Revsbech NP (1989) An oxygen microsensor with a guard cathode. Limnology and Oceanography 34: 474–478.
32. Fan XR, Jia LJ, Li YL, Smith SJ, Miller AJ, et al. (2007) Comparing nitrate storage and remobilization in two rice cultivars that differ in their nitrogen use efficiency. Journal of Experimental Botany 58: 1729–1740.
33. Zhang CF, Peng SB, Peng XX, Chavez A Q, Bennett J (1997) Response of glutamine synthetase isoforms to nitrogen sources in rice (*Oryza sativa* L.) roots. Plant Science 125: 163–170.
34. Ogawa T, Fukuoka H, Yano H, Ohkawa Y (1999) Relationships between nitrite reductase activity and genotype-dependent callus growth in rice cell cultures. Plant Cell Reports 18: 576–581.
35. Jensen CR, Luxmoore RJ, Van Gundy SD, Stolzy LH (1969) Root air-space measurements by a pycnometer method. Agronomy Journal 61: 474–475.
36. Kim JD, Jugsujinda A, Carbonell-Barrachina AA, DeLaune RD, Patrick WH (1999) Physiological functions and methane and oxygen exchange in Korean rice cultivars grown under controlled soil redox potential. Botanical Bulletin of Academia Sinica 40: 185–191.
37. Ratering S, Schnell S (2001) Nitrate-dependent iron (II) oxidation in paddy soil. Environmental. Microbiology 3: 100–109.
38. Frenzel P, Rothfuss F, Conrad R (1992) Oxygen profiles and methane turnover in a flooded rice microcosm. Biology and Fertility of Soils 14: 84–89.
39. Wang YF, Yu ZW, Li SX, Yu SL (2003) Effects of nitrogen rates and soil fertil ity levels on root nitrogen uptake and assimilation and grain protein content of winter wheat. Plant Nutrition and Fertilizer Science 9: 39–44.
40. Mugnai S, Azzarello E, Baluška F, Mancuso S (2012) Local root apex hypoxia induces NO-mediated hypoxic acclimation of the entire root. Plant and Cell Physiology 53: 912–920.
41. Baluška F, Mancuso S, Volkmann D, Barlow PW (2010) Root apex transition zone: a signaling-response nexus in the root. Trends in Plant Science 15: 402–408.
42. Sorrell BK (1999) Effect of external oxygen demand on radial oxygen loss by *Juncus* roots in titanium citrate solutions. Plant, Cell and Environment 22: 1587–1593.
43. Chu HY, Fujii T, Morimoto S, Lin XG, Yagi K, et al. (2007) Community structure of ammonia-oxidizing bacteria under long-term application of mineral fertilizer and organic manure in a sandy loam soil. Applied and Environmental Microbiology 73: 485–491.
44. Colmer TD, Bloom AJ (1998) A comparison of NH_4^+ and NO_3^- net fluxes along roots of rice and maize. Plant, Cell and Environment 21: 240–246.

2

Association Mapping for Important Agronomic Traits in Core Collection of Rice (*Oryza sativa* L.) with SSR Markers

Peng Zhang[1,2], Xiangdong Liu[1], Hanhua Tong[2], Yonggen Lu[1]*, Jinquan Li[1,3]*

1 State Key Laboratory for Conservation and Utilization of Subtropical Agro-bioresources, South China Agricultural University, Guangzhou, China, **2** State Key Laboratory of Rice Biology, China National Rice Research Institute, Hangzhou, China, **3** Department of Plant Breeding and Genetics, Max Planck Institute for Plant Breeding Research, Cologne, Germany

Abstract

Mining elite genes within rice landraces is of importance for the improvement of cultivated rice. An association mapping for 12 agronomic traits was carried out using a core collection of rice consisting of 150 landraces (Panel 1) with 274 simple sequence repeat (SSR) markers, and the mapping results were further verified using a Chinese national rice micro-core collection (Panel 2) and a collection from a global molecular breeding program (Panel 3). Our results showed that (1) 76 significant ($P<0.05$) trait-marker associations were detected using mixed linear model (MLM) within Panel 1 in two years, among which 32% were identical with previously mapped QTLs, and 11 significant associations had >10% explained ratio of genetic variation; (2) A total of seven aforementioned trait-marker associations were verified within Panel 2 and 3 when using a general linear model (GLM) and 55 SSR markers of the 76 significant trait-marker associations. However, no significant trait-marker association was found to be identical within three panels when using the MLM model; (3) several desirable alleles of the loci which showed significant trait-marker associations were identified. The research provided important information for further mining these elite genes within rice landraces and using them for rice breeding.

Editor: Jauhar Ali, International Rice Research Institute, Philippines

Funding: This work was supported by Fund of the National Natural Science Foundation of China grant 30700494. The funder had no role in study design, data collection and analysis, decision to publish, or preparation of the manuscript.

Competing Interests: The authors have declared that no competing interests exist.

* Email: yglu@scau.edu.cn (YL); lijinquan@scau.edu.cn (JL)

Introduction

As a staple cereal crop, rice (*Oryza sativa* L.) feeds more than 50% of the world's population [1] and is one of the most important components of human diet in many regions of the world. Thus, genetic improvement of rice for yield is important to the meet food demand of a growing global population. Rice landraces have a greater genetic diversity than elite cultivars (or commercial cultivars) and represent an intermediate stage in domestication between wild rice and elite cultivars [2], which make it easier to be used in rice breeding than wild rice and at the same time still keeping most of the diversity in rice germplasm resource. Therefore, mining elite genes within the germplasm of rice landraces is of importance for the improvement of cultivated rice.

Linkage mapping and association mapping based linkage disequilibrium (LD) are two main methods for locating genes or QTLs. The major limitations of linkage mapping are that only two alleles at any given locus can be studied in bi-parental crosses and a low mapping resolution [3], whereas association mapping promises to overcome the limitations of linkage mapping [4]. Moreover, association mapping identifies QTLs by examining the trait-marker associations and enables researchers to use modern genetic technologies to exploit natural diversity and locate valuable genes in the genome [5].

Association mapping has been widely used in plant research since it was firstly reported in maize [6,7]. In recent years, association mapping has been applied in *Arabidopsis*, maize, barley, durum wheat, spring wheat, sorghum, sugarcane, sugar beet, soybean, grape, forest tree species and forage grasses [8] as well as rice [9,10,11,12]. For example, an association mapping was performed with 60 simple sequence repeat (SSR) markers and 114 restriction fragment length polymorphism (RFLP) markers for 12 agronomic traits within 218 inbred lines of rice originating from United States of America (USA) and Asia [13]. An association mapping was performed for five agronomic traits in a population of 103 cultivars using 123 SSRs [14] as well as for grain shape using a collection of 293 accessions of Asian cultivated rice [15]. An association mapping for starch quality traits using both candidate gene-based association mapping and genome-wide association study (GWAS) strategies was performed [16]. More than 3.6 million SNPs were detected by sequencing 517 rice landraces and applied for GWAS for 14 agronomic traits [17]. However, to our knowledge, an association mapping with a high number of SSR markers was seldom performed in the previous studies. Moreover, no earlier research performed an association mapping in one population and at the same time verified the association mapping results in other populations.

The choice of appropriate germplasm to maximize the number of historical recombinations and mutation events (and thus reduce LD) within and around the gene of interest is critical for the success of association analysis [18]. One of the methods to obtain most of the phenotypes is to construct a core collection. A core collection is a subset chosen to represent most genetic diversity of an initial collection with a minimum of redundancies [19,20,21].

Core collections facilitate the users to access useful samples of small sizes while still keeping most of the genetic variability contained within the gene pool of a specific crop [22]. The construction of a core collection was widely applied in rice as well as other crops. Thus, a core collection might be an ideal mapping population for association mapping. Some rice core collections have been used as association mapping populations in previous studies [23,24]. However, the mapping population in the studies mentioned above were two subsets consisting of 547 and 203 accessions chosen randomly from United States Department of Agriculture (USDA) rice core collection which consists of 1790 rice entries, which cannot effectively maintain the genetic diversity in the original collections. Moreover, the number of SSR markers for genotyping was low (72 and 155) in the studies. As far as we know, no earlier research on association mapping based on a core collection of rice landraces was available.

Population structure may cause false positives in association mapping. To overcome this problem, an approach using a mixed-model was proposed for association mapping, which take both population structure (Q) and kinship (K) into account for the reduction of false positives [25]. In recent years, comparisons of different statistical models e.g. Q, Q+K and P+K have been conducted for *Arabidopsis* [26], sweet sorghum [27], maize [28] and rice [23]. However, false positive might not be absolutely avoided through the aforementioned models. To avoid them, it required that the significant associations identified within one population should be verified in another population [29].

In our previous studies, a rice core collection (Ting's rice core collection) consisting of 150 accessions of rice landraces has been constructed based on 15 quantitative traits and 34 qualitative traits from 2262 accessions of rice landraces of the Ting's collection with an optimal sampling strategy [30]. Moreover, population structure and LD of the rice core collection had been examined in details [31]. In this study, an association mapping was performed for 12 agronomic traits in the Ting's core collection assessed with 274 SSR markers. Moreover, the significant trait-marker associations identified in the population were verified within a Chinese national rice micro-core collection and a collection from a global molecular breeding program. The study aimed to (1) perform association mapping for 12 important agronomic traits in the Ting's core collection and verify some of the mapping results in another two core collections, (2) compare the effectiveness of different statistical models and different significant thresholds for association mapping, and (3) identify desirable alleles of the loci which showed significant trait-marker associations for rice breeding.

Materials and Methods

Plant material

Three rice collections, i.e. Ting's core collection (Panel 1), the Chinese national micro-core collection (Panel 2), and a collection from the core collection of a global molecular breeding program (Panel 3) were used in this study. Panel 1 was collected by the researcher Ying Ting during 1920–1964 from all over China as well as from Korea, Japan, Philippines, Brazil, Celebes, Java, Oceania, and Vietnam. The original collection comprises 7128 rice landraces [32]. The core collection (Panel 1) with 150 accessions was constructed from 2262 accessions of 7128 based on a strategy of stepwise clustering and preferred sampling on adjusted Euclidean distances and weighted pair-group average method using integrated qualitative and quantitative traits [30]. Panel 2 with 197 accessions was provided by China Agricultural University, and Panel 3 with 122 accessions was offered by the International Rice Research Institute (IRRI). The information for each variety is shown in Table S1 in File S1.

Phenotyping

All of the three panels were cultivated at the farm of South China Agricultural University, Guangzhou (23°16N, 113°8E), during the late season (July-November) for two consecutive years (2008 and 2009). A randomized complete block design with three replications was used during each season. The space between rows and between plants was set to 20 and 16.5 cm, respectively. Thirty plants of each variety were grown in three rows with 10 plants per row. For each block, the five plants in the middle position of the second row of each variety were selected so that the marginal effect was avoided. 12 agronomic traits for these plants were investigated. Heading date (HD) was recorded as days from sowing to flowering time when 30% of the individuals of one variety started flowering. Plant height (PH), panicle length (PL), grain length (GL), grain width (GW), flag leaf length (FLL), and flag leaf width (FLW) were measured in centimeters. Seed set rate (SS, %) was the percentage of filled grains divided by the total grains per plant. For 1000-grain weight (1000GW), 100 grains were measured in grams with three replicates and then its average was multiplied by 10. For grain length (GL) and width (GW), ten grains were randomly selected and measured with a digital vernier caliper.

Genotyping

274 SSR markers evenly distributed across the 12 chromosomes of rice were selected to genotype all varieties in Panel 1 (Table S2 in File S1). A total of 23, 25, 24, 22, 21, 22, 21, 25, 23, 24, 23, and 21 of these markers were mapped to chromosomes 1 to 12, respectively. The average distance between the loci in chromosomes 1 to 12 is 7.5 cM, 8.2 cM, 9.4 cM, 7.4 cM, 7.1 cM, 6.3 cM, 5.8 cM, 5.4 cM, 5.2 cM, 4.7 cM, 5.6 cM and 5.3 cM, respectively. Markers which prefix RM were summarized in [33,34,35,36] and those with prefix PSM were summarized in [37]. DNA was extracted using a modified SDS method [38]. The volume of the polymerase chain reaction (PCR) was 10 µl. The profile of the PCR program was as follows: 94°C for 5 mins followed by 29 cycles of 94°C for 1 min, 55°C for 1 min, 72°C for 1 min with a final extension of 5 minutes at 72°C. PCR products were separated in size by 6% polyacrylamide gel electrophoresis and detected by silver staining [39]. A standard marker (100–600 bp, produced by Shanghai Biocolor BioScience & Technolgy Company) was added on each gel as control during the gel run. The size of PCR products were detected by BIO Imagine System with software Genetools from SynGene and were manually re-checked twice [31]. The length of each allele was compared to the standard bands of the standard marker and scored.

Data analysis

Means and standard deviation (SD) for 12 traits were calculated using Excel software. The percentage of phenotypic variation explained by population structure was calculated using a General Linear Model (GLM) with software SPSS 17.0 for Windows (SPSS Inc. Chicago, IL, USA). The broad-sense heritability (H^2) was calculated as $H^2 = \sigma_g^2 / \left(\sigma_g^2 + \sigma_e^2\right)$, where σ_g^2 is the genetic variance, σ_e^2 is the environmental variance. They were calculated using software QGA Station 1.0 (Zhu Jun, Zhejiang University, China). Correlation coefficients between traits were calculated using the software SPSS.

Polymorphism information content (PIC) which measures the extent of polymorphism for marker gene(s) or marker sequence(s) was calculated using the program POWERMARKER V3.25.

Software Structure V2.3.1 was used to infer population structure and get Q matrices [40,41]. During the running, a range of genetic clusters from $K = 1$ to 15 with the admixture model was examined, and for each K it was replicated 5 times. Each run implemented with a burn-in period of 100,000 steps followed by 100,000 Monte Carlo Markov Chain replicates. Due to the distribution of L(K) did not show a clear cutoff point for the true K, an ad hoc measure ΔK was used to detect the numbers of subgroup. That run with the maximum likelihood was applied to subdivide the varieties into different subgroups based on the maximum membership probability. A Q-matrix was obtained from the membership probability of each variety. Our previous study indicated that there were two distinct subgroups in Panel 1, which were in accordance with the germplasm types of *indica* and *japonica* rice [31]. The Q-matrix was used for further association mapping. The Loiselle algorithm was chosen for calculating kinship matrix (K) by software SPAGeDi [42]. Rare alleles with frequency of less than 10% in population were filtered as missing data in association analysis. Quantile–quantile plots were generated for observed against expected $-\log_{10}(P)$ using software SAS version 9.0 (SAS Institute 2002), where observed P values were obtained from association mapping and expected P values from the assumption that no associations happened between marker and trait.

Association analysis was performed using the software TASSEL (www.maizegenetics.net/tassel). For the mixed linear model (MLM) method, both K and Q matrices were incorporated, whereas for the GLM method, only population structure information (Q-matrix) was used as a covariate. Significance of associations between loci and traits were determined by their P values ($P<0.05$) which were calculated by the statistical models, and the phenotypic variance explained by the significant loci was calculated through analysis of variance (ANOVA). Since MLM method performs better in controlling spurious associations than GLM method [43], we first ranked the significant ($P<0.05$) association from MLM and then compared the significance of these markers ($P<0.05$) in the permutation based on GLM association tests. For the comparison, we calculated and used other two significant thresholds (i.e. Minimum Bayes factor (BF) and Bonferroni threshold) besides the P value. BF was calculated using the following formula: $BF = -e*P*\ln(P)$ [44,45]. The Bonferroni threshold [46] was $1/274 = 0.00365$, where 274 is the number of association tests for each traits in this study. Duncan multiple comparisons was implemented in SPSS for comparisons of performance of agronomic traits relevant to different alleles of the significant trait-marker associations.

Results

Phenotypic variation

The rice landraces in Panel 1 revealed a wide range of phenotypic variation in 12 agronomic traits (Table 1). Heading date, plant height, 1000-grain weight, flag leaf length, flag leaf length/width, and panicle numbers per plant showed similar distributions in both two years (Figures S1–S6 in File S1). On average about 12.4% of phenotypic variation was influenced by population structure. The broad-sense heritability ranged from 74.8% (1000GW) to 99.8% (GW) for these traits.

Phenotypic correlation analysis

Extremely significant ($P<0.01$) positive correlations both in 2008 and 2009 were found between HD and PH, PH and PL, FLL and FLL/FLW, PL and FLL, PL and FLW, GL and GL/GW, GW and 1000GW, GL and 1000GW, HD and FLW, PH and FLL, SS and 1000GW, PH and FLW (Table 2). Extremely significant ($P<0.01$) negative correlations in both two years were found between HD and 1000GW, GW and GL/GW, FLW and FLL/FLW, FLW and PN.

Relative kinship among individuals in the three panels

In Panel 1, about 55% of pairwise kinship estimates were zero and only 4.73% of pairwise kinship coefficient were larger than 0.5, indicating that these varieties were unrelated (Figure 1). In Panel 2 and 3, 55.9% and 60.4% of pairwise kinship coefficient were larger than 0.5, respectively (Figure S7 in File S1), indicating that these varieties have certain kinship relationship.

The effect of controlling type I error using MLM

Observed versus expected P values for each trait-marker association were plotted to assess the control of type I errors. Uniform distributions between the observed and expected P values for all traits were observed, and were demonstrated by similar distributions in two years (Figures 2 and 3). As the deviations from the expectation demonstrated that the statistical analysis may cause spurious associations [28], our result indicated that the false positives were well controlled in the MLM method in this study.

Trait-marker associations

152 significant ($P<0.05$) trait-marker associations were found using the GLM model for the 12 agronomic traits both in 2008 and 2009, and 15 (~10%) of 152 trait-marker associations were detected in the previous studies (Table 3). Furthermore, 184 and 217 significant ($P<0.05$) trait-marker associations were identified using MLM in 2008 and 2009, respectively. Among them, 76 trait-marker associations were significant ($P<0.05$) both in 2008 and 2009. The number of significant loci associated with each agronomic trait in two years ranged from 0 (seed set rate) to 13 (plant height). Moreover, 24 (~32%) of the 76 trait-marker associations were in the same or similar genomic regions where QTLs were detected in previous studies (http://www.gramene.org/), and the other 52 trait-marker associations were new associations which were not previously identified.

Eleven of the 76 trait-marker associations had 10% or more explained percentage of the total variation (R^2), i.e. HD (PSM184), PH (RM530, RM590), PL (PSM184), GL/GW (RM447), FLL (RM287), FLW (RM235), 1000GW (RM7, RM538 and RM206), and PN (RM311) both in 2008 and 2009 (Table 4). When using BF and the Bonferroni threshold as significance thresholds, there were 15 and 3 trait-marker associations out of the 76 significant associations which still showed significant associations, respectively. Moreover, the three trait-marker significant associations shown by Bonferroni threshold were also significant when using BF as significant threshold. Furthermore, 59 of the 76 trait-marker associations were found to be significant when using the GLM model in two years.

Impact of allele frequency on the power to detect a QTL

We further investigated the relationship between the P values of significant trait-marker associations and the PIC values of related markers. For all trait-marker associations, only 3.5% of markers had a PIC value lower than 0.2 (Figure 4). Most of the markers which showed significant associations with related traits had a PIC value larger than 0.2, which meant that these markers showed a higher power to detect a QTL.

Table 1. Descriptive statistics, percentage of phenotypic variation explained by population structure (R^2), and heritability in broad sense (h^2) for 12 agronomic traits in Panel 1.

Trait	Year	Mean±S.D.	Range	R^2(%)	h^2(%)
Heading days (day)	2008	71.0±7.6	61.0–95.0	1.2	78.1
	2009	66.7±9.9	52.0–92.0	7.5	
Plant height (cm)	2008	144.5±26.4	66.0–209.5	25.1	97.4
	2009	150.8±30.3	72.8–229.0	28.1	
Seed set rate (%)	2008	79.1±11.7	24.4–98.0	1.2	76.8
	2009	84.3±11.5	25.3–98.3	10.8	
Panicle length (cm)	2008	24.8±2.9	15.8–31.5	24.9	94.9
	2009	25.6±3.3	15.7–35.2	31.5	
Grain length(GL) (mm)	2008	7.9±0.6	6.2–9.6	9.6	76.5
	2009	8.0±0.6	6.6–10.5	5.3	
Grain width(GW) (mm)	2008	3.1±0.4	2.3–4.1	8.2	99.4
	2009	3.1±0.3	2.4–3.7	12.0	
GL/GW	2008	2.6±0.4	1.9–3.9	10.5	99.5
	2009	2.6±0.4	1.9–3.7	11.1	
1000-grain weight (g)	2008	21.5±3.8	11.0–34.1	1.8	74.8
	2009	23.0±3.9	11.8–35.7	2.9	
Flag leaf length(FLL) (cm)	2008	43.2±8.6	23.0–75.0	34.1	88.7
	2009	39.6±6.6	23.6–56.1	17.5	
Flag leaf width(FLW) (cm)	2008	1.7±0.3	1.0–2.2	2.6	99.8
	2009	1.6±0.3	0.9–2.2	6.7	
FLL/FLW	2008	26.3±6.4	13.5–50.2	28.4	97.2
	2009	25.1±6.2	13.6–49.0	2.5	
Panicles number per plant	2008	7.9±2.6	3.0–20.0	12.3	94.6
	2009	8.7±2.5	4.6–18.2	0.6	

Verification of association mapping results in Panel 2 and Panel 3

For the 76 significant trait-marker associations in Panel 1, because some SSR markers show more than one significant associations with related traits, the number of related SSR markers is less than 76, i.e. 55 SSR markers in this study. All these 55 SSR markers were further used to genotype Panel 2 and 3. Based on these genotyping data, the population structure of both Panel 2 and 3 indicated two distinct subgroups (Figure S8 in File S1).

Association analysis was performed within the two Panels using both MLM and GLM approaches with the 55 SSR markers. A total of 20 and 31 significant trait-marker associations were detected using MLM within Panel 2 and Panel 3, respectively. Seven significant trait-marker associations which were detected in Panel 1 using MLM model were identical with those in Panel 2 and Panel 3 using the GLM model, respectively. However, there was no identical trait-marker association within the three Panels when using the MLM model (Table 5). In Panel 2, RM219 [47], RM469 [48] and RM204 [49] showed significant associations with plant height and they were also reported by previous researches. Among them, the association for marker RM469 with plant height had the highest R^2 (10.08%). Similarly, in Panel 3, the association for marker RM590 with plant height had the highest R^2 (39.96%). RM339 which showed significant associations with heading days, were reported by previous researches [50] (Table 6).

Performance of traits relevant to different alleles of significant loci

Seven markers, i.e. PSM184, RM447, RM469, RM235, RM206, RM311, and RM277, were selected for analysis of trait performance relevant to different alleles of significant loci based on their high explained percentage of genetic variation and supported by several significant thresholds (Table 4). For PSM184, the individuals carrying the allele 222 bp (the size of PCR product for the SSR markers, the same as below) had a significantly ($P<0.01$) lower plant height and panicle length than those carrying other two alleles 205 bp and 215 bp (Table 7). For RM447, the individuals carrying the allele 109 bp had a significantly ($P<0.01$) higher grain width and significantly ($P<0.01$) lower grain length/width ratio than those carrying other two alleles 100 bp and 117 bp. For RM469, the individuals carrying the allele 94 bp had a significantly ($P<0.01$) lower flag leaf length than those carrying other two other alleles 83 bp and 88 bp. For RM206, the individuals carrying the allele 162 bp had a significantly ($P<0.01$) higher 1000-grain weight than those carrying the other four alleles 123 bp, 125 bp, 130 bp and 143 bp. For RM311, the individuals carrying the allele 143 bp, 143 bp and 153 bp showed a significantly ($P<0.05$) higher panicle number per plant than those carrying other two alleles 147 bp and 157 bp. For RM235, the individuals carrying the allele 108 bp showed a significantly ($P<0.05$) higher flag leaf width than those carrying the alleles 115 bp, 117 bp, 121 bp and 123 bp, whereas the individuals carrying the allele 123 bp had a significantly ($P<0.05$) lower flag leaf width

Table 2. Correlation coefficients for 12 agronomic traits in 2008 and 2009.

	HD	PH	SS	PL	GL	GW	GL/GW	1000GW	FLL	FLW	FLL/FLW	PN
HD		0.263**	−0.085	0.072	−0.017	0.028	−0.040	−0.316**	0.168	0.497**	0.034	−0.103
PH	0.425**		0.107	0.544**	0.104	0.112	−0.092	0.189*	0.477**	0.240**	0.224**	−0.124
SS	−0.176*	0.005		0.074	−0.140	−0.005	−0.083	0.301**	0.074	0.146	−0.044	−0.151
PL	0.364**	0.648**	0.013		0.146	0.014	0.024	0.143	0.378**	0.294**	0.116	−0.074
GL	0.031	0.018	−0.022	0.254**		0.014	0.515**	0.247**	0.153	−0.014	0.116	−0.080
GW	−0.076	0.120	0.007	−0.035	−0.363**		−0.836**	0.233**	−0.010	0.000	0.015	−0.040
GL/GW	0.064	−0.112	−0.016	0.135	0.758**	−0.864**		−0.103	0.043	−0.010	0.015	−0.006
1000GW	−0.325**	−0.003	0.255**	0.098	0.280**	0.514**	−0.263**		0.307**	0.089	0.168	−0.157
FLL	0.064	0.243**	0.014	0.337**	0.070	0.072	−0.012	0.149		0.098	0.732**	−0.202*
FLW	0.216**	0.291**	0.016	0.282**	−0.014	−0.042	0.035	−0.073	0.097		−0.569**	−0.324**
FLL/FLW	−0.129	−0.034	0.021	−0.037	−0.022	0.133	−0.106	0.129	0.591**	−0.681**		0.042
PN	−0.079	−0.120	0.116	−0.133	0.167*	−0.107	0.160	0.021	−0.051	−0.404**	0.286**	

Note: Above the diagonal is the Pearson correlation coefficient in 2008 and below the diagonal there is the Pearson correlation coefficient in 2009. HD: Heading days, PH: Plant height, SS: Seed set rate, PL: Panicle length, GL: Grain length, GW: Grain width, 1000GW: 1000-grain weight, FLL: Flag leaf length, FLW: Flag leaf width and PN: Panicles number per plant.
**,* represents significant correlation at $\alpha = 0.01$ and 0.05, respectively.

than those carrying the alleles 91 bp, 108 bp, and 115 bp. For RM277, the individuals carrying the allele 117 bp had a higher grain length than those carrying the allele 111 bp (Duncan multiple comparisons was not been performed due to it had only two alleles).

Discussion

Comparison of different mapping populations for association mapping

An appropriate population with maximized phenotypic variation is critical for the success of an association analysis [18,51]. Rice landraces represent an intermediate stage in domestication between wild and elite cultivars [2], which possess high genetic diversity and many exotic genes, and therewith provide useful germplasm resources for rice breeding. Moreover, association mapping based on a core collection of rice landraces would help to catch as much phenotypic variation as possible.

China is well known as one of the origin center of cultivated rice with abundant genetic resources for rice. As early as in 1920–1964s, Professor Ying Ting collected more than 7128 accessions of rice landraces from all over China as well as some countries which grow rice as a major crop. The collection is one of the earliest collections for rice germplasm resources and therefore was named Ting's rice germplasm collection [30]. Our previous results based on the core collection from it indicated that (1) the percentage of SSR loci pairs in significant ($P<0.05$) LD was 46.8%; (2) LD decayed rapidly to the threshold, i.e. the 95% quantile of r^2 between unlinked loci pairs, at 1.03 cM in the entire collection; and (3) there were many LD blocks. These previous results indicated that Panel 1 was an appropriate population for association mapping. Therefore, our association mapping was performed based on Panel 1.

The populations in previous studies for association analysis in rice included populations from the USDA core collection [14,16,24], landraces [16,17], elite cultivars [16], and mini-core collection [23]. The mapping populations in the researches of Agrama et al. [14,24,52] and Li et al. [23] were subsets chosen randomly from the USDA core collection, which consisted of 92, 547 and 203 accessions, respectively. Moreover, the number of SSR markers was 123, 72 and 155, which was rather low for association mapping. In the study of Zhao et al. [11], 416 rice accessions including only two landraces were randomly selected and only 100 SSR markers were used.

Our results indicated that there is a wide-range of phenotypic variation for 12 agronomic traits in Panel 1. For heading days, flag leaf length, flag leaf width, grain length, grain width, grain length/width and panicle length, there was less phenotypic variations than described in the research of Jin et al. [16], while for plant height and 1000 grains weight, more phenotypic variation was found than reported in the research of Jin et al. [16]. The comparison with the results of Li et al. [23] indicated that less phenotypic variation was found in this study for heading days, 1000-grain weight and panicle length, while more was found for plant height, panicle number per plant and seed set rate. More phenotypic variation was found than reported in the research of Agrama et al. [14] for grain length, grain width and 1000-grain weight.

Choice of statistical models and statistical parameters to control type I error

There are two frequently used models (i.e. MLM and GLM) which were implemented in the software TASSEL for association analysis [17,23,28]. In this study, we used the MLM (Q+K) [25] which accounted for population structure and kinship relationship

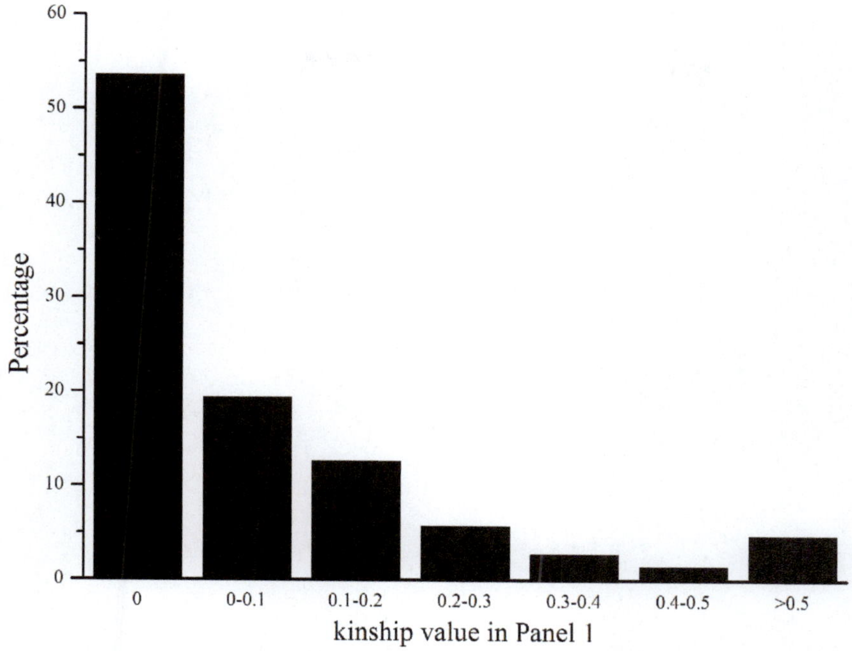

Figure 1. Distribution of pairwise relative kinship values in Panel 1. The height of the black bar represents the percentage of varieties in different ranges of kinships.

to minimize spurious associations. For comparison, GLM was also used. In our study, 137 (~90%) trait-marker associations were possibly new loci when using GLM model, whereas 52 (~68%) trait-marker associations were possibly new loci when using MLM model. The ratio of possibly new significant loci detected using GLM model was much higher than that using MLM model. However, the new significant loci might be false positive because GLM model did not account for kinship.

Furthermore, the significance threshold (*P* value) must be set considerately in the association mapping. Using a smaller *P* value as threshold might lose more minor QTLs, while using a higher *P* value as threshold might get more false positive QTLs. To reliably

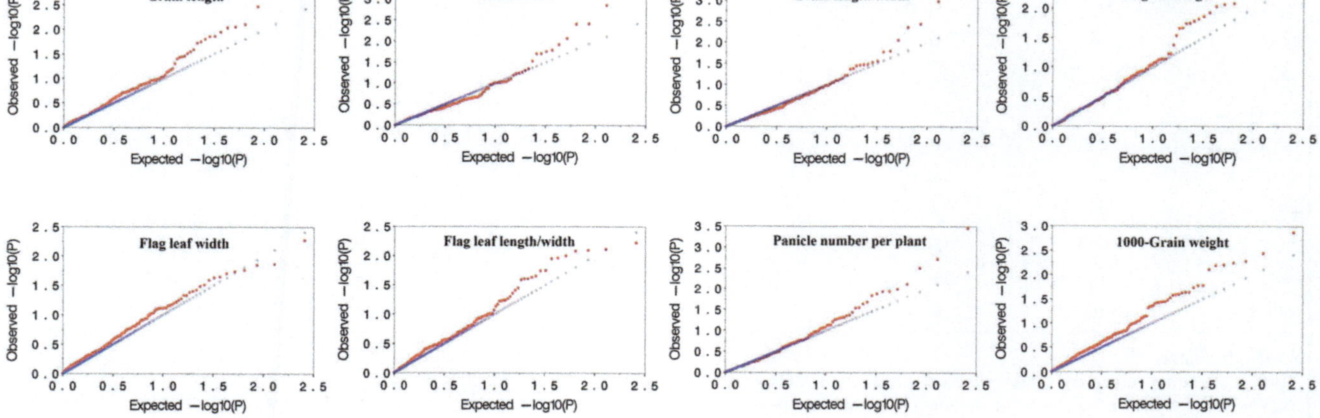

Figure 2. Plots of observed versus expected *P*-values using MLM (Q+K) model for 12 agronomic traits in 2008. The blue symbol the represents expected *P*-values, and the red symbol represents the observed *P*-values.

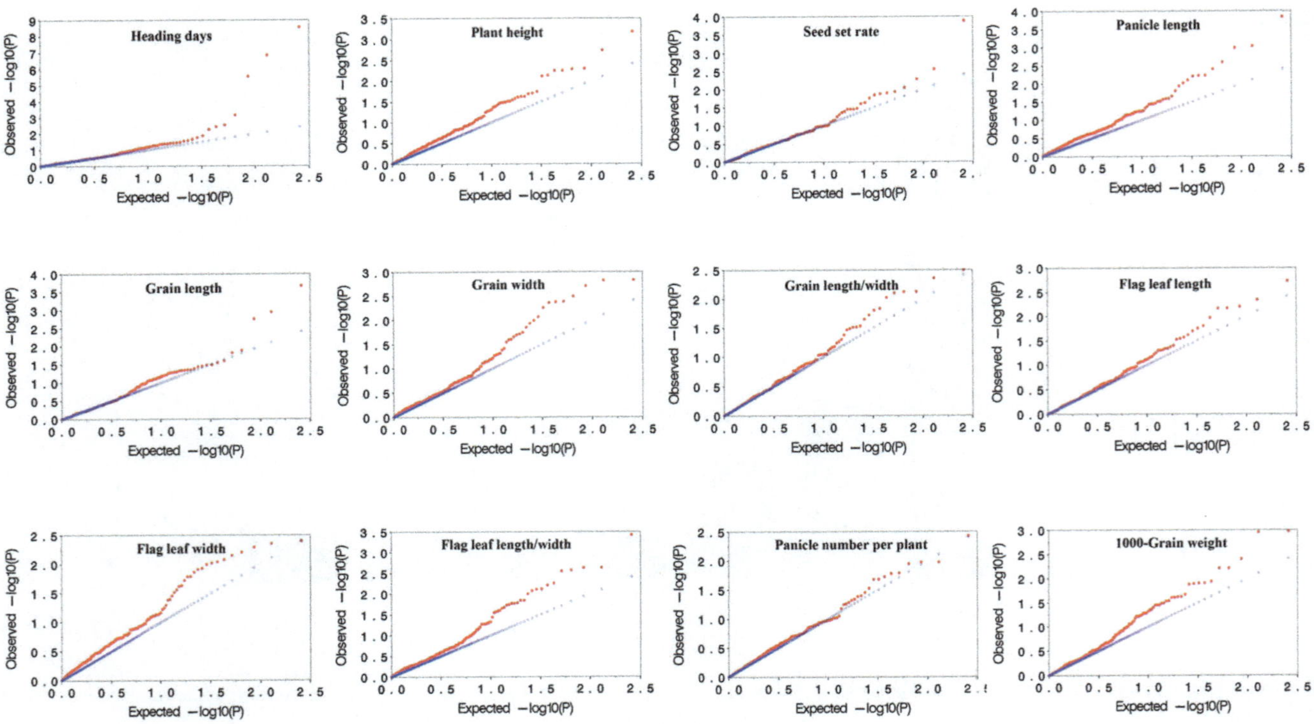

Figure 3. Plots of observed versus expected *P*-values using MLM (Q+K) model for 12 agronomic traits in 2009. Blue symbol represents expected *P*-values, and red symbol represents observed *P*-values.

interpret the MLM-derived significant associations in our study, we also used minimum BF estimation [44] for the MLM association results. Minimum BF estimates over *P* values of MLM approach may help to understand the overall impact of the associations [45]. We also used a Bonferroni threshold for identifying the associations derived from MLM analysis. The

Table 3. Summary of association mapping results for 12 agronomic traits using MLM model in Panel 1.

Traits	No.[a]	No. of significant associations using MLM		
		2008	2009	No.[c]
HD	4(2)[b]	11(6)	15(8)	10(0)
PH	13(10)	20(14)	16(12)	21(5)
SS	0(0)	9(2)	15(4)	1(0)
PL	2(2)	18(11)	24(13)	11(2)
GL	10(3)	19(4)	13(2)	9(1)
GW	6(0)	13(1)	18(4)	2(0)
GL/GW	9(2)	14(3)	14(3)	9(2)
1000GW	8(3)	22(12)	24(12)	15(1)
FLL	5(0)	14(2)	16(4)	4(0)
FLW	10(1)	13(2)	22(4)	38(2)
FLL/FLW	8(0)	15(1)	25(0)	23(1)
PN	1(1)	16(7)	15(9)	9(1)
Total	76(24)	184(65)	217(75)	152(15)

Note: In this table,
[a]number of SSR loci shows the same trait-marker association (MLM, $P<0.05$) in the both years;
[b]number in parentheses represents the number of trait-marker associations which is located in the same or similar genomic region where QTLs were detected in previous studies;
[c]the number of SSR loci showing the same trait-marker association (GLM, $P<0.05$) in both years.
HD: Heading days, PH: Plant height, SS: Seed set rate, PL: Panicle length, GL: Grain length, GW: Grain width, 1000GW: 1000-grain weight, FLL: Flag leaf length, FLW: Flag leaf width and PN: Panicles number per plant.

Table 4. Association mapping results for 12 agronomic traits in two years using MLM model in Panel 1.

No.	Marker names	Chr. No	Chr. position(cM)	2008		2009		QTL Accession ID[f]	QTL region (cM)
				P value (MLM)	R^2 (%)	*P* value (MLM)	R^2 (%)		
Heading days									
1	RM341	2	82.7	0.0382[e]	3.29	0.0037[ace]	6.18	AQFW189	81.0–129.6
2	RM339	8	72.2	0.0469[ce]	4.72	0.0291[e]	5.16	AQS010	45.4–73.0
3	RM224	11	120.1	0.0246[c]	8.72	0.0176[c]	8.81	-	-
4	PSM184	12	26.0	0.0289	11.08	0.0001[bcd]	35.4	-	-
Plant height									
1	RM530	2	170.1	0.0015[bcde]	12.06	0.0007[bcde]	12.16	AQCU198	165.4–189.4
2	RM138	2	196.8	0.0074[ace]	5.74	0.0018[bcde]	6.93	CQV7	189.9–202.8
3	PSM130	3	130.7	0.0395[ce]	4.88	0.0340[ce]	4.62	CQAB22	94.3–132.8
4	RM469	6	2.2	0.0094[ace]	10.86	0.0241[e]	7.98	AQFW082	0–17.3
5	RM204	6	25.1	0.0146[ce]	9.51	0.0050[ace]	10.37	AQHR045	24.1–37.9
6	RM225	6	26.2	0.0449[e]	6.10	0.0497[e]	5.35	AQHR045	24.1–37.9
7	RM18	7	90.4	0.0010[bcd]	8.26	0.0314[c]	3.16	-	-
8	RM219	9	11.7	0.0029[bcde]	10.82	0.0360[ce]	5.86	AQGS003	0–23.8
9	PSM167	10	55.6	0.0478[c]	2.93	0.0074[ac]	4.92	-	-
10	RM147	10	99.8	0.0462[e]	7.20	0.0392[e]	8.27	AQFF070	90.4–106.0
11	RM590	10	117.2	0.0078[a]	14.89	0.0056[a]	14.08	-	-
12	RM21	11	85.7	0.0396[e]	7.66	0.0439[e]	6.82	AQHR060	80.8–87.7
13	PSM184	12	26.0	0.0015[bcde]	16.95	0.0424[e]	9.01	AQFW086	14.9–44.1
Panicle length									
1	RM228	10	96.3	0.0404[ce]	3.40	0.0313[ce]	3.20	AQCU037	88.0–129.0
2	PSM184	12	26.0	0.0424[ce]	10.80	0.0009[bcde]	16.44	AQFJ050	0–28.1
Grain length									
1	RM81A	1	77.5	0.0307[c]	5.26	0.0357[c]	4.75	-	-
2	RM341	2	82.7	0.0078[ac]	5.36	0.0017[bcd]	7.08	-	-
3	RM156	3	125.7	0.0010[bcd]	8.30	0.0002[bcd]	10.10	-	-
4	RM127	4	150.1	0.0352[ce]	5.05	0.0480[e]	4.33	CQAL24	149.7–173.2
5	PSM158	9	33.0	0.0112[ce]	6.84	0.0372[ce]	4.69	AQCA004	0–63.4
6	PSM171	11	4.8	0.0284[ce]	5.38	0.0420[ce]	4.52	AQT016	0–72.7
7	RM144	11	123.2	0.0012[bcd]	8.01	0.0288[c]	3.40	-	-
8	RM4A	12	5.2	0.0399[c]	6.33	0.0450[c]	5.77	-	-
9	RM277	12	57.2	0.0190[c]	9.01	0.0011[bcd]	13.42	-	-
10	PSM191	12	99.7	0.0359[c]	5.05	0.0470[c]	4.37	-	-

Table 4. Cont.

No.	Marker names	Chr. No	Chr. position(cM)	2008		2009		QTL Accession ID[f]	QTL region (cM)
				P value (MLM)	R^2 (%)	*P* value (MLM)	R^2 (%)		
Grain width									
1	RM237	1	115.2	0.0162[c]	4.40	0.0019[bcd]	6.95	-	-
2	RM276	6	40.3	0.0013[bcd]	8.14	0.0015[bcd]	7.50	-	-
3	RM557	6	41.9	0.0038[ac]	8.67	0.0032[acd]	8.53	-	-
4	RM223	8	80.5	0.0005[bcd]	9.50	0.0086[ac]	4.96	-	-
5	RM447	8	124.6	0.0195[c]	9.12	0.0014[bcd]	13.13	-	-
6	RM19	12	20.9	0.0085[ac]	9.05	0.0044[ac]	9.58	-	-
GL/GW									
1	RM237	1	115.2	0.0036[acd]	6.38	0.0158[c]	4.06	-	-
2	RM208	2	186.4	0.0299[c]	8.07	0.0191	9.54	-	-
3	RM559	4	155.8	0.0169[c]	7.69	0.0343[c]	6.04	-	-
4	RM276	6	40.3	0.0161[ce]	4.39	0.0421[ce]	2.91	CQAL18	18.5–49.0
5	RM557	6	41.9	0.0043[ace]	8.26	0.0307[ce]	4.89	CQAL18	18.5–49.0
6	RM223	8	80.5	0.0006[bcd]	9.11	0.0075[ac]	4.99	-	-
7	RM447	8	124.6	0.0087[a]	10.40	0.0032[acd]	11.41	-	-
8	RM224	11	120.1	0.0357	7.73	0.0044[ac]	10.80	-	-
9	RM277	12	57.2	0.0329[c]	7.89	0.0296[c]	7.54	-	-
Flag leaf length									
1	RM138	2	196.8	0.0092[ac]	5.34	0.0401	3.14	-	-
2	RM469	6	2.2	0.0182[c]	9.40	0.0287[c]	8.14	-	-
3	PSM340	9	90.1	0.0046[ac]	6.02	0.0272[c]	3.55	-	-
4	RM167	11	20.3	0.0085[ac]	7.17	0.0063[ac]	7.49	-	-
5	RM287	11	68.6	0.0032[acd]	12.22	0.0045[ac]	11.28	-	-
Flag leaf width									
1	PSM128	3	96.6	0.0326[c]	3.52	0.0242[c]	3.70	-	-
2	RM571	3	205.4	0.0417[ce]	4.92	0.0091[ace]	6.91	AQFW011	190.5–209.0
3	RM225	6	26.2	0.0172[c]	7.95	0.0353[c]	6.28	-	-
4	RM584	6	26.2	0.0397[c]	3.26	0.0123[c]	4.57	-	-
5	RM182	7	61.0	0.0224	5.90	0.0277[c]	5.25	-	-
6	RM407	8	5.7	0.0257[c]	8.67	0.0095[ac]	11.53	-	-
7	RM339	8	72.2	0.0197[c]	6.11	0.0040[ac]	8.18	-	-
8	RM222	10	11.3	0.0053[ac]	6.07	0.0062[ac]	5.49	-	-
9	PSM170	10	68.6	0.0181[c]	8.68	0.0138[c]	8.62	-	-

Table 4. Cont.

No.	Marker names	Chr. No	Chr. position(cM)	2008		2009		QTL Accession ID[f]	QTL region (cM)
				P value (MLM)	R^2 (%)	*P* value (MLM)	R^2 (%)		
10	RM235	12	91.3	0.0140[c]	16.67	0.0308[c]	13.71	-	-
FLL/FLW									
1	RM571	3	205.4	0.0246[c]	5.78	0.0028[acd]	8.58	-	-
2	RM348	4	137.9	0.0231[c]	7.45	0.0056[ac]	9.20	-	-
3	RM559	4	155.8	0.0499	6.07	0.0004[bcd]	13.68	-	-
4	RM153	5	3.0	0.0388[c]	5.05	0.0172[c]	5.88	-	-
5	RM469	6	2.2	0.0108[c]	10.66	0.0225[c]	8.34	-	-
6	PSM340	9	90.1	0.0078[ac]	5.51	0.0146[c]	4.29	-	-
7	PSM172	11	7.6	0.0110[c]	8.77	0.0027[bcd]	10.40	-	-
8	RM167	11	20.3	0.0058[ac]	8.10	0.0023[bcd]	8.86	-	-
1000GW									
1	RM341	2	82.7	0.0036[acd]	6.89	0.0128[c]	4.49	-	-
2	PSM374	2	83.6	0.0078[ac]	7.92	0.0400[c]	4.69	-	-
3	RM7	3	64	0.0052[ace]	13.92	0.0125[ce]	10.77	AQDV056	51.3–86.7
4	RM252	4	99	0.0256[c]	7.53	0.0011[bcd]	12.00	-	-
5	RM538	5	132.7	0.0366	8.51	0.0321[c]	12.38	-	-
6	RM447	8	124.6	0.0331[c]	9.03	0.0355[c]	7.83	-	-
7	RM239	10	25.2	0.0253[ce]	8.51	0.0295[ce]	7.85	AQEY016	21.1–44.4
8	RM206	11	102.9	0.0013[bcde]	9.03	0.0115[ce]	12.19	AQGP076	102.9–102.9
Panicle number per plant									
1	RM311	10	25.2	0.0031[bcde]	17.52	0.0190[ce]	12.42	AQDY128	20.9–25.9

Note:
[a]BFmin with moderate to strong evidence for association (>0.05–0.13);
[b]BFmin with strong to very strong evidence for association (≤0.05);
[c]supported by the GLM in TASSEL (≤0.05);
[d]the Bonferroni threshold (<0.0036);
[e]supported by previous literature; R^2 represents the genetic variants explained by the marker; QTLs detected in previous studies (http://www.gramene.org/).

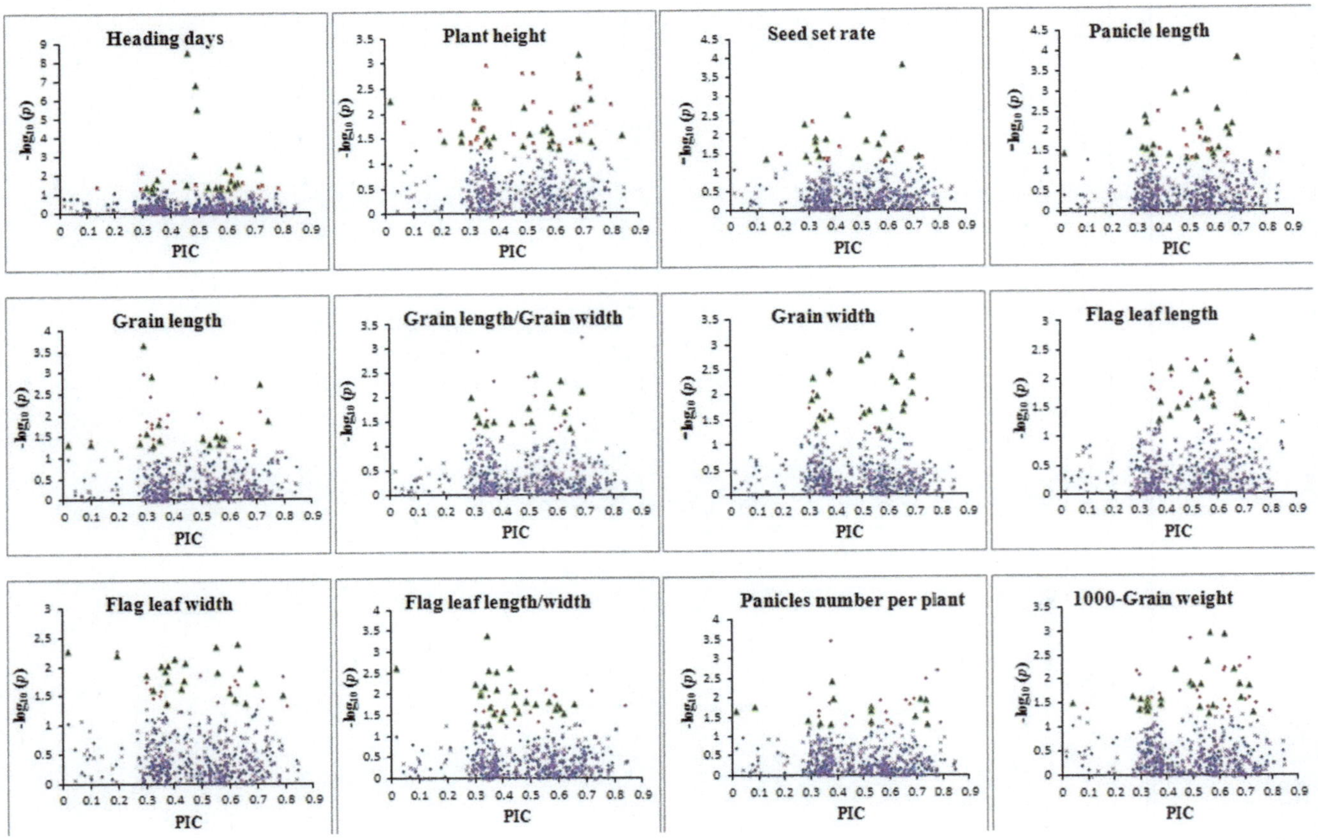

Figure 4. Relationship between PIC and *P*-value for marker–trait associations for 12 agronomic traits in two years. Green asterisk refers to the total markers used in traits in 2008. A red asterisk refers to the markers significantly associated with traits in 2008. A purple asterisk refers to the total markers used in traits in 2009. A green triangle refers to the markers significantly associated with traits in 2009.

statistical parameters had been used successfully in association mapping of cotton [8]. Our results indicated that three significant trait-marker associations (i.e. plant height-RM530, grain length-RM156 and grain width-RM276) reached simultaneously the three thresholds (i.e. $P<0.05$, minimum BF, and the Bonferroni), which should be emphasized in future studies.

Moreover, molecular markers can be used to calculate the relative kinship between pairs of individuals in a study, which provides useful information for quantitative inheritance studies. Relative kinship reflects the approximate identity between two given individuals over an average probability of identity between two random individuals [25]. Our results indicated that most varieties had no or weak relationship with each other in the Ting's core collection, which might be due to the fact that these varieties were chosen from a diverse rice cultivating region including all over China, East Asia, and Southeast Asia. The quantile-quantile plot indicated that MLM (Q+K) performed well in association mapping on 12 agronomic traits, which could correct false positive trait-marker associations (Figure 2 and 3).

Association analysis within Ting's core collection

Using Ting's rice core collection genotyped with 274 SSR markers, we performed association mapping for 12 agronomics traits with two years data using the MLM and GLM models implemented in TASSEL. In this study, most (~80%) of the significant associations found using the MLM approach were also supported by the GLM approach in both years. The percentage of associations identical to previous reported QTLs was about 32%, which was higher than those in the research of Li et al. [23], but

Table 5. Summary of trait-marker associations within the three Panels.

Population	MLM[(1)]	GLM[(2)]	GLM[(3)]
Panel 2	0	7(3)	3(1)
Panel 3	0	7(2)	2(1)

Note:
[(1)]Number of the same trait-marker associations using MLM found both in Panel 1 and Panel 2 or Panel 3;
[(2)]Number of the same trait-marker associations using GLM ($P<0.05$) found both in Panel 1 and Panel 2 or Panel 3.
[(3)]Number of the same trait-marker associations using GLM ($P<0.01$) found both in Panel 1 and Panel 2 or Panel 3.
In parentheses, the number of trait-marker associations which are identical with the published mapping results in previous literature is given.

Table 6. The same trait-marker associations in Panel 2 and 3 using GLM model compared with those in Panel 1.

Trait	Panel 2						Panel 3					
	Marker names	Chr. No	Genetic distance (cM)	R^2 (%)	QTL Accession ID	QTL region (cM)	Marker names	Chr. No	Genetic distance (cM)	R^2 (%)	QTL Accession ID	QTL region (cM)
Plant height	RM204[a]	6	25.1	4.79	AQHR045	24.1–37.9	RM590	10	117.2	39.96	AQEX006	113.5–117.0
	RM219[a]	9	11.7	6.46	AQGS003	0–23.8						
	RM469[a]	6	2.2	10.08	AQFW082	0–17.3						
Grain length	PSM158[a]	9	33.0	4.69	AQCA004	0–63.4	PSM191	12	99.7	11.42	AQCV009	47.0–51.5
GL/GW	RM559	4	155.8	3.73	AQFA016	47.8–47.8	RM208	2	186.4	6.74	AQGB055	55.9–77.9
							RM277	12	57.2	5.44	AQCV021	30.0–39.7
Flag leaf length	RM138	2	196.8	7.34	AQFW023	205.8–231.5						
Heading days							RM339[a]	8	72.2	9.00	AQS010	45.4–73.0
							PSM184	12	26.0	3.92	AQAX020	17.5–22.2
FLL/FLW	RM469	6	2.2	9.90	AQEJ014	33.7–38.3						
1000-GW							PSM374	2	83.6	4.47	AQT007	0–81.4

Note: [a]represents the locus identical with previous mapping results, R^2 represents the genetic variance explained by the marker.

lower than those in the research of Agrama et al. [14]. The 76 significant trait-marker associations which were detected in both years were potential markers for effective marker-assisted selection programs in rice. Moreover, 52 of the 76 significant associations which were not detected in previous studies might be some new potential loci. For instance, the trait-marker associations for heading days with PSM184, plant height with RM590, grain length/width with RM447, flag leaf length with RM287, flag leaf width with RM235, 1000-grain weight with RM538, and 1000-grain weight with RM206, explained more than 10% of genetic variations both in 2008 and 2009.

For heading days, two of the four significant trait-marker associations were identical to previous reported QTLs, i.e. RM341 and RM339, were identical to previous reported QTLs in the research of Mei et al. [48] and Kunihiro et al. [50], respectively. Moreover, RM339 was also significantly associated with heading days in Panel 2 and 3. For heading days, ten of 13 significant trait-marker associations were identical to previous reported QTLs, i.e. RM530 in the research of Mei et al. [53], RM138 in the research of Fang et al. [51], PSM130 in the research of Cao et al. [54], RM469 (which also showed significant association in Panel 2 and 3) and PSM184 in the research of Mei et al. [48], RM204 (which also showed significant association in Panel 2 and 3) and RM225 in the research of Yang et al. [49], RM219 (which also showed significant association in Panel 2 and 3) in the research of Xiao et al. [47], RM21 and RM147 in the research of Lanceras et al. [55]. For panicle length, the two significant trait-marker associations were also identical to previous reported QTLs, i.e. RM228 and PSM184 in the research of Mei et al. [53] and Jiang et al. [56], respectively. For grain length, three of ten significant trait-marker associations were identical to previous reported QTLs in the previous researches, i.e. RM127 in the research of Tan et al. [57], PSM158 in the research of Xing et al. [58], and PSM171 in the research of Yoshida et al. [59]. For grain length/width, two of nine significant trait-marker associations were identical to previous reported QTLs in the previous researches, i.e. RM276 and RM557 reported by Tan et al. [57]. For flag leaf width, one of nine significant trait-marker associations were identical to previous reported QTLs, i.e. RM571 in the research of Mei et al. [48]. For 1000-grain weight, there of eight significant trait-marker associations were identical to previous reported QTLs in the previous researches, i.e. RM7 in the research of Hittalmani et al. [60], RM239 in the research of Gao et al. [61], and RM206 in the research of Cho et al. (this reference cannot be found, but QTL ID can be found in GRAMENE website). For panicle number per plant, the only one significant trait-marker association was also identical to previous reported QTL, i.e. RM311 in the research of Kobayashi et al. [62].

Verification association mapping results within Panel 2 and Panel 3

It is worthwhile to further verify the significant associations identified within one population in a different population [29]. In this study, 55 SSR markers for the 76 trait-marker associations identified in Panel 1 were used to genotype two other populations, i.e. Panel 2 and Panel 3, and an association mapping was performed using both MLM and GLM approaches. When using the GLM approach, seven significant trait-marker associations were identical within Panel 1 and Panel 2 or Panel 3. Moreover, three of the seven identical significant trait-marker associations in the two panels were reported by previous studies. Although the GLM would bring more false positive results than the MLM when it was used alone, however, some significant trait-marker associations were first detected Panel 1 in our research and

Table 7. Duncan multiple comparisons for different allelic effects on traits.

Trait	Marker	Allele (bp)	Mean±SD	Trait	Marker	Allele (bp)	Mean±SD
Plant height (cm)	PSM184	205	148.96 ± 26.94^{Aa}	1000-grain weight (g)	RM206	123	21.76 ± 4.48^{Aa}
		215	152.34 ± 28.01^{Aa}			125	21.80 ± 2.38^{Aa}
		222	134.11 ± 32.91^{Bb}			130	22.39 ± 3.94^{Aa}
Panicle length (cm)	PSM184	205	25.52 ± 3.14^{Aa}			143	20.71 ± 3.51^{Aa}
		215	25.37 ± 3.13^{ABa}			162	26.09 ± 5.30^{Bb}
		222	23.89 ± 4.12^{Bb}	Panicle number per plant	RM311	143	8.14 ± 2.77^{Aa}
Grain width (mm)	RM447	100	3.06 ± 0.31^{Aa}			145	8.66 ± 2.32^{Aa}
		109	3.24 ± 0.40^{Bb}			147	7.86 ± 2.18^{Ab}
		117	3.05 ± 0.36^{Aa}			153	8.75 ± 3.18^{Aa}
Grain length/width	RM447	100	2.66 ± 0.32^{Aa}			157	7.54 ± 2.10^{Ab}
		109	2.44 ± 0.44^{Bb}	Flag leaf width (cm)	RM235	91	1.79 ± 0.19^{Aab}
		117	2.69 ± 0.45^{Aa}			108	1.82 ± 0.21^{Aa}
Flag leaf length (cm)	RM469	83	42.02 ± 7.47^{Aa}			115	1.70 ± 0.18^{Ab}
		88	43.08 ± 8.13^{Aa}			117	1.63 ± 0.35^{Abc}
		94	39.10 ± 7.84^{Bb}			121	1.59 ± 0.28^{Abc}
						123	1.57 ± 0.25^{Ac}

Note: Capital and lower letters represent significant difference at $\alpha = 0.01$ and 0.05, respectively. Allele (bp) is PCR product amplified by SSR markers.

proved by several statistical thresholds as well as by previous mapping results. After that, we used the GLM to verify our mapping results in Panel 2 and 3. Therefore, it makes sense for verification of association mapping results by the fact that some common trait-marker associations were detected by the GLM approach.

We observed that there were no overlapping QTLs among the three panels with the GLM approach. The reasons might be (1) different compositions and origins of the varieties in three panels, where Panel 1 only consists of original rice landraces from China and some other rice growing countries which were collected during 1920–1964 before the emergence of hybrid rice, while Panel 2 consists of rice landraces as well as modern rice cultivars and maintainer lines in hybrid rice breeding from China, and Panel 3 is a worldwide collection and consists of modern rice cultivars including cytoplasmic sterile line, maintainer lines, and some landraces; (2) that different allelic frequencies might exist for the three panels which consist of different compositions and origins. The explanations were supported by our observations that (1) frequency of some alleles was different in the three panels and some alleles only exist in one panel (Table S3 in File S1), and (2) in our another experiment some alleles associated with aluminum tolerance were different for different germplasm types (data not shown).

When using the MLM approach, no identical significant trait-marker associations were found among the three panels. Previous studies on linkage mapping and association mapping also found that different mapping populations detected different QTL regions [14,48,63,64,65]. The reasons might be due to that (1) a much lower number of SSR markers (55 SSRs) was used in Panel 2 and Panel 3 than in Panel 1 (274 SSRs); (2) the 55 SSR markers are associated with relevant traits which were not randomly distributed across the genome, which might reduce the exactness of measurement for population structure and kinship; (3) the relative kinship calculated by 274 SSRs in Panel 1 was quite different than those calculated by the 55 SSRs in Panel 2 and 3, where in Panel 1 only 4.73% of pairwise kinship coefficient were larger than 0.5 and most of them were zero, whereas 55.9% and 60.4% of pairwise kinship coefficient in Panel 2 and 3 were larger than 0.5, respectively (Figure S8 in File S1); and (4) the degree of association might be reduced in MLM compared to those in GLM [50], which meant that when using much less SSR markers, the weak significant trait-marker associations in GLM might be not significant in MLM. As verification experiments were rarely performed in previous association studies, it is required to find an efficient solution for verification in future as well as to check the repeatability in different association mapping populations.

Prospects for association mapping based on core collections

Association mapping has become a promising approach to mine elite genes within germplasm populations compared to traditional linkage mapping. Association mapping based on a core collection would help to capture as much phenotypic variation as possible. Compared to a natural population or a breeding population with a broad genetic basis, the LD level in a core collection might be low due to its diverse origin. Therefore, more markers might be required for association mapping. However, due to the quick LD decay, fine mapping using association analysis might be possible with a core collection. As quick, automated, economic genotyping technologies (such as genotyping by sequencing) have been developed, genotyping large germplasm resources with high density markers and GWAS in such mapping populations has become possible. Because such an association could be further applied in rice breeding by molecular marker assisted selection, it would be promising to make use of the elite genes in the diverse germplasm resources by the current strategy.

Supporting Information

File S1 **Table S1,** Accessions, variety names, origin, germplasm types of 150 rice varieties in Panel 1. **Table S2,** Summary statistics of the 274 SSR markers used in this study. **Table S3,** Allele frequency of the 55 significant markers in three panels. **Figure S1,** Frequency distribution of heading days, plant height, seed set rate and panicle length in Panel 1 in 2008. The height of black bar represents the number of varieties in different range of traits. **Figure S2,** Frequency distribution of grain length, grain width, grain length/width and 1000 grain weight in Panel 1 in 2008. The height of black bar represents the number of varieties in different range of traits. **Figure S3,** Frequency distribution of flag leaf length, flag leaf width, flag leaf length/width and panicle number per plant in Panel 1 in 2008. The height of black bar represents the number of varieties in different range of traits. **Figure S4,** Frequency distribution of heading days, plant height, seed set rate and panicle length in Panel 1 in 2009. The height of black bar represents the number of varieties in different range of traits. **Figure S5,** Frequency distribution of grain length, grain width, grain length/width and 1000 grain weight in Panel 1 in 2009. The height of black bar represents the number of varieties in different range of traits. **Figure S6,** Frequency distribution of flag leaf length, flag leaf width, flag leaf length/width and panicle number per plant in Panel 1 in 2009. The height of black bar represents the number of varieties in different range of traits. **Figure S7,** Distribution of pairwise relative kinship values in Panel 2 and 3. The height of black bar represents the percentage of varieties in different range of kinships. **Figure S8,** Delta K change according to different K among Panel 2 and Panel 3 identified by STRUCTURE under Admixture model.

Acknowledgments

We are grateful to Dr. Guoyou Ye from International Rice Research Institute, Dr. Xiaoling Li, Dr. Lan Wang, Dr. Zhixiong Chen, Dr. Xuelin Fu, Dr. Youxin Yang, Ms Xingjuan Zhao and Ms. Shuhong Yu from South China Agricultural University for their assistance in the experiment, and thank Miss Anja Bus from Max Planck institute for plant breeding research for the improvement of English writing.

Author Contributions

Conceived and designed the experiments: JL PZ. Performed the experiments: PZ. Analyzed the data: PZ JL. Contributed reagents/materials/analysis tools: JL XL HT. Wrote the paper: PZ JL YL.

References

1. Mather KA, Caicedo AL, Polato NR, Olsen KM, McCouch S, et al. (2007) The extent of linkage disequilibrium in rice (*Oryza sativa* L.). Genetics 177: 2223–2232.
2. Londo JP, Chiang YC, Hung KH, Chiang TY, Schaal BA (2006) Phylogeography of Asian wild rice, *Oryza rufipogon*, reveals multiple independent domestications of cultivated rice, *Oryza sativa*. Proc Natl Acad Sci U S A 103: 9578–9583.
3. Flint-Garcia SA, Thornsberry JM, Buckler ES (2003) Structure of linkage disequilibrium in plants. Annu Rev Plant Biol 54: 357–374.

4. Kraakman ATW, Niks RE, Van den Berg PMMM, Stam P, Van Eeuwijk FA (2004) Linkage disequilibrium mapping of yield and yield stability in modern spring barley cultivars. Genetics 168: 435–446.
5. Zhu C, Gore M, Buckler ES, Yu J (2008) Status and Prospects of association mapping in plants. The Plant Genome 1: 5–20.
6. Remington DL, Thornsberry JM, Matsuoka Y, Wilson LM, Whitt SR, et al. (2001) Structure of linkage disequilibrium and phenotypic associations in the maize genome. Proc Natl Acad Sci U S A 98: 11479–11484.
7. Huang X, Han B (2014) Natural variations and genome-wide association studies in crop plants. Annu Rev Plant Biol 65: 531–551.
8. Abdurakhmonov IY, Abdukarimov A (2008) Application of association mapping to understanding the genetic diversity of plant germplasm resources. Int J Plant Genomics 2008: 574927.
9. Han B, Huang X (2013) Sequencing-based genome-wide association study in rice. Curr Opin Plant Biol 16: 133–138.
10. Huang X, Zhao Y, Wei X, Li C, Wang A, et al. (2012) Genome-wide association study of flowering time and grain yield traits in a worldwide collection of rice germplasm. Nat Genet 44: 32–39.
11. Zhao K, Tung CW, Eizenga GC, Wright MH, Ali ML, et al. (2011) Genome-wide association mapping reveals a rich genetic architecture of complex traits in *Oryza sativa*. Nat Commun 2: 467.
12. Famoso AN, Zhao K, Clark RT, Tung CW, Wright MH, et al. (2011) Genetic architecture of aluminum tolerance in rice (*Oryza sativa*) determined through genome-wide association analysis and QTL mapping. Plos Genetics 7.
13. Zhang N, Xu Y, Akash M, McCouch S, Oard JH (2005) Identification of candidate markers associated with agronomic traits in rice using discriminant analysis. Theoretical and Applied Genetics 110: 721–729.
14. Agrama HA, Eizenga GC, Yan W (2007) Association mapping of yield and its components in rice cultivars. Molecular Breeding 19: 341–356.
15. Iwata H, Ebana K, Uga Y, Hayashi T, Jannink JL (2010) Genome-wide association study of grain shape variation among *Oryza sativa* L. germplasms based on elliptic Fourier analysis. Molecular Breeding 25: 203–215.
16. Jin L, Lu Y, Xiao P, Sun M, Corke H, et al. (2010) Genetic diversity and population structure of a diverse set of rice germplasm for association mapping. Theoretical and Applied Genetics 121: 475–487.
17. Huang X, Wei X, Sang T, Zhao Q, Feng Q, et al. (2010) Genome-wide association studies of 14 agronomic traits in rice landraces. Nat Genet 42: 961–967.
18. Yan JB, Warburton M, Crouch J (2011) Association Mapping for Enhancing Maize (*Zea mays* L.) Genetic Improvement. Crop Science 51: 433–449.
19. Frankel OH (1984) Genetic perspectives of germplasm conservation. In: Arber W, Llimensee K, Peacock WJ (Eds.), Genetic Manipulation: Impact on Man and Society. Cambridge University Press, UK, pp.161–170.
20. Frankel OH, Brown AHD (1984a) Current plant genetic resources—a critical appraisal. In: Genetics: new frontiers, vol 4. Oxford and IBH Publ, New Delhi, India, pp. 1–11.
21. Frankel OH, Brown AHD (1984b) Plant genetic resources today: acritical appraisal. In: Hoden HW, Williams JT (eds) Crop genetic resources: conservation and evaluation. George Allen and Urwin, London, pp. 249–257.
22. Brown AHD (1995) The core collection at the crossroads. In: Hodgkin T, Brown AHD, van Hintum TJL, Morales EAV (Eds.), Core Collections of Plant Genetic Resources. John Wiley and Sons, Chichester, UK, pp. 3–19.
23. Li XB, Yan WG, Agrama H, Jia LM, Shen XH, et al. (2011) Mapping QTLs for improving grain yield using the USDA rice mini-core collection. Planta 234: 347–361.
24. Agrama HA, Yan W (2009) Association mapping of straighthead disorder induced by arsenic in *Oryza sativa*. Plant Breeding 128: 551–558.
25. Yu JM, Pressoir G, Briggs WH, Bi IV, Yamasaki M, et al. (2006) A unified mixed-model method for association mapping that accounts for multiple levels of relatedness. Nature Genetics 38: 203–208.
26. Zhu CS, Yu JM (2009) Nonmetric multidimensional scaling corrects for population structure in association mapping with different sample types. Genetics 182: 875–888.
27. Wang ML, Zhu CS, Barkley NA, Chen ZB, Erpelding JE, et al. (2009) Genetic diversity and population structure analysis of accessions in the US historic sweet sorghum collection. Theoretical and Applied Genetics 120: 13–23.
28. Yang XH, Yan JB, Shah T, Warburton ML, Li Q, et al. (2010) Genetic analysis and characterization of a new maize association mapping panel for quantitative trait loci dissection. Theoretical and Applied Genetics 121: 417–431.
29. Wray NR, Yang J, Hayes BJ, Price AL, Goddard ME, et al. (2013) Pitfalls of predicting complex traits from SNPs. Nat Rev Genet 14: 507–515.
30. Li XL, Lu YG, Li JQ, Xu HM, Muhammad QS (2011) Strategies on sample size determination and qualitative and quantitative traits integration to construct core collection of rice (*Oryza sativa*). Rice Science 18: 46–55.
31. Zhang P, Li JQ, Li XL, Liu XD, Zhao XJ, et al. (2011) Population structure and genetic diversity in a rice core collection (*Oryza sativa* L.) investigated with SSR markers. PLoS One 6 (12).
32. Li JQ, Zhang P (2012) Genetic diversity in plants. In: Çalişkan M, editor. Chapter5: assessment and utilization of the genetic diversity in rice. Hard cover: InTech-Open Access Publisher.
33. Chen X, Temnykh S, Xu Y, Cho YG, McCouch SR (1997) Development of a microsatellite framework map providing genome-wide coverage in rice, *Oryza sativa* L. Theoretical and Applied Genetics: 553–567.
34. Temnykh S, Park WD, Ayes N, Cartinhour S, Hauck N, et al. (2000) Mapping and genome organization of microsatellite sequences in rice, *Oryza sativa* L. Theoretical and Applied Genetics: 697–712.
35. Temnykh S, Declerck G, Lukashova A, Lipovich L, Cartinhour S, et al. (2001) Computational and experimental analysis of microsatellites in rice (*Oryza sativa* L.): frenquency, length variation, transposon associations, and genetic marker potential. Genetics Research: 1441–1452.
36. McCouch SR, Teytelman L, Xu Y, Lobos KB, Clare K, et al. (2002) Development and mapping of 2240 new SSR markers for rice (*Oryza sativa* L.). DNA Res 9: 199–207.
37. Huang CF (2003) Development of position-specific microsatellite markers and molecular mapping lf insect resistant genes in rice (*Oryza sativa* L.). M.Sc. Thesis, South China Agricultural University.
38. Zheng KL, Huang N, Bennett J, Khush GS (1995) PCR-based phylogenetic analysis of wide compatibility varieties in *Oryza sativa* L. Theoretical and Applied Genetics: 65–69.
39. Panaud O, Chen X, McCouch SR (1996) Development of microsatellite markers and characterization of simple sequence length polymorphism (SSLP) in rice (*Oryza sativa* L.). Mol Gen Genet 252: 597–607.
40. Pritchard JK, Stephens M, Donnelly P (2000a) Inference of population structure using multilocus genotype data. Genetics 155: 945–959.
41. Pritchard JK, Stephens M, Rosenberg NA, Donnelly P (2000b) Association mapping in structured populations. Am J Hum Genet 67: 170–181.
42. Hardy OJ, Vekemans X (2002) SPAGeDi: a versatile computer program to analyse spatial genetic structure at the individual or population levels. Molecular Ecology Notes 2: 618–620.
43. Yu J, Buckler ES (2006) Genetic association mapping and genome organization of maize. Curr Opin Biotechnol 17: 155–160.
44. Goodman SN (2001) Of P-values and Bayes: a modest proposal. Epidemiology 12: 295–297.
45. Katki HA (2008) Invited commentary: Evidence-based evaluation of p values and Bayes factors. American Journal of Epidemiology 168: 384–388.
46. Moran MD (2003) Arguments for rejecting the sequential Bonferroni in ecological studies. Oikos 100: 403–405.
47. Xiao J, Li J, Grandillo S, Ahn SN, Yuan L, et al. (1998) Identification of trait-improving quantitative trait loci alleles from a wild rice relative, *Oryza rufipogon*. Genetics 150: 899–909.
48. Mei HW, Li ZK, Shu QY, Guo LB, Wang YP, et al. (2005) Gene actions of QTLs affecting several agronomic traits resolved in a recombinant inbred rice population and two backcross populations. Theoretical and Applied Genetics 110: 649–659.
49. Yang GH, Xing YZ, Li SQ, Ding JZ, Yue B, et al. (2006) Molecular dissection of developmental behavior of tiller number and plant height and their relationship in rice (*Oryza sativa* L.). Hereditas 143: 236–245.
50. Kunihiro Y, Qian Q, Sato H, Teng S, Zeng DL, et al. (2002) QTL analysis of sheath blight resistance in rice (*Oryza sativa* L.). Yi Chuan Xue Bao 29: 50–55.
51. Flint-Garcia SA, Thuillet AC, Yu JM, Pressoir G, Romero SM, et al. (2005) Maize association population: a high-resolution platform for quantitative trait locus dissection. Plant J 44: 1054–1064.
52. Agrama HA, Eizenga GC (2008) Molecular diversity and genome-wide linkage disequilibrium patterns in a worldwide collection of *Oryza sativa* and its wild relatives. Euphytica 160: 339–355.
53. Mei HW, Luo LJ, Ying CS, Wang YP, Yu XQ, et al. (2003) Gene actions of QTLs affecting several agronomic traits resolved in a recombinant inbred rice population and two testcross populations. Theoretical and Applied Genetics 107: 89–101.
54. Cao G, Zhu J, He C, Gao Y, Yan J, et al. (2001) Impact of epistasis and QTL×environment interaction on the developmental behavior of plant height in rice (*Oryza sativa* L.). Theoretical and Applied Genetics 103: 153–160.
55. Lanceras JC, Pantuwan G, Jongdee B, Toojinda T (2004) Quantitative trait loci associated with drought tolerance at reproductive stage in rice. Plant Physiology 135: 384–399.
56. Jiang GH, Xu CG, Li XH, He YQ (2004) Characterization of the genetic basis for yield and its component traits of rice revealed by doubled haploid population. Yi Chuan Xue Bao 31: 63–72.
57. Tan YF, Xing YZ, Li JX, Yu SB, Xu CG, et al. (2000) Genetic bases of appearance quality of rice grains in Shanyou 63, an elite rice hybrid. Theoretical and Applied Genetics 101: 823–829.
58. Xing YZ, Tan YF, Xu CG, Hua JP, Sun XL (2001) Mapping quantitative trait loci for grain appearance traits of rice using a recombinant inbred line population. Acta Botanica Sinica 43: 840–845.
59. Yoshida S, Ikegami M, Kuze J, Sawada K, Hashimoto Z, et al. (2002) QTL analysis for plant and grain characters of sake-brewing rice using a doubled haploid population. Breeding Science 52: 309–317.
60. Hittalmani S, Huang N, Courtois B, Venuprasad R, Shashidhar HE, et al. (2003) Identification of QTL for growth- and grain yield-related traits in rice across nine locations of Asia. Theoretical and Applied Genetics 107: 679–690.
61. Gao Y, Zhu J, Song Y, He C, Shi C, et al. (2004) Analysis of digenic epistatic effects and QE interaction effects QTL controlling grain weight in rice. Journal of Zhejiang University Science 5: 371–377.
62. Kobayashi S, Fukuta Y, Sato T, Osaki M, Khush GS (2003) Molecular marker dissection of rice (*Oryza sativa L.*) plant architecture under temperate and tropical climates. Theoretical and Applied Genetics 107: 1350–1356.

63. Huang N, Courtois B, Khush GS, Lin HX, Wang GL, et al. (1996) Association of quantitative trait loci for plant height with major dwarfing genes in rice. Heredity 77: 130–137.
64. Septiningsih EM, Prasetiyono J, Lubis E, Tai TH, Tjubaryat T, et al. (2003) Identification of quantitative trait loci for yield and yield components in an advanced backcross population derived from the *Oryza sativa* variety IR64 and the wild relative *O-rufipogon*. Theoretical and Applied Genetics 107: 1419–1432.
65. Thomson MJ, Tai TH, McClung AM, Lai XH, Hinga ME, et al. (2003) Mapping quantitative trait loci for yield, yield components and morphological traits in an advanced backcross population between *Oryza rufipogon* and the *Oryza sativa* cultivar Jefferson. Theoretical and Applied Genetics 107: 479–493.

3

Expression of Cry1Ab and Cry2Ab by a Polycistronic Transgene with a Self-Cleavage Peptide in Rice

Qichao Zhao[1]◑, Minghong Liu[1]◑, Miaomiao Tan[1], Jianhua Gao[2], Zhicheng Shen[1]*

1 State Key Laboratory of Rice Biology, Institute of Insect Sciences, Zhejiang University, Hangzhou, China, **2** College of Life Science, Shanxi Agricultural University, Taigu, China

Abstract

Insect resistance to *Bacillus thuringiensis* (Bt) crystal protein is a major threat to the long-term use of transgenic Bt crops. Gene stacking is a readily deployable strategy to delay the development of insect resistance while it may also broaden insecticidal spectrum. Here, we report the creation of transgenic rice expressing discrete Cry1Ab and Cry2Ab simultaneously from a single expression cassette using 2A self-cleaving peptides, which are autonomous elements from virus guiding the polycistronic viral gene expression in eukaryotes. The synthetic coding sequences of Cry1Ab and Cry2Ab, linked by the coding sequence of a 2A peptide from either foot and mouth disease virus or porcine teschovirus-1, regardless of order, were all expressed as discrete Cry1Ab and Cry2Ab at high levels in the transgenic rice. Insect bioassays demonstrated that the transgenic plants were highly resistant to lepidopteran pests. This study suggested that 2A peptide can be utilized to express multiple Bt genes at high levels in transgenic crops.

Editor: Mario Soberón, Instituto de Biotecnología, Universidad Nacional Autónoma de México, Mexico

Funding: This research was supported by The National Natural Science Foundation of China (31021003, http://www.nsfc.gov.cn/) received by Zc Shen. The funders had no role in study design,data collection and analysis, decision to publish, or preparation of the manuscript.

Competing Interests: The authors have declared that no competing interests exist.

* Email: zcshen@zju.edu.cn

◑ These authors contributed equally to this work.

Introduction

Since first demonstrated in transgenic tobacco in 1987, Bt crystal toxin genes have been widely utilized in transgenic crops for pest management [1,2]. However, due to the widespread application of Bt toxins, several major insect pests, including fall armyworm (*Spodoptera frugiperda*), dimondback moth (*Plutella xylostella*), maize stalk borer (*Busseola fusca*), cotton bollworm (*Helicoverpa armigera*), western corn rootworm (*Diabrotica virgifera*) and cabbage looper (*Trichoplusia ni*) have developed resistance to Bt toxins in field or greenhouse, which threatens the long-term utilization of Bt crops in the future [3–5].

Several strategies have been proposed and/or deployed to cope with the development of insect resistance to Bt toxins, including high-dose/refuge strategy, discovery of novel insecticidal genes with novel modes of actions, modification of used Bt genes and Bt gene stacking [6–8]. Bt gene stacking strategy introduces more than one Bt genes into plant and has been demonstrated to be an effective way to delay the development of insect resistance to Bt toxins [9–11]. Expression of multiple genes using conventional approaches has several potential limitations, most notably imbalanced expression among different genes and a large T-DNA size required to include multiple genes [12]. Efficient and stoichiometric expression of discrete proteins may be achieved by a polycistronic system involving self-cleaving peptides such as the 2A peptide from foot and mouth disease virus (F2A) or porcine teschovirus-1 (P2A) [12,13]. When expressed in eukaryotic system, nascent 2A peptide can interact with the ribosome exit tunnel to dictate a stop-codon-independent termination at the final proline codon of 2A peptide [14]. Subsequently translation is reinitiated on the same proline codon. When linked by a 2A peptide coding sequence, different genes are co-expressed from a single open reading frame.

2A self-cleaving peptides have been extensively studied previously [15]. In this study, the DNA encoding F2A or P2A was used to link two potent insecticidal Bt genes, the truncated *Cry1Ab* encoding the N-terminal 648 amino acids of active Cry1Ab endotoxin (Genbank:AAG16877.1) and the full-length *Cry2Ab* encoding Cry2Ab endoxin (Genbank:AAA22342.1) with 634 amino acid residues, to generate a polycistronic gene for co-expression in transgenic rice. Analysis of the obtained transgenic rice lines revealed that discrete Cry1Ab and Cry2Ab were indeed co-expressed at a level approximately comparable to transgenic plants expressing traditional monocistronic Bt genes. Insect bioassays demonstrated that the transgenic rice generated was highly resistant to its target insects.

Materials and Methods

Rice cultivar

Elite rice (*Oryza sativa spp. Japonica*) cultivar Xiushui 134 originated from the Jiaxing Academy of Agricultural Science in Zhejiang Province was used for *Agrobacterium*-mediated transformation. The homozygous transgenic rice line KMD1, containing a synthetic truncated *Cry1Ab* gene under control of maize ubiquitin promoter [16,17], was kindly provided by Dr. Gongyin Ye from Institute of Insect Sciences, ZheJiang University.

Construction of binary vector for rice transformation

pCambia1300 (Cambia, Canberra, Australia) was used for construction of binary vectors for plant transformation. This vector was first modified by substituting the hygromycin-resistant gene expression cassette with a glyphosate-tolerant *5-enolpyruvyl-shikimate-3-phosphate synthase* (*EPSPS*) gene expression cassette. The modified vector was named as pCambia1300–GLY, and was further used to clone an expression cassette for a polycistronic gene encoding Cry1Ab and Cry2Ab linked by a 2A peptide (Fig. 1).

Four polycistronic genes of *Cry1Ab* and *Cry2Ab* linked by the coding sequence of F2A or P2A were generated. They were named as *Cry1Ab-F2A-Cry2Ab* (*Cry1-F-Cry2*), *Cry1Ab-P2A-Cry2Ab* (*Cry1-P-Cry2*), *Cry2Ab-F2A-Cry1Ab* (*Cry2-F-Cry1*), and *Cry2Ab-P2A-Cry1Ab* (*Cry2-P-Cry1*), respectively (Fig. 1). *Cry1-F-Cry2* is a fusion gene, in the order from 5' to 3', of coding sequences of Cry1Ab, F2A, and Cry2Ab (GenBank: KJ716232); *Cry1-P-Cry2* is a fusion gene of Cry1Ab, P2A, and Cry2Ab (GenBank: KJ716233); *Cry2-F-Cry1* is a fusion gene of Cry2Ab, F2A, and Cry1Ab (GenBank: KJ716234); and *Cry2-P-Cry1* is a fusion gene of Cry2Ab, P2A, and Cry1Ab (GenBank: KJ716235).

Maize ubiquitin-1 promoter (pUbi) was obtained by PCR from maize genome with primers pUbi-F (5' attaagcttagcttgcatgcctacagtg 3', with *Hin*dIII restriction site underlined) and pUbi-R (5' taaggatccctctagagtcgacctgca 3', with *Bam*HI restriction site underlined). The pUbi fragment digested with *Hin*dIII and *Bam*HI, the polycistronic gene fragment digested with *Bam*HI and *Kpn*I and the vector 1300-GLY predigested with *Hin*dIII and *Kpn*I were ligated to generate plasmid p1300-Cry1-F-Cry2, p1300-Cry1-P-Cry2,p1300-Cry2-F-Cry1 and p1300-Cry2-P-Cry1, respectively (Fig. 1).

Agrobacterium-mediated transformation

The T-DNA plasmids were separately transformed into *Agrobacterium tumefaciens* LBA4404 by electroporation. The *Agrobacterium*-mediated transformation of rice was carried out according to Hiei Y. *et al.*, except that 2 mM glyphosate (Sigma-Aldrich, St. Louis, MO, USA) was used for selection [18].

RT-PCR analysis of transgenic rice

Total RNA was extracted from rice leaf with Trizol reagent (Invitrogen, Carlsbad, CA,USA). Concentration of RNA was determined by a spectrophotometer (Thermo Scientific, DE, USA). One μg RNA was treated with Dnase I and used as template to synthesize the first strand of cDNA using cDNA synthesis Kit (Thermo Scientific, DE, USA). Finally the synthesized cDNA was used as PCR template. The primers P1-2-F (5' CAGCGGCAACGAGGTGTACA 3') and P1-2-R (5' TAGGCGTCGCAGATGGTGGT 3') were used to amplify the joint parts in Cry1-F-Cry2 and Cry1-P-Cry2 (Fig. 2**a**). The PCR products were expected to be 321 bp. The primers P2-1-F (5' GCCGCTCGACATCAACGTGA 3') and P2-1-R (5' TCAGGCTGATGTCGATGGGG 3') were used to amplify the joint parts in Cry2-F-Cry1 and Cry2-P-Cry1 (Fig. 2**a**). The PCR products were expected to be 314 bp. The procedure for both PCRs was pre-denaturation at 95°C for 3 min; then 30 cycles of denaturation at 95°C for 40 s, annealing at 50°C for 30 s, and extension at 72°C for 30 s; finally followed with extension at 72°C for 5 min. The PCR products were analyzed by electrophoresis in 1% (w/v) agarose gel.

Western blot analysis

Leaf sample of 0.01 g collected from 1-month-old rice was frozen in liquid nitrogen for 45 s and disrupted with TissueLyserII (Qiagen, Hilden, Germany) at 40 Hz for 45 s. The sample was then homogenized in 200 μL 1×PBS Buffer by shaking vigorously for 30 s and centrifuged at 12000 rpm for 15 min at 4°C. The supernatants were collected as protein samples for western blot analysis. These samples were separated by electrophoresis on 8% polyacrylamide gel and transferred to PVDF membrane (Pall, Ann Arbor, MI, USA). The blotted membrane was blocked in 5% (w/v) skim milk in Tris Buffer Saline with 0.1% Tween-20 (TBST) for 1 h at room temperature. The membrane was then incubated with primary antibody against either Cry1Ab or Cry2Ab (diluted 1:1000 in 1% skim milk/TBST) for 1 h at room temperature with gentle shaking. The polyclonal antisera against Cry1Ab and Cry2Ab were prepared from New Zealand white rabbits immunized with purified recombinant Cry1Ab and Cry2Ab from *Escherichia coli*, respectively. After three times of 5 minute wash in TBST, the membrane was incubated with HRP-conjugated secondary antibody (Promega, Madison, WI, USA) diluted at

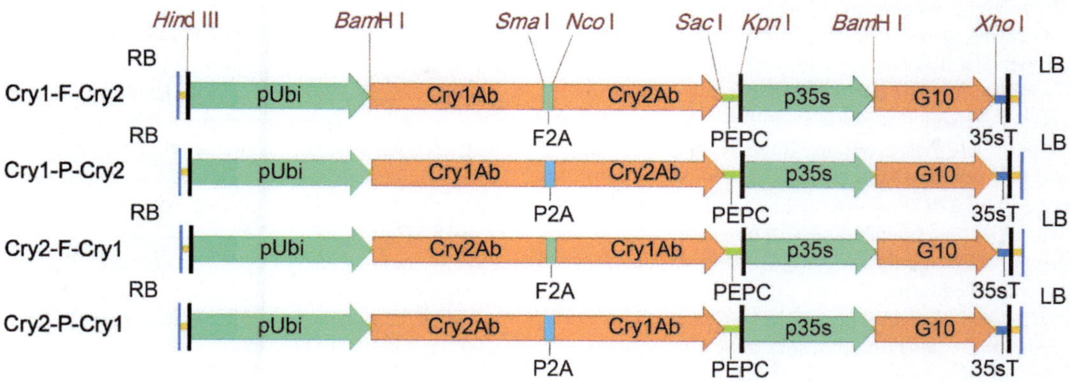

Figure 1. Schematic diagram of T-DNAs for transgenic expression of 2A linked polycistronic genes. Each T-DNA contained two expression cassettes, one for insecticidal polycistronic gene under control of maize ubiquitin-1 promoter (pUbi), the other for glyphosate-tolerant *5-enolpyruvylshikimate-3-phosphate synthase (EPSPS)* gene (G10) under control of Cauliflower mosaic virus 35s promoter (p35s). Cry1Ab and Cry2Ab,synthetic Bt inseciticidal gene; F2A and P2A, synthetic DNAs encoding Foot and Mouth Disease Virus 2A and Porcine teschovirus-1 2A, respectively; RB and LB, right border and left border of T-DNA; PEPC, maize *phosphoenolpyruvate carboxylase* gene terminator; 35sT, Cauliflower mosaic virus 35S gene terminator.

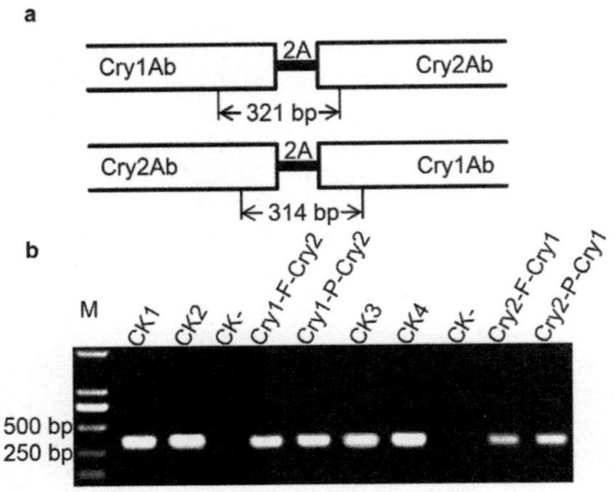

Figure 2. RT-PCR detection of mRNA integrity of the polycistronic genes. a, schematic diagram of the fragments amplified by RT-PCR; **b,** electrophoresis analysis of RT-PCR products. CK1, plasmid containing *Cry1-F-Cry2*; CK2, plasmid containing *Cry1-P-Cry2*; CK3, plasmid containing *Cry2-F-Cry1*; CK4, plasmid containing *Cry2-P-Cry1*; CK-, cDNA prepared from non-transgenic rice; M, DNA marker.

1:5000 with 1% skim milk/TBST for 1 h at room temperature with gentle shaking. After another three times of 5 minute wash in TBST, signals were visualized with DAB substrate (Sigma-Aldrich).

Protein quantification of Cry1Ab and Cry2Ab

Concentrations of Cry1Ab and Cry2Ab in the leaf of transgenic rice at tillering stage were determined by enzyme-linked immunosorbent assay (ELISA) using Envirologix kit AP003 and AP005 (Envirologix, Portland, ME, USA), respectively, according to the manufacturer's recommendation. For each construct, 3 transgenic lines were selected, and the samples prepared with the same method from non-transgenic rice were used to eliminate the basal absorption at 450 nm. After quantification of Cry1Ab and Cry2Ab, the molar ratio of the upstream protein to downstream protein of each transgenic line selected was calculated and analyzed by One-Way ANOVA [19] in SPSS.

Insect bioassay

Insect bioassays were conducted with cotton bollworm (Bt-susceptible and Cry1Ac-resistant), beet armyworm (*Spodoptera exigua*), striped stem borer (*Chilo suppressalis*) and rice leaf roller (*Cnaphalocrocis medinalis*). For beet armyworm and striped stem borer, detached leaf bioassay was carried out [20]. Two leaf blades from the same line were collected at 2-3 cm length at seedling stage and placed in a 70-mm-diameter petri dish lined with a pre-moistened filter paper. The leaf samples in the petri dish were infested with 10 newly hatched neonates. The petri dishes were then sealed with parafilm membrane and placed in dark at 28°C. The result was recorded after 3 days incubation. For cotton bollworm, detached leaf bioassay was carried out except that one leaf blade was placed in a petri dish and infested with one neonate to avoid cannibalism of cotton bollworm. Each line was infested with a total of 10 neonates. Eggs of Bt-susceptible cotton bollworm (CB-S), beet armyworm and striped stem borer were obtained from Genralpest Biotech (Genralpest Biotech, Bejing, China). Eggs of Cry1Ac-resistant cotton bollworm (CB-RR) were kindly provided by Dr. Kongming Wu from Institute of Plant Protection, Chinese Academy of Agricultural Sciences. For rice leaf roller, whole plant assay was carried out. Tillering-stage rice grown in greenhouse was infested with 10 newly hatched neonates. The result was recorded 2 weeks after infestation. Eggs of rice leaf roller used in the assays were obtained from caged moths collected from rice field. Each assay was repeated for 3 times.

Results

Transgene transcribed as a long intact mRNA

About 30 independent transgenic lines were generated via *Agrobacterium*-mediated transformation for each of the four constructs. To investigate whether the transcript of each polycistronic transgene is a long intact mRNA with sequences encoding both Cry1Ab and Cry2Ab, RT-PCR was used to detect the sequences in the area connecting the two Bt genes, which include sequences from both *Cry1Ab* and *Cry2Ab*, and the 2A peptide coding sequence in between (Fig. 2**a**). The RT-PCR products with expected sizes were detected clearly in the transgenic lines from all of the four constructs (Fig. 2**b**), suggesting that each of the polycistronic genes was transcribed into a long intact mRNA as expected.

Co-expression of discrete Cry1Ab and Cry2Ab from polycistronic transgenes

Transgenic lines from each construct were analyzed by western blot analysis using antisera against Cry1Ab and Cry2Ab,respectively. When detected with antiserum against Cry1Ab, a band of approximately 72 kD was detected among different transgenic lines, indicating that the Cry1Ab protein was expressed as a discrete protein rather than a fusion protein (Fig. 3 Upper panel). When detected by antiserum against Cry2Ab, a band of about 68~70 kD was detected among all the transgenic lines (Fig. 3 Lower panel), indicating a discrete Cry2Ab protein was expressed. A possible fusion protein of "Cry-2A-Cry" would be at a size of approximately 147 kD. However, no significant signal of proteins at such size was detected by antiserum against either Cry1Ab or Cry2Ab, suggesting that there was no significant amount of fusion protein expressed in the transgenic rice. The upstream Cry2Ab detected in the Cry2-F-Cry1 and the Cry2-P-Cry1 transgenic lines was slightly larger than the downstream Cry2Ab in the Cry1-F-Cry2 and the Cry1-P-Cry2. This increased size was likely due to the 2A peptide residues attached to the upstream protein during translation,which has been demonstrated in other 2A polycistronic transgenes [21].

Determination of gene expression level and 2A cleavage efficiency

To determine the expression levels of Bt genes with 2A polycistronic transgene and evaluate the cleavage efficiency of 2A peptide in each construct, the amounts of Cry1Ab and Cry2Ab expressed in the transgenic rice leaves were determined by ELISA. Three transgenic lines at tillering stage from each construct were selected. The expression levels of Cry1Ab and Cry2Ab were different among different lines as expected. The concentrations of the soluble Cry1Ab and Cry2Ab in the leaf were in the range of 0.67 to 1.82 μg/g and 0.69 to 2.31 μg/g of leaf fresh weight (LFW), respectively (Fig. 4**a**).

The molar ratio (R) of the upstream protein to the downstream protein of each construct was calculated to estimate the cleavage efficiency. The R value for polycistronic gene using F2A as 2A cleavage peptide was lower than that using P2A (Fig. 4**b**), indicating that the cleavage efficiency of F2A peptide was higher than P2A. The R value of polycistronic gene with Cry1Ab at the upstream was lower than that with Cry2Ab at the upstream (Fig. 4**b**), suggesting that the Cry1Ab at the upstream was likely

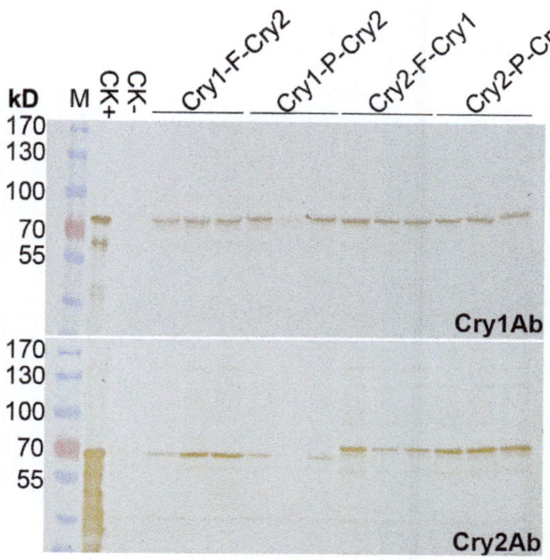

Figure 3. Western bolt analysis of Cry1Ab and Cry2Ab in transgenic rice. Three transgenic lines from each construct were selected for the analysis. Each sample was detected with antiserum against Cry1Ab (upper) and Cry2Ab (bottom) respectively. Cry1Ab or Cry2Ab protein expressed by *E.coli* was used as positive control (CK+). Sample prepared from non-transgenic rice was used as negative control (CK-). M, prestained protein ladder.

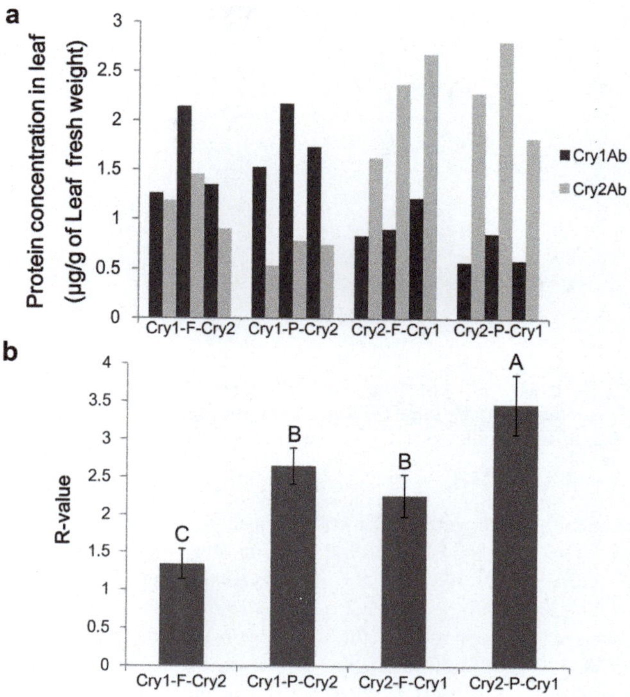

Figure 4. Cry1Ab and Cry2Ab concentrations in different transgenic rice lines and comparison of 2A cleavage efficiency. **a**, concentrations of Cry1Ab and Cry2Ab in leaves of 3 selected transgenic lines from each construct; **b**, R values of different 2A constructs. R value is the molar ratio of the upstream protein to the downstream protein. Different capital letters on each bar indicate extreme significant differences ($p<0.01$, Fisher's least-significant difference and Duncan's multiple range test).

more efficient in cleavage than the Cry2Ab at the upstream, regardless of whether F2A or P2A was the cleavage peptide in the synthetic construct. The transgenic lines from the construct of Cry1-F-Cry2 had the R value close to 1, indicating that the two Bt proteins were expressed almost equally (Fig. 4**a**).

Insect-resistant activity of transgenic rice

The transgenic rice lines were assayed for their insecticidal activity using neonates of CB-S, striped stem borer and rice leaf roller. Three transgenic lines from each construct were selected and age-matched non-transgenic rice plants were used as control. Mortalities of all the 3 insect species feeding on transgenic rice were 100% while those on the non-transgenic rice were much lower (Fig. 5). In the assays with CB-S and striped stem borer, the transgenic leaves were only slightly bitten by the insects while non-transgenic controls were severely consumed (Fig. 6**a**). In the assays with rice leaf roller, no rolled leaf was observed in the transgenic plant while several leaves were rolled in the non-transgenic plant (Fig. 6**b**). The bioassays demonstrated that transgenic rice plants generated from the 2A constructs were highly insect-resistant.

The transgenic rice lines were also assayed for their activity against beet armyworm and CB-RR. The KMD1 transgenic rice at the same growing stage was used as control. The mortalities of beet armyworm feeding on the 2A transgenic rice lines were obvious higher than that feeding on KMD1 ($p<0.01$) (Figure 7). The KMD1 showed a rather low activity to CB-RR, while the 2A transgenic rice showed much higher activity ($p<0.01$) (Figure 7).

Discussion

In this study, Cry1Ab and Cry2Ab, two potent Bt crystal proteins with different receptors in the insect midgut [22], were expressed by the polycistronic transgene using 2A peptide. To our best knowledge, this is the first application of 2A peptide for expressing two Bt genes in a transgenic crop. Utilization of two or more Bt genes simultaneously is desirable for management of insect resistance to Bt toxins as well as for broadening insecticidal spectrum of Bt crops. Self-cleaving 2A peptides and 2A-like sequences have also been found in insect virus, and some of them have high self-cleavage activity [23]. While there is no scientific base for any safety concern for using 2A peptide from mammalian virus, it may be better for public perception to use 2A peptides from insect virus for the development of transgenic rice.

Introduction of multiple genes into plant permits complex and sophisticated manipulation of traits in transgenic crops [2]. The number of genes being introduced into crops for genetic engineering is increasing steadily. For instance, the recently released commercial transgenic corn "Genuity SmartStax" (Monsanto, St Louis, MS, USA) has a total of 8 transgenes for insect-resistance and herbicide-tolerance. With more traits under research and development for transgenic improvement, more genes are expected to be utilized for future transgenic crops. Clearly polycistronic strategy using 2A peptide can simplify the process of gene stacking significantly, as it enables us to introduce multiple genes into plants using a single T-DNA with a single promoter [2]. Moreover, polycistronic strategy alleviates the concern of gene silencing induced by the insertion of homologous promoters or multiple T-DNA sequences into recipient genome [2,24]. Additionally, equal expressions of different genes could be achieved by the 2A polycistronic strategy [12,24].

The expression levels of Cry1Ab and Cry2Ab by 2A polycistronic transgene in this study were comparable to traditional monocistronic transgene. The Cry1Ab concentration of transgenic rice KMD1, which is highly resistant to striped stem borer and rice leaf roller [16], was 2.95 μg/g of LFW [25]. The Cry1C concentration in homozygous rice line T1c-19 harboring a

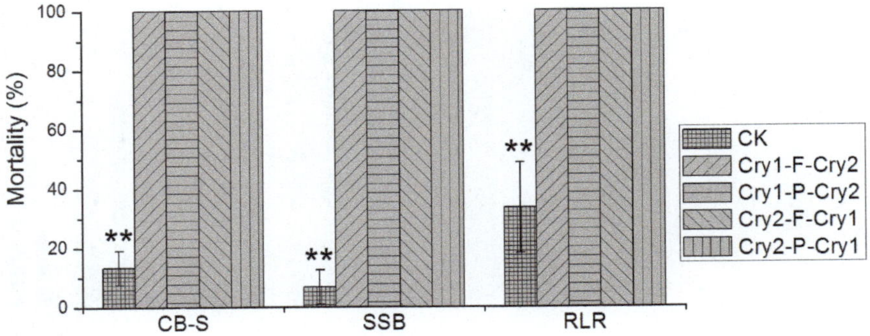

Figure 5. Mortality of Bt-susceptible cotton bollworm, striped stem borer and rice leaf roller feeding on transgenic rice plants. Non-transgenic rice at the same growing stage was used as control (CK). ** on the bar indicate extreme significant differences ($p<0.01$, Kruskal Wallis test). CB-S, Bt-susceptible cotton bollworm; SSB, striped stem borer; RLR, rice leaf roller.

synthetic Cry1C gene under control of maize ubiquitin promoter was 1.38 μg/g of LFW at the heading stage [26]. The Cry2A protein concentrations of the homozygous transgenic rice lines harboring a synthetic Cry2 gene driven by maize ubiquitin promoter was reported at the range from 9.65 to 12.11 μg/g of LFW [27]. In the transgenic *indica* rice line harboring both *Cry1Ac* and *Cry2Ab* gene, the Cry2Ab concentration was reported to be around 1 μg/g LFW [28]. Thus, the expression levels of the Cry1Ab and Cry2Ab in the transgenic rice lines obtained by this study were well within the range of the expression levels of the monocistronic transgene reported previously.

Although cotton bollworm was not a pest of rice, it has been long a good target insect for evaluating Bt toxicity. Moreover, cotton bollworm is a good model insect for the study of insect resistance to Bt toxins because of the availabilities of both sensitive and resistant lines to Cry1Ac. The 2A transgenic lines obtained in this study showed high insect-resistant activity against cotton bollworm, either sensitive or resistant to Cry1Ac, while the KMD1 transgenic rice had little activity toward Cry1Ac-resistant cotton bollworm. This suggested that little or no cross resistance existed between Cry1Ac and Cry2Ab and the 2A peptide based transgenic Bt rice will be useful for the management of Bt resistance developed by insect pests.

The 2A transgenic lines obtained in this study conferred significantly higher activity toward beet armyworm than the KMD1 transgenic rice containing Cry1Ab only [16], suggesting that the Cry2Ab expressed in the 2A transgenic lines indeed contributed to the enhancement of insecticidal activity. A similar phenomenon was also observed in transgenic cotton BollgardII. While Bollgard cotton only had moderate toxic activity to beet armyworm, BollgardII cotton showed greatly improved activity due to the addition of Cry2Ab [29].

The upstream proteins expressed with 2A polycistronic strategy are usually attached with the 2A peptide residues at the C-terminus [30]. The western blot analysis in this study appeared to agree with it. However, due to the limitation of the western blot analysis in determination of protein size, C-terminal amino acid sequencing and mass spectrum are required to confirm. While no adverse impact was observed for the utilization of 2A peptide in gene stacking [12], the Bt toxins attached with a 2A peptide will be considered as different proteins according to the regulation of transgenic crops, and thus additional safety studies will be required for commercialization of transgenic crops using 2A cleaving peptide.

Figure 6. Insect bioassay for transgenic rice with a polycistronic Bt transgene. a, bioassay with Bt-susceptible cotton bollworm (CB-S) and striped stem borer (SSB). **b,** bioassay with rice leaf roller. For each assay, age-matched non-transgenic rice was used as control (CK).

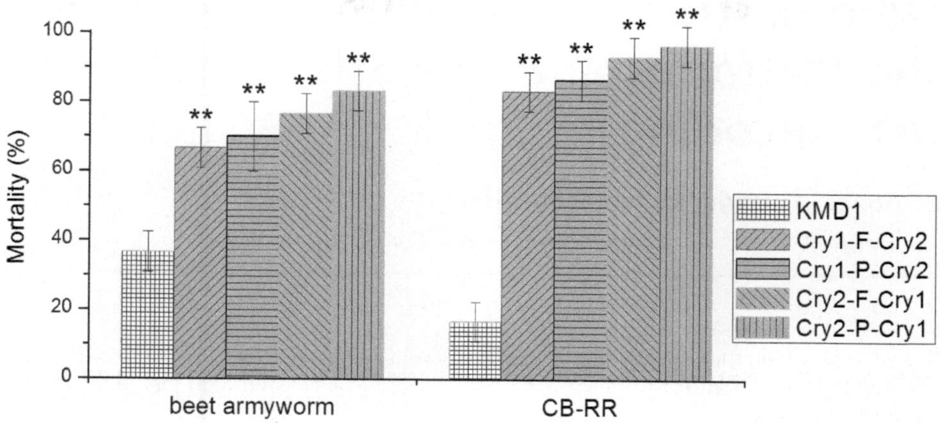

Figure 7. Mortality of beet armyworm and Cry1Ac-resistant contton bollworm feeding on KMD1 and various 2A transgenic rice lines. The transgenic rice KMD1 was used as control. ** on the bar indicate extreme significant difference ($p<0.01$, Fisher's least-significant difference and Duncan's multiple range test). CB-RR, Cry1Ac-resistant cotton bollworm.

Acknowledgments

We thank Kongming Wu and Gemei Liang for the supply of Cry1Ac-resistant cotton bollworm. We thank Gongyin Ye for the supply of KMD1 transgenic rice.

Author Contributions

Conceived and designed the experiments: QcZ MhL JHG ZcS. Performed the experiments: QcZ MhL MmT. Analyzed the data: QcZ MhL MmT. Contributed reagents/materials/analysis tools: QcZ MhL JHG. Wrote the paper: QcZ MhL JHG ZcS.

References

1. Romeis J, Meissle M, Bigler F (2006) Transgenic crops expressing Bacillus thuringiensis toxins and biological control. Nature Biotechnology 24: 63–71.
2. Halpin C (2005) Gene stacking in transgenic plants – the challenge for 21st century plant biotechnology. Plant Biotechnology Journal 3: 141–155.
3. Gassmann AJ, Petzold-Maxwell JL, Keweshan RS, Dunbar MW (2011) Field-Evolved Resistance to Bt Maize by Western Corn Rootworm. PLoS ONE 6.
4. Tabashnik BE, Van Rensburg JBJ, Carriere Y (2009) Field-Evolved Insect Resistance to Bt Crops: Definition, Theory, and Data. Journal of Economic Entomology 102: 2011–2025.
5. Wang P, Zhao J-Z, Rodrigo-Simon A, Kain W, Janmaat AF, et al. (2006) Mechanism of resistance to Bacillus thuringiensis toxin Cry1Ac in a greenhouse population of cabbage looper, Trichoplusia ni. Applied and Environmental Microbiology: 01834–01806.
6. Bates SL, Zhao JZ, Roush RT, Shelton AM (2005) Insect resistance management in GM crops: past, present and future. Nature Biotechnology 23: 57–62.
7. Shelton AM, Tang JD, Roush RT, Metz TD, Earle ED (2000) Field tests on managing resistance to Bt-engineered plants. Nature Biotechnology 18: 339–342.
8. Tabashnik BE, Carriere Y, Dennehy TJ, Morin S, Sisterson MS, et al. (2003) Insect resistance to transgenic Bt crops: Lessons from the laboratory and field. Journal of Economic Entomology 96: 1031–1038.
9. Cao J, Zhao JZ, Tang JD, Shelton AM, Earle ED (2002) Broccoli plants with pyramided cry1Ac and cry1C Bt genes control diamondback moths resistant to Cry1A and Cry1C proteins. Theoretical and Applied Genetics 105: 258–264.
10. Yang Z, Chen H, Tang W, Hua HX, Lin YJ (2011) Development and characterisation of transgenic rice expressing two Bacillus thuringiensis genes. Pest Management Science 67: 414–422.
11. Zhao JZ, Cao J, Li YX, Collins HL, Roush RT, et al. (2003) Transgenic plants expressing two Bacillus thuringiensis toxins delay insect resistance evolution. Nature Biotechnology 21: 1493–1497.
12. Szymczak-Workman AL, Vignali KM, Vignali DAA (2012) Design and Construction of 2A Peptide-Linked Multicistronic Vectors. Cold Spring Harbor Protocols 2012: pdb.ip067876.
13. Ha S-H, Liang YS, Jung H, Ahn M-J, Suh S-C, et al. (2010) Application of two bicistronic systems involving 2A and IRES sequences to the biosynthesis of carotenoids in rice endosperm. Plant Biotechnology Journal 8: 928–938.
14. Brown JD, Ryan MD (2010) Ribosome "Skipping": "Stop-Carry On" or "StopGo" Translation Recoding: Expansion of Decoding Rules Enriches Gene Expression. InAtkinsJFGestelandRFeditors: New York:Springerpp101–121.
15. Luke G, Escuin H, De Felipe P, Ryan M (2010) 2A to the fore - research, technology and applications. Biotechnology and Genetic Engineering Reviews 26: 223–260.
16. Shu QY, Ye GY, Cui HR, Cheng XY, Xiang YB, et al. (2000) Transgenic rice plants with a synthetic cry1Ab gene from Bacillus thuringiensis were highly resistant to eight lepidopteran rice pest species. Molecular Breeding 6: 433–439.
17. Ye GY, Shu QY, Yao HW, Cui HR, Cheng XY, et al. (2001) Field evaluation of resistance of transgenic rice containing a synthetic cry1Ab gene from Bacillus thuringiensis Berliner to two stem borers. Journal of Economic Entomology 94: 271–276.
18. Hiei Y, Komari T, Kubo T (1997) Transformation of rice mediated by Agrobacterium tumefaciens. Plant Molecular Biology 35: 205–218.
19. Heiberger R, Neuwirth E (2009) One-Way ANOVA. R Through Excel: New YorkSpringerpp165–191.
20. Chen Y, Tian J-C, Shen Z-C, Peng Y-F, Hu C, et al. (2010) Transgenic Rice Plants Expressing a Fused Protein of Cry1Ab/Vip3H Has Resistance to Rice Stem Borers Under Laboratory and Field Conditions. Journal of Economic Entomology 103: 1444–1453.
21. Halpin C, Cooke SE, Barakate A, Amrani AE, Ryan MD (1999) Self-processing 2A-polyproteins – a system for co-ordinate expression of multiple proteins in transgenic plants. The Plant Journal 17: 453–459.
22. Hernandez-Rodriguez CS, Hernandez-Martinez P, Van Rie J, Escriche B, Ferre J (2013) Shared Midgut Binding Sites for Cry1A. 105, Cry1Aa, Cry1Ab, Cry1Ac and Cry1Fa Proteins from Bacillus thuringiensis in Two Important Corn Pests, Ostrinia nubilalis and Spodoptera frugiperda. Plos One 8.
23. Ryan M (2014) 2A and 2A like sequences. Available: http://www.st-andrews.ac.uk/ryanlab/2A_2Alike.pdf/. Accessed 4 September 2014.
24. Sainsbury F, Benchabane M, Goulet M-C, Michaud D (2012) Multimodal Protein Constructs for Herbivore Insect Control. Toxins 4: 455–475.
25. Tian J-C, Chen Y, Li Z-L, Li K, Chen M, et al. (2012) Transgenic Cry1Ab Rice Does Not Impact Ecological Fitness and Predation of a Generalist Spider. PLoS ONE 7: e35164.
26. Tang W, Chen H, Xu C, Li X, Lin Y, et al. (2006) Development of insect-resistant transgenic indica rice with a synthetic cry1C* gene. Molecular Breeding 18: 1–10.
27. Chen H, Tang W, Xu C, Li X, Lin Y, et al. (2005) Transgenic indica rice plants harboring a synthetic cry2A* gene of Bacillus thuringiensis exhibit enhanced resistance against lepidopteran rice pests. Theoretical and Applied Genetics 111: 1330–1337.
28. Bashir K, Husnain T, Fatima T, Latif Z, Mehdi SA, et al. (2004) Field evaluation and risk assessment of transgenic indica basmati rice. Molecular Breeding 13: 301–312.
29. Perlak FJ, Oppenhuizen M, Gustafson K, Voth R, Sivasupramaniam S, et al. (2001) Development and commercial use of Bollgard cotton in the USA – early promises versus today's reality. The Plant Journal 27: 489–501.
30. Szymczak-Workman AL, Vignali KM, Vignali DAA (2012) Verification of 2A Peptide Cleavage. Cold Spring Harbor Protocols 2012: pdb.prot067892.

4

Direct Chloroplast Sequencing: Comparison of Sequencing Platforms and Analysis Tools for Whole Chloroplast Barcoding

Marta Brozynska, Agnelo Furtado, Robert James Henry*

Queensland Alliance for Agriculture and Food Innovation, The University of Queensland, Brisbane, Queensland, Australia

Abstract

Direct sequencing of total plant DNA using next generation sequencing technologies generates a whole chloroplast genome sequence that has the potential to provide a barcode for use in plant and food identification. Advances in DNA sequencing platforms may make this an attractive approach for routine plant identification. The HiSeq (Illumina) and Ion Torrent (Life Technology) sequencing platforms were used to sequence total DNA from rice to identify polymorphisms in the whole chloroplast genome sequence of a wild rice plant relative to cultivated rice (cv. Nipponbare). Consensus chloroplast sequences were produced by mapping sequence reads to the reference rice chloroplast genome or by *de novo* assembly and mapping of the resulting contigs to the reference sequence. A total of 122 polymorphisms (SNPs and indels) between the wild and cultivated rice chloroplasts were predicted by these different sequencing and analysis methods. Of these, a total of 102 polymorphisms including 90 SNPs were predicted by both platforms. Indels were more variable with different sequencing methods, with almost all discrepancies found in homopolymers. The Ion Torrent platform gave no apparent false SNP but was less reliable for indels. The methods should be suitable for routine barcoding using appropriate combinations of sequencing platform and data analysis.

Editor: Hikmet Budak, Sabanci University, Turkey

Funding: The research was funded by Rural Industries Research and Development Corporation (Australia). The funders played no role in study design, data collection, decision to publish or preparation of the manuscript.

Competing Interests: This research was supported by a grant from Rural Industries Research and Development Corporation, Australian Government.

* Email: robert.henry@uq.edu.au

Introduction

A universal plant barcoding method based upon sequencing a specific gene or combination of specific genes has proven elusive. Analysis of different plant groups has required the use of different genetic loci. The emergence of next generation sequencing (NGS) technology allowing whole chloroplast genomic DNA sequencing has provided the opportunity to use the whole chloroplast genome sequence as a barcode [1]. The chloroplast has unique features found in all green plants, conserved sufficiently to be readily aligned for comparison of different samples and large enough to contain variation, which allow species or sub species distinction across the seed plants [2]. Analysis of complex mixtures of plants has applications in many ecological studies [3].

A generic method for rapid and cost effective DNA-based identification of plants at this level (species or sub-species) will find wide application in industry and research [4,5]. Industrial applications will include the identification of plant components in whole and processed foods and the management of food processing to ensure food safety and authenticity of labelling [6]. This will complement the use of next generation sequencing to screen for the presence of pathogenic microbes [7]. Protection of intellectual property rights associated with plant varieties will also be simplified by a standard approach to plant identification. These tools will also support protection of biosecurity and management of wild plant populations of rare or threatened plants.

Early analysis relied on either specific amplification of the chloroplast genome or separation of chloroplasts from nuclear and mitochondrial DNA before chloroplast DNA isolation and sequencing the amplified or cloned product derived from PCR amplification [8]. However, the universal amplification of chloroplast sequences from all species has proven difficult and chloroplast purification is laborious and not perfect [9,10]. Recently a simple approach involving NGS of a total genomic DNA preparation was proposed [11]. This method relies on successful extraction of chloroplast genome sequence reads from total genomic DNA reads and their assembly to form a consensus sequence for the sample that can be used as a barcode. Reference-guided mapping of sequence reads and *de novo* assembly have been used to generate these whole chloroplast barcodes [12].

DNA sequencing platforms differ in their throughput and accuracy [13]. Accuracy of a DNA sequence assembled using NGS reads is dependent on the read length, sequencing depth, sequence coverage or width and evenness of coverage but also on the accuracy of the sequencing platform and the assembly and analysis pipeline [14]. We compared two different sequencing technologies, the Illumina platform that uses sequencing-by-syntheses (SBS) chemistry and the Ion Torrent that is based on semiconductor sequencing technology. Sequencing on both platforms commences with library preparation, which involves fragmentation of total genomic DNA, purifying to uniform and desired fragment size and ligation to sequencing adapters specific to the platform [15,16]. In Illumina systems the fragments are subsequently denatured and fixed on the surface of a proprietary glass flowcell, followed by solid-phase amplification (bridge amplification). As a result clusters are created which contain clonally amplified DNA templates. Next the clusters are sequenced in parallel using four fluorescently labelled nucleotides with reversible dye terminators. After every sequencing cycle and base incorporation to the template the polymerization terminates, a CCD (charge-coupled device) captures the fluorescent signal and identifies the base. After the base call the dye is cleaved and the sequencing process continues. This technology was first implemented in an instrument called Genome Analyzer (GA) with subsequent releases of improved series up to GAIIx. The next major system launched by Illumina was the HiSeq2000 with improved output and read length. Data from both GAIIx and HiSeq was used in the current study.

In Ion Torrent systems the prepared libraries are immobilized to beads and amplified using emulsion PCR which takes place within microdroplets of aqueous solution and oil [17]. During the sequencing individual bases are incorporated by the action of a DNA polymerase. As a result of this reaction a proton is released and the resulting change in pH is measured. The reaction occurs in a proprietary chip, which acts as a pH meter. Unlike Illumina, this system does not use fluorescent dyes and light detection, shortening the sequencing time. The Ion Torrent instrument used in the study was a Personal Genome Machine (PGM).

We examined the reliability of these two different sequencing platforms coupled with different data analysis approaches when applied to obtaining the chloroplast consensus sequences of a wild rice genotype and a cultivated rice genotype. Although the Ion Torrent platform is widely used for sequencing and limited published data is available on its application for transcript [18] and amplicon [19] sequencing in plants, it has not previously been applied to plant genomic DNA analysis. The most common applications of the Ion Torrent system have been to small genomes (bacterial samples) and targeted sequencing of amplicons or transcripts, whereas the Illumina has been widely used for applications including whole human genome re-sequencing, *de novo* plant and animal genome sequencing, exome, transcriptome (RNA-Seq) and metagenomics investigations. The chloroplast genome represents a specific sequencing challenge with the presence of an inverted repeat. This study compares total DNA sequencing with the Illumina and Ion Torrent platforms and options for analysis of the sequences generated to produce a whole chloroplast genome sequence as a barcode for plant identification. Whole genome NGS was carried out on two rice genotypes: *Oryza sativa* spp. *japonica* var. Nipponbare referred to here as the reference-rice-genotype (R-rice-genotype) as its chloroplast sequence (reference-sequence) is publicly available and an Australian wild rice genotype sample (W-rice-genotype).

Materials and Methods

Plant materials

Seeds of rice (*Oryza sativa* cv. Nipponbare), referred to as reference-rice-genotype (R-rice-genotype), were germinated and entire seedlings were used for DNA extraction. Leaf tissue from a wild rice plant, referred to as wild-rice-genotype (W-rice-genotype) collected from a field containing a population of perennial wild rice (*Oryza rufipogon*-like) at Abattoir Creek in North Queensland, was used for DNA extraction [20].

DNA extraction

DNA from leaf tissue of the W-rice-genotype was extracted using modification of the CTAB method [21]. DNA from seedlings of the R-rice-genotype was extracted as described for the W-rice-genotype but with further purification as described by [11].

Sequencing

The Illumina and the Ion Torrent sequencing platforms were used for the shotgun sequencing of total genomic DNA of both rice samples. The wild rice sample was sequenced on the Ion Torrent (200 bp reads) and the Illumina HiSeq platform (100 bp paired end reads). The cultivated rice genotype was subjected to sequencing using the Ion Torrent (200 bp reads) platform while Illumina reads (36 bp paired-end), generated on the GAII, were sourced from available archived data [11]. Sequencing on the Illumina HiSeq 2000 (Illumina, San Diego, CA, USA), to generate 100 bp paired end reads with an average library insert size of 500–600 bp, was outsourced to the Australian Genome Research Facility (AGRF, Melbourne, Australia). For sequencing using the Ion Torrent platform, genomic DNA was sheared using the Covaris S220 instrument (www.covarisinc.com) and used for preparing sequencing libraries according to the standard Ion Torrent PGM protocol. The resulting individual DNA libraries were quality checked and quantified on the Agilent 2100 Bioanalyzer using the High Sensitivity DNA kit (Agilent). Following template amplification and enrichment on the Ion OneTouch2 (Ion OneTouch 200 Template Kit v2; #4478316) and OneTouch2 ES instruments, each sample was loaded onto one PGM #318 chip and sequenced using Ion PGM Sequencing 200 Kit v.2 (#4482006) according to manufacturer's protocol.

Data analysis – consensus

Raw reads from both sequencing platforms, PGM Ion Torrent and Illumina, were imported to CLC Genomics Workbench 6.0 (CLC-GW) (www.clcbio.com) and read statistics assessed using sequencing data quality control, followed by read trimming for quality, length and presence of ambiguous bases. Ion Torrent reads were trimmed with the quality score limit set to 0.05 (which corresponded to PHRED quality value >15) and a minimum read length of 20 bp. Reads from both Illumina platforms, GAII and HiSeq, were trimmed with a quality score limit of 0.01 (which corresponded to PHRED >22) and the same minimum read length of 30 bp. Ion Torrent data was trimmed at a slightly lower quality value due to the lower average quality compared to the Illumina reads (Table S1; avg. PHRED for Ion Torrent reads: 25 and 26, and avg. PHRED for Illumina: 29 and 32). The Illumina HiSeq platform generates sequencing data fifty to sixty times more in order of magnitude as compared to the Ion Torrent data. Thus, in order to compare Ion Torrent and Illumina HiSeq platforms at similar coverage, a subset of HiSeq Illumina reads for the wild rice sample and matching the Ion Torrent read numbers for that same sample, was randomly extracted from the whole reads set. Ion Torrent raw reads were also alternatively trimmed using the Torrent Suite Software version 3.6 using default analysis settings with the following modifications: the quality-based trimming of the sequencing reads was adjusted to predicted-PHRED-score 17 over the sliding window of 20 bp (default values 15, and 30 respectively) and the minimal length filter for reported reads was set to 20 (default 8).

Consensus sequence generation using the mapping tool was carried out as follows. The CLC-GW was used to map trimmed reads to the chloroplast genome sequence of *Oryza sativa* spp. *japonica* var. Nipponbare (GenBank accession – GU592207) used as the reference-sequence. The mapping process involved the following parameters: mismatch cost 2, insertion cost 3, deletion cost 3, length fraction 0.9, similarity fraction 0.9, and global alignment setting. Non-specific matches were also mapped randomly to the reference genome. Consensus sequences were extracted from all mapping runs giving five sequences in total; two consensus sequences for *O. sativa* Nipponbare each derived from the Illumina GAII reads and the Ion Torrent reads, and three consensus sequences for the wild rice sample each derived from the Illumina HiSeq sequence reads, Illumina HiSeq sequence subset-reads and the Ion Torrent reads. Conflicts between reads were resolved by voting for the majority of the reads at the given position. Consensus sequence were aligned to the reference sequence and analysed for SNPs and indels.

Consensus sequence generation using the assembly tool was carried out as follows using both the CLC-GW and the Torrent Suite Software. The CLC trimmed reads from both platforms were assembled in CLC-GW using the *de novo* assembly tool with the following parameters: automatic word and bubble sizes with 200 bp for minimum contig length. Moreover the reads were mapped back to contigs (mismatch cost 2, insertion cost 3, deletion cost 3, length fraction 0.9, and similarity fraction 0.9) and contigs were updated based on the results. The trimmed Ion Torrent reads were also assembled in the Torrent Suite Software using the Assembler plugin v.3.4.2 utilizing the MIRA assembly algorithm (http://sourceforge.net/projects/mira-assembler/). The assembly was performed in the reference-assisted mode using the reference-sequence. All contigs generated were aligned to the reference-sequence using NUCmer alignment generator from MUMmer package [22]. Owing to the chloroplast genome structure (two long inverted repeats) the option to use all matches regardless of the uniqueness was enabled in the script. All other parameters were left as default. Based on the MUMer output the longest contigs with the best mapping results (query coverage, precent identity and alignment length) were chosen as scaffolds and manually joined together by overlapping the remaining short contigs. In cases when there was not enough contigs to reconstruct the chloroplast genome sequence, *de novo* assembly was repeated with different parameter for word size in the CLC-GW and subjected again to the alignment tool. Consensus sequences from contigs mapping and joining were also screened for SNPs and indels based on the reference-sequence.

Data analysis – variant detection

The consensus sequences generated by both mapping and *de novo* assembly of the reads were imported into Geneious 6.1.6 (www.geneious.com) and each of them was separately mapped to reference-sequence. To generate the alignments the Geneious Map to Reference tool was used with the default parameters. Subsequently the alignments were screened to find variations/SNPs using the tool available in Geneious package.

Validation of the most probable variant

The wild rice chloroplast consensus sequences derived from reads mapping from Illumina and Ion Torrent platforms were aligned and scanned for SNPs and indels (as described above). If a variant was found in one consensus but was not present in the other it was reported as a discrepancy and closely studied. To confirm which of the disagreeing variants was more likely to be true the following procedure was adopted: (1) visual investigation of mapping results in CLC-GW and (2) aligning the conflicting regions from mapping derived consensus with contigs sequences originated from *de novo* assemblies.

Data deposition

The chloroplast genome for Australian wild rice genotype sample (W-rice-genotype) has been deposited in GenBank with the sequence accession number KF428978.

Results

Mapping and *de novo* assembly of sequence reads

The CLC Genomics Workbench was used to trim sequence reads from both the Illumina and Ion Torrent platforms and then processed, using the mapping and *de novo* assembly mode, to generate a mapping-consensus or *de novo*-consensus chloroplast sequences, respectively. The Torrent Suite Software, a dedicated analysis tool for Ion Torrent reads, was used to trim sequence reads from the Ion Torrent platform and then processed using the reference-guided *de novo* assembly mode (the only available mode) to generate a reference-guided *de novo*-consensus chloroplast sequence. The published chloroplast sequence of the R-rice-genotype (*Oryza sativa* spp. *japonica* var. Nipponbare) was used as a reference chloroplast genome sequence (reference-sequence) for mapping of reads or contigs. Sequencing data of the two rice genotypes, generated by both sequencing platforms, was of good quality with average PHRED scores ranging above 25 (Table S1).

In case of the R-rice-genotype, CLC trimmed-reads from both sequencing platforms mapped to the entire reference-sequence indicating complete reference-sequence coverage. However, the mapping-consensus chloroplast sequences varied in length based on sequencing platform reads. Similar results were obtained for the W-rice-genotype (Table S2).

Using the *de novo* mode in the CLC-GW and reads from both sequencing platforms, a number of contigs were generated with a subset of these mapping to the reference-sequence (Table S3). Similar results were obtained with the Ion Torrent reads when using the reference-assisted *de novo* mode on the Torrent Suite Software (Table S3). For both rice genotypes, contigs, which mapped to the reference-sequence, were checked if contiguous/overlapping to generate a *de novo*-consensus sequence. Using the *de novo* mode in the CLC-GW, respective *de novo*-consensus chloroplast sequences were generated for the reference-rice-genotype and the wild-rice-genotype but only from their respective Ion Torrent reads and not their Illumina reads. Using the reference-assisted *de novo* mode, reference-guided *de novo*-consensus chloroplast sequences were generated for both genotypes.

Thus, all contigs generated from Ion Torrent reads using both assembly algorithms (in the CLC-GW and Torrent Suite Software) could be joined together based on their position and overlapping information creating full length consensus sequences for the chloroplast genome (Table S3). Consensus sequences could not be created from contigs generated in the CLC-GW from Illumina reads for either of the rice genotypes, possibly because of the limitations of read length (Table S3). Although these contigs were well mapped to the reference, there were gaps between some of them that prevented their joining.

Analysis of reference-rice-genotype

The mapping-consensus chloroplast genome sequence (134,551 bp) generated by mapping the Illumina reads from [11] to the reference-sequence in the CLC-GW was, as expected, identical to the reference-sequence (Table 1). The mapping-

Table 1. Comparison of chloroplast consensus sequences of the cultivated reference rice genotype (*Oryza sativa* Nipponbare).

Sequencing platform	Source	Sequence length (bp)	Variants	Deletions	Insertions	Mismatches
Used available reference	GenBank (GU592207)	134,551	–	–	–	–
Illumina [11]	mapping- consensus	134,551	0	0	0	0
Ion Torrent	mapping- consensus	134,520	31	30	1	0
	de novo- consensus (CLC-GW	134,503	54	48	6	0
	reference-assisted *de novo*- consensus (Torrent Suite)	134,581	50	9	41	0

consensus sequence generated by mapping of the Ion Torrent reads to the reference-sequence in the CLC-GW was shorter than the reference-sequence by 31 nucleotides with no base mismatches, but diverged at 31 positions comprising of 30 deletions and 1 insertion and where all but one of the polymorphisms were found in homopolymer regions of length from 2 to 9 bases. The *de novo*-consensus sequence built by mapping and joining the contigs from *de novo* assembly of Ion Torrent reads (performed in CLC-GW) was shorter than the reference-sequence by 48 bp with no base mismatches, but diverged at 54 positions comprising of 48 deletions and 6 insertion and where all but two of the polymorphisms (indels) were found in homopolymer regions of length from 2 to 17 identical bases.

The reference-guided *de novo*-consensus chloroplast sequence obtained from contigs generated in Torrent Suite Software using Ion Torrent reads was longer than the reference-sequence by 30 bases with no base mismatches, but diverged at 50 positions comprising of 9 deletions and 41 insertions and where all polymorphisms (indels) were found in homopolymer regions of length from 2 to 12 identical bases.

The variants in the mapping-derived consensus sequences generated by the CLC-GW and using the Illumina or Ion Torrent reads were predominantly deletions, which is in contrast to the reference-guided *de novo*-consensus chloroplast sequence generated by the Torrent Suite Software using Ion Torrent reads, where the variants were predominantly insertions. Overall, 135 variants were detected using these different approaches. Among them only 8 were common for all of the methods (the same position, length and type). The rest of the variants were either present in two of the approaches or were unique to only one of them.

The analysis of regions in the chloroplast genome where the indels were found in the Ion Torrent data (both mapping and assembly consensus) revealed biases in terms of polymorphism nature, nucleotide type and location. Data from all approaches were merged and duplicated regions were removed. The number of indels that involved C/G bases was slightly higher than A/T bases (53 and 44, respectively). It is worth noting that the GC content of the chloroplast genome is 39%. Also, more deletions than insertions were observed (except for the consensus from the Torrent Suite Software assembly) and the majority were of G/C nucleotides in short homopolymeric regions (2–5 bp long with the vast majority of these variants occurring in 3 bp long homopolymer) (Figure 1). On the other hand, insertions were found mostly in longer homopolymers (5–17 bp long) and involved almost only A/T nucleotides.

Analysis of wild-rice-genotype

The reference-sequence was used to generate consensus chloroplast sequences using mapping or *de novo* approaches. For variant analysis, consensus sequences were compared to the reference-sequence unless indicated otherwise. The mapping-consensus sequence generated by mapping the Illumina reads of the wild rice plant using the CLC-GW was shorter than the reference-sequence by 20 bases (Table 2). In comparison, there were 128 variants found between the two sequences, which included 18 deletions, 13 insertions and 97 mismatches. Among mismatches 92 were single-nucleotide variants (SNPs) (48 transitions and 44 transversions) and 5 were multi-nucleotide variants (MNVs). The mapping-consensus obtained by mapping a subset of Illumina reads differed from the previous sequence by one deletion of 2 bases (Table 2), although, this variation was not found in the reference-sequence.

The consensus sequence generated by mapping Ion Torrent reads of the wild-rice-genotype using the CLC-GW was shorter than the reference-sequence by 26 bases and shorter by 6 bases than the consensus sequence from Illumina sequencing. The two sequences differed by 139 variants, which included 30 deletions, 14 insertions and 95 mismatches. Among the mismatches 90 were SNPs (47 transitions and 43 transversions) and 5 were MNVs.

As was observed for the R-rice-genotype, the de novo-consensus sequence differed from the reference-sequence at more positions than the mapping-consensus sequence (Table 1 and Table 2). In addition, a higher number of deletions were observed in consensus sequences obtained from the CLC-GW analysis as compared to the Torrent Suite Software analysis where more insertions were observed (Table 1 and Table 2).

Comparison of sequencing platforms

The mapping-consensus chloroplast sequences of the wild-rice-genotype generated in the CLC-GW using the Illumina and Ion Torrent reads were used to compare sequence platforms. When comparing variations found in the wild-rice-genotype by both sequencing technologies, 20 discrepancies (variant in one consensus not present in the other) were observed (Table 3). However, in the Ion Torrent consensus sequence there were fifteen variants and in the Illumina consensus there were only three. Moreover, two variants found at the same position differed in nucleotide composition in both consensus sequences.

Read mapping files derived from both mapping-consensus sequences were visually inspected to check for reads mapped to all 20 positions to determine sequence read error or for mapping error. The region in the Ion Torrent mapping-consensus sequence with the substitution of TT → AA (Table 3, variant #1) had several mapped reads with many mismatches. Reads from this region when extracted and blasted against the nucleotide sequence collection (nr/nt) at NCBI website (http://blast.ncbi.nlm.nih.gov) gave hits to the chloroplast but also to the nuclear DNA of *Oryza* species. In addition, this variant was not observed in the corresponding region of the contig sequences obtained from *de novo* assembly (from both analysis software), indicating that the

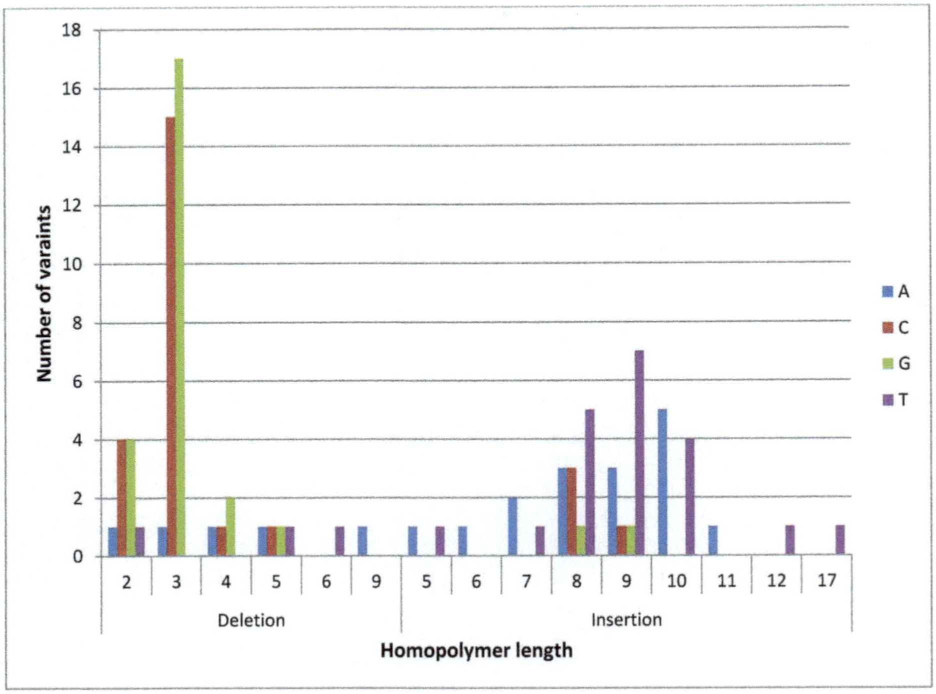

Figure 1. Variants in indels in cultivated (cv. Nipponbare) rice chloroplast consensus. Sequences generated by mapping and assembly of Ion Torrent reads to the available chloroplast sequence in GenBank for this genotype. The number of variants is shown with respect to its type (deletion or insertion) and position (the length of homopolymer region where the variants were found).

variant was not due to sequencing error but due to mapping artefact. Similarly, mapping error and not sequence read error was the reason for variants detected at several positions in one or the other mapping-consensus sequences (Table 3: for # 2 see Figure 2, for #3 and #4 see Figure 3), as these variants were not observed in the corresponding region of the contig sequences obtained from *de novo* assembly (from both analysis software). One of the variants (#7), an insertion of an A in the Ion Torrent consensus and AA in the Illumina consensus, was found in a long homopolymer stretch of 10 A's. The location of the variation suggested that in both cases it could be an error. Variations at this position in contigs sequences varied from 2 to 3 A's insertions. This polymorphism was not called in the consensus from the subset of Illumina data. Comparison of other chloroplast genomes, known to have been sequenced on the Illumina platform (GAII) [11], showed some to have this insertion, namely Australian *Oryza rufipogon* (GenBank accession – JN005833), Asian *Oryza rufipogon* (GenBank accession – JN005832) and *Oryza meridionalis* (GenBank accession – JN005831). Interestingly, the insertion was not present in the chloroplast genome sequence of *Oryza sativa* spp. *japonica* var. Nipponbare (reference-sequence), *Oryza sativa* ssp. *indica* isolate 93-11 (GenBank accession – AY522329) and *Oryza nivara* (GenBank accession – AP006728) which were sequenced by Sanger technology. Thus for this variant we cannot conclude with certainty if this discrepancy is due to mapping artefact or due to read error. However, all other errors found in Ion Torrent sequence (Table 3, #8 to #20) were not due mapping errors but read errors from either deletions or insertions in homopolymer regions.

Based on the analysis and all the findings using various approaches outlined above, an amended wild-rice-genotype chloroplast consensus sequence was created incorporating the most probable variants from Table 3. To assess the correctness of this amended chloroplast sequence paired-end reads from Illumina

Table 2. Comparison of chloroplast consensus sequences of the wild rice (*Oryza rufipogon*-like).

Sequencing platform	Source	Sequence length (bp)	Variants	Deletions	Insertions	Mismatches
Used available reference	GenBank (GU592207)	134,551	–	–	–	–
Illumina	mapping- consensus	134,531	128	18	13	97
	mapping-consensus (subset of reads)	134,529	129	19	13	97
Ion Torrent	mapping consensus	134,525	139	30	14	95
	de novo- consensus (CLC-GW)	134,521	155	43	17	95
	reference-assisted *de novo*- consensus (Torrent Suite)	134,554	147	23	28	96

Table 3. Inconsistent variations found in wild rice chloroplast mapping-consensus sequences and their validation.

Variations		Reference-sequence		Mapping-consensus sequence		The most probable variant
No	Type	Position	Allele	HiSeq	Ion Torrent	
#1	MNV	57,036	TT	TT	AA	TT
#2	ins	65,465^65,466	–	–	TCCTATTTAATA	TTCCTATTTAATA
#3	MNV	66,897	CGAT	TAGA	CGAT	TAGAAATAAAAAATTCTAA
#4	SNP	66,902	C	A	C	TAGAAATAAAAAATTCTAA
#5	SNP	17,366	T	A	T	T
#6	SNP/del	17,368	C	A	–	–
#7	ins	3,545^3,546	–	AA	A	–
#8	del	21,808	C	C	–	C
#9	del	57,027	T	T	–	T
#10	del	81,342	G	G	–	G
#11	del	91,427	C	C	–	C
#12	del	91,589	C	C	–	C
#13	del	97,135	G	G	–	G
#14	del	111,639	G	G	–	G
#15	del	116,139	C	C	–	C
#16	del	118,025	C	C	–	C
#17	del	119,245	G	G	–	G
#18	del	122,914	C	C	–	C
#19	del	123,568	G	G	–	G
#20	del	133,816	C	C	–	C

Variations derived by Illumina and Ion Torrent sequencing.
SNP – single-nucleotide variant, MNV – multi-nucleotide variant, ins – insertion, del – deletion.

sequencing platform were re-mapped to this new wild rice chloroplast genome. Illumina reads as against Ion Torrent reads of the W-rice-genotype were selected for remapping as even short Illumina reads (36 bases) of the R-rice-genotype provided a consensus matching the reference-sequence (Table 1) [11], as against an inaccurate consensus when Ion Torrent reads were used (Table 1) due to indel-associated errors. The new mapping results for the wild rice show only one discrepancy described earlier (Table 3, #7) in a long homopolymer region. All other sites were identical giving an additional assurance of the correctness of the final consensus.

Discussion

Various NGS technologies are now available for the rapid sequencing of whole genomes [17,23,24]. The choice of selecting one or more NGS technologies depends on the yield of data required, read length required, cost per data point and accuracy of the sequence data [25]. Systematic as against random errors can compromise the use of sequence data [26,27]. In this study we analysed NGS reads of two rice genotypes obtained from the Ion Torrent platform and compared it for accuracy to those obtained from the Illumina platform. A comparison of the three sequencing systems is presented in Table 4.

The Ion Personal Genome Machine (PGM) is characterised by long read length, very short run times and inexpensive consumables [28]. The drawbacks include low output, high cost per Gb (gigabase) of sequence [25] and the longer time required for the library preparation relative to that required for the sequencing run. Moreover, the PGM produces biased sequences from high homopolymer regions, which result in indel errors [27]. However, this is a developing platform with high potential for improvement in accuracy, read length and cost per read and per Gb. As the machine is the chip, it can be easily upgraded by the release of new chips. The PGM was the first next-generation sequencer, which price that dropped below $100 K. At the time of the study the PGM cost was around $80 K, which included the instrument, temple preparation system, enrichment system and server. The main advantages of HiSeq2000 system are the high throughput, large output data volumes per run and low cost of reagents per Gb [29]. The shortcomings of the platform embrace short read length and the need for advanced computational resources to process and store the enormous data volumes created from each run. Furthermore, the run time is considerably longer than that for the PGM, and the Illumina has a much higher initial capital cost. The specification of the GAIIx system was very similar to the HiSeq2000 with smaller output per run, higher cost per Gb and significantly lower instrument cost. The sequencing accuracy on both Illumina platforms is high with the most commonly encountered errors being substitutions. A wide range of Illumina instruments are now available with differing data volumes, running times and costs.

We selected the cultivated rice genotype (cv. Nipponbare) as one of the samples as its chloroplast sequence, derived by Sanger sequencing and considered as accurate, was available and could be used as reference sequence to check for errors in consensus sequences derived from NGS sequencing on both platforms. The aim was to generate both mapping and *de novo*-consensus sequences to determine read accuracy of both platforms.

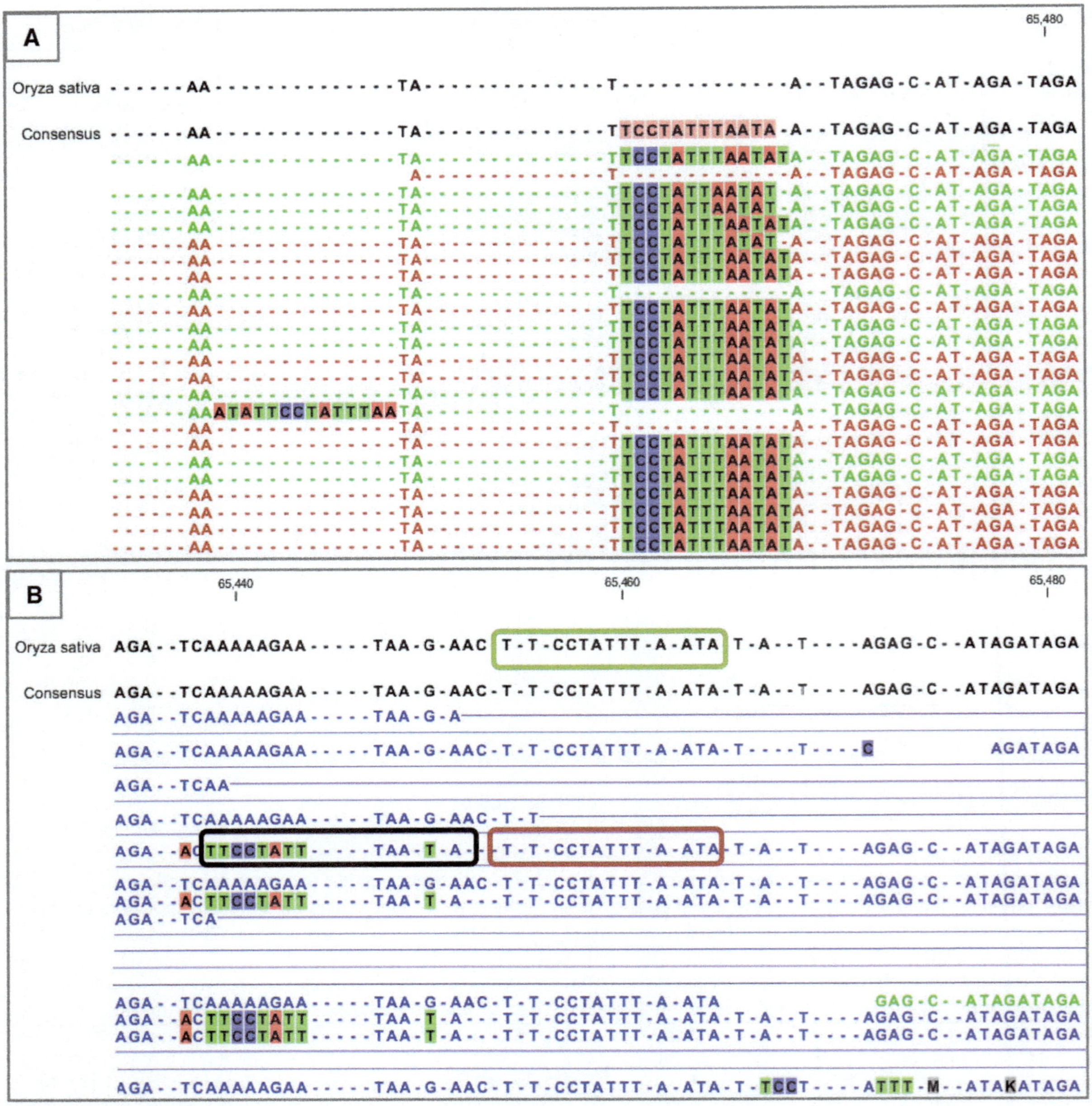

Figure 2. Snapshot of mapping results of wild rice Ion Torrent (A) and Illumina (B) reads. Reads were mapped to the chloroplast reference of *Oryza sativa* cv. Nipponbare. In the mapping of Ion Torrent reads there was a long insertion (TCCTATTTAATA) reported in the consensus sequence of wild rice chloroplast ((A), marked with orange background colour). This insertion was missed in the mapping of Illumina reads, although it was present in the reads ((B), example of the read sequence marked in black rectangle). The nucleotides in the insertion were duplicated in wild rice (sequence marked in red rectangle), and not in the reference genome where only one copy of these nucleotides was present (marked in green rectangle). The duplicated region was a probable cause of the misalignment of reads. Oryza sativa – fragment of chloroplast sequence of *Oryza sativa* spp. *japonica* var. Nipponbare; Consensus – consensus sequence of wild rice chloroplast sequence derived by mapping reads from Illumina (A) and Ion Torrent (B) platforms to the reference. Nucleotides with background colours represent the mismatches between reads and the reference sequence; paired end reads are shown in blue; single reads are shown in green and red (in forward and reverse orientation, respectively).

We were unable to obtain *de novo*-consensus sequences for both genotypes with Illumina reads when processed by CLC-GW as the contigs generated were non-contiguous when mapped to the reference-sequence. In the case of reference-genotype, read coverage depth and read length contributed to non-contiguous contigs as the read depth in regions corresponding to gaps was between 7 and 30, when examined for base calling and read mapping depth to the mapping-consensus, which is well below the average mapping read coverage depth of 127.7 (Table S2). In the case of wild-rice-genotype with longer read length (100 bases), read length but not read coverage depth may have contributed to non-contiguous contigs as the read depth in regions corresponding to the gaps was between 7,000 to 11,000, when examined for base calling and read mapping depth to the in the mapping-consensus, which is well above the average mapping read coverage depth of 8,143.2 (Table S2). Interestingly, *de novo*-consensus, although not accurate (Table 1) was generated for the reference-genotype using CLC-GW and the longer (200 bases) Ion Torrent reads. We can

Oryza sativa (reference)	CCTTCAGTTA TTTCGCA--- ---------- ------CGAT ACTAAAAAAT
Illumina reads mapping	CCTTCAGTTA TTTCGCA--- ---------- ------TAGA AATAAAAAAT
Ion Torrent reads mapping	CCTTCAGTTA TTTCGCA--- ---------- ------CGAT ACTAAAAAAT
Illumina reads assembly (CLC)	CCTTCAGTTA TTTCGCATAG AAATAAAAAA TTCTAACGAT ACTAAAAAAT
Ion Torrent reads assembly (CLC)	CCTTCAGTTA TTTCGCATAG AAATAAAAAA TTCTAACGAT ACTAAAAAAT
Ion Torrent reads assembly (Suite)	CCTTCAGTTA TTTCGCATAG AAATAAAAAA TTCTAACGAT ACTAAAAAAT
Final consensus	CCTTCAGTTA TTTCGCATAG AAATAAAAAA TTCTAACGAT ACTAAAAAAT

Figure 3. Alignment of regions #3 and #4 from Table 3 showing discrepancies in consensus sequences. The fragment circled in red shows false called SNPs (#3 and #4, Table 3, Illumina consensus); these SNPs were incorrect because of the long insertion present in wild rice sequence but not in the reference. The fragments circled in green illustrate this long insertion found in wild rice chloroplast genome by means of reads assembly from both platforms and both assembly tools. Final sequence was created based on this information. Oryza sativa (reference) – region 66860.66940 from chloroplast sequence of *Oryza sativa* spp. *japonica* var. Nipponbare; Illumina reads mapping and Ion Torrent reads mapping – regions from consensus sequence generated by mapping wild rice Illumina and Ion Torrent reads, respectively, to the reference sequence; Illumina reads assembly and Ion Torrent reads assembly – regions from contigs generated by assembly of reads from Illumina and Ion Torrent platforms, respectively; CLC – assembly performed in CLC Genomic Workbench; Suite – assembly performed in Torrent Suite Software; Final consensus – final wild rice chloroplast genome sequence (GenBank accession – KF428978).

thus conclude that read lengths of more than 100 bases would be required to successfully obtain a *de novo*-consensus of the chloroplast genome.

In case of the reference-genotype, mapping-consensus sequences were generated using reads from both platforms, with no mismatches. However, mapping-consensus sequences showed several indels due to inaccuracies in the Ion Torrent reads mainly in homopolymer regions. In case of the short Illumina reads (36 bases) of the R-rice-genotype, neither the read length (36 bases) nor the sequence yield (minimum reference-sequence read mapping coverage of 7) was a limitation in generating an accurate mapping-consensus when processed by the CLC-GW, indicating the accuracy of the Illumina sequencing platform. In case of the longer Ion Torrent reads (200 bases), when processed by the CLC-GW and the Torrent Suite Software, consensus-sequences with no mismatches were generated but with several variants all of which were indels and predominantly in homopolymer regions (Table 1). These results indicate the actual presence of these anomalous indels in the Ion Torrent reads which contributed to the inaccuracy in the consensus sequence generated and to the unreliability of any indels calls made when using these reads. It is known that the Ion Torrent sequencing platform generates anomalous indels [27]. Interestingly, the preference of the CLC-GW and the Torrent Suite analysis tool in filtering out predominantly either insertions or deletion, respectively, as observed in both rice genotypes (Tables 1 and 2), indicates the

Table 4. The comparison between the tree sequencing systems utilised in the study.

	Illumina GAIIx	Illumina HiSeq2000	Ion Torrent PGM –318 chip
Sequencing method	Synthesis (light detection)	Synthesis (light detection)	Synthesis (proton detection)
Amplification method	Bridge PCR	Bridge PCR	Emulsion PCR
Read length[a]	Up to 2×150 bp	1×50 bp, 2×50 bp, 2×100 bp	~200 bp, ~400 bp
Paired reads	Yes	yes	Yes
Insert size	Up to 700 bp	Up to 700 bp	Up to 250 bp
Output data/run	30 Gb	600 Gb	Up to 2 Gb
Time/run[b]	10–14 days	8–11 days	4-7 hours
Cost/Gb[c]	$148	$41	$1000
Instrument cost	$256 K	$654 K	$80 K
Accuracy	>99.9%	>99.9%	99%
Error rate[d]	≥0.1%	≥0.1%	~1%
Primary errors	Substitutions	Substitutions	Indels
DNA requirements	0.05–1 ug	0.05–1 ug	0.1–1 ug

Gb – gigabase, bp – base pair, K – thousand, uq - microgram.
[a] Annotation '2 x' refers to paired end reads and '1 x' to single reads.
[b] Run time from minimum to maximum read lengths.
[c] Includes one sample and one sequencing kit per run.
[d] percentage of errors per base in single read.

possibility of these analysis tools to be amended to deal with anomalous indels in the Ion Torrent reads. We also observed a bias for indels in homopolymer regions, where deletions occurred in shorter regions while insertions occurred in longer regions. These findings are inconsistent with a recent study on Ion Torrent sequencing bias where it was reported that an increased deletion rate was positively correlated with increased lengths of homopolymer regions and that insertions were independent of homopolymer length [27,30]. This is probably because of the different genomes used or that the chloroplast genome is AT-rich and the longest C/G stretches found are 9 bp long (2 regions), 8 bp (4 regions) and 7 bp (9 regions).

The systematic pattern of bias, specific types of nucleotide variants in long versus short repeats, in the single base indels in homopolymers (Figure 1) is likely to be a feature of chloroplasts with their specific patterns of homopolymer repeat frequency. However, systematic errors reported in the Ion Torrent data in earlier studies [27] were not found in this analysis of the chloroplast possibly because we used a more recent version of the technology. We can conclude that both sequencing platforms produce high quality data with relatively low rates of discrepancy in the calling of polymorphisms, especially SNPs, but the Ion Torrent data would be less reliable for indel calls.

We compared the two sequencing platforms in generating an accurate consensus chloroplast sequence by comparing the mapping-consensus sequences generated for both genotypes. For the R-rice genotype, no mismatches were observed indicating the two platforms to be comparable for SNP calls as has been reported for microbial genomes [25]. When comparing the mapping-consensus sequences of the W-rice-genotype we identified twenty discrepancies. It must be noted that some of the discrepancies were caused by misalignment of reads and as has been previously recognized [31] there is an important issue with mismatches in close proximity of indel events. Some read mapping tools can have problems with these regions resulting in false SNPs calling. However, for both the rice genotypes we observed erroneous indel calls with Ion Torrent reads mainly in homopolymer regions.

The importance of obtaining a *de novo* sequence was clear from our study as discrepancies observed in the mapping sequences of the W-rice-genotype were curated using the *de novo* contigs. The *de novo* assembly of chloroplast genomes using the tools applied in this study was easier with the longer Ion Torrent data than with the Illumina data. Although increasing number of reference genome sequences are being generated to support chloroplast genome analysis, the ability to generate whole chloroplast genome sequences *de novo* will find wide application. The first report of total plant DNA analysis for chloroplast sequencing [11] used short Illumina reads (36 bp) and relied on a reference genome for successful assembly. More recent studies based upon longer reads [12,23] have reported more success with *de novo* assembly.

Conclusions

This study demonstrates that *de novo* assembly of an accurate whole chloroplast genome sequence will be possible for routine plant barcoding. Analysis of plants based upon appropriate and careful analysis of shotgun sequencing of total DNA promises to provide a barcode that will have wide application. Having a well curated and reliable consensus chloroplast sequence for a plant sample provides greater certainty of obtaining reliable results in many critical studies such as plant identification, purity assessment, phylogenetic analysis and heteroplasmy analysis. Continuing improvements in sequencing platforms and analysis tools will make this method more reliable and cost effective for a wide range of research and industrial applications. The significant advances in these competing sequencing platforms that have been foreshadowed by the manufactures promise dramatic reductions in cost in the near future. This would make barcoding by sequencing the whole chloroplast in this way the preferred option for plant identification in many research and industrial applications. A comparison of the platforms at the same level of coverage is useful in providing a guide to likely comparative performance as the number of reads on these platforms increases in the future. Costs will change for each platform as read length and read volume are increased.

However cost is not the only advantage of this approach. The use of total DNA from the sample without amplification simplifies the analysis but also provides greater opportunity to measure in an unbiased way the contribution of different genotypes in complex mixtures. The risks of preferential amplification or enrichment of chloroplasts from a specific genotype that complicated earlier analysis will be avoided.

Supporting Information

Table S1 Summary statistics of raw reads obtained from PGM Ion Torrent and Illumina platforms for rice (*Oryza sativa* spp. *japonica* var. Nipponbare) and wild rice (*Oryza rufipogon*-like plant from Australia). Quality distribution is represented as PHRED scores.

Table S2 Mapping statistics of CLC trimmed reads from Illumina and PGM Ion Torrent platforms.

Table S3 Assembly results performed on CLC Genomic Workbench and Torrent Suite Software of reads from Illumina and Ion Torrent platforms and mapping results of generated contigs to chloroplast reference-sequence.

Acknowledgments

We thank the local support team from Life Technologies for technical assistance with this project.

Author Contributions

Conceived and designed the experiments: MB AF RH. Performed the experiments: MB. Analyzed the data: MB AF. Contributed to the writing of the manuscript: MB AF RH.

References

1. Straub SCK, Parks M, Weitemier K, Fishbein M, Cronn RC, et al. (2012) Navigating the tip of the genomic iceberg: Next-generation sequencing for plant systematics. Am J Bot 99: 349–364.
2. Parks M, Cronn R, Liston A (2009) Increasing phylogenetic resolution at low taxonomic levels using massively parallel sequencing of chloroplast genomes. BMC Biol 7: 84–84.
3. Taberlet P, Coissac E, Pompanon F, Brochmann C, Willerslev E (2012) Towards next-generation biodiversity assessment using DNA metabarcoding. Mol Ecol 21: 2045–2050.
4. Henry RJ (2001) Plant genotyping: the DNA fingerprinting of plants. New York: CABI Pub.
5. Henry RJ (2013) Molecular Markers in Plants. Hoboken, NJ: John Wiley & Sons.

6. Galimberti A, De Mattia F, Losa A, Bruni I, Federici S, et al. (2013) DNA barcoding as a new tool for food traceability. Food Res Int 50: 55–63.
7. Diaz-Sanchez S, Hanning I, Pendleton S, D'Souza D (2013) Next-generation sequencing: the future of molecular genetics in poultry production and food safety. Poult Sci 92: 562–572.
8. Jansen RK, Raubeson LA, Boore JL, dePamphilis CW, Chumley TW, et al. (2005) Methods for Obtaining and Analyzing Whole Chloroplast Genome Sequences. In: Elizabeth AZ, Eric HR, editors. Methods in Enzymology: Academic Press. pp. 348–384.
9. Lang BF, Burger G (2007) Purification of mitochondrial and plastid DNA. Nat Protoc 2: 652–660.
10. Cronn R, Liston A, Parks M, Gernandt DS, Shen R, et al. (2008) Multiplex sequencing of plant chloroplast genomes using Solexa sequencing-by-synthesis technology. Nucleic Acids Res 36: e122.
11. Nock CJ, Waters DLE, Edwards MA, Bowen SG, Rice N, et al. (2011) Chloroplast genome sequences from total DNA for plant identification. Plant Biotechnol J 9: 328–333.
12. McPherson H, Rossetto M, van der Merwe M, Delaney SK, Edwards MA, et al. (2013) Capturing chloroplast variation for molecular ecology studies: a simple next generation sequencing approach applied to a rainforest tree. BMC Ecol 13: 8–18.
13. Loman NJ, Misra RV, Dallman TJ, Constantinidou C, Gharbia SE, et al. (2012) Performance comparison of benchtop high-throughput sequencing platforms. Nat Biotechnol 30: 434–434.
14. Sims D, Sudbery I, Ilott NE, Heger A, Ponting CP (2014) Sequencing depth and coverage: key considerations in genomic analyses. Nat Rev Genet 15: 121–132.
15. Illumina Inc website. Available: http://res.illumina.com/documents/products/techspotlights/techspotlight_sequencing.pdf. Accessed 2014 September 8.
16. Illumina Inc website. Available: http://res.illumina.com/documents/products/illumina_sequencing_introduction.pdf. Accessed 2014 September 8.
17. Rothberg JM, Edwards M, Hoon J, Simons JF, Marran D, et al. (2011) An integrated semiconductor device enabling non-optical genome sequencing. Nature 475: 348–352.
18. Lee KH, Park J, Williams DS, Xiong Y, Hwang I, et al. (2013) Defective chloroplast development inhibits maintenance of normal levels of abscisic acid in a mutant of the Arabidopsis RH3 DEAD-box protein during early post-germination growth. Plant J 73: 720–732.
19. Hartwig B, James GV, Konrad K, Schneeberger K, Turck F (2012) Fast isogenic mapping-by-sequencing of ethyl methanesulfonate-induced mutant bulks. Plant Physiol 160: 591–600.
20. Sotowa M, Ootsuka K, Kobayashi Y, Hao Y, Tanaka K, et al. (2013) Molecular relationships between Australian annual wild rice, Oryza meridionalis, and two related perennial forms. Rice 6: 26.
21. Furtado A (2014) DNA extraction from vegetative tissue for next-generation sequencing. In: Henry RJ, Furtado A, editors. Cereal Genomics. Methods in Molecular Biology. New York: Springer Science+Business Media. pp. 1.
22. Kurtz S, Phillippy A, Delcher AL, Smoot M, Shumway M, et al. (2004) Versatile and open software for comparing large genomes. Genome Biol 5: 12–R12.
23. Ferrarini M, Cestaro A, Sargent DJ, Moretto M, Ward JA, et al. (2013) An evaluation of the PacBio RS platform for sequencing and de novo assembly of a chloroplast genome. BMC Genomics 14: 670–670.
24. Shendure J, Ji H (2008) Next-generation DNA sequencing. Nat Biotechnol 26: 1135–1145.
25. Quail MA, Smith M, Coupland P, Otto TD, Harris SR, et al. (2012) A tale of three next generation sequencing platforms: comparison of Ion Torrent, Pacific Biosciences and Illumina MiSeq sequencers. BMC Genomics 13: 341–341.
26. Meacham F, Boffelli D, Dhahbi J, Martin DIK, Singer M, et al. (2011) Identification and correction of systematic error in high-throughput sequence data. BMC Bioinformatics 12: 451–451.
27. Bragg LM, Stone G, Butler MK, Hugenholtz P, Tyson GW (2013) Shining a light on dark sequencing: characterising errors in Ion Torrent PGM data. Plos Comput Biol 9: e1003031.
28. Liu L, Li Y, Li S, Hu N, He Y, et al. (2012) Comparison of next-generation sequencing systems. J Biomed Biotechnol 2012: 251364.
29. Glenn TC (2011) Field guide to next-generation DNA sequencers. Molecular ecology resources 11: 759–769.
30. Ross MG, Russ C, Costello M, Hollinger A, Lennon NJ, et al. (2013) Characterizing and measuring bias in sequence data. Genome Biol 14: R51.
31. Albers CA, Lunter G, MacArthur DG, McVean G, Ouwehand WH, et al. (2011) Dindel: accurate indel calls from short-read data. Genome Res 21: 961–973.

5

Reflectance Variation within the In-Chlorophyll Centre Waveband for Robust Retrieval of Leaf Chlorophyll Content

Jing Zhang[1], Wenjiang Huang[2], Qifa Zhou[1]*

1 College of Life Sciences, Zhejiang University, Hangzhou, China, **2** Key Laboratory of Digital Earth Science, Institute of Remote Sensing and Digital Earth, Chinese Academy of Sciences, Beijing, China

Abstract

The in-chlorophyll centre waveband (ICCW) (640–680 nm) is the specific chlorophyll (Chl) absorption band, but the reflectance in this band has not been used as an optimal index for non-destructive determination of plant Chl content in recent decades. This study develops a new spectral index based solely on the ICCW for robust retrieval of leaf Chl content for the first time. A glasshouse experiment for solution-culture of one chlorophyll-deficient rice mutant and six wild types of rice genotypes was conducted, and the leaf reflectance (400–900 nm) was measured with a high spectral resolution (1 nm) spectrophotometer and the contents of chlorophyll a (Chl*a*), chlorophyll b (Chl*b*) and chlorophyll a+b (Chl*t*) of the rice leaves were determined. It was found that the reflectance curves from 640 nm to 674 nm and from 675 nm to 680 nm of the low-chlorophyll mutant leaf were drastically steeper than that of the wild types in the ICCW. The new index based on the reflectance variation within ICCW, the difference of the first derivative sum within the ICCW (DFDS_ICCW), was highly sensitive ($r = -0.77$, $n = 93$, $P < 0.01$) to Chl*t* while the mean reflectance (R_ICCW) in the ICCW became insensitive ($r = -0.12$, $n = 93$, $P > 0.05$) to Chl*t* when the leaf Chl*t* was higher than 200 mg/m^2. The best equations of R-ICCW and DFDS_ICCW yielded an RMSE of 78.7, 32.9 and 107.3 mg/m^2, and an RMSE of 37.4, 16.0 and 45.3 mg/m^{-2}, respectively, for predicting Chl*a*, Chl*b* and Chl*t*. The new index could rank in the top 10 for prediction of Chl*a* and Chl*t* as compared with the 55 existing indices. Additionally, most of the 55 existing Chl-related VIs performed robustly or strongly in simultaneous prediction of leaf Chl*a*, Chl*b* and Chl*t*.

Editor: Jinxing Lin, Beijing Forestry University, China

Funding: This work was funded by the National Natural Science Foundation of China (grant numbers 41271363 and 41271412). QFZ received 41271363 and WJH received 41271412. The funders had no role in study design, data collection and analysis, decision to publish, or preparation of the manuscript.

Competing Interests: The authors have declared that no competing interests exist.

* Email: zqifa2002@yahoo.com

Introduction

Chlorophyll (Chl) a and Chl b are major constituents of the photosynthetic apparatus in higher plants. Chl a and Chl b are interconverted in the chlorophyll cycle [1]. Leaf Chl a concentration (Chl*a*) and Chl b concentration (Chl*b*) indicate a plant's photosynthetic capacity and health status, and determination of Chl*a*, Chl*b* and ratios of Chl*a* to Chl*b* are also helpful for understanding the light acclimation mechanisms in higher plants [2]. Conventionally, leaf Chl*a* and Chl*b* are determined with a traditional wet extraction analysis based on measuring the extinction of the extract at the major red absorption maxima of Chl a (~664 nm) and b (~647 nm) in the in-chlorophyll centre waveband (640–680 nm), and by inserting these values into simultaneous equations [2,3]. In recent decades, there has been an increasing interest in non-destructively determining leaf and canopy Chl content by measuring leaf and canopy spectral reflectance. Particular efforts have been devoted to the development of robust algorithms for Chl*t* determination from the leaf to canopy scale [4–10]. Contrastingly, studies conducted for determination of individual Chl*a* or individual Chl*b* with spectral vegetation indices (VIs) are much less frequent [4,6,11]. Reflectance in the ICCW had been used for a long time as an indicator of chlorophyll content of leaves, but has not been used as an optimal index since Thomas and Gausman (1977) [12] found that reflectance near 675 nm became saturated at medium to high chlorophyll concentrations [6]. In recent decades, many studies have found that reflectance in the green and red-edge spectral regions was optimal for non-destructive estimation of leaf Chl content in a wide range of its variation [13–16]. The results of Féret et al. (2011) [17] showed that the reflectance in the red-edge and near infrared spectral regions simulated with the Prospect 5 radiative transfer model provided an accurate estimation of leaf Chl content. Recently, Main et al. (2011) [11] assessed the performance of 73 published VIs for leaf Chl estimation and also found that the indices using off-chlorophyll absorption centre wavebands (OCCW) performed better than those using ICCW. To our best knowledge, no VIs based solely on ICCW for Chl estimation have been developed since Thomas and Gausman (1977) [12] found the saturated reflection of plant leaves. Plant leaves have a reflectance minima around 675 nm, and there are substantial differences in reflectance among different wavelengths in the ICCW. Is the reflectance difference within the ICCW

Table 1. The existing vegetation indices used in this study.

Index	Formulation	Reference
log(1/R737)	log(1/R737)	Yoder, Pettigrew-Crosby (1995)
SIPI	(R800-R445)/(R800-R680)	Peñuelas et al. (1995)
Ratcart	R695/R760	Carter et al. (1996)
PSSRa	R800/R680	Blackburn (1998)
PSSRb	R800/R635	Blackburn (1998)
PSNDa	(R800-R675)/(R800+R675)	Blackburn (1998)
PSNDb	(R800-R650)/(R800+R650)	Blackburn (1998)
PSSRchla	R810/R676	Blackburn (1999)
PSRI	(R680-R500)/R750	Merzlyak et al. (1999)
SR705	R750/R705	Sims, Gamon (2002)
ND705	(R750-R705)/(R750+R705)	Sims, Gamon (2002)
mND705	(R750-R445)/(R700-R445)	Sims, Gamon (2002)
mSR705	(R750-R705)/(R750+R705-2×R445)	Sims, Gamon (2002)
Readone	R415/R695	Read et al. (2002)
RGRcan	(R612+R660)/(R510+R560)	Steddom et al. (2003)
NDVIcanste	(R760-R708)/(R760+R708)	Steddom et al. (2003)
Red edge Model	(R800/R700)-1	Gitelson et al. (2005)
Green Model	(R800/R550)-1	Gitelson et al. (2005)
OSAVI	1.16×(R800-R670)/(R800+R670+0.16)	Rondeaux et al. (1996)
$CI_{red\ edge}$	(R800/R700)-1	Gitelson et al. (2005)
EVI2	2.5×(R800-R660)/(1+R800+2.4×R660)	Jiang et al. (2008)
CARI	$R700\times(\mathrm{sqrt}(a\times670+R670+b)^2)/R670\times(a^2+1)^{0.5}$ a=(R700-R550)/150 b=R550-a×550	Kim et al. (1994)
Carter[A]	R695/R420	Carter (1994)
Carter2[A]	R695/R760	Carter (1994)
Carter3[A]	R605/R760	Carter (1994)
Carter4[A]	R710/R760	Carter (1994)
Carter5[A]	R695/R670	Carter (1994)
Carter6[A]	R550	Carter (1994)
DD	(R749-R720)-(R701-R672)	Le Maire et al. (2004)
Datt[A]	(R850-R710)/(R850-R680)	Datt (1999)
Datt2[A]	R850/R710	Datt (1999)
Datt4[A]	R672/(R550×R708)	Datt (1998)
Datt5[A]	R672/R550	Datt (1998)
Datt6[A]	R860/(R550×R708)	Datt (1998)
Gitelson2[A]	(R750-R800/R695-R740)-1	Gitelson et al. (2003)
Gitelson[A]	1/R700	Gitelson et al. (1999)
mNDVI	(R800-R680)/(R800+R680-2×R445)	Sims, Gamon (2002)
Maccioni[A]	(R780-R710)/(R780-R680)	Maccioni et al. (2001)
mSR	(R800-R445)/(R680-R445)	Sims, Gamon (2002)
SRPI	R430/R680	Peñuelas et al. (1995)
NDVI2[A]	(R750-R705)/(R750+R705)	Gitelson, Merzlyak (1994)
NPCI	(R680-R430)/(R680+R430)	Penuelas et al. (1994)
REP_LE[A]	700+40×(Rre-R700)/(R740-R700) Rre=(R670+R780)/2	Cho, Skidmore (2006)
REP_Li[A]	700+40×((R670+R780/2)/(R740-R700))	Guyot, Baret (1988)
SR1[A]	R750/R700	Gitelson, Merzlyak (1997)
SR2[A]	R752/R690	Gitelson, Merzlyak (1997)
SR3[A]	R750/R550	Gitelson, Merzlyak (1997)
SR4[A]	R700/R670	McMurtey et al. (1994)
SR5[A]	R675/R700	Chappelle et al. (1992)
SR6[A]	R750/R710	Zarco-Tejada, Miller (1999)

Table 1. Cont.

Index	Formulation	Reference
SR7[A]	R440/R690	Lichtenthaler et al. (1996)
Sum_Dr2[A]	sum of first derivative reflectance between R680 and R780	Filella, Penuelas (1994)
Vogelmann[A]	R740/R720	Vogelman et al. (1993)
Vogelmann2[A]	(R734-R747)/(R715+R726)	Vogelman et al. (1993)
SPAD reading	Based on the transmittance at 650 nm and 940 nm	Konica Minota, Japan

associated with the Chl content? This study has two objectives. The first is to examine the robustness of simultaneous estimation of Chl*a*, Chl*b* and Chl*t* with the existing Chl-related VIs and commercial chlorophyll meter readings by using a dataset of measured reflectance, Chl*a*, Chl*b* and Chl*t* of rice leaves of different genotypes including low-chlorophyll mutants (low in Chl content) at different stages. Second, we test if the reflectance difference within the ICCW is associated with the Chl content by using the constructed dataset and then solely using ICCW to develop a new VI simultaneously sensitive to Chl*a*, Chl*b* and Chl*t*.

Materials and Methods

2.1. Plant materials and growth conditions

A pot experiment was conducted in a greenhouse with natural light (mean daily photosynthetically active radiation 130 μmol $m^{-2}\ s^{-1}$ during the whole growth period) and controlled temperature (daily maximum 27.6°C, daily minimum 16.2°C during the rice growing period) and humidity (24.5–85.1% average daily relative humidity, RH, throughout the whole rice growing period) at Zhejiang University Experimental Farm, Hangzhou, China (30°14′ N, 120°10′ E). Six wild types of rice genotypes (IG1, IG23, IG24, DJ, NIP and ZH11) and one chlorophyll-deficient mutant (IG20) were solution-cultured according to the IRRI prescription [18], but the nitrogen level was designed as 1/5×40 mg l^{-1} (low N) and 40 mg l^{-1} (normal N), respectively, for two nitrogen treatments. The mutant 'IG 20' is an isogenic line of the recurrent parent "Zhefu 802" bred by China National Rice Research Institute. A completely random design with four replications was used. Each pot contained a 6.0-L nutrient solution and three seedlings. The nutrient solution was

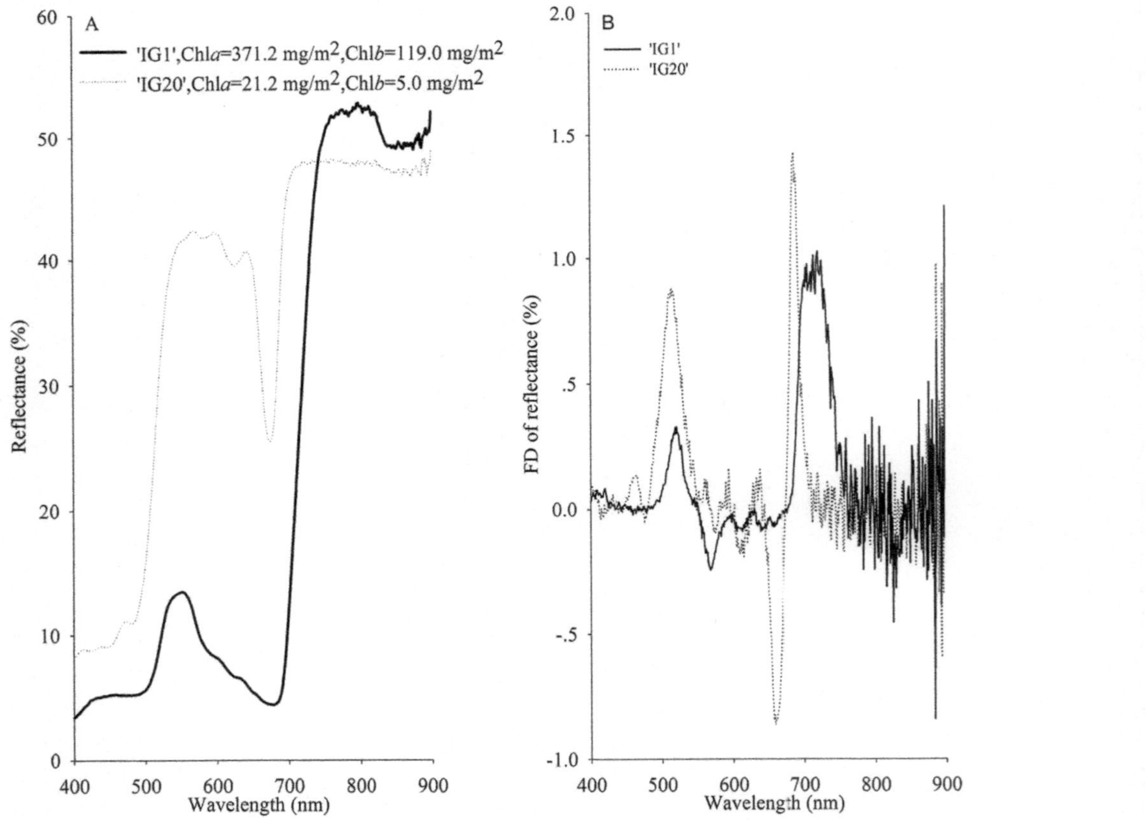

Figure 1. The reflectance curve (A) and the first derivative (FD) of reflectance curve (B) in the mutant (IG20) and wild type (IG1). Chl*a* and Chl*b* represent the leaf chlorophyll a content and chlorophyll b content, respectively.

Table 2. The best prediction equations of the existing vegetation indices.

Index	Prediction target	r	Prediction equation	R^2	RMSE (mg/m^2)	Rank
Log(1/R737)	Chla	0.34	$y=-34230x^2-110191x-88409$	0.25	73.8	a52
	Chlb	0.40	$y=-13672x^2-43899x-35148$	0.29	29.0	b47
	Chlt	0.37	$y=-47901x^2-154090x-123557$	0.28	99.0	t52
SIPI	Chla	−0.65	$y=221.3x^{-6.194}$	0.78	59.5	a45
	Chlb	−0.51	$y=63.261x^{-6.7392}$	0.50	29.6	b49
	Chlt	−0.62	$y=288.46x^{-6.2375}$	0.76	84.1	t45
Ratcart	Chla	−0.83	$y=577.68e^{-4.297x}$	0.94	37.8	a11
	Chlb	−0.70	$y=2.4669x^{-2.057}$	0.77	15.8	b25
	Chlt	−0.82	$y=768.46e^{-4.3833x}$	0.94	50.9	t19
PSSRa	Chla	0.81	$y=4.3255x^{1.6601}$	0.90	50.4	a34
	Chlb	0.72	$y=1.7069e^{0.327x}$	0.72	23.2	b36
	Chlt	0.81	$y=5.2238x^{1.6927}$	0.89	68.1	t34
PSSRb	Chla	0.90	$y=14.01x^{1.5063}$	0.93	41.2	a19
	Chlb	0.90	$y=1.5702x^2+0.682x+0.6033$	0.84	13.6	b6
	Chlt	0.99	$y=16.707x^{1.556}$	0.95	46.9	t10
PSNDa	Chla	0.74	$y=1.3021e^{6.2414x}$	0.87	52.1	a37
	Chlb	0.61	$y=0.1751e^{7.1639x}$	0.62	26.0	b43
	Chlt	0.72	$y=1.5724e^{6.335x}$	0.86	72.5	t40
PSNDb	Chla	0.83	$y=9.6049e^{4.182x}$	0.94	38.8	a15
	Chlb	0.73	$y=717.58x^2-555.53x+77.434$	0.80	15.5	b19
	Chlt	0.83	$y=11.591e^{4.2866x}$	0.95	49.0	t13
PSSRchla	Chla	0.81	$y=3.9395x^{1.6948}$	0.90	50.4	a33
	Chlb	0.72	$y=1.6415e^{0.3287x}$	0.72	23.2	b37
	Chlt	0.81	$y=4.744x^{1.7285}$	0.89	68.0	t33
PSRI	Chla	−0.52	$y=152.13e^{-13.23x}$	0.61	85.8	a55
	Chlb	−0.34	$y=43.635e^{-12.77x}$	0.31	37.9	b55
	Chlt	−0.48	$y=198.91e^{-13.064x}$	0.57	120.3	t55
SR705	Chla	0.91	$y=23.775x^{2.5135}$	0.89	45.8	a26
	Chlb	0.88	$y=19.518x^2-22.118x+8.0188$	0.81	15.2	b9
	Chlt	0.93	$y=28.788x^{2.5989}$	0.91	54.2	t27
ND705	Chla	0.91	$y=572.06x^{0.9776}$	0.94	37.5	a7
	Chlb	0.83	$y=724.6x^2-161.79x+13.25$	0.80	15.3	b13
	Chlt	0.91	$y=758.62x^{0.9945}$	0.93	51.6	t21
mND705	Chla	0.90	$y=22.471x^{1.5336}$	0.89	47.9	a29
	Chlb	0.89	$y=0.8471x^2+15.357x-14.561$	0.80	15.5	b21
	Chlt	0.92	$y=27.138x^{1.5862}$	0.91	57.3	t29
mSR705	Chla	0.91	$y=494.39x^{0.994}$	0.94	36.8	a6
	Chlb	0.83	$y=517.31x^2-133.21x+13.164$	0.80	15.4	b15
	Chlt	0.91	$y=654.1x^{1.0094}$	0.93	50.7	t18
Readone	Chla	0.88	$y=1720.4x^{2.5357}$	0.85	54.9	a40
	Chlb	0.84	$y=838.41x^{3.1792}$	0.73	20.9	b34
	Chlt	0.89	$y=2403.7x^{2.619}$	0.87	69.2	t35
RGRcan	Chla	−0.68	$y=6638.5e^{-5.523x}$	0.82	68.7	a51
	Chlb	−0.53	$y=2736.3e^{-6.1144x}$	0.55	32.2	b54
	Chlt	−0.66	$y=8855.4e^{-5.5601x}$	0.79	96.6	t51
NDVIcanste	Chla	0.91	$y=609.94x^{0.925}$	0.94	36.6	a5
	Chlb	0.83	$y=783.43x^2-128.31x+11.471$	0.80	15.5	b16
	Chlt	0.91	$y=809.92x^{0.9412}$	0.93	50.3	t17
Red edge Model	Chla	0.91	$y=117.36x^{0.821}$	0.95	35.5	a3
	Chlb	0.88	$y=6.683x^2+11.629x+4.7987$	0.80	15.5	b17

Table 2. Cont.

Index	Prediction target	*r*	Prediction equation	R^2	RMSE (mg/m^2)	Rank
	Chlt	0.92	$y=151.08x^{0.8386}$	0.95	45.6	t8
Green Model	Chla	0.91	$y= 118.66x^{1.0178}$	0.94	37.6	a9
	Chlb	0.93	$y=4.6913x^2+28.539x-2.6421$	0.87	12.5	b2
	Chlt	0.94	$y=151.82x^{1.0515}$	0.96	41.2	t2
OSAVI	Chla	0.75	$y=1.556e^{5.2403x}$	0.88	50.2	a31
	Chlb	0.62	$y=0.2085e^{6.0466x}$	0.64	25.2	b40
	Chlt	0.73	$y=1.8751e^{5.324\ x}$	0.87	69.7	t37
CI red edge	Chla	0.91	$y=117.36x^{0.821}$	0.95	35.5	a4
	Chlb	0.88	$y=6.683x^2+11.629x+4.7987$	0.80	15.5	b18
	Chlt	0.92	$y=151.08x^{0.8386}$	0.95	45.6	t9
EVI2	Chla	0.82	$y=7.4037e^{1.9921x}$	0.93	41.2	a20
	Chlb	0.71	$y=1.084e^{2.3895x}$	0.73	19.9	b32
	Chlt	0.81	$y=8.9479e^{2.0371x}$	0.93	53.9	t26
CARI	Chla	−0.87	$y=0.0159x^2-5.7648x+540.52$	0.79	39.2	a16
	Chlb	−0.82	$y=0.0141x^2-3.6719x+247.15$	0.80	15.2	b10
	Chlt	−0.88	$y=0.0299x^2-9.4367x+787.67$	0.83	48.0	t11
Carter[A]	Chla	−0.87	$y=1418.9e^{-0.839x}$	0.91	39.4	a17
	Chlb	−0.76	$y=676.39x^{-3.0899}$	0.74	19.2	b31
	Chlt	−0.86	$y=1941.8e^{-0.86x}$	0.92	49.8	t15
Carter2[A]	Chla	−0.83	$y=577.68e^{-4.297x}$	0.94	37.8	a12
	Chlb	−0.70	$y=2.4669x^{-2.057}$	0.77	15.8	b26
	Chlt	−0.82	$y=768.46e^{-4.3833x}$	0.94	50.9	t20
Carter3[A]	Chla	−0.85	$y=579.04e^{-4.3x}$	0.95	35.4	a2
	Chlb	−0.74	$y=2.4748\ x^{-2.051}$	0.82	13.0	b4
	Chlt	−0.84	$y=774.39e^{-4.4086}$	0.96	43.9	t5
Carter4[A]	Chla	−0.90	$y=2561.5e^{-4.845x}$	0.92	38.2	a14
	Chlb	−0.81	$y=593.69x^2-1001x+423.27$	0.79	15.8	b24
	Chlt	−0.90	$y=3589.7e^{-4.9847x}$	0.93	45.4	t7
Carter5[A]	Chla	−0.43	$y=-98.296x^2+341.52x34.281$	0.21	75.8	a53
	Chlb	−0.48	$y=-57.997x+199.2$	0.23	30.2	b51
	Chlt	−0.45	$y=-99.659x^2+290.02x157.36$	0.22	102.6	t53
Carter6[A]	Chla	−0.88	$y=1248.3e^{-0.102x}$	0.91	43.6	a23
	Chlb	−0.83	$y=0.2785x^2-18.344x+299.57$	0.86	12.9	b3
	Chlt	−0.89	$y=1738.4e^{-0.106x}$	0.94	49.0	t12
DD	Chla	0.91	$y=171.95e^{0.0753x}$	0.85	41.7	a22
	Chlb	0.83	$y=0.1316x^2+4.8546x+52.571$	0.80	15.5	b20
	Chlt	0.91	$y=0.2558x^2+15.278x+255.86$	0.87	42.2	t3
Datt[A]	Chla	0.90	$y=18.526e^{4.5459x}$	0.90	44.5	a25
	Chlb	0.83	$y=443.78x^2-139.88x+14.677$	0.81	15.0	b8
	Chlt	0.91	$y=22.272e^{4.6979x}$	0.92	51.9	t23
Datt2[A]	Chla	0.89	$y=29.472x^{2.8339}$	0.83	57.3	a42
	Chlb	0.90	$y=17.484x^2+14.947x-30.522$	0.81	14.9	b7
	Chlt	0.92	$y=35.395x^{2.9529}$	0.86	69.4	t36
Datt4[A]	Chla	0.69	$y=-237156x^2+25959x-55.707$	0.48	61.4	a46
	Chlb	0.81	$y=66027x^2+7841x-38.216$	0.65	20.3	b33
	Chlt	0.75	$y=-171128x^2+33800x-93.923$	0.56	77.3	t42
Datt5[A]	Chla	−0.27	$y=-5518x^2+3806.5x-375.93$	0.44	63.9	a49
	Chlb	−0.09	$y=-2482x^2+1820.7x-232.67$	0.46	25.3	b41
	Chlt	−0.23	$y=-8000x^2+5627.2x-608.59$	0.46	85.3	t46

Table 2. Cont.

Index	Prediction target	r	Prediction equation	R^2	RMSE (mg/m^2)	Rank
Datt6[A]	Chla	0.86	$y=2546.3x^{1.2194}$	0.85	54.5	a39
	Chlb	0.92	$y=-563.98x^2+748.31x-18.395$	0.86	13.0	b5
	Chlt	0.91	$y=3709.6x^{1.2735}$	0.89	64.0	t32
Gitelson2[A]	Chla	−0.75	$y=5.2141e^{-1.03x}$	0.63	83.9	a54
	Chlb	−0.74	$y=0.4714e^{-1.3512x}$	0.59	31.4	b53
	Chlt	−0.77	$y=5.7811e^{-1.0753x}$	0.66	109.9	t54
Gitelson[A]	Chla	0.88	$y=50890x^{2.0381}$	0.89	50.3	a32
	Chlb	0.87	$y=15333x^2-5.1781x-4.1117$	0.78	16.2	b28
	Chlt	0.90	$y=80087x^{2.1079}$	0.91	60.4	t31
mNDVI	Chla	0.71	$y=1.1004e^{5.2579x}$	0.84	56.5	a41
	Chlb	0.56	$y=0.1657e^{5.8958x}$	0.58	28.5	b45
	Chlt	0.68	$y=1.3561\ e^{5.3138x}$	0.82	80.0	t43
Maccioni[A]	Chla	0.90	$y=468.03x^{1.1116}$	0.93	37.6	a10
	Chlb	0.81	$y=524.36x^2-195.48x+19.343$	0.79	15.8	b23
	Chlt	0.90	$y=22.432e^{4.6798x}$	0.93	44.9	t6
mSR	Chla	−0.32	$y=-0.0202x^2-3.6039x+105.02$	0.47	62.1	a48
	Chlb	−0.14	$y=-0.0073x^2-1.1513x+37.45$	0.31	28.5	b46
	Chlt	−0.28	$y=-0.0275x^2-4.7551x+142.47$	0.44	87.0	t47
NDVI2[A]	Chla	0.91	$y=572.06x^{0.9776}$	0.94	37.5	a8
	Chlb	0.83	$y=724.6x^2-161.79x+13.25$	0.80	15.3	b14
	Chlt	0.91	$y=758.62x^{0.9945}$	0.93	51.6	t22
NPCI	Chla	−0.76	$y=185.21e^{-5.54x}$	0.90	51.3	a35
	Chlb	−0.63	$y=51.691e^{-6.487x}$	0.67	25.0	b39
	Chlt	−0.75	$y=240.87e^{-5.6355x}$	0.89	70.6	t39
REP_LE[A]	Chla	0.73	$y=2E\text{-}06e^{0.0261x}$	0.85	47.0	a27
	Chlb	0.59	$y=8E\text{-}08e^{0.0288x}$	0.56	25.5	b42
	Chlt	0.71	$y=2E\text{-}06e^{0.0263x}$	0.83	74.6	t41
REP_Li[A]	Chla	−0.62	$y=7E+18x^{-5.741}$	0.76	57.8	a43
	Chlb	−0.49	$y=3E+19x^{-6.1545}$	0.48	30.0	b50
	Chlt	−0.60	$y=1E+19x^{-5.7714}$	0.74	95.2	t50
SR1[A]	Chla	0.91	$y=20.424x^{2.0298}$	0.91	44.5	a24
	Chlb	0.88	$y=7.965x^2-5.2604x+2.1228$	0.80	15.3	b11
	Chlt	0.92	$y=24.674x^{2.0062}$	0.92	52.5	t25
SR2[A]	Chla	0.88	$y=10.593x^{1.5411}$	0.94	41.6	a21
	Chlb	0.83	$y=1.7565x^2-4.8351x+9.0096$	0.76	17.0	b29
	Chlt	0.89	$y=12.789\ x^{1.5807}$	0.94	52.0	t24
SR3[A]	Chla	0.91	$y=21.529x^{2.2107}$	0.85	51.5	a36
	Chlb	0.93	$y=5.8111x^2+17.204x-25.878$	0.88	12.2	b1
	Chlt	0.94	$y=-15.607x^2+238.91x-238.84$	0.89	38.6	t1
SR4[A]	Chla	−0.06	$y=-114.37x^2+719.91x-871.51$	0.38	67.0	a50
	Chlb	−0.17	$y=-37.435x^2+226.79x-260.71$	0.28	29.3	b48
	Chlt	−0.10	$y=-151.81x^2+946.7x-1132.2$	0.37	92.5	t48
SR5[A]	Chla	−0.09	$y=-9120.2x^2+6433.1x-859.07$	0.48	61.7	a47
	Chlb	0.05	$y=-3456.2x^2+2521.7x-367.62$	0.41	26.4	b44
	Chlt	−0.05	$Y=-12576x^2+8954.8x-1226.7$	0.48	83.9	t44
SR6[A]	Chla	0.92	$y=26.484x^{3.1156}$	0.87	47.7	a28
	Chlb	0.88	$y=41.718x^2-58.098x+22.191$	0.80	15.3	b12
	Chlt	0.93	$y=32.113x^{3.2249}$	0.90	57.0	t28
SR7[A]	Chla	0.90	$y=429.79x^{2.2355}$	0.93	39.6	a18

Table 2. Cont.

Index	Prediction target	*r*	Prediction equation	R^2	RMSE (mg/m^2)	Rank
	Chlb	0.83	$y = 288.79x^2-165.83x+30.259$	0.75	17.1	b30
	Chlt	0.90	$y = 570.74x^{2.2932}$	0.94	49.3	t14
SRPI	Chla	0.78	$y = 4.2726e^{3.6528}$	0.90	52.4	a38
	Chlb	0.67	$y = 0.577e^{4.3543x}$	0.70	23.9	b38
	Chlt	0.77	$y = 5.141e^{3.7278x}$	0.90	70.4	t38
Sum_Dr2[A]	Chla	0.75	$y = 1E\text{-}05x^{4.4928}$	0.80	59.1	a44
	Chlb	0.59	$y = 0.1296e^{0.143x}$	0.53	30.4	b52
	Chlt	0.72	$y = 1E\text{-}05x^{4.532}$	0.78	94.5	t49
Vogelmann[A]	Chla	0.92	$y = 29.72x^{6.135}$	0.86	45.6	a30
	Chlb	0.87	$y = 313.02x^2-580.26x+274.16$	0.79	15.9	b27
	Chlt	0.93	$y = 36.19x^{6.3497}$	0.88	60.3	t30
Vogelmann2[A]	Chla	−0.91	$y = -10448x^2-3992.6x+21.527$	0.84	33.8	a1
	Chlb	−0.88	$y = 4359.1x^2-660.68x+5.9525$	0.79	15.6	b22
	Chlt	−0.93	$y = -6088.5x^2-4653.3x+27.479$	0.87	42.5	t4
SPAD	Chla	0.90	$y = 1.9176x^{1.3184}$	0.94	37.8	a13
	Chlb	0.82	$y = 2.9234e^{0.0794x}$	0.76	22.2	b35
	Chlt	0.90	$y = 2.2727\ x^{1.3451}$	0.93	50.1	t16

replaced as the electric conductivity decreased to half of the original. The plants were transplanted on October 1, 2013.

2.2. Chlorophyll meter and spectral measurements

The second uppermost leaves of each treatment were measured in situ with a SPAD 502 model chlorophyll meter (Konica Minota Inc., Japan) around the midpoint at tillering, booting and heading. After the measurement of the chlorophyll meter, the leaves were immediately sampled and stored in an ice box, and transported to the lab for leaf reflectance measurements. The reflectance of the single leaf was measured with an integrating sphere (model LISR-3100, Shimadzu Scientific Instruments Inc., Japan) coupled to a UV-3600 UV-VIS-NIR spectrophotometer (Shimadzu Scientific Instruments Inc., Japan) in the wavelength range of 400–900 nm around the midpoint of each leaf. The spectral meter has a 1-nm resolution in the region of 400–900 nm.

2.3. Determination of leaf Chl contents

After spectral measurements, 15 leaf discs of 0.5 cm^2 from each leaf were sampled for determination of leaf Chl content. The Chl a and Chl b contents per unit area were measured spectrophotometrically using a solution of alcohol, acetone and water (4.5:4.5:1, V/V/V) as a solvent, employing the equations of Lichtenhaler and Wellburn (1983) [19]. The total Chl content was calculated as Chl*a* plus Chl*b*. The leaves that appeared evidently desiccative were not used in this study. We measured a total of 108 leaves across tillering, booting and heading stages, which included 12 leaves of the mutant and 96 leaves of the wild types.

2.4. Data analysis

The scatterplots of the reflectance and the first derivative (FD) reflectance *vs* Chl*a*, Chl*b* and Chl*t* were plotted, and the curves were visually analysed for extraction of spectral signatures of interest including shape, peak position, trough position and inflection point. FD was calculated with the following equation:

$$FD(\lambda) = R(\lambda+1) - R(\lambda) \quad (1)$$

where FD(λ), R(λ) and R(λ+1) represent the first derivative reflectance at wavelength λ (nm), reflectance at λ and reflectance at λ+1, respectively.

The existing published Chl-related VIs selected in this study and their formulations were summarized in Table 1 [4–7,20–46]. Only leaf-scale indices were collected. Among the 55 indices, none were solely based on the ICCW, although 21 indices used the ICCW.

The sensitivity of the VIs to Chl contents were tested with the correlation coefficients between the VIs and the Chl content, and the correlation coefficients were computed with Excel 10.0 (Microsoft).

The relationship between the VIs and the Chl content (Chl*a* or Chl*b* or Chl*t*) were fitted with linear, power, exponential, logarithmic and polynomial equations and the equation with the highest determination coefficients (R^2) was selected as the best equation. The root mean square error (RMSE) was computed for each best equation, and the predictive performance of the VIs was assessed by ranking the RMSE values in ascending order. The relationships were fitted with Excel.

Results

3.1. Rice leaf Chl content

All the leaves of both the normal N treatment and the low N treatment of the mutant 'IG 20' were yellow-green in color during the whole growth period. The leaves of the wild types were green in colour, although the low N treatments were shallower in leaf colour than the normal N treatments. The means and ranges of Chl content (mg/m^2) for the 96 leaf samples of the conventional genotypes as well as Chl*a*/Chl*b* were 260.5 (148.7–378.5) for Chl*a*, 81.8 (31.9–135.3) for Chl*b*, 342.3 (209.4–497.7) for Chl*t* and 3.76 (1.99–6.55) for Chl*a*/Chl*b*. The values for the 12 leaf samples of the low-chlorophyll mutant were 52.2 (11.9–157.5) for Chl*a*,

14.7 (0.2–40.5) for Chl*b*, 66.8 (16.9–198.0) for Chl*t* and 11.08 (1.05–114.35) for Chl*a*/Chl*b*. The leaves of the wild types had an evidently higher Chl*a*, Chl*b* and Chl*t* and a much lower ratio of Chl*a* to Chl*b* than the leaves of the mutant. As compared with the previous study [6] for constructing VIs for Chl*a*, Chl*b* and Chl*t* estimation, this study had a similar mean Chl content, a lower minimum Chl content, a lower maximum Chl content, and a significantly larger variation of ratios of Chl*a* to Chl*b*.

3.2. Leaf spectral reflectance signatures and construction of the new VI

As shown in Figure 1A, a profound difference in leaf spectral reflectance was observed between the conventional rice genotypes and the mutant. The reflectance curves from 640 nm to 674 nm and from 675 nm to 680 nm of the mutant leaf of a low Chl content were drastically steeper than those of the wild types in the ICCW. For both the wild types and the mutant, the inflection point of the reflectance spectra in the ICCW was 645 nm, where the FD value of reflectance started to be positive (Figure 1B). Additionally, the reflectance trough around 620 nm became evident, and the green peak around 550 nm was broadened and deformed in the reflectance spectra of the mutant as compared with that of the wild types. The reflectance spectra of all the leaves of the mutant were visually similar in shape and reflection band positions.

Based on the spectral signatures in the ICCW we observed, we found that the reflectance variation within the ICCW was sensitive to the Chl content, and constructed a new VI—the difference of

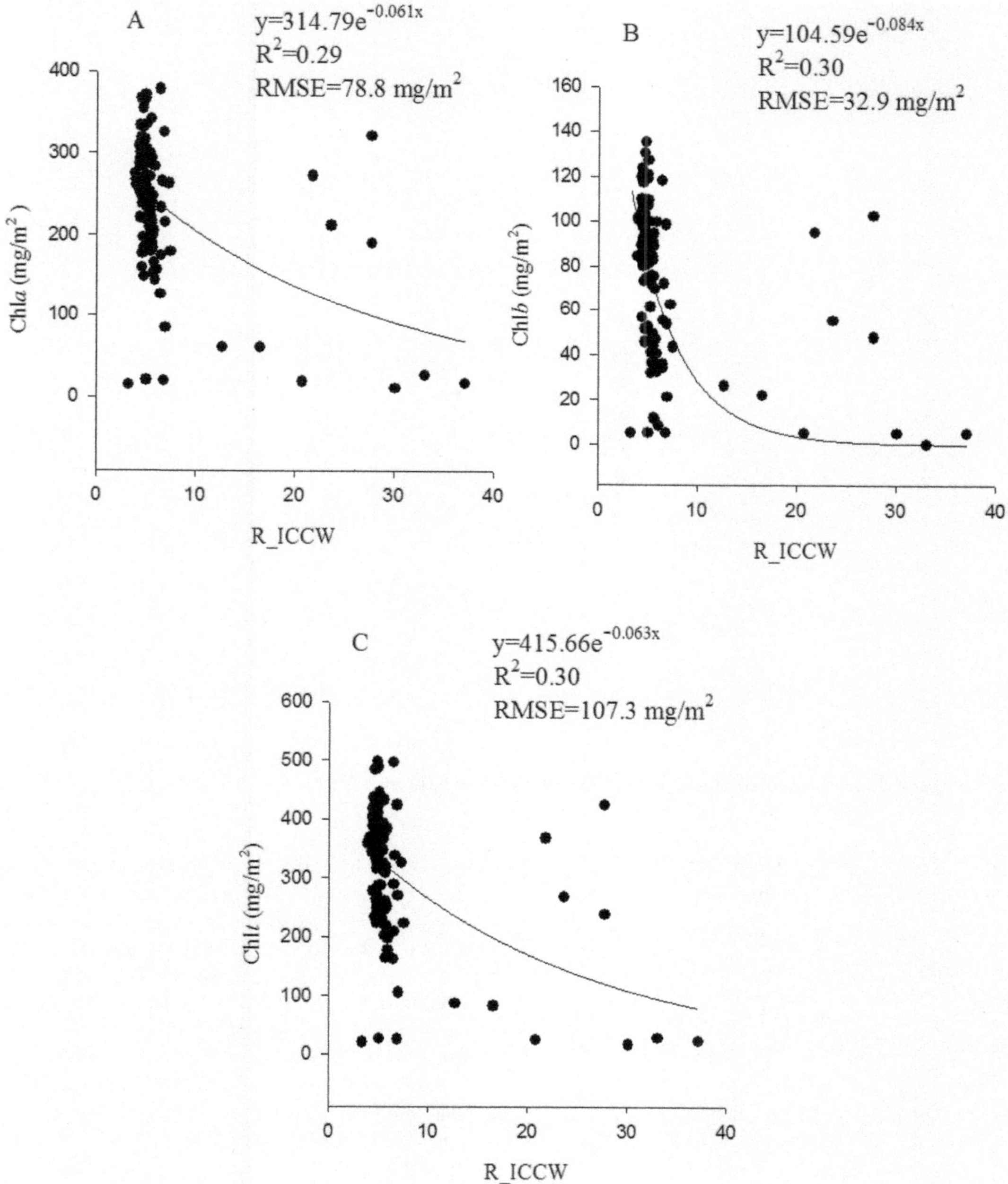

Figure 2. The best prediction models of R_ICCW for Chl*a* (A), Chl*b* (B) and Chl*t* (C).

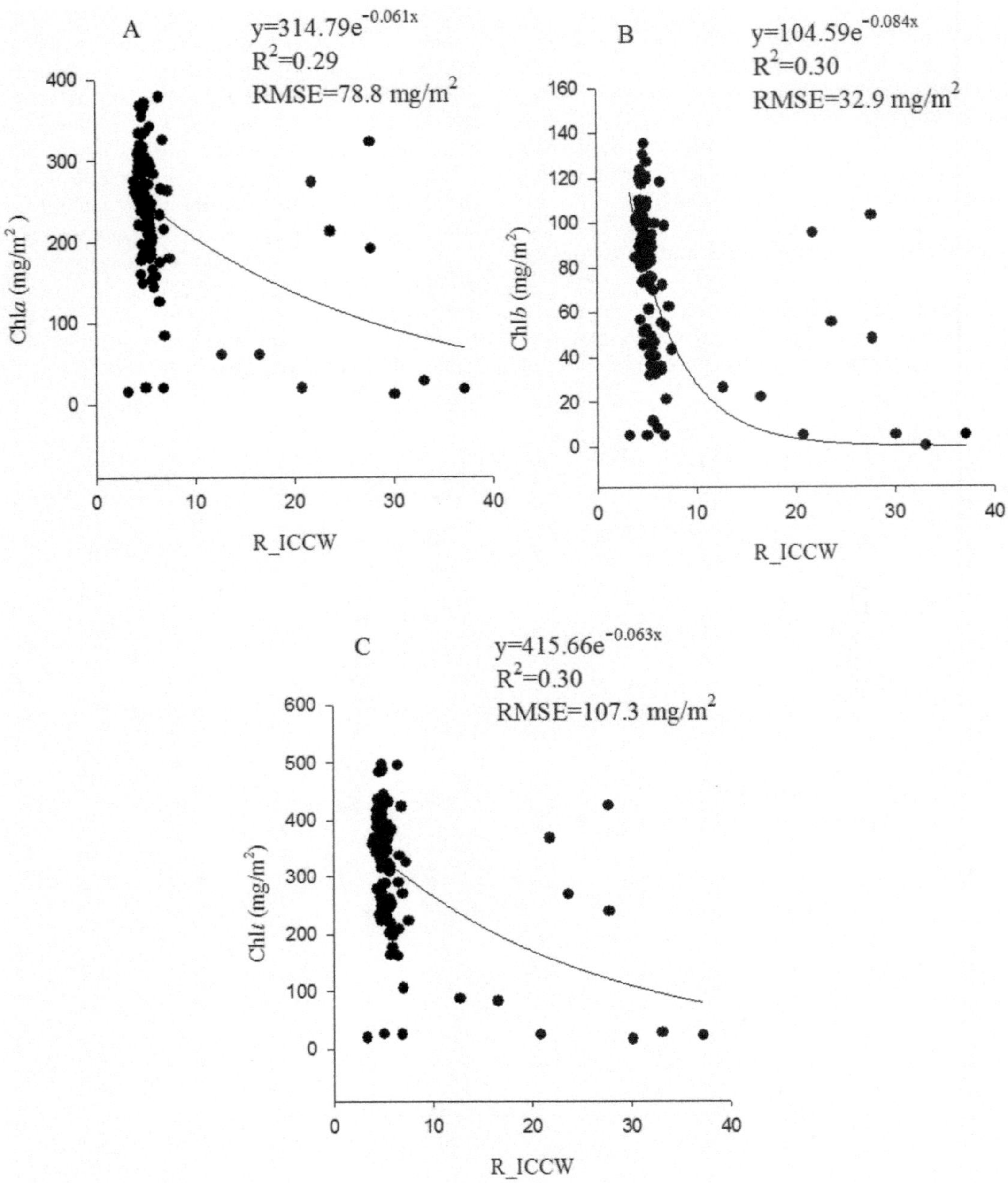

Figure 3. The best prediction models of DFDS_ICCW for Chl*a* (A), Chl*b* (B) and Chl*t* (C).

first derivative sum within the ICCW (DFDS_ICCW)—for simultaneous retrieval of Chl*a*, Chl*b* and Chl*t*:

$$\text{DFDS\_ICCW} = \text{sum of } FD_{675\text{–}680} - \text{sum of } FD_{640\text{–}674} \quad (2)$$

where the sum of $FD_{675\text{–}680}$ and the sum of $FD_{640\text{–}674}$ represent the sum of the first derivative reflectance between R675 and R680 and that between R640 and R674, respectively. R640, R674, R675 and R680 are the reflectance at 640 nm, 674 nm, 675 nm and 680 nm, respectively.

3.3. Sensitivity of the VIs to Chl*a*, Chl*b* and Chl*t*

Of the 55 VIs tested (Table 2), 24 were robustly sensitive to the leaf Chl*t* ($r2 \geq 0.81$, $n = 108$), 19 were strong ($0.49 \leqq r2 < 0.81$, $n = 108$), 5 were moderate ($0.25 \leqq R2 < 0.49$) and 5 were weak ($0.04 \leqq R2 < 0.25$). Only 2 indices, $SR4^A$ and $SR5^A$, were insignificantly ($P > 0.05$, $n = 108$) related to the leaf Chl*a*, Chl*b* and Chl*t*. Generally, the sensitivity of the indices to Chl*t* was similar to that of Chl*a*, and the sensitivity of the indices to Chl*b* was slightly lower than Chl*t* or Chl*a*. The results showed that most of the tested indices were highly sensitive to Chl*a*, Chlb and Chl*t*.

The mean reflectance in the ICCW (R-ICCW) was significantly ($P < 0.05$) related to Chl*a*, Chl*b* and Chl*t* with a low correlation strength, yielding an r ($n = 108$) of −0.45, −0.40 and −0.45, respectively. In contrast, the new VI, DFDS_ICCW, had an r ($n = 108$) of −0.86, −0.76 and −0.85 as correlated with Chl*a*, Chl*b* and Chl*t*, respectively, indicating that this index was highly sensitive to Chl*t*, Chl*a* and Chl*b*. When leaf Chl*t* was higher than 200 mg/m^2, the r value was −0.77 ($n = 93$, $P < 0.01$) and −0.12 ($n = 93$, $P > 0.05$) respectively between DFDS-ICCW and Chl*t* and between R_ICCW and Chl*t*. The results demonstrated that DFDS-ICCW was still highly sensitive, but R_ICCW became

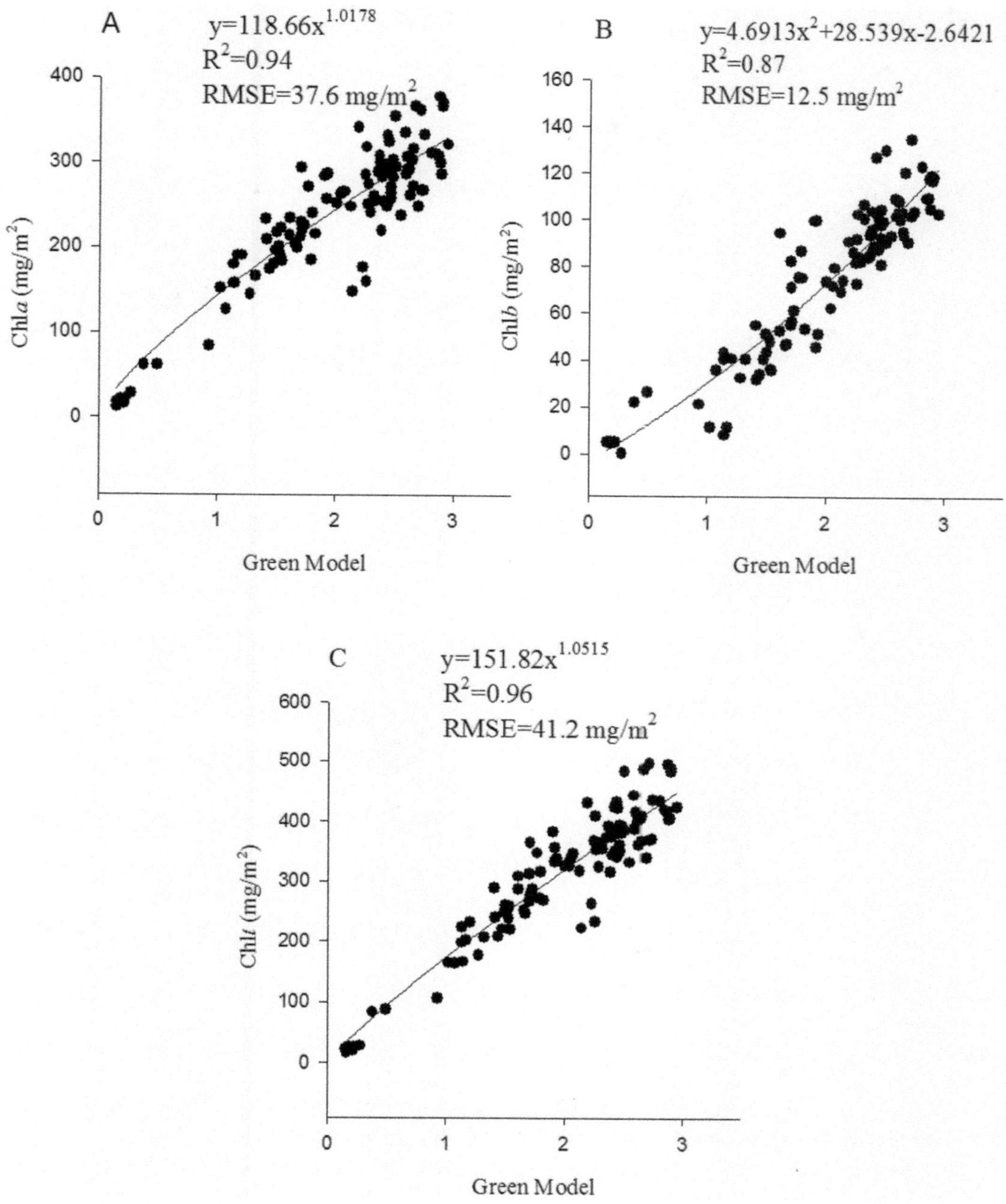

Figure 4. The best prediction models of Green Model for Chl*a* (A), Chl*b* (B) and Chl*t* (C).

insensitive to Chl*t* when Chl*t* was at medium and high levels. As shown in Figure 2C, the R-ICCW tended to be saturated when leaf Chl*t*>200 mg/m^2. Contrastingly, DFDS_ICCW decreased sensitively with the Chl*t* even when Chl*t* was higher than 200 mg/m^2 (Figure 3C). This result confirmed the saturated reflection of the leaves at medium to high Chl content.

3.4. Prediction of Chl*a*, Chl*b* and Chl*t* with the best-fit equations

The best equations of R-ICCW (Figure 2) and DFDS_ICCW (Figure 3) were all exponential equations. For R-ICCW, the exponential equations yielded an RMSE (mg/g^2) of 78.7 for Chl*a*, 32.9 for Chl*b* and 107.3 for Chl*t*. The DFDS_ICCW equations yielded an RMSE of 37.4 for Chl*a*, 16.0 for Chl*b* and 45.3 for Chl*t*. The results indicated that DFDS_ICCW had a drastically higher prediction accuracy for Chl*a*, Chl*b* and Chl*t* than R_ICCW. The prediction accuracy of DFDS_ICCW was slightly lower for Chl*b* than Chl*a* or Chl*t*.

The prediction performance with a best prediction equation for all of the 55 existing indices are presented in Table 2. Interestingly, none of the best equations were linear; they were exponential, polynomial and power. The RMSE (mg/m^2) ranged from 33.8 to 85.8 for Chl*a*, from 12.2 to 37.9 for Chl*b* and from 38.6 to 120.3 for Chl*t*, which demonstrated that there was a large difference of prediction accuracy between the best index and the last index. However, the RMSE (mg/m^2) of the top 30 indices ranged from 33.8 to 49.6 for Chl*a*, from 12.2 to 17.1 for Chl*b* and from 38.6 to 60.3 for Chl*t*, indicating that the differences in the RMSE were not large in the top 30 indices. An index of high

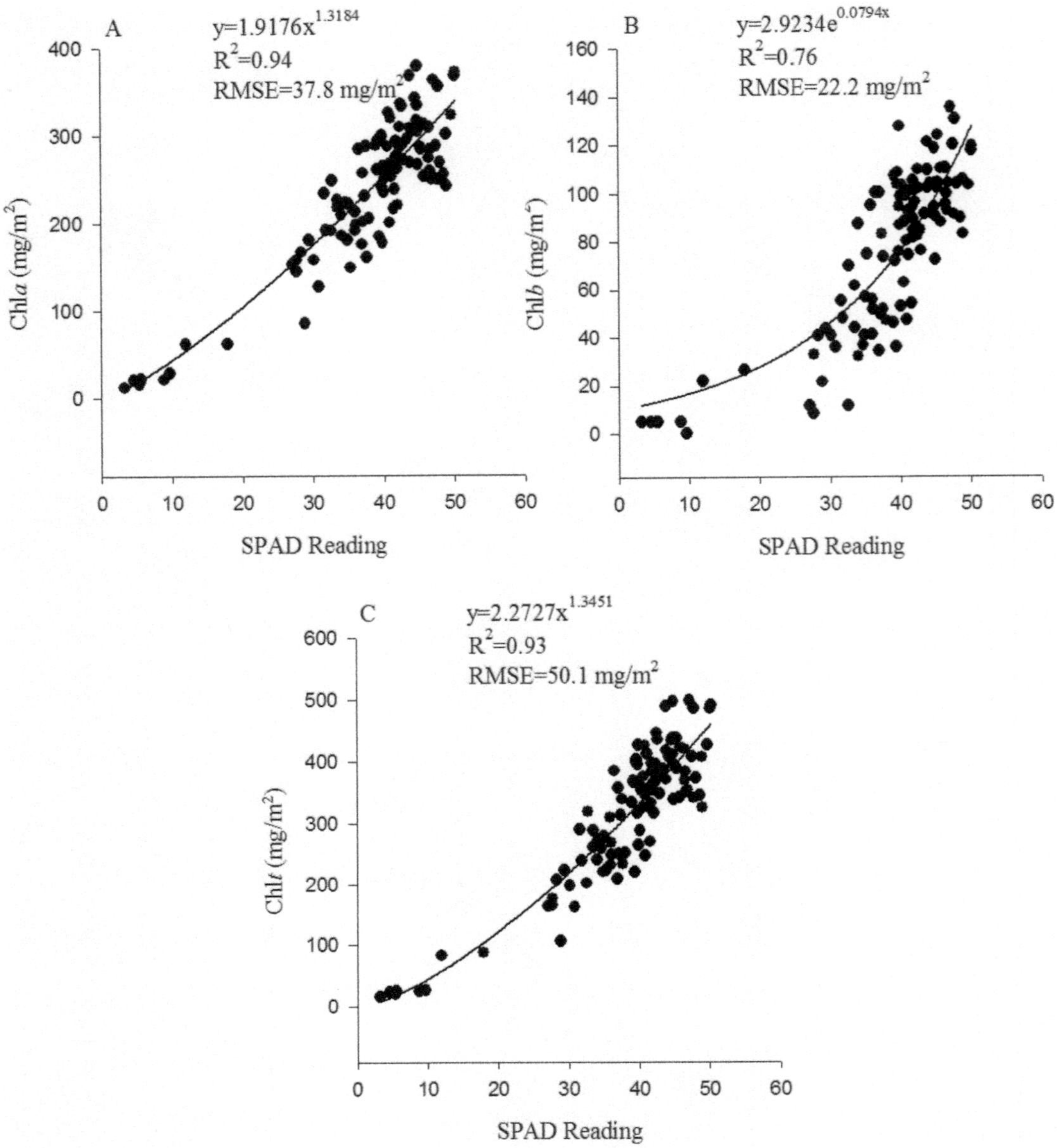

Figure 5. The best prediction models of SPAD readings for Chl*a* (A), Chl*b* (B) and Chl*t* (C).

predictive ability for Chl*t* (e.g. Green Model) generally also performed well for prediction of Chl*a* or Chl*b*, although the prediction accuracy for Chl*b* was generally and slightly lower than that for Chl*a* or Chl*t*, and an index of low predictive ability for Chl*t* (e.g. PSRI) was also weak for prediction of Chl*a* or Chl*b*. The SPAD reading ranked 13th, 35th and 16th among the 55 indices, respectively for prediction of Chl*a*, Chl*b* and Chl*t* with the polynomial equations, which indicated that it was also a strong index for predicting the leaf Chl contents.

The prediction results of the best VI, Green Model, together with the SPAD reading are also presented in Figure 4 and Figure 5, which confirm their high accuracy for prediction of Chl*a*, Chl*b* and Chl*t*.

The results in this study demonstrated that most of the existing indices could be used for simultaneous retrieval of Chl*a*, Chl*b* and Chl*t*.

As compared with the 55 indices, the prediction accuracy of DFDS_ICCW was similar to Datt2^A, ranking 7th for Chl*a* prediction, similar to SR6^A ranking 28th for Chl*b* prediction and similar to Carter4^A ranking 7th for Chl*t* prediction. The results indicated that DFDS_ICCW could simultaneously and robustly predict Chl*a*, Chl*b* and Chl*t*.

Discussion

Most of the existing VIs as well as the SPAD reading were simultaneously and robustly or strongly related to Chl*a*, Chl*b* and Chl*t*, and achieved a high accuracy for Chl*a*, Chl*b* and Chl*t* prediction. As most of the indices were originally sought for prediction of Chl*t*, the results in this study suggested that the indices could be extended for simultaneous retrieval of Chl*a*, Chl*b* and Chl*t*. None of the best-fit equations for prediction of Chl*a*, Chl*b* and Chl*t* were linear equations; therefore, the ranking of the existing indices in this study was not in agreement with that of Main et al. (2011) [11], who used a linear equation for all indices. The VIs based on red edge (e.g. REP_LE^A and REP_Li^A) ranked high for leaf Chl prediction in the previous study, but ranked low in this study. The indices excluding the ICCW generally performed better than those including the ICCW in this study, which is consistent with the previous study [11]. Particularly, both the best index for Chl*a* and Chl*t*, SR3^A, and the best index for Chl*b*, Vogelmann2^A, did not use ICCW. The simple ratio indices—SR4^A based on 670 nm in the ICCW and 700 nm and S5^A based on 675 nm in the ICCW and 700 nm—were the only indices that were insignificantly ($P>0.05$) related to the Chl contents. In contrast, another simple ratio index, SR3^A based on 550 nm and 750 nm in the OCCW, was the best index for prediction of Chl*a* and Chl*t*. In the ICCW, the reflectance curves from 640 nm to 674 nm and from 675 nm to 680 nm of the mutant leaf of a low Chl content were drastically steeper than that of the wild types of medium to high Chl content. This spectral signature could enlighten us to use the reflectance variation within the ICCW for retrieval of plant Chl content, although further studies are needed for understanding the mechanisms causing this signature. The successful detection of the reflectance variation within the ICCW in this study could be attributed to the high spectral resolution (1 nm) of the spectral photometer, as the current widely-used spectral meter, the Field Spec spectroradiometer (Analytical Spectral Devices, Boulder, CO, USA), has a spectral resolution of 3 nm in the red band.

Plant leaves tend to have saturated reflectance in the ICCW [6,12] when leaf Chl*t* is medium to high, which has limited the use of this spectral region for non-destructive determination of leaf Chl. The results in this study also showed that the R-ICCW tended to be saturated when leaf Chl*t* was higher than 200 mg/m^2. However, the new spectral index based on the reflectance variation within the ICCW decreased sensitively with the Chl*t* even when Chl*t* was greater than 200 mg/m^2. The new index could rank in the top 10 for prediction of Chl*a* and Chl*t* as compared with the 55 tested indices, and also achieved a promising accuracy for Chl*b* prediction. Therefore, the results suggested that ICCW could also be used for development of robust VIs for retrieval of plant Chl contents. Unlike the existing VIs, the new index is solely based on the specific Chl adsorption band. Therefore, the retrieval of Chl by using this index may not be confounded by non-Chl factors, e.g. other pigments and leaf structure. Further studies are needed for confirmation of this finding at different scales (e.g. canopy and region) and for different plant species

Conclusions

Most of the 55 existing VIs could robustly or strongly and simultaneously predict Chl*a*, Chl*b* and Chl*t* in the rice leaves of a large variation of ratios of Chl*a* to Chl*b* in this study. It was found that the reflectance curves from 640 nm to 674 nm and from 675 nm to 680 nm of the mutant leaf were drastically steeper than those of the wild types in the ICCW, which implied that the reflectance variation within ICCW could be used for retrieval of Chl content The new index based solely on the reflectance variation within the ICCW were simultaneously and strongly sensitive to Chl*a*, Chl*b* and Chl*t* and achieved a high accuracy for prediction of Chl*a*, Chl*b* and Chl*t*. The results suggested that ICCW could also be of potential for development of robust VIs for retrieval of plant Chl content with non-destructive reflectance measurement approaches.

Acknowledgments

We thank Dr Wang Danying at the China National Rice Research Institute for providing the rice seeds. We also thank the reviewers and the editor for their valuable comments.

Author Contributions

Conceived and designed the experiments: QFZ JZ WJH. Performed the experiments: JZ QFZ. Analyzed the data: QFZ JZ. Contributed reagents/materials/analysis tools: JZ QFZ. Wrote the paper: QFZ WJH JZ.

References

1. Tanaka R, Tanaka A (2011) Chlorophyll cycle regulates the construction and destruction of the light-harvesting complexes. Biochimica et Biophysica Acta 1807: 968–976.
2. Porra RJ (2002) The chequered history of the development and use of simultaneous equations for the accurate determination of chlorophylls *a* and *b*. Photosynth Res 73: 149–156.
3. Arnon DI (1949) Copper enzymes in isolated chloroplasts. Polyphenoloxidase in *Beta vulgaris*. Plant Physiol 24: 1–15.
4. Chappelle EW, Kim MS, McMurtrey JE III (1992) Ratio analysis of reflectance spectra (RARS): an algorithm for the remote estimation of the concentrations of chlorophyll a, chlorophyll b and carotenoids in soybean leaves. Remote Sens Environ 39: 239–247
5. Blackburn GA (1998) Quantifying chlorophylls and carotenoids at leaf and canopy scales: an evaluation of some hyper-spectral approaches. Remote Sens Environ 66: 273–285.
6. Datt B (1998) Remote sensing of chlorophyll a, chlorophyll b, chlorophyll a + b and total carotenoid content in Eucalyptus leaves. Remote Sens Environ 66: 111–121.
7. Sims DA, Gamon JA (2002) Relationship between leaf pigment content and spectral reflectance across a wide range species, leaf structures and development stages. Remote Sens Environ 81: 337–354.
8. Gitelson AA, Gritz Y, Merzlyak MN (2003) Relationships between leaf chlorophyll content and spectral reflectance and algorithms for nondestructive chlorophyll assessment in higher plant leaves. J Plant Physiol 160: 271–282.
9. Gitelson AA, Vina A, Ciganda V, Rundquist DC, Arkebauer TJ (2005) Remote estimation of canopy chlorophyll content in crops. Geophys Res Lett 32: L08403. doi: 10.1029/2005 GL022688.
10. Schlemmer M, Gitelson A, Schepers J, Ferguson R, Peng Y, et al. (2013) Remote estimation of nitrogen and chlorophyll contents in maize at leaf and canopy levels. Int J Appl Earth Obs 25: 47–54.
11. Main R, Cho MA, Mathieu R, O'Kennedy RM, Ramoelo A, et al. (2011) An investigation into robust spectral indices for leaf chlorophyll estimation. ISPRS J Photogramm 66: 751–761.
12. Thomas JR, Gausman HW (1977) Leaf reflectance versus leaf chlorophyll and carotenoids concentration for eight crops. Agron J 63: 845–847.
13. Blackburn GA (2006) Hyperspectral remote sensing of plant pigments. J Exp Bot 58: 855–867.
14. Gitelson AA, Gritz Y, Merzlyak MN (2003) Relationships between leaf chlorophyll content and spectral reflectance and algorithms for non-destructive chlorophyll assessment in higher plant leaves. J Plant Physiol 160: 271–282.
15. Hatfield JL, Gitelson AA, Schepers JS, Walthall CL (2008) Application of spectral remote sensing for agronomic decisions. Agron J 100: S117–S131.
16. le Maire G, Francois C, Dufrene E (2004) Towards universal broad leaf chlorophyll indices using PROSPECT simulated database and hyperspectral reflectance measurements. Remote Sens Environ 89: 1–28.
17. Féret JB, Francois C, Gitelson AA, Barry KM, Panigada C, et al. (2011) Optimizing spectral indices and chemometric analysis of leaf chemical properties using radiative transfer modeling. Remote Sens Environ 115: 2742–2750.
18. He QX, Zhou QF, Sun XM (2005) Strikingly high content of grain protein in solution-cultured rice. J Sci Food Agr 85: 1197–1202.
19. Lichtenhaler HK, Wellburn AR (1983) Determination of total carotenoids and chlorophylls a and b of leaf extracts in different solvents. Biochem Soc T 11: 591–592.
20. Yoder BJ, Pettigrew-Crosby RE (1995) Predicting nitrogen and chlorophyll content and concentrations from reflectance spectra (400–2500 nm) at leaf and canopy scales. Remote Sens Environ 53: 199–211.
21. Peñuelas J, Baret F, Filella I (1995) Semiempirical indices to assess carotenoids chlorophyll-a ratio from leaf spectral reflectance. Photosynthetica 31: 221–230.
22. Carter GA, Cibula WG, Miller RL (1996) Narrow-band reflectance imagery compared with thermal imagery for early detection of plant stress. J Plant Physiol 148: 515–522.
23. Blackburn GA (1998) Quantifying chlorophylls and carotenoids at leaf and canopy scales: an evaluation of some hyper-spectral approaches. Remote Sens Environ 66: 273–285.
24. Blackburn GA (1999) Relationships between spectral reflectance and pigment concentrations in stacks of deciduous broadleaves. Remote Sens Environ 70: 224–237.
25. Merzlyak MN, Gitelson AA, Chivkunova OB, Rakitin VY (1999) Non-destructive optical detection of pigment changes during leaf senescence and fruit ripening. Physiol Plantarum 106: 135–141.

26. Sims DA, Gamon JA (2002) Relationship between leaf pigment content and spectral reflectance across a wide range species, leaf structures and development stages. Remote Sens Environ 81: 337–354.
27. Read JJ, Tarpley L, McKinion JM, Reddy KR (2002) Narrow-waveband reflectance ratios for remote estimation of nitrogen status in cotton. J Environ Qual 31: 1442–1452.
28. Steddom K, Heidel G, Jones D, Rush CM (2003) Remote detection of rhizomania in sugar beets. Phytopathology 93: 720–726.
29. Gitelson AA. Vina A, Ciganda V, Rundquist DC, Arkebauer TJ (2005) Remote estimation of canopy chlorophyll content in crops. Geophys Res Lett 32: L08403. doi: 10.1029/2005 GL022688.
30. Rondeaux G, Steven M, Baret F (1996) Optimization of soil adjusted vegetation indices. Remote Sens Environ 55: 95–107.
31. Jiang Z, Huete AR, Didan K, Miura T (2008) Development of a two-band enhanced vegetation index without a blue band. Remote Sens Environ 112: 3833–3845.
32. Kim MS, Daughtry CST, Chappelle EW, McMurtrey III JE, Walthall CL (1994) The use of high spectral resolution bands for estimating absorbed photosynthetically active radiation (Apar). In: Proc. Sixth Symposium on Physical Measurements and Signatures in Remote Sensing, Val D'Isere, France, January 17–21, pp. 299–306.
33. Gitelson AA, Buschmann C, Lichtenthaler HK (1999) Thechlorophyll fluorescence ratio F735/F700 as an accurate measure of the chlorophyll content in plants. Remote Sens Environ 69: 296–302.
34. Maccioni A, Agati G, Mazzinghi P (2001) New vegetation indices for remote measurement of chlorophylls based on leaf directional reflectance spectra. Journal of Photochemistry and Photobiology 61: 52–61.
35. Gitelson A, Merzlyak MN (1994) Quantitative estimation of chlorophyll-a using reflectance spectra: experiments with autumn chestnut and maple leaves. Journal of Photochemistry and Photobiology B: Biology 22: 247–252.
36. Peñuelas J, Gamon JA, Fredeen AL, Merino J, Field CB (1994) Reflectance indices associated with physiological changes in nitrogen and water limited sunflower leaves. Remote Sens Environ 48: 135–146.
37. Cho MA, Skidmore AK (2006) A new technique for extracting the red-edge position from hyperspectral data: the linear extrapolation method. Remote Sens Environ 101: 181–193.
38. Guyot G, Baret F (1988) Utilisation de la haute résolution spectrale pour suivrel'état des couverts végétaux. In: Guyenne TD, Hunt, JJ (Eds.), Proc. Fourth International Colloquium on Spectral Signatures of Objects in Remote Sensing, ESA SP-287, Assois, France, 18–22 January, pp. 279–286.
39. Gitelson AA, Merzlyak MN (1997) Remote estimation of chlorophyll content in higher plant leaves. Int J Remote Sens 18: 2691–2697.
40. McMurtey III JE, Chappelle EW, Kim MS, Meisinger JJ, Corp LA (1994) Distinguish nitrogen fertilization levels in field corns (Zea maysL.) with actively induced fluorescence and passive reflectance measurements. Remote Sens Environ 47: 36–44.
41. Zarco-Tejada PJ, Miller JR (1999) Land cover mapping at BOREAS using red-edge spectral parameters from CASI imagery. J Geophys Res 104: 27921–27933.
42. Lichtenthaler HK, Lang M, Sowinska M, Heisel F, Miehe JA (1996) Detection of vegetation stress via a new high resolution fluorescence imaging system. J Plant Physiol 148: 599–612.
43. Peñuelas J, Filella I, Lloret P, Munoz F, Vilajeliu M (1995) Reflectance assessment of mite effects on apple trees. Int J Remote Sens 16: 2727–2733.
44. Filella I, Peñuelas J (1994) The red-edge position and shape as indicators of plant chlorophyll content, biomass and hydric status. Int J Remote Sens 15: 1459–1470.
45. Vogelman JE, Rock BN, Moss DM (1993) Red-edge spectral measurements from sugar maple leaves. Int J Remote Sens 14: 1563–1575.
46. Jin XL, Diao WY, Xiao CH, Wang FY, Chen B, et al. (2013) Estimation of wheat agronomic parameters using new spectral indices. PLoS ONE 8: e72736. doi: 10.1371/journal.pone.0072736.

6

QTLs for Tolerance of Drought and Breeding for Tolerance of Abiotic and Biotic Stress: An Integrated Approach

Shalabh Dixit[1], B. Emma Huang[2], Ma. Teresa Sta Cruz[1], Paul T. Maturan[1], Jhon Christian E. Ontoy[1], Arvind Kumar[1]*

1 International Rice Research Institute (IRRI), Los Baños, Laguna, Philippines, **2** Computational Informatics, CSIRO, Dutton Park, Queensland, Australia

Abstract

Background: The coupling of biotic and abiotic stresses leads to high yield losses in rainfed rice (*Oryza sativa* L.) growing areas. While several studies target these stresses independently, breeding strategies to combat multiple stresses seldom exist. This study reports an integrated strategy that combines QTL mapping and phenotypic selection to develop rice lines with high grain yield (GY) under drought stress and non-stress conditions, and tolerance of rice blast.

Methodology: A blast-tolerant BC_2F_3-derived population was developed from the cross of tropical *japonica* cultivar Moroberekan (blast- and drought-tolerant) and high-yielding *indica* variety Swarna (blast- and drought-susceptible) through phenotypic selection for blast tolerance at the BC_2F_2 generation. The population was studied for segregation distortion patterns and QTLs for GY under drought were identified along with study of epistatic interactions for the trait.

Results: Segregation distortion, in favour of Moroberekan, was observed at 50 of the 59 loci. Majority of these marker loci co-localized with known QTLs for blast tolerance or NBS-LRR disease resistance genes. Despite the presence of segregation distortion, high variation for DTF, PH and GY was observed and several QTLs were identified under drought stress and non-stress conditions for the three traits. Epistatic interactions were also detected for GY which explained a large proportion of phenotypic variance observed in the population.

Conclusions: This strategy allowed us to identify QTLs for GY along with rapid development of high-yielding purelines tolerant to blast and drought with considerably reduced efforts. Apart from this, it also allowed us to study the effects of the selection cycle for blast tolerance. The developed lines were screened at IRRI and in the target environment, and drought and blast tolerant lines with high yield were identified. With tolerance to two major stresses and high yield potential, these lines may provide yield stability in rainfed rice areas.

Editor: Keqiang Wu, National Taiwan University, Taiwan

Funding: The work was funded by Global Rice Science Partnership (GRiSP), Generation Challenge Program (GCP), Federal ministry for economic cooperation and development (BMZ), Germany, and the Bill and Melinda Gates Foundation (BMGF). The funders had no role in study design, data collection and analysis, decision to publish, or preparation of the manuscript.

Competing Interests: The authors have declared that no competing interests exist.

* Email: a.kumar@irri.org

Introduction

Developing crop varieties with higher yield potential is a challenging task in modern-day plant breeding. The Green Revolution led to a historic breakthrough in increasing the yield potential of crops, such as rice and wheat, through the development of semi-dwarf varieties that has fed the increasing population till now. However, it is predicted that a 60% increase in agricultural production needs to be achieved by 2050 to feed the ever-growing population [1].

Unlike other cereals, rice is cultivated across a much wider range of ecosystems, including irrigated, rainfed upland, rainfed lowland, flooded, and deepwater ecosystems across tropical and temperate climate conditions. Despite this diversity of cultivation systems and environments, the major focus of research in increasing rice yield potential has been for the irrigated ecosystem in the post-Green Revolution period. However, no major breakthrough in breaking the yield potential barrier has been achieved until now.

The low-yielding rainfed lowland rice ecosystem has great potential for yield improvement. This ecosystem occupies about 38% of the total rice area, but contributes only 21% to total rice production [2]. In Asia alone, 34 million ha of rainfed lowland rice and 8 million ha of rainfed upland rice [3] experience drought stress of varying intensity almost every year. These drought events are often coupled with increased incidence of other biotic stresses, such as blast, brown spot, and bacterial blight, leading to a further

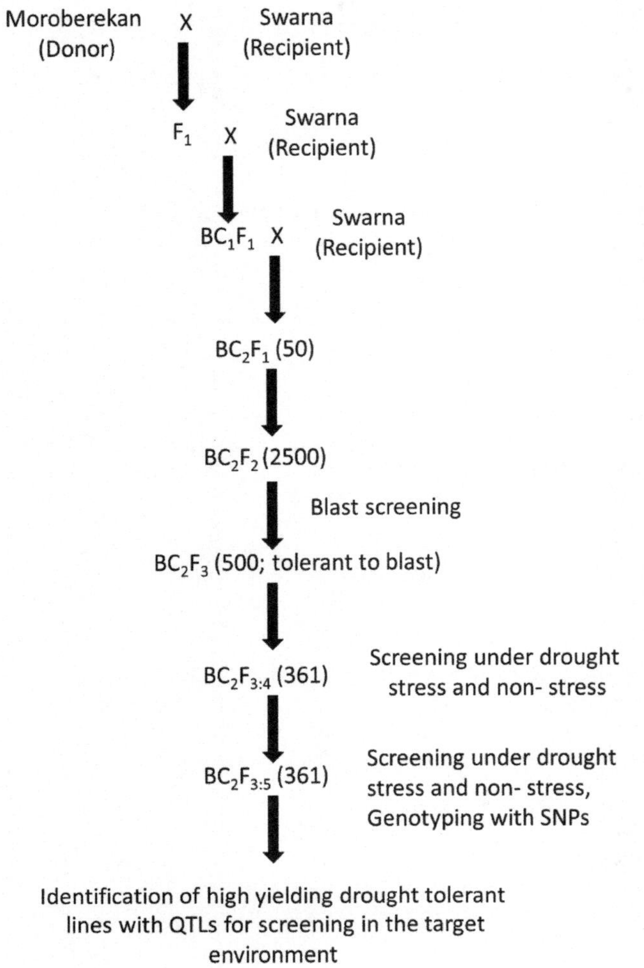

Figure 1. Population development and phenotypic screening strategy used to develop a blast-tolerant BC_2F_3-derived mapping population for mapping QTLs for high grain yield under drought stress and non-stress conditions.

decline in rice yield. Rice yields in these ecosystems remain low at 1.0–2.5 t ha^{-1} [4].

Despite the importance of drought as a major factor in yield reduction in rainfed ecosystems, few efforts have been made to develop high-yielding drought-tolerant cultivars. As a result, a large proportion of these rainfed areas are covered with high-yielding varieties such as Swarna, Samba Mahsuri, IR64, and MTU1010 that were specifically developed for irrigated ecosystems and suffer heavy yield losses under drought. On the other hand, several drought-tolerant cultivars, such as N22 and Moroberekan, have evolved in these rainfed ecosystems through natural or artificial selection over the years. These cultivars carry genes for tolerance of a wide range of biotic and abiotic stresses, but lack farmers' and consumers' preference because of their low yield potential and poor grain quality. In many cases, a single cultivar shows tolerance of a combination of biotic and abiotic stresses. This provides a unique opportunity for rice breeders to develop high-yielding varieties with tolerance of multiple abiotic and biotic stresses.

The lack of effective selection criteria and low heritability of grain yield (GY) under drought have been reported as the two major reasons for the slow progress in breeding drought-tolerant

Table 1. Details of experiments conducted for identification of QTLs and screening of drought-tolerant lines identified from the mapping population at IRRI and in IRRI-SA.

Experiment	Season	Ecosystem	Population size	Location	Experimental design	Replications	Plot size
1	DS2011	LSS	361	IRRI	38×10 AL	2	1.0
2	DS2012	LMS	361	IRRI	38×10 AL	2	2.0
3	DS2011	LNS	361	IRRI	38×10 AL	2	1.0
4	DS2012	LNS	361	IRRI	38×10 AL	2	2.0
5	WS2012	USS	99	IRRI	12×10 AL	2	1.0
6	DS2013	LSS	77	IRRI	10×10 AL	2	2.0
7	DS2013	LNS	77	IRRI	10×10 AL	2	2.0
8	WS2013	LMiS	95	IRRI-SA	8×13 AL	2	2.4
9	WS2013	LNS	95	IRRI-SA	8×13 AL	2	3.2

DS = dry season, WS = wet season, LSS = lowland severe stress, LMS = lowland moderate stress, LNS = lowland non-stress, UMS = upland moderate stress.

varieties [5]. With increased precision in large-scale phenotyping, many studies have shown moderate to high heritability for GY under drought [6–8]. These studies used direct selection for GY in progenies derived from high-yielding drought-susceptible varieties and drought-tolerant donors as a successful strategy to improve yield under drought. This strategy has also proven beneficial to combining high yield potential with drought tolerance and has resulted in the development and release of several high-yielding rice varieties with good yield under drought in countries such as India, Nepal, Bangladesh, Indonesia, and the Philippines [9]. Along with the release of varieties through conventional breeding approaches, several large-effect QTLs for GY under drought have also been identified [2,10–15]. Similarly, several studies have also reported independent and epistatic QTLs for GY and other traits of agronomic importance [16–19]. However, only a few studies have reported QTLs leading to a yield advantage under both drought stress and non-stress environments, taking grain yield as the primary selection criterion. Occurrence of drought is normally coupled with increased incidence of diseases such as rice blast, brown spot, and bacterial blight. However, very few studies have been undertaken to understand the genetics of these abiotic and biotic stresses simultaneously in a mapping population.

In our study, an advanced backcross (AB) QTL approach was used on a large backcross population developed from a cross between tropical japonica, drought- and rice blast-tolerant donor Moroberekan with highly popular recipient *indica* rice variety Swarna. The objectives were to: (1) develop a strategy to combine breeding approaches for tolerance of blast and drought with QTL mapping, (2) understand the effect of a selection cycle for blast tolerance during population development on segregation distortion and population structure, (3) identify QTLs for GY under drought stress and non-stress conditions, and (4) develop high-yielding blast- and drought-tolerant breeding lines from the mapping population and evaluate them in the target environment.

Materials and Methods

Experiments for this study were conducted at the Experiment Station of the International Rice Research Institute (IRRI), Los Baños, Laguna, Philippines, during the dry seasons (DS) of 2011, 2012, and 2013 and the wet season (WS) of 2012. IRRI is located at 14°13′ N and 121°15′ E, at an elevation of 21 m above mean sea level. The study also presents results from two experiments conducted at the IRRI South Asia breeding hub (IRRI-SA) located at the International Crop Research Institute for Semi-Arid Tropics (ICRISAT), Patancheru, Hyderabad (AP), India in WS2013. ICRISAT is located at 17°31′ N and 78°15′ E at an elevation of 516 m above sea level.

Plant material

A BC_2F_3-derived population developed from the cross of Moroberekan and Swarna was used in this study. Moroberekan, the tolerant donor, is an upland adapted tropical japonica [20] cultivar from New Guinea. It is a long-duration line with sturdy plant type, deep roots, and tolerance of drought and rice blast. However, this variety has poor yield potential due to its low tillering ability and lower number of grains per panicle. On the other hand, Swarna (MTU7029), the drought-susceptible recipient parent used in this study, is a lowland-adapted high-yielding indica [2,21] variety derived from the cross Vashishtha × Mahsuri. It is a long-duration semi-dwarf variety with high tillering ability and grain yield but is highly susceptible to diseases such as rice blast and bacterial blight. This variety is grown on a large area in rainfed and irrigated ecologies across India, Nepal, and Bangladesh and is regarded as a mega-variety of rice.

Figure 1 presents the strategy used to develop the mapping population. The BC_2F_1 plants developed through the cross of the two parents were advanced through the single seed descent (SSD) method to develop a large BC_2F_2 population. This population was screened for tolerance of rice blast. Blast-tolerant plants identified through this screening were advanced to the BC_2F_3 generation through SSD. Seeds from BC_2F_3 single plants were used to develop a $BC_2F_{3:4}$ mapping population to screen for GY under drought stress and non-stress conditions. Furthermore, single plant selections were made from the mapping population under drought stress. These selections were advanced, purified, and screened at IRRI, Philippines and IRRI-SA, India to identify lines that perform the best in both environments.

Stress ecosystems

Field experiments for mapping QTLs were conducted in upland and lowland ecosystems. Throughout the study, the term "upland" refers to field experiments conducted under dry direct-seeded, unpuddled, aerobic conditions, whereas "lowland" refers to field experiments conducted under flooded, puddled, transplanted, and anaerobic conditions. Experiments were conducted under both drought stress and non-stress conditions. In this study, fully irrigated experiments with no drought stress imposed throughout the season are referred to as non-stress experiments; whereas, those experiments on which drought stress was imposed during the reproductive stage (starting at 40 and 50 days after seeding, respectively, under upland and lowland conditions) are referred to as drought stress experiments. In order to further classify the stress experiments into mild, moderate, or severe stress, the percentage reduction in yield as compared to non-stress conditions was used as a criterion [6].

Phenotypic evaluation and data collection

Details of all the experiments, including population size, experimental design, number of replications, and plot size are presented in Table 1. A population of 361 BC_2F_3-derived lines was screened under lowland stress and non-stress conditions in DS2011 and DS2012. A subset of 100 lines from this population was screened under upland stress conditions in WS2012, along with parents Moroberekan and Swarna. Breeding lines developed through plant selection from this population were screened under lowland stress and non-stress conditions in DS2013 at IRRI and WS2013 at IRRI-SA. All experiments were conducted as replicated field trials in α lattice design in continuous plots. The experimental setup has proved effective for screening large rice mapping populations for grain yield under drought using smaller plot sizes. The method has been used in several QTL identification studies, which have led to the identification of large and consistent QTLs now being used in marker-assisted breeding (MAB) programs [2,10–15]. Field and crop management were carried out using the protocol presented in File S1.

In all experiments, data on days to flowering (DTF), plant height (PH) at maturity, and grain yield (GY) were recorded. DTF was recorded as the number of days from seeding to the day that 50% of the plants had flowering tillers. The PH of three plants from each plot were measured at maturity from the ground to the tip of the tallest tiller, and then averaged to get the mean PH for analysis. GY from each plot was harvested at physiological maturity, dried to a moisture content of 14%, and then weighed [11]. This data set was then used to calculate the quantity used in QTL analysis, namely GY of the genotypes in kg ha^{-1}.

Generation of genotypic data

Fresh leaves for all lines were collected and freeze-dried. DNA was extracted from freeze-dried leaf samples by a modified CTAB method in deep-well plates. The DNA was then quantified and purified and lines were genotyped using KASPar SNP assays. These SNPs were selected as subsets from the set of 1,536 and 44K SNP chips [22,23] converted to SNP assays and made available through the integrated breeding platform (https://www.integratedbreeding.net/snp-marker-conversion). A total of 2,015 SNP markers were screened for polymorphism between the two parents. Out of the 2,015 SNPs screened for polymorphism, 591 polymorphic SNP loci were identified. A total of 200 polymorphic markers spaced uniformly across the genome were used to generate the genotypic data of the population. To ensure data quality, lines having percentages higher than expected of heterozygote alleles (>40%), Moroberekan alleles (>40%), or missing data (>30%) were filtered before analysis. Furthermore, markers were removed if the frequency of genotypes deriving from Moroberekan (homozygote and heterozygote) was <2%, as they were essentially monomorphic.

Statistical analysis

ANOVA and means. Data from all experiments for computation of means and standard error of difference (SED) were analyzed using CROPSTAT version 7.2.3. Mixed model analysis of data from individual years was carried out using the model

$$y_{ijk} = \mu + g_i + r_j + b_k(r_j) + e_{ijk}$$

where y_{ijk} is the measurement recorded in a plot, μ is the overall mean, g_i is the effect of the ith genotype, r_j is the effect of the jth replicate, $b_k(r_j)$ is the effect of the kth block within the jth replicate, and e_{ijk} is the error. Genotypic effects were considered fixed and the replicates and block effects were random.

A combined analysis was conducted for lowland experiments to obtain line means across years (2011 and 2012) under stress and non-stress conditions. The model used was

$$y_{ijkl} = \mu + g_i + m_l + r_j(m_l) + b_k[r_j(m_l)] + g_i m_l + e_{ijkl}$$

where y_{ijkl} is the measurement recorded in a plot, μ is the overall mean, g_i is the effect of the ith genotype, m_l is the effect of the lth year, $r_j(m_l)$ is the effect of the jth replicate within the lth year, $b_k[r_j(m_l)]$ is the effect of the kth block within the jth replicate of the lth year, $g_i m_l$ is the effect of the interaction between the ith genotype and lth year, and e_{ijkl} is the error. The effects of replicate within year and block within replicate of year were considered random, whereas the other effects were considered fixed. Due to the high variability for DTF in the population, adjusted mean GY were also calculated by taking DTF as covariate using the above models for QTL mapping. This was done to estimate the mean GY of lines free from the variation caused due to their difference in flowering time (File S3).

Heritability. Broad-sense heritability (*H*) of the traits for single years was calculated as shown below:

$$H = \frac{\sigma^2_{G'}}{\sigma^2_{G'} + \left[\frac{\sigma^2_e}{r}\right]}$$

where $\sigma^2_{G'}$ is the genetic variance, σ^2_E is the plot residual variance and r is the number of replications. The formula used to calculate *H* across years (2011 and 2012) was

$$H = \frac{\sigma^2_G}{\sigma^2_G + \left[\frac{\sigma^2_{GM}}{M} + \frac{\sigma^2_e}{rM}\right]}$$

[24], where σ^2_G is the genetic variance, σ^2_{GY} is the genotype × year variance, σ^2_E is the plot residual variance, and r and M are the number of replicates and years, respectively. Variance components were calculated using the REML algorithm of PROC VARCOMP of SAS V.9.1 [25]. A class analysis was conducted to understand the effect of DTF and PH on GY. Lines were divided into five DTF and GY classes and heat maps were constructed showing GY response in all possible combinations of these classes under drought stress and non-stress conditions. 'Levelplot' in R was used to construct the heat maps.

QTL mapping

Genetic analysis of the population included estimation of segregation distortion, composite interval mapping (CIM), and estimation of epistatic effects. The level of segregation distortion was assessed using the function 'geno.table' in the R package qtl [26]. In order to understand the co-location of markers showing significant distortion to blast QTLs and known genes for disease resistance in rice, a physical map of SNP markers (used in this study), QTLs for blast resistance (reported in Moroberekan-derived mapping populations), and known disease resistance genes in rice was constructed. Information about disease resistance genes as well as those within the identified QTLs and their locations was obtained from the MSU Rice Genome Annotation Project, whereas QTL information was obtained from www.gramene.org and used along with the SNP markers from this study to construct a map using GGT 2 [27]. To map QTLs for GY, DTF, and PH, the genome was scanned for QTLs at a step size of 5 cM, using composite interval mapping ('cim') in R/qtl [26] with 5 cofactors. The genome-wide significance level was determined based on 1,000 permutations. A two-dimensional scan was also performed to identify epistatic QTLs using 'scantwo' in R/qtl with 1,000 permutations.

Results

Population development and segregation patterns

Our study reports a population development strategy using donors tolerant of more than one biotic and/or abiotic stress for simultaneous QTL mapping and pureline development in elite genetic background (Fig. 1). In this study, tropical japonica cultivar Moroberekan was used as the donor parent for tolerance of drought and rice blast, whereas *indica* variety Swarna was used as the recipient. Our strategy involved developing a large BC_2F_2 population and screening this population for blast tolerance. Only blast-tolerant plants were advanced further to develop the BC_2F_3-derived mapping population, which was used to detect QTLs for grain yield under drought and to develop purelines by plant selection in the mapping population. Since selection pressure was applied for blast tolerance, segregation distortion was expected within the mapping population. The genotypic data generated from BC_2F_3-derived lines were used to conduct the test for segregation distortion. A total of 50 and 9 marker loci showed segregation distortion in favour of the donor parent Moroberekan and recipient parent Swarna, respectively (Table S1). The largest segregation distortions were seen on chromosomes 3, 4, and 7 (Table S1). Segregation distortion in favour of Swarna was seen at

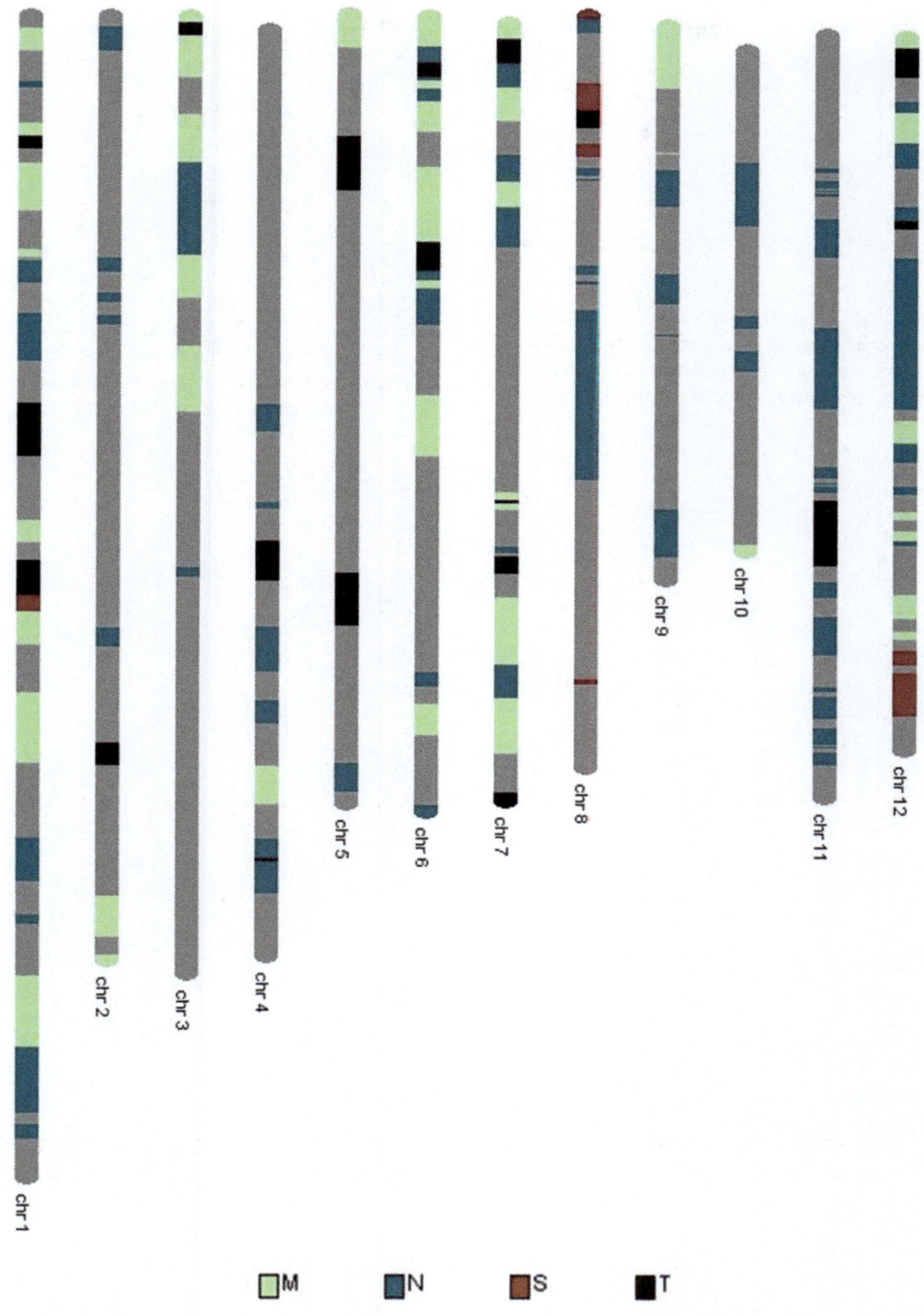

Figure 2. Physical map of markers showing segregation distortion, known genes for disease resistance in rice, and QTLs reported for blast resistance in the Moroberekan/CO39 population. M = markers showing distortion in favour of Moroberekan, S = markers showing distortion in favour of Swarna, N = genes coding for NBS-LRR proteins, and T = QTLs reported for blast resistance in the Moroberekan/CO39 population.

chromosomes 1, 8, and 12. The physical map of SNP markers from this study, known disease-resistance genes, and meta QTLs for blast resistance along the rice genome, highlighted NBS-LRR disease resistance genes close to the markers showing significant segregation distortion in favour of Moroberekan on chromosomes 3, 6, and 7 (Fig. 2). Since the majority of blast resistance genes code for NBS-LRR type proteins, the presence of these genes close to the markers showing distortion may be expected as a result of phenotypic selection for blast tolerance.

Phenotypic variation

Table 2 presents the results of ANOVA, means, heritability, and percentage yield reduction (under drought stress) in the experiments conducted at IRRI and IRRI-South Asia (IRRI-SA)

Table 2. Means ± SED (M), broad-sense heritability (H) and P values for grain yield, days to flowering, and plant height and percentage yield reduction (YR) of trial mean grain yield under stress and non-stress conditions.

	Grain yield (kg ha^{-1})					Days to flowering					Plant height (cm)					
Experiment	P1	P2	M	H	*P*	P1	P2	M	H	*P*	P1	P2	M	H	*P*	YR
1	88	87	677±563	0.60	****	109	110	99±4	0.85	****	72	56	68±9	0.8	****	93
2	954	2326	2115±688	0.61	****	100	102	92±5	0.83	****	102	76	82±8	0.87	****	45
Combined	384	1211	1390±626	0.55	****	104	106	95±5	0.87	****	87	66	75±9	0.70	****	66
3	2234	4552	4370±868	0.78	****	97	98	91±5	0.72	****	126	91	102±8	0.84	****	
4	1248	4012	3859±1049	0.54	****	93	93	88±7	0.78	****	118	92	99±10	0.69	****	
Combined	1701	4282	4118±676	0.71	****	95	95	90±5	0.72	****	122	90	101±6	0.80	****	
5	691	89	827±357	0.92	****	95	112	100±5	0.88	****	107	75	89±7	0.85	****	80
6	0	0	1274±433	0.93	****	105	105	87±2	0.98	****	84	58	74±6	0.86	****	83
7	2893	6857	7380±1628	0.67	****	94	95	86±2	0.95	****	128	91	100±7	0.82	****	
8	NA	NA	4757±856	0.83	****	NA	NA	94±2	0.85	****	NA	NA	97±6	0.59	***	40
9	NA	NA	7895±1027	0.31	****	NA	NA	93±3	0.92	****	NA	NA	116±7	0.61	****	

[a]P1 = Moroberekan, P2 = Swarna, M = population mean, H = heritability, **** = significant at 0.01% *P* level.

Table 3. Genetic correlation between grain yield (GY), days to flowering (DTF), and plant height (PH) under lowland drought stress and non-stress conditions.

	DTF(LSS)	PH(LSS)	GY(LSS)	DTF(LMS)	PH(LMS)	GY(LMS)	DTF (LNS I)	PH (LNS I)	GY(LNS I)	DTF (LNS II)	PH(LNS II)	GY (LNS II)
DTF(LSS)	1											
PH(LSS)	−0.597***	1										
GY(LSS)	−0.813***	0.551***	1									
DTF(LMS)	0.936***	−0.495***	−0.716***	1								
PH(LMS)	−0.042	0.727***	−0.018	−0.040	1							
GY(LMS)	−0.519***	0.168***	0.711***	−0.679***	−0.145**	1						
DTF (LNS I)	0.998***	−0.699***	−0.794***	0.869***	−0.166***	−0.568***	1					
PH (LNS I)	−0.238***	0.831***	0.185***	−0.170***	0.969***	−0.063	−0.297***	1				
GY(LNS I)	−0.297***	0.348***	0.308***	−0.318***	0.264***	0.579***	−0.403***	0.317***	1			
DTF (LNS II)	0.877***	−0.426***	−0.672***	0.845***	−0.137*	−0.664***	0.810***	−0.166**	−0.226***	1		
PH(LNS II)	0.114*	0.668***	−0.204***	0.095	0.801***	−0.238***	−0.086	0.894***	0.123*	0.037	1	
GY (LNS II)	−0.401***	0.245***	0.471***	−0.457***	0.084	0.949****	−0.482***	0.058	0.861***	−0.368***	−0.095	1

LSS = lowland severe stress, LMS = lowland moderate stress, LNS = lowland non-stress. *, **, ***, **** = significant at 5%, 1%, 0.1% 0.01% *P* levels, respectively.

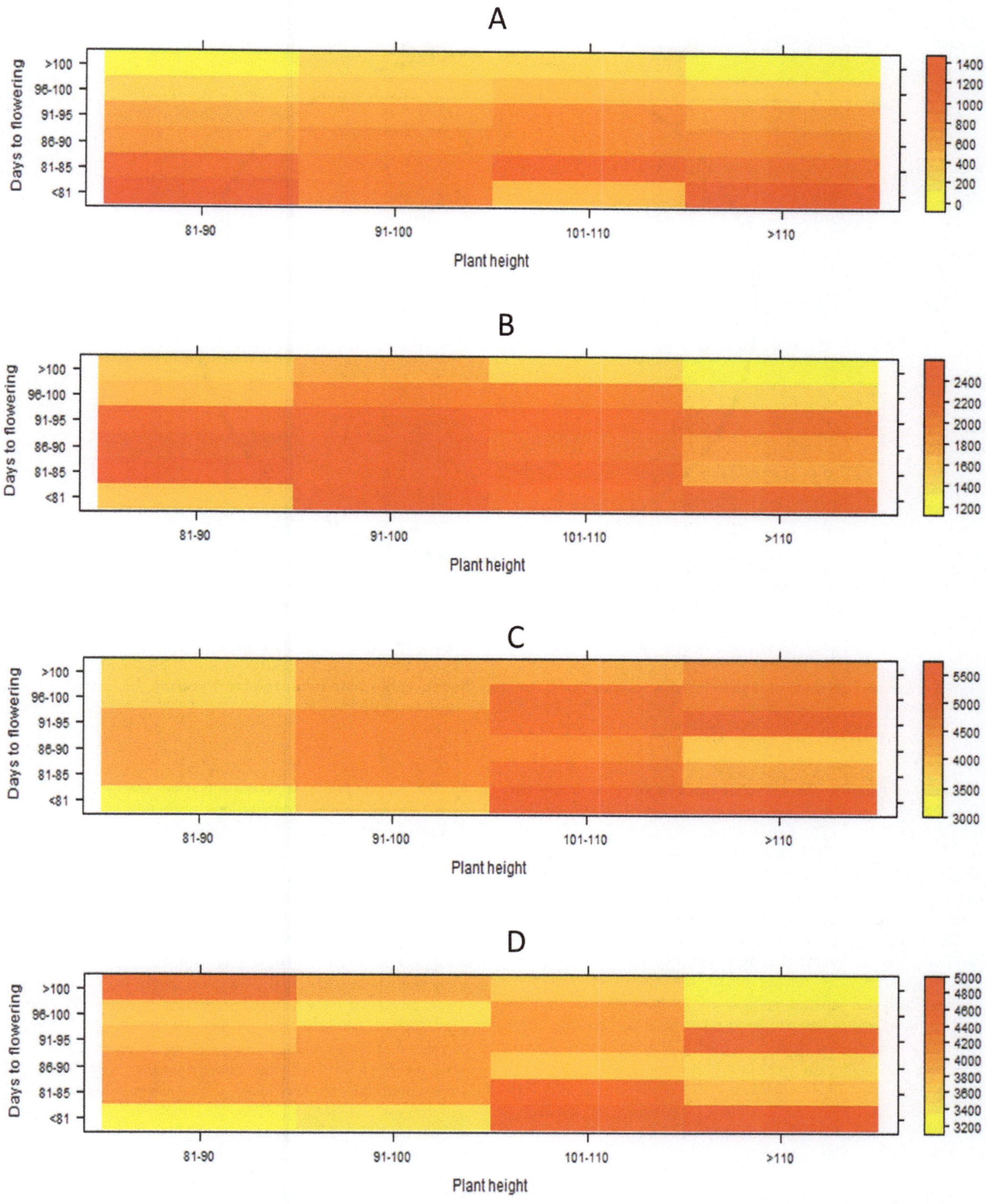

Figure 3. Heat maps showing the effect of DTF and PH on yield performance of lines under (A) lowland severe stress (LSS), (B) lowland moderate stress (LMS), (C) lowland non-stress (LNS I) and (D) LNS II conditions. Increasing intensity of colors from yellow to orange indicates increasing yield levels while white color shows the unavailable PH class.

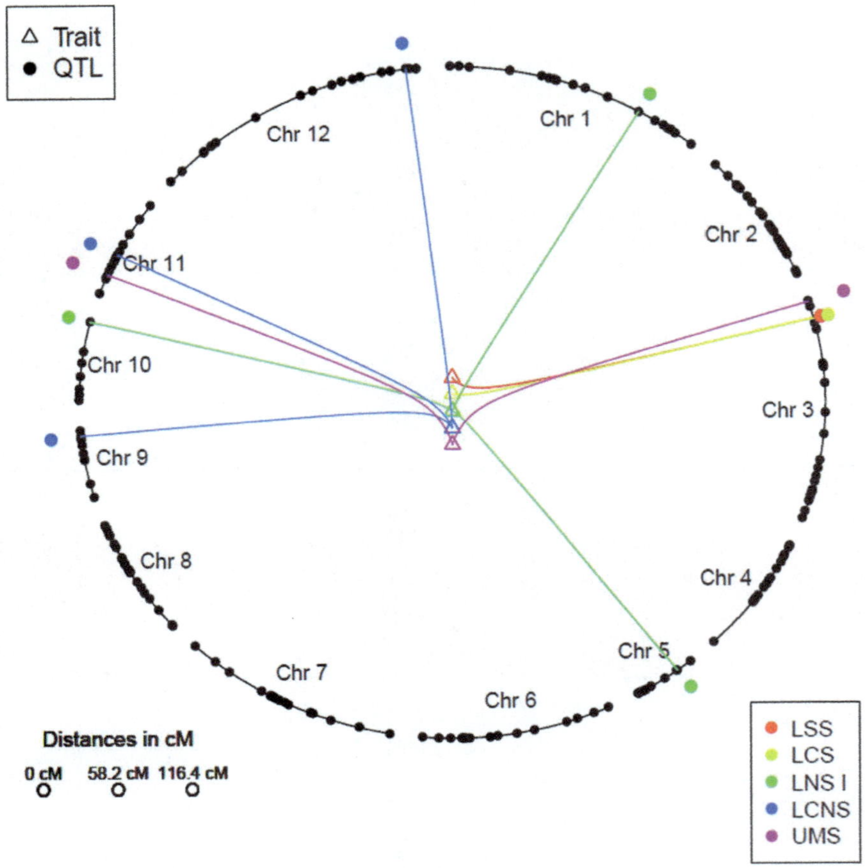

Figure 4. Circular genome plot showing the positions of QTLs identified for grain yield (adjusted for flowering time) under drought stress and non-stress conditions. LSS = lowland severe stress, LNS = lowland non-stress, LCS = lowland combined stress, LCNS = lowland combined non-stress, UMS = upland moderate stress.

under upland and lowland conditions. The full population was screened in Experiments 1-4 under stress and non-stress conditions. Yield reduction in Experiments 1 and 2 varied with stress levels in the respective seasons from severe for Experiment 1 to moderate in Experiment 2. Experiments 6 and 7 followed similar trends as Experiments 1 and 2 had in terms of yield. Both parents had similar yield under severe stress, however, Swarna (P2), being the high-yielding parent, had higher yield under moderate stress conditions in the lowland. The high yield of Swarna under moderate stress was also due to the poor adaptation of Moroberekan, an upland cultivar to lowland conditions. The population means remained higher than those for parents in both stress experiments, suggesting the role of high yield and drought tolerance in deciding the yield under drought.

The advantage of Moroberekan (P1) over Swarna was seen more clearly under upland conditions (Experiment 5), in which a large difference was recorded in the yield of both parents. Mild stress levels were recorded in Experiment 8 as this experiment was conducted in the wet season under naturally occurring drought stress. Moderate to high heritability estimates were recorded for yield in all experiments.

In all DS experiments, both parents had similar DTF under stress and non-stress conditions, but their progeny showed a wide range of DTF values. While Swarna is a well-known long-duration variety, there was a possibility of Moroberekan contributing to the earliness of the progeny. While, this earliness was not seen in the cultivar itself in DS experiments, this difference became clear in the WS experiment (5) when Moroberekan flowered much earlier than Swarna. Similar observations were also recorded in seed multiplication trials under lowland conditions in WS2013 (data not presented).

Trait correlation and effect

Genetic correlations between DTF, PH, and GY under lowland stress and non-stress conditions (DS2011 and DS2012) are presented in Table 3. These four experiments were conducted using the whole population, and hence, were used to understand trait correlations. Clear correlation patterns between DTF and GY were seen in all experiments. The two traits were negatively correlated in both stress and non-stress experiments. Regardless of stress severity levels, earliness of lines related to higher yield under stress environments. Also, under non-stress conditions, a similar pattern was observed, indicating that the reduced growth period did not lead to reduced yield potential. The degree of correlation however was much higher in stress experiments than in non-stress experiments. In contrast to DTF, no clear correlation patterns between PH and GY were observed in both stress and non-stress conditions.

In order to further understand the relationships between DTF, PH, and GY, a class analysis was conducted to measure GY for varying combinations of DTF and PH levels (Fig. 3). At any level of PH, greater differences were seen in GY for varying levels of DTF than vice versa. This agrees with the higher correlation observed between DTF and GY than between PH and GY. In

Table 4. QTLs for grain yield adjusted for flowering time (GY), plant height (PH), and flowering time (DTF) identified under severe and/or moderate reproductive-stage drought stress and non-stress conditions in DS2011 and DS2012.

Trait	QTL	Ecosystem	Peak marker	Chromosome	cM	R^2	A	*P*
GY	$qDTY_{3.2}$	LSS	Id3000757	3	15.0	16.0	281.6	<0.001
	$qDTY_{3.2}$	LCS	Id3000757	3	15.0	14.0	211.3	<0.001
	$qDTY_{1.1}$	LNS I	id1021344	1	154.0	5.5	552.5	0.009
	$qDTY_{5.1}$	LNS I	wd5002636	5	13.5	17.0	1743.0	0.009
	$qDTY_{10.2}$	LNS I	id10006250	10	69.3	4.9	−350.0	0.047
	$qDTY_{9.2}$	LCNS	id9007259	9	54.1	4.7	−353.4	0.012
	$qDTY_{11.2}$	LCNS	id11009456	11	36.7	10.0	1223.0	0.025
	$qDTY_{12.2}$	LCNS	ud12000118	12	212.0	4.8	620.7	0.050
	$qDTY_{3.2}$	UMS	id3000019	3	0.0	19.0	417.6	0.014
	$qDTY_{11.1}$	UMS	id11002801	11	17.4	25.0	1967.0	0.001
PH	$qDTH_{1.1}$	LSS	Id1021344	1	154.0	4.2	5.1	0.003
	$qDTH_{2.1}$	LMS	Id2006643	2	70.9	6.3	5.6	<0.001
	$qDTH_{8.1}$	LMS	Id8007896	8	95.0	4.9	4.2	<0.001
	$qDTH_{10.1}$	LMS	Id10004477	10	43.0	4.4	2.0	<0.001
	$qDTH_{1.1}$	LCS	Id1021344	1	150.0	31.0	9.5	<0.001
	$qDTH_{1.1}$	LNS I	Id1021344	1	154.0	9.1	9.1	<0.001
	$qDTH_{2.1}$	LNS I	Id2007213	2	75.0	4.5	6.5	0.008
	$qDTH_{10.1}$	LNS I	Id10004477	10	43.0	3.6	4.4	0.002
	$qDTH_{2.1}$	LNS II	Id2007213	2	74.3	9.1	6.6	0.008
	$qDTH_{3.1}$	LNS II	Id3002556	3	58.2	4.9	−3.3	0.031
	$qDTH_{1.1}$	LCNS	Id1021344	1	154.0	9.1	7.6	<0.001
	$qDTH_{2.1}$	LCNS	Id2007213	2	74.3	7.4	6.5	<0.001
	$qDTH_{7.1}$	LCNS	Id7005631	7	107.0	2.8	4.6	0.024
DTF	$qDTF_{3.1}$	LSS	Id3000757	3	20.0	30	−5.0	<0.001
	$qDTF_{7.1}$	LSS	Id7003271	7	91.0	8.4	−5.6	<0.001
	$qDTF_{10.1}$	LSS	Id10006250	10	69.3	3.4	1.9	0.040
	$qDTF_{11.1}$	LSS	Id11000855	11	5.0	4.9	−2.5	0.028
	$qDTF_{3.1}$	LMS	Id3000757	3	20.0	29.0	−4.9	<0.001
	$qDTF_{8.1}$	LMS	Id8004692	8	60.4	13.0	−6.7	<0.001
	$qDTF_{3.1}$	LCS	Id3000757	3	20.0	34.0	−5.2	<0.001
	$qDTF_{8.1}$	LCS	Id8004692	8	60.4	13.0	−6.6	<0.001
	$qDTF_{1.1}$	LNS I	Id1014783	1	130.0	7.9	−2.7	0.031
	$qDTF_{3.1}$	LNS I	Id3000946	3	25.0	17.0	−3.4	<0.001
	$qDTF_{3.1}$	LNS II	Id3000019	3	0.0	14.0	−2.6	<0.001

Table 4. Cont.

Trait	QTL	Ecosystem	Peak marker	Chromosome	cM	R^2	A	*P*
	$qDTF_{1.1}$	LCNS	Id1014783	1	125.0	3.9	−1.5	0.023
	$qDTF_{3.1}$	LCNS	Id3000946	3	25.0	21.0	−3.3	<0.001
	$qDTF_{6.1}$	LCNS	Id6010576	6	118.0	7.8	−4.0	0.044
	$qDTF_{2.2}$	UMS	Id2008112	2	90.0	11.0	−7.0	<0.001
	$qDTF_{3.1}$	UMS	Id3000019	3	0.0	19.0	−4.2	<0.001

qDTY, *qDTH*, and *qDTF* are QTLs for GY, PH, and DTF, respectively. Numbers before decimals denote the chromosome number, whereas those after are the order of their detection. A = additive effect of the peak marker, R^2 = percentage of phenotypic variance explained by the QTL, LSS = lowland severe stress, LMS = lowland moderate stress, LCS = lowland combined stress, LNS = lowland non-stress, LCNS = lowland combined non-stress, USS = upland moderate stress.

general, lines with medium maturity (DTF = 81–95 days) were the most consistent at all levels of plant height across all four experiments. However, lines with plant height from 101–110 cm showed the most consistent performance across the four experiments as compared to those beyond these levels.

Genetic variation

Analysis of GY identified several QTLs for each trait significant within and across environments (Fig 4.,Table 4). The most consistent QTL identified for GY was $qDTY_{3.2}$, which had an effect under drought stress across both upland and lowland conditions. The QTL explained a phenotypic variance effect of 16.0% and 19.0% under lowland severe stress (LSS) and upland moderate stress (UMS), respectively (Table 4). Apart from this, $qDTY_{11.1}$ was identified was identified under UMS condition. The QTL explained a phenotypic variance of 25.0%. To the best of our knowledge, this is the first time $qDTY_{11.1}$ is being reported for GY under upland drought stress.

In this study, large QTLs for DTF were identified, which reflect the negative correlations between DTF and GY under drought. A QTL for DTF ($qDTF_{3.1}$) was also identified at the $qDTY_{3.2}$ locus in this study. Several other QTLs for DTF were also observed at chromosomes 2, 6, 7, 8, 10 and 11 although $qDTF_{3.1}$ remained to be the most consistent QTL across drought stress and non-stress conditions (Table 4). Similarly, for PH, several QTLs were identified across the genome with $qDTH_{1.1}$ as the largest and most consistent QTL for the trait (Table 4). Majority of the GY QTLs under stress and non-stress were contributed by Moroberekan; however, two QTLs ($qDTY_{9.2}$ and $qDTY_{10.2}$) for yield under non-stress were also contributed by Swarna. Similarly, for PH and DTF, majority of the QTLs were contributed by the donor parent Moroberekan, but one QTL for each trait ($qDTH_{3.1}$ for PH and $qDTF_{10.1}$ for DTF) was also contributed by the recipient Swarna.

While trangressive segregants were observed in the population for all three traits, the contribution of both parents to higher performance of progeny lines and the presence of epistatic effect have become clear. In order to further explain the phenotypic variance present for grain yield under drought stress and non-stress conditions, tests for epistatic interactions were done. Epistatic interactions for GY under non-stress were observed between several loci under non-stress conditions. However, $eQTL_{10.1}$ consistently showed epistatic interactions with other loci across the genome (Table 5). Other consistant loci include $eQTL_{1.1}$ and $eQTL_{8.1}$ at chromosomes 1 and 8 respectively (Table 5). The high and consistent epistatic interactions of these loci suggest their importance in explaining the genetic basis of GY in lines under non-stress conditions. In this study, lines with high yield under both drought stress and non-stress conditions were identified (Table S2). Despite being more early-maturing than the recipient parent Swarna, these lines were able to maintain their yield potential equal to or more than Swarna under non-stress conditions.

Lines identified from the mapping population were evaluated in the target environment (IRRI-SA) for GY under naturally occurring drought stress and non-stress conditions to understand the consistency of performance of the lines. Figure 5 presents the scatter plots between the GY of selected lines at IRRI and IRRI-SA. Although the stress levels were very mild, the high-yielding drought-tolerant lines selected in DS2013 at IRRI were able to hold similar yield patterns at IRRI-SA. The R^2 values were non-significant, but no negative correlations were observed (Fig. 5). More significant correlations may have been apparent under higher severity of stress in the target environment. Lines finally

Table 5. Epistatic QTLs affecting grain yield (kg ha^{-1}) adjusted for flowering time under drought stress and non-stress conditions in DS2011 and DS2012.

Experiment	*QTL 1*	Chr (1)	cM (1)	Marker (1)	*QTL 2*	Chr 2	cM (2)	Marker (2)	*P*	R^2
LNS I	$eQTL_{8.1}$	8	75	id8006997	$eQTL_{10.1}$	10	55	id10004477	0.001	11.9
LNS II	$eQTL_{1.1}$	1	200	id1010697	$eQTL_{10.1}$	10	45	id10004477	0.015	13.6
LNS II	$eQTL_{1.1}$	1	200	id1010697	$eQTL_{12.1}$	12	175	id12008700	0.037	14.7
LCNS	$eQTL_{4.1}$	4	55	id4001882	$eQTL_{9.1}$	9	40	id9002357	0.039	10.8
LCNS	$eQTL_{7.1}$	7	80	id7002907	$eQTL_{10.1}$	10	50	id10004477	0.021	8.8
LCNS	$eQTL_{8.1}$	8	75	id8006997	$eQTL_{10.1}$	10	55	id10004477	<0.001	12.1
LCNS	$eQTL_{10.1}$	10	50	id10004477	$eQTL_{8.1}$	11	25	id11006588	0.024	8.9

LNS = lowland non-stress, LCNS= lowland combined non-stress, A = additive effect, R^2 = phenotypic variance explained.

identified from these experiments will undergo another testing for tolerance to blast to check for any loss of tolerance and grain quality testing to shortlist a final set of lines for deployment in the target environment.

Gene content of $qDTY_{3.2}$ and $qDTY_{11.1}$

Gene content of $qDTY_{3.2}$ **and** $qDTY_{11.1}$ was classified into 7 classes and percentage of genes falling into each class was calculated (Fig 6.). A large proportion of the genes in bot QTLs were genes coding for known proteins other than those known to be directly related to stress response. Apart from this, large proportion of genes were also related to other expressed proteins in both QTLs. $qDTY_{3.2}$ however showed a much higher percentage of genes related to enzyme synthesis and function while $qDTY_{11.1}$ showed a much higher percentage of retrotransposon proteins as compared to $qDTY_{3.2}$. 8% of the genes within $qDTY_{3.2}$ 6% of genes for $qDTY_{11.1}$ were genes known to be directly associated to stress response or those which could be crucial for maintenance of plant function under stress. These included genes coding for heat shock proteins, no apical meristem proteins, membrane, cell cycle related proteins and zinc finger proteins (File S2). $qDTY_{3.2}$ also showed the presence of flowering related genes such as AP2 domain containing proteins and MADS box family genes while $qDTY_{11.1}$ showed the presence of a large number of nucleotide-binding site leucine- rich repeats (NBS-LRR) genes which are known to be major candidates for blast resistance.

Discussion

Our study reports an integrated strategy to develop drought- and blast-tolerant lines along with simultaneous identification of QTLs for GY under drought. The study also attempts to explain the genetics behind the selection cycle for blast tolerance and its effect on drought tolerance. Several previous studies have used different kinds of mapping populations and strategies to identify QTLs for tolerance of rice blast and grain yield under drought. However, to the best of our knowledge, there has been no attempt to identify QTLs for high grain yield under drought stress on a mapping population tolerant of rice blast while simultaneously developing high-yielding, drought-tolerant breeding lines. Ballini et al. [28] conducted a meta-analysis to pinpoint the location of major genes reported for blast tolerance in rice. Similarly, Swamy et al. [29] conducted a meta-analysis on QTLs for grain yield under drought and reported several meta-QTLs controlling the trait across the rice genome. Both studies provide an extensive coverage of literature, indicating the large number of studies targeting these traits independently. However, reports of the two traits targeted in the same study seldom exist. In our study, the test for segregation distortion in the population showed most of the distortion to be toward the blast-tolerant parent Moroberekan, showing that it was the major contributor for blast tolerance in the population. Interestingly, distortion in favour of Swarna was also observed at some loci. While some of these marker loci co-locate with NBS-LRR genes, it is likely that these regions play a role in imparting tolerance to blast. The breeding strategy not only allowed us to select for but also to detect loci with small and large effects contributed by both parents. With observations of increased incidence of diseases such as blast in drought conditions, it becomes important to develop lines tolerant to both traits to have more stable yields in rainfed environments. Identifying QTLs independently affecting two different traits and pyramiding them together is a two-step process that is normally used to accomplish this task. The alternative mapping strategy presented in Figure 1 allows the simultaneous achievement of both these goals,

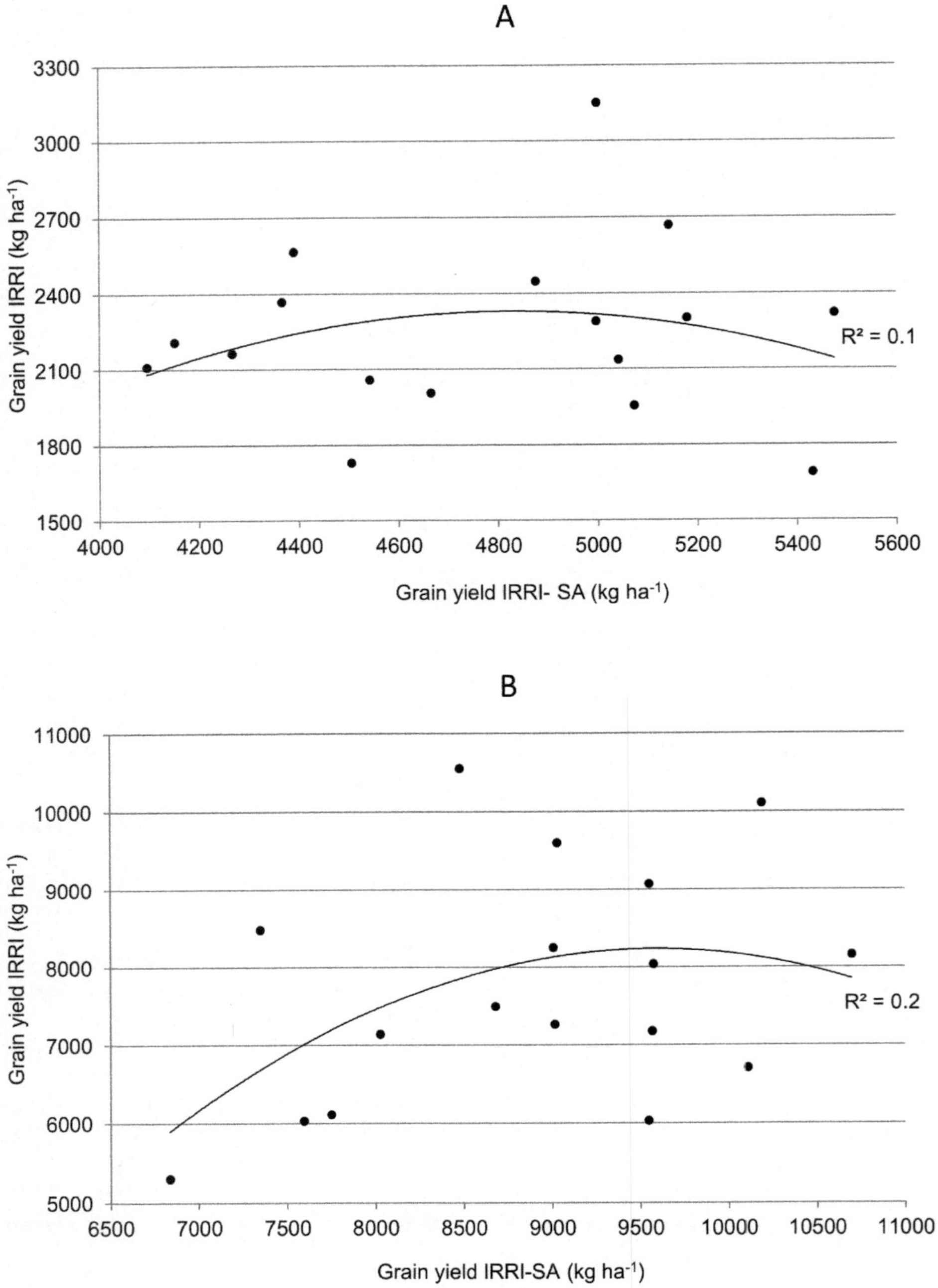

Figure 5. Correlation of GY performance of selected lines under drought stress and non-stress conditions in IRRI and IRRI-SA.

particularly for traits where QTLs/genes conferring tolerance are unknown. We observed that, through systematic early-generation screening of large populations, lines with higher tolerance of one of the two targeted traits (blast in this study) may be obtained. These lines can then be used as mapping populations for QTL identification for the second target trait (high GY under drought stress and non-stress in this study), thus enabling the identification of QTLs and development of high-yielding lines with tolerance of multiple stresses in a reduced time span. Despite the segregation distortion due to screening for blast tolerance, high genetic variation could be seen for DTF, PH, and GY under drought and non-stress conditions. Clear pattern of genetic correlations between the three traits were observed and specific DTF and PH classes showed GY advantage under different stress and non-stress conditions (Table 3, Fig. 3).

Advanced backcross (AB) lines were particularly suitable in such cases where a traditional cultivar is used as a donor for an elite recipient background. Moroberekan has long been known as an important donor for tolerance of drought. In addition, being a tropical japonica accession, it is likely to carry genes that may provide opportunity to increase yield potential under normal situations. Despite these advantages, it has not been successfully used in the breeding program because of its poor combining ability and probable adverse linkage drag. In this study, the earlier segregating generations, such as F_1- and BC_1-derived populations, were not found suitable for the study because of the high percentage of lines with poor performance and poor plant type.

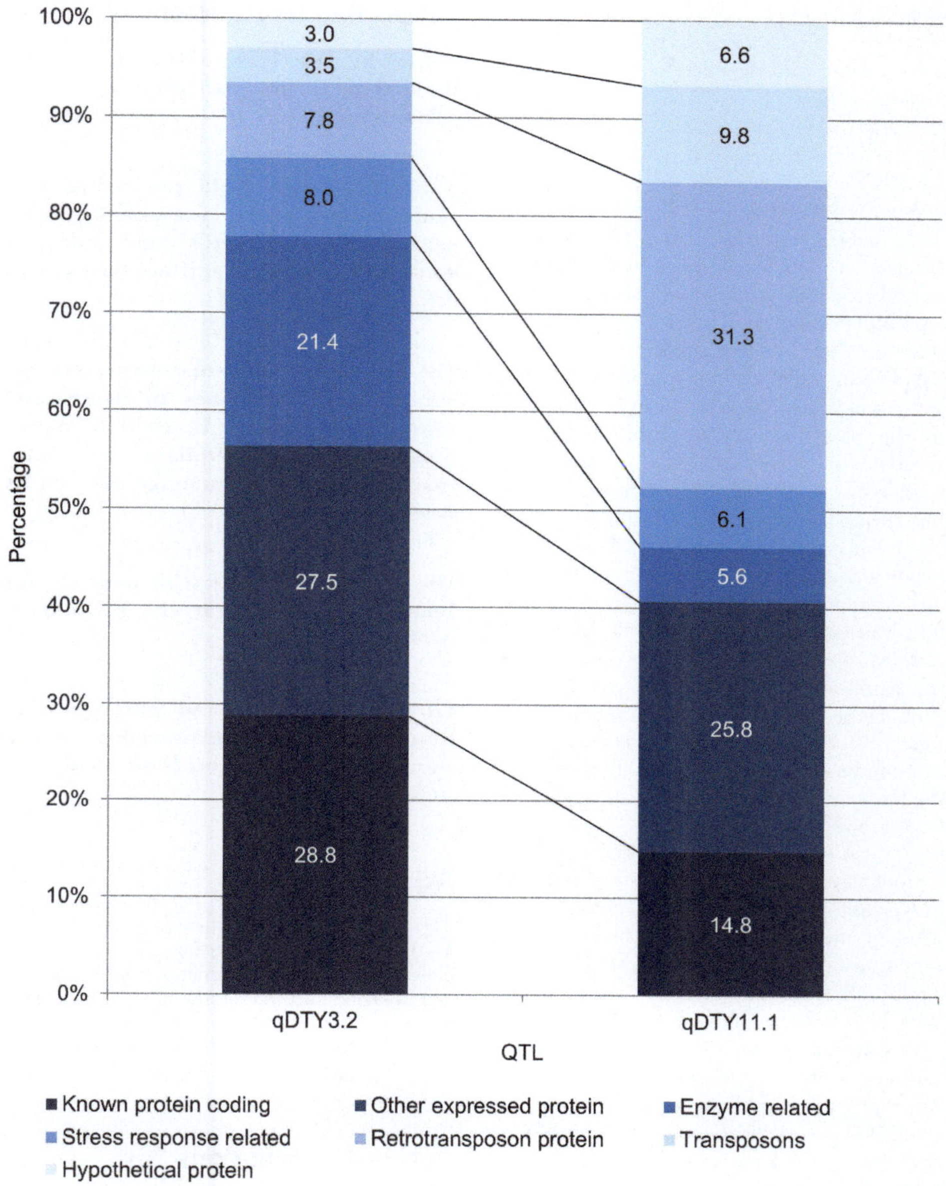

Figure 6. Comparison of gene content of *qDTY*$_{3.2}$ and *qDTY*$_{11.1}$ based on percentage of genes belonging to seven gene classes within the two QTLs.

However, the BC_2-derived lines showed a very high recovery of recipient parent Swarna. In our evaluation, several lines with 10–20 days earlier duration than Swarna, but that had higher yield than Swarna under both irrigated control and drought conditions, were identified (Table S2). DS screening at IRRI has proven highly beneficial in uniform screening of large segregating populations to identify high-yielding drought-tolerant lines. Moderate to high correlation has been reported previously between yield in DS screening at IRRI and WS screening in India [30]. Similar patterns were also seen for the lines selected in this study (Figure 5).

qDTY$_{3.2}$ was the most consistent QTL identified for GY under drought in this study (Table 4, Figure 4). This QTL has been reported in several previous studies for showing independent or epistatic effects on GY under upland and lowland conditions [11,13,31,32]. The QTL region coincides with the *HD9* locus, which controls DTF in rice. While *qDTF*$_{3.1}$ was found to collocate with this locus in the present study, it is likely that the Moroberekan allele at this locus causes earliness, leading to yield advantage under drought.

The role of drought escape mechanism coming into play due to this locus cannot be denied. However, the large variety of genes directly or indirectly related to stress response at this locus (File S2) shows that this locus could affect other traits related to stress response apart from flowering. Several existing reports also suggest the effect of this locus on a wide range of traits. Yadav et al. [13] reported the effect of this locus on grain yield under drought while Bernier et al. [10] reported its effect on plant height and biomass. Similarly, the region has been reported to affect traits including chlorophyll content, root volume under drought, low-temperature germination, and spikelet fertility [32–34]. The wide range of QTLs detected at this locus for traits related to yield and adaptation to abiotic stresses indicates the existence of higher genetic diversity on this chromosomal region. These results are suggestive of either multiple-linked QTLs or pleiotropic effects of one or more than one QTL on a wide range of traits.

The high genic diversity along with the effect of this region on flowering time may be the reason of its effect on yield under drought under wider range of environments as compared to $qDTY_{11.1}$ which showed effect under upland conditions only.

Similar to $qDTY_{3.2}$, $qDTY_{11.1}$ also showed the presence of a wide range of stress response related genes (File S2). However, NBS-LRR genes were by far the most dominating group within this region, while these genes are well known candidates for major blast resistance genes [28]; it is likely that the QTL may affect tolerance to blast along with affecting yield under upland drought.

Epistatic interactions are one of the major factors that determine the phenotypic expression of a mapping population [35]. It has also been seen in a large number of QTL mapping studies conducted in the past that, despite nearly complete coverage of the chromosomes, the identified QTLs do not explain the entire phenotypic variance observed for a trait [35]. This potentially indicates the presence of epistatic interactions contributing to phenotypic variation in a segregating population along with other factors contributing the same. The presence of transgressive segregants in mapping populations is also suggestive of epistasis. In this study, transgressive segregants for all three traits were observed. The study of epistatic interactions showed the presence of loci having significant interactions with several other loci across the genome. These loci could be very important in understanding the genetic basis of yield under drought and non-stress conditions. Understanding the genetic composition of these loci and their effect on plant function may allow us to use these loci for maximum advantage in breeding programs.

Our study involved a two-way approach to develop lines tolerant of blast and drought with high yield potential. Some of these lines show plant type and grain type similar to Swarna with added tolerance of drought and blast. These lines may act as a replacement of the variety where Swarna is highly preferred. On the other hand, lines with different plant and grain types can be tested as prospective new high-yielding cultivars. In addition, lines with the identified QTLs are also being used in other marker-assisted breeding programs as improved donors. The study has allowed us to demonstrate the combination of conventional breeding approaches with QTL identification and to understand the effect of selection pressure on QTL mapping. This strategy can be very effective in cases where traits are likely to be controlled by multiple genes/QTLs across the genome. In such cases, phenotypic selection for one trait is less likely to affect the other, leaving sufficient variation in the population for QTL mapping.

Supporting Information

Table S1 Markers showing significant segregation distortion across the genome with the direction of distortion.

Table S2 Grain yield performance of lines selected under severe stress and non-stress conditions in dry-season screening at IRRI, and under naturally occurring wet-season drought and non-stress conditions at IRRI-SA.

File S1 Standard evaluation system used for large scale screening of rice lines to determine their tolerance reaction rice blast (A), field management of lowland experiments with details of crop cultivation, fertilizer management, and management of insect pests and weeds (B).

File S2 List of genes with possible function related to biotic and abiotic stress response within $qDTY_{3.1}$, $qDTY_{11.1}$.

File S3 Phenotypic and genotypic data used for QTL mapping including trait values, marker names, positions and genotypes for individual lines in the mapping population.

Acknowledgments

The authors are thankful for the support received from Dr. H. Leung, Dr. M. Thomson, during the process of genotyping and development of the manuscript, Dr Nickolai Alexandrov, Dr Casiana Vera Cruz, Dr Bo Zhou and Ms Rose Imee Zhella Morantte for the support received during the process of manuscript development and Teodoro Cuevas, Marcial Panting, Jocelyn O. Guevarra, Danilo De Ocampo, Carmela Malabanan, Ariel Formarejo, and Ruth Erica Carpio for the technical support received in the completion of this study.

Author Contributions

Conceived and designed the experiments: SD AK. Performed the experiments: SD MTSC PTM. Analyzed the data: SD BEH. Contributed reagents/materials/analysis tools: MTSC JCEO. Wrote the paper: SD BEH AK.

References

1. Alexandratos N, Bruinsma J (2012) World agriculture towards 2030/2050. ESA Working Paper No. 12–03.
2. Vikram P, Mallikarjuna Swamy BP, Dixit S, Ahmed HU, Sta Cruz MT, et al. (2011) $qDTY_{1.1}$, a major QTL for rice grain yield under reproductive-stage drought stress with a consistent effect in multiple elite genetic backgrounds. BMC Genet 12: 89.
3. Huke RE, Huke EH (1997) Rice area by type of culture: South, Southeast, and East Asia. IRRI, Los Baños, Philippines.
4. Serraj R, Bennett J, Hardy B (2008) Drought frontiers in rice: crop improvement for increased rainfed production. Singapore: World Scientific Publishing and Los Baños (Philippines): International Rice Research Institute.
5. Ouk M, Basnayake J, Tsubo M, Fukai S, Fischer KS, et al. (2006) Use of drought response index for identification of drought tolerant genotypes in rainfed lowland rice. Field Crops Res 99: 48–58.
6. Kumar A, Bernier J, Verulkar S, Lafitte HR, Atlin GN (2008) Breeding for drought tolerance: direct selection for yield, response to selection and use of drought-tolerant donors in upland and lowland-adapted populations. Field Crops Res 107: 221–231.
7. Venuprasad R, Lafitte HR, Atlin GN (2007) Response to direct selection for grain yield under drought stress in rice. Crop Sci 47: 285–293.
8. Venuprasad R, Sta Cruz MT, Amante M, Magbanua R, Kumar A, et al. (2008) Response to two cycles of divergent selection for grain yield under drought stress in four rice breeding populations. Field Crops Res 107: 232–244.
9. Kumar A, Verulkar SB, Mandal NP, Variar M, Shukla VD, et al. (2012) High-yielding, drought-tolerant, stable rice genotypes for the shallow rainfed lowland drought-prone ecosystem. Field Crops Res 133: 37–47.
10. Bernier J, Kumar A, Venuprasad R, Spaner D, Atlin GN (2007) A large-effect QTL for grain yield under reproductive-stage drought stress in upland rice. Crop Sci 47: 507–516.
11. Venuprasad R, Dalid CO, Del Valle M, Zhao D, Espiritu M, et al. 2009 Identification and characterization of large-effect quantitative trait loci for grain yield under lowland drought stress in rice using bulk-segregant analysis. Theor Appl Genet 120: 177–190.
12. Ghimire KH, Quiatchon LA, Vikram P, Mallikarjuna Swamy BP, Dixit S, et al. (2012) Identification and mapping of a QTL ($qDTY_{1.1}$) with a consistent effect on grain yield under drought. Field Crops Res 131: 88–96.
13. Yadav RB, Dixit S, Raman A, Mishra KK, Vikram P, et al. (2013) A QTL for high grain yield under lowland drought in the background of popular rice variety Sabitri from Nepal. Field Crops Res 144: 281–287.

14. Mishra KK, Vikram P, Yadaw RB, Swamy BPM, Dixit S, et al. (2013) *qDTY*$_{12.1}$: a locus with a consistent effect on grain yield under drought in rice. BMC Genet 14: 12.
15. Dixit S, Singh A, Sta Cruz MT, Maturan PT, Amante M, et al. (2014) Multiple major QTLs lead to stable yield performance of rice cultivars across variable drought intensities. BMC Genet 15: 16.
16. You A, Lu X, Jin H, Ren X, Liu K, et al. (2006) Identification of quantitative trait loci across recombinant inbred lines and testcross populations for traits of agronomic importance in rice. Genetics 172: 1287–1300.
17. Yamagishi J, Miyamoto N, Hirotsu S, Laza RC, Nemoto K (2004) QTLs for branching, floret formation, and pre-flowering floret abortion of rice panicle in a temperate *japonica* × tropical *japonica* cross. Theor Appl Genet 109: 1555–1561.
18. Xing YZ, Tan YF, Hua JP, Sun XL, Xu CG, et al. (2002) Characterization of the main effects, epistatic effects and their environmental interactions of QTLs on the genetic basis of yield traits in rice. Theor Appl Genet 105: 248–257.
19. Thomson MJ, Tai TH, McClung AM, Lai X-H, Hinga ME, et al. (2003) Mapping quantitative trait loci for yield, yield components and morphological traits in an advanced backcross population between *Oryza rufipogon* and the *Oryza sativa* cultivar Jefferson. Theor Appl Genet 107: 479–493.
20. McNally KL, Childs KL, Bohnert R, Davidson RM, Zhao K, et al. 2009 Genomewide SNP variation reveals relationships among landraces and modern varieties of rice. Proc Nat Acad Sci USA 106(30): 12273–12278.
21. Neeraja CN, Maghirang-Rodriguez R, Pamplona A, Heuer S, Collard B, et al. (2007) A marker-assisted backcross approach for developing submergence-tolerant rice cultivars. Theor Appl Genet 115: 767–776.
22. Zhao K, Wright M, Kimball J, Eizenga G, McClung A, et al. (2010) Genomic diversity and introgression in *O. sativa* reveal the impact of domestication and breeding on the rice genome. PLOS One 5 (5): e10780.
23. Zhao K, Tung C-W, Eizenga GC, Wright MH, Ali ML, et al. (2011) Genome-wide association mapping reveals a rich genetic architecture of complex traits in *Oryza sativa*. Nat Comm 2: 467/DOI: 10.1038/ncomms1467.
24. Cooper M, DeLacy IH, Basford KE (1996) Relationship among analytical methods used to analyse genotypic adaptation in multi-environment trials, pp-193–224. In M. Cooper and G.L. Hammer editors Plant adaptation and crop improvement University Press, Cambridge, UK.
25. SAS Institute Inc (2004) SAS online doc 9.1.3. SAS Institute Inc., Cary, NC, USA.
26. Broman KW, Wu H, Sen S. Churchill GA (2003) R/qtl: QTL mapping in experimental crosses. Bioinformatics 19: 889–890.
27. Berloo RV (2008) GGT 2.0: versatile software for visualization and analysis of genetic data. J Heredity 99(2): 232–236.
28. Ballini E, Morel J-B, Droc G, Price A, Courtois B, et al. (2008) A genome-wide meta-analysis of rice blast resistance genes and quantitative trait loci provides new insights into partial and complete resistance. Mol. Plant Microbe Interact 21(7): 859–868.
29. Swamy BPM, Vikram P, Dixit S, Ahmed HU, Kumar A (2011) Meta-analysis of grain yield QTL identified during agricultural drought in grasses showed consensus. BMC Genomics 12: 319.
30. Verulkar SB, Mandal NP, Dwivedi JL, Singh BN, Sinha PK, et al. (2010) "Breeding resilient and productive rice genotypes adapted to drought-prone rainfed ecosystems of India", Field Crops Res 117: 197–208.
31. Dixit S, Mallikarjuna Swamy BP, Vikram P, Bernier J, Sta Cruz MT, et al. (2012) Increased drought tolerance and wider adaptability of *qDTY12.1* conferred by its interaction with *qDTY2.3* and *qDTY3.2*". Mol Breed 30: 1767–1779.
32. Ding X, Li X, Xiong L (2011) Evaluation of near-isogenic lines for drought resistance QTL and fine mapping of a locus affecting flag leaf width, spikelet number, and root volume in rice Theor Appl Genet 123: 815–826.
33. Ishimaru K Yano M, Aoki N, Ono K, Hirose T, et al. (2001) Toward the mapping of physiological and agronomic characters on a rice function map: QTL analysis and comparison between QTLs and expressed sequence tags. Theor Appl Genet 102: 793–800.
34. Fujino K, Sekiguchi H, Sato T, Kiuchi H, Nonoue Y, et al. (2004) Mapping of quantitative trait loci controlling low-temperature germinability in rice (*Oryza sativa* L.). Theor Appl Genet 108: 794–799.
35. Li Z, Pinson SRM, Park WD, Paterson AH, Stansel JW (1997) Epistasis for three grain yield components in rice (*Oryza sativa* L.). Genetics 145: 453–465.

7

Cropping Systems and Cultural Practices Determine the *Rhizoctonia* Anastomosis Groups Associated with *Brassica* spp. in Vietnam

Gia Khuong Hoang Hua[☯], **Lien Bertier**[☯], **Saman Soltaninejad**, **Monica Höfte***

Laboratory of Phytopathology, Department of Crop Protection, Faculty of Bioscience Engineering, Ghent University, Gent, Belgium

Abstract

Ninety seven *Rhizoctonia* isolates were collected from different *Brassica* species with typical *Rhizoctonia* symptoms in different provinces of Vietnam. The isolates were identified using staining of nuclei and sequencing of the rDNA-ITS barcoding gene. The majority of the isolates were multinucleate *R. solani* and four isolates were binucleate *Rhizoctonia* belonging to anastomosis groups (AGs) AG-A and a new subgroup of A-F that we introduce here as AG-Fc on the basis of differences in rDNA-ITS sequence. The most prevalent multinucleate AG was AG 1-IA (45.4% of isolates), followed by AG 1-ID (17.5%), AG 1-IB (13.4%), AG 4-HGI (12.4%), AG 2-2 (5.2%), AG 7 (1.0%) and an unknown AG related to AG 1-IA and AG 1-IE that we introduce here as AG 1-IG (1.0%) on the basis of differences in rDNA-ITS sequence. AG 1-IA and AG 1-ID have not been reported before on *Brassica* spp. Pathogenicity tests revealed that isolates from all AGs, except AG-A, induced symptoms on detached leaves of several cabbage species. In *in vitro* tests on white cabbage and Chinese cabbage, both hosts were severely infected by AG 1-IB, AG 2-2, AG 4-HGI, AG 1-IG and AG-Fc isolates, while under greenhouse conditions, only AG 4-HGI, AG 2-2 and AG-Fc isolates could cause severe disease symptoms. The occurrence of the different AGs seems to be correlated with the cropping systems and cultural practices in different sampling areas suggesting that agricultural practices determine the AGs associated with *Brassica* plants in Vietnam.

Editor: Mark Gijzen, Agriculture and Agri-Food Canada, Canada

Funding: This work was supported by a scholarship from the Special Research Fund of Ghent University (BOF) given to Gia Khuong Hoang Hua. Lien Bertier was funded by a PhD grant of the Agency of Innovation by Science and Technology in Flanders (IWT). The funders had no role in study design, data collection and analysis, decision to publish, or preparation of the manuscript.

Competing Interests: The authors have declared that no competing interests exist.

* Email: monica.hofte@ugent.be

☯ These authors contributed equally to this work.

Introduction

Vietnam is a country in Southeast Asia in which the agricultural sector accounts for more than 22% of the GDP, 30% of export and 52% of all employment. Vietnam is not only one of the world leaders in rice and coffee export, but also the third world's largest vegetable producer. *Brassicas* are among the main vegetables produced for both local consumption and export [1]. Vegetables in Vietnam are mainly produced by poor households living in the Red River and Mekong River delta (see Figure 1) in intensive cultivation systems or in rotation with other crops. Due to the lack of knowledge in crop management, limited availability of technology and land fragmentation, farmers are suffering heavy yield losses year after year. Among the limiting factors in vegetable production is the occurrence of *Rhizoctonia* diseases, which has been recognized as one of the most important threats.

Rhizoctonia is a genus of basidiomycete fungi causing many important plant diseases. Based on differences in the number of nuclei per cell, *Rhizoctonia* isolates have been differentiated into uninucleate *Rhizoctonia*, binucleate *Rhizoctonia* (teleomorphs: *Ceratobasidium* spp. and *Tulasnella* spp.) and multinucleate *Rhizoctonia* (teleomorphs: *Thanatephorus* spp. and *Waitea* spp.) [2]. *Rhizoctonia* species can also be classified using biochemical and molecular techniques. Among those, rDNA-ITS sequence analysis appears to be the most convenient and reliable method [3]. Currently, isolates of *R. solani*, the most widely recognized species within the multinucleate *Rhizoctonia* group, have been divided into 13 anastomosis groups (AGs), while 16 AGs of binucleate *Rhizoctonia* have been recognized [2–7]. Due to the considerable genetic diversity, several AGs have been further divided into subgroups based on phylogenetic differences. These phylogenetic differences can be associated with differences in morphology, ecology, pathogenicity and biochemical characteristics, although this is not necessarily the case [8].

Compared to binucleate *Rhizoctonia*, multinucleate *R. solani* AGs usually have a wider host range and higher virulence. *R. solani* can survive for a long period in plant debris, contaminated seeds, or infested soils as mycelium or sclerotia [9,10]. Under favorable conditions, sclerotia geminate and form delicate hyphae that will grow toward the host plants [11]. *Brassica* vegetables can be attacked by several different AGs of *R. solani* resulting in the development of various diseases such as foliar blight, wirestem and damping off. In previous studies, AGs 1-IB, 1-IC, 2-1, 2-2 IIIB, 3, 4-HGI, 4-HGII, 4-HGIII, 5, 7, 9 and 10 were shown to be pathogenic on *Brassica* crops grown in Canada [12,13], Australia [14], Japan [15,16], North America [17–19], Brazil [20], China [21], Belgium [22], and the UK [23]. Although *Rhizoctonia*

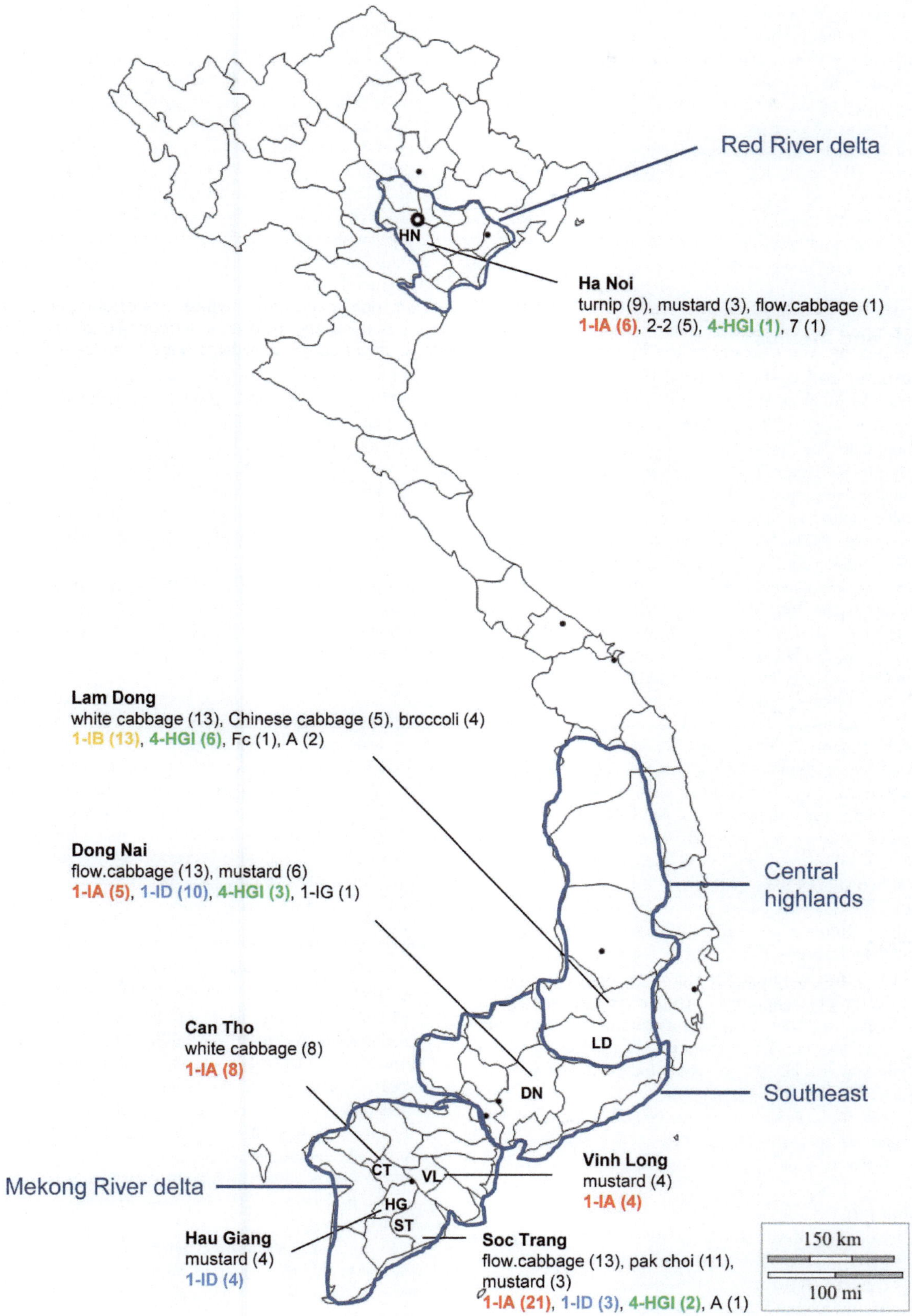

Figure 1. Location of sites for collection of *Rhizoctonia* isolates from *Brassica* spp. in Vietnam. The seven provinces sampled are: Ha Noi (districts of Gia Lam, Thanh Tri and Dong Anh), Lam Dong (Da Lat city and Duc Trong district), Dong Nai (Bien Hoa city), Vinh Long (Binh Tan district), Can Tho (Cai Rang district), Hau Giang (Phung Hiep district) and Soc Trang (Soc Trang city and My Xuyen district). In each city or district of one province, one to two wards were surveyed and these wards are marked with a start✦. Different colors are used to highlight the most important AGs found in our survey including AG 1-IA, AG 1-IB, AG 1-ID and AG 4-HGI.

diseases occur severely and frequently on leafy vegetables cultivated in Vietnam, there have been no reports about the AGs and subgroups that attack *Brassica* crops. Therefore, our research aimed at (i) identifying the species and AGs of *Rhizoctonia* present on *Brassica* plants in different vegetable producing regions in Vietnam, and (ii) verifying the susceptibility of *Brassica* vegetables to *Rhizoctonia* isolates collected. Our story revealed that *Rhizoctonia* AGs such as AG 1-IA and AG 1-ID, which have not been reported before on *Brassica* spp., are predominant in Vietnam, which is presumably linked with the cultural practices and cropping systems in the different sampling areas. Moreover, we describe two new AGs that were previously unknown.

Materials and Methods

Field sampling and pathogen isolation

Sampling was done on private farms by Gia Khuong Hoang Hua (Vietnamese citizen) with permission from the farmer. In general, permission by the authorities was not required since the sampling studies were carried out on private farms. Only cabbage plants showing symptoms of the undesired *Rhizoctonia* fungus were sampled, hence the field studies did not involve endangered or protected species. The GPS coordinates of the locations where samples were collected are presented in Table S1 in File S1.

The survey was conducted from September to October 2011 on various fields in the Red River delta (Ha Noi; an important vegetable production area of the North), the Mekong River delta (Vinh Long, Can Tho, Hau Giang and Soc Trang; main vegetable production areas of the South), the Central Highlands (Lam Dong; vegetables are mainly produced for export) and the Southeast (Dong Nai; vegetables are mainly produced for domestic consumption) (Figure 1). These regions were chosen because of their importance in vegetable production and because they offer a good representation of the different climatic and topographic conditions in Vietnam and, therefore, a good representation of the distribution of *Rhizoctonia* spp. on *Brassica* spp. in Vietnam can be obtained. Total production area, climatic conditions and main agricultural activities of these regions are listed in Table 1.

A total of 142 *Brassica* plants with *Rhizoctonia*-like symptoms were sampled. Infected root and leaf tissues were washed in running tap water, surface-disinfected in 1% sodium hypochlorite solution for two min and then rinsed twice in sterile water before placing on 1% water agar medium supplemented with streptomycin (0.05 g L^{-1}). After 24 h of incubation, *Rhizoctonia*-like hyphal tips growing out of these tissues were transferred to fresh potato dextrose agar (PDA; Difco) plates and incubated for two to four days at 28°C.

Ninety seven *Rhizoctonia* isolates were recovered and subjected to nuclei staining and sequencing of the ITS-rDNA region. All isolates are listed in Table 2.

Nuclei staining

Rhizoctonia isolates were cultured on sterile glass slides covered by PDA for two days at 28°C. Actively growing fungal hyphae were stained with 10 µg mL^{-1} 4, 6-diamino-2-phenyl indole (DAPI; Sigma-Aldrich) and the number of nuclei per hyphal cell was determined using an Olympus BX51 microscope [22].

DNA extraction, PCR and sequencing of the rDNA-ITS region

The rDNA-ITS region of all collected isolates was sequenced for identification to the AG and subgroup level. The usefulness of the rDNA-ITS region for identification of unknown *Rhizoctonia* spp. has been clearly shown by Sharon et al. [2,3].

Rhizoctonia isolates were grown on potato dextrose broth at 28°C for one week. Mycelial mats were harvested by filtration and ground in liquid nitrogen to produce a fine powder. Total genomic DNA was extracted using the DNeasy Plant Mini Kit (Qiagen). The rDNA-ITS fragment including the 5.8 S gene was amplified using primers ITS4 (5′-TCCTCCGCTTATTGATATGC-3′) and ITS5 (5′-GGAAGTAAAAGTCGTAACAAGG-3′) [24]. The PCR amplification reactions were performed by adding 2 µL genomic DNA (5–10 ng μL^{-1}) to 23 µL of reaction mixture containing 2.5 µL PCR buffer (10×; Qiagen), 5 µL Q-solution (Qiagen), 0.5 µL dNTPs (10 mM; Fermentas GmbH), 1.75 µL of each primer (10 µM), 0.15 µL Taq DNA polymerase (5 units μL^{-1}; Fermentas GmbH) and 11.35 µL ultrapure sterile water. Amplification was performed using a Flexcycler PCR Thermal Cycler (AnalytikJena) programmed for an initial denaturation step at 94°C for 10 min followed by 35 cycles at 94°C for 1 min, 55°C for 1 min and 72°C for 1 min. Cycling ended with a final extension step at 72°C for 10 min. Amplification products were separated in 1% agarose gels in TAE-buffer at 100 V for 30 min and visualized by ethidium bromide staining on a UV transilluminator. The sequences of both strands were determined by LGC Genomics GmbH (Berlin, Germany) using Sanger sequencing.

Identification using BLAST and phylogenetic analysis

Consensus sequences for all 97 isolates were created with BioEdit version 7.1.11. To determine the AG of the isolates, the rDNA-ITS consensus sequences obtained were compared to those in Genbank using the BLASTn tool.

However, since Genbank is an uncurated database, it can contain inaccurately designated *Rhizoctonia* spp., as has been previously shown by Sharon et al. [3]. Therefore, comparison of rDNA-ITS sequences of unknown isolates to a curated database of sequences containing representative rDNA-ITS sequences of all known uninucleate, binucleate and multinucleate *Rhizoctonia* AG and subgroups provides a more reliable identification. Such a database of representative sequences is available from Sharon et al. [2] and was provided by Michal Sharon to us. However, not all known AGs were present in this database, therefore we added representative isolates of the following AGs: AG 1-1E, AG 1-1F, AG 2-2 WB [25] and AG 13 [4]. The total number of sequences in the database was 129 and the Genbank accession numbers of all these sequences can be found in Table S2 in File S2.

Multiple alignments for multinucleate and binucleate *Rhizoctonia* isolates were constructed using MUSCLE which is implemented in MEGA 6 [26] and checked manually afterwards. The resulting alignments had a length of 717 bp (multinucleate *Rhizoctonia* isolates) and 768 bp (binucleate *Rhizoctonia* isolates).

Separate phylogenetic trees were constructed for multinucleate *Rhizoctonia* and binucleate *Rhizoctonia* isolates. For the binucleate tree, one representative isolate for each known binucleate AG was included together with the binucleate *Rhizoctonia* isolates obtained in this work.

For the multinucleate tree, reference isolates from the curated database (Table S2 in File S2) were added for all AGs present in our collection from Vietnam. Another 32 isolates from a characterization study in Vietnam that has not been published (Thuan et al., unpublished; Genbank accession numbers with prefix 'EF' in Table 3), and found on a range of crops and belonging to AGs AG 1-IA, AG 1-ID and AG 4-HGI were also added to the multinucleate *Rhizoctonia* alignment for phylogenetic analysis. Phylogenetic trees were built using the neighbour joining algorithm with 1000 bootstrap repeats using MEGA 6 [26]. Model

Table 1. Sampling locations and their relevant characteristics [53–59].

Sampling location	Agricultural area (1000 ha)	Average temperature (°C)	Main crop
Red River Delta			
Ha Noi	152.24	24	- Cabbage, tomato, cucumber, radish
			- Rice
Central highlands region			
Lam Dong	279.00	High land: 14	- Lettuce, cabbage, carrot, potato
		Low land: 21	- Coffee, tea, cashew-nut tree, cotton
Southeast region			
Dong Nai	289.02	27	- Coffee, cotton, black pepper
			- Durian, grapefruit, mango
Mekong River Delta			
Vinh Long	116.18	27	- Rice
			- Mango, orange, grapefruit, durian
			- Cabbage, cucumber, bean
Can Tho	115.00	27	- Rice
			- Cabbage, cucumber
Hau Giang	139.07	27	- Durian, pineapple, grapefruit
			- Rice
Soc Trang	278.15	27	- Rice
			- Grapefruit, mango, durian

testing was done using the software implemented in MEGA 6 and the K2+ G DNA substitution model was chosen.

Aggressiveness of *Rhizoctonia* isolates towards detached leaves of cabbages, rice and water spinach

Nine isolates (representing the nine different *Rhizoctonia* AGs collected) were randomly selected and tested for virulence towards several *Brassica* crops in two independent experiments (Figures 2, 3 and 4). To confirm their pathogenicity, tests were repeated for AGs 1-IA, 1-IB, 1-ID, 2-2, 4-HGI and A since each of these AGs consists of more than one isolate. Results of additional tests are shown in Tables S3, S4 and S5 in File S1.

Leaves of white cabbage (*Brassica oleracea*), Chinese cabbage (*B. chinensis*), pak choi (*B. chinensis*), mustard cabbage (*B. juncea*) and Chinese flowering cabbage (*B. parachinensis*) were cut into pieces (3×3 cm). The soil substrate used in our experiments was a mixture (w/w) of 50% potting soil (Structural; Snebbout, Kaprijke, Belgium) and 50% sand (Cobo garden; Belgium). Sets of six leaf discs were placed in a plastic box (16×11×6 cm) containing 400 g of soil substrate. Inoculum of *Rhizoctonia* spp. was produced according to the method described by Scholten et al. [27]. Briefly, water-soaked wheat kernels were autoclaved for 25 min on two successive days and then inoculated with three fungal discs (diameter 5 mm) cut at the edge of a 3-day-old *Rhizoctonia* colony cultured on PDA. Flasks containing the inoculated kernels were incubated for 14 days at 28°C and shaken every 3–4 days to avoid coagulation. Two *Rhizoctonia*-infected kernels which were comparable in size were buried 2 cm below each leaf disc. Leaf discs inoculated with sterile wheat kernels served as a control. All boxes were incubated in a growth chamber at 22°C.

The detached leaf bio-assay was also conducted to investigate the pathogenicity of the *Rhizoctonia* isolates on rice and water spinach, two important hosts of *Rhizoctonia* spp. in tropical countries. Surface-sterilized seeds of rice (*Oryza sativa* cv. CO39) and water spinach (*Ipomoea aquatic* cv. Trang Nong) were sown in plastic trays (45×45×10 cm) filled with 4 kg of soil substrate and kept in a growth chamber at 28°C for four weeks before their leaves were detached. Rice leaves were then cut into pieces (8 cm long) and six rice leaf pieces or six water spinach leaves were put in one square Petri dish containing a sterile filter paper moistened with sterile water. Two sterile glass slides were placed in the middle of each Petri dish to keep the leaves away from water. A 5-mm plug harvested from 3-day-old cultures of *Rhizoctonia* spp. on PDA was placed at the center of each leaf or leaf piece and the Petri dishes were incubated at 28°C.

After four days of incubation, disease severity was scored based on the following disease scale: 0 = no symptoms observed; 1 = lesions covered less than 25% of leaf surface; 2 = lesions covered 25–50% of leaf surface; 3 = lesions covered 50–75% of leaf surface; 4 = lesions covered more than 75% of leaf surface or dead leaf. The experiment had a completely randomized design. Each treatment consisted of 12 leaves or leaf pieces equally divided into two boxes or Petri dishes.

In vitro pathogenic potential of *Rhizoctonia* spp. on seedlings of white cabbage and Chinese cabbage

The same nine *Rhizoctonia* isolates were studied for their *in vitro* pathogenic potential using the method described by Keijer et al. [28]. Six surface-sterilized seeds of white cabbage (*B. oleracea* cv. TN180) or Chinese cabbage (*B. chinensis* cv. Elton) were germinated on Gamborg B5 medium (Gamborg B5 medium including vitamins; Duchefa) in a square Petri dish. Two mycelial disks (5 mm in diameter) from 3-day-old *Rhizoctonia* cultures grown on PDA were placed between seeds. In the control dishes, sterile PDA discs were used for inoculation. The Petri dishes were incubated at 22°C in the dark for two days for seed germination. Then, the second halves of the Petri dishes were covered with aluminum foil to protect the roots from light and placed in an

Table 2. Characterization of *Rhizoctonia* isolates collected from diseased *Brassica* crops grown in Vietnam by sequencing the ITS-region.

AG/Subgroup	Host plant	Isolate[ab]	Genbank accession numbers
1-IA	*B. parachinensis* (Chinese flowering cabbage)	STST03-1, STST03-3, STST03-4, STST04-2, **STMX04-1, STMX04-2, STMX04-3, STMX04-4**, STMX04-5	KF907702, KF907703, KF907704, KF907705
	B. juncea (Mustard cabbage)	DNBH01-1, DNBH01-2, DNBH01-3, DNBH02-2, DNBH02-3	
		HNGL01-1, HNGL01-2, **HNGL01-3**	KF907706
		STST02-1, STST02-2	
		VLBT01-1, VLBT01-2, VLBT01-3, VLBT01-4	
	B. chinensis (Pak choi)	**STMX01-1**, STMX01-2, **STMX01-4**, STMX01-5, STMX02-1, STMX02-2, **STMX02-3**, **STMX03-1**, **STMX03-2**, STMX03-3	KF907707, KF907708, KF907709, KF907710, KF907711
	B. oleraceae (Turnip cabbage)	HNDD01-1, HNDD01-2, **HNDD01-3**	KF907712
	B. oleraceae (White cabbage)	CTCR01-1, CTCR01-2, **CTCR01-3**, **CTCR02-1**, **CTCR02-2**, CTCR02-3, CTCR03-1, CTCR03-2	KF907713, KF907714, KF907715
1-IB	*B. chinensis* (Chinese cabbage)	LDDT03-1	
	B. oleraceae (Broccoli)	LDDL01-1, **LDDL01-2**, LDDL01-3, LDDL01-4	KF907716
	B. oleraceae (White cabbage)	**LDDL04-1**, LDDL04-2, LDDL04-3, LDDL04-4, LDDL04-5, **LDDL05-1**, LDDL05-2, **LDDL05-3**	KF907717, KF907718, KF907719
1-ID	*B. parachinensis* (Chinese flowering cabbage)	**DNBH03-1, DNBH03-2, DNBH03-3**, DNBH03-4, DNBH03-5, DNBH05-1-1, DNBH05-1-3, DNBH05-2-2, **DNBH05-3-1**, **DNBH05-4**	KF907720, KF907721, KF907722, KF907723, KF907724
		STST04-1, STST04-3	
	B. juncea (Mustard cabbage)	HGPH01-1, HGPH01-2, **HGPH01-3**, HGPH01-4	KF907725
		STST02-3	
1-IG	*B. parachinensis* (Chinese flowering cabbage)	**DNBH05-1-2**	KF907730
2-2	*B. parachinensis* (Chinese flowering cabbage)	**HNTT01-1**	KF907726
	B. oleraceae (Turnip cabbage)	**HNDA01-1**, HNDA01-2, **HNDA01-3**, **HNDA01-4**	KF907727, KF907728, KF907729
4-HGI	*B. chinensis* (Chinese cabbage)	LDDT01-1, LDDT01-2, LDDL02-2	KF907731
	B. parachinensis (Chinese flowering cabbage)	DNBH05-2-1, DNBH05-3-2	
		STST01-1, **STST03-2**	KF907732
	B. juncea (Mustard cabbage)	DNBH02-1	
	B. oleraceae (Turnip cabbage)	**HNDD01-4**	KF907733
	B. oleraceae (White cabbage)	LDDT02-1, LDDT02-2, LDDT02-3	
7	*B. oleraceae* (Turnip cabbage)	**HNDA02-1**	KF907734
A	*B. chinensis* (Pak choi)	STMX01-3	
	B. oleraceae (White cabbage)	**LDDL03-1**, LDDL03-2	KF907735
Fc	*B. chinensis* (Chinese cabbage)	**LDDL02-1**	KF907736

[a]The first two letters represent provinces in which the samples were collected (i.e. CT: Can Tho, VL: Vinh Long, HG: Hau Giang, ST: Soc Trang, DN: Dong Nai, LD: Lam Dong and HN: Ha Noi).
[b]Isolates in bold (unique sequences) are submitted to Genbank.

upright position in a growth chamber (22°C, 12 h light). The disease severity was recorded for root and hypocotyl or for leaves after six days of incubation using the following disease scale: 0 = healthy, no symptoms; 1 = lesions covering less than 25% of the root, hypocotyl or leaf surface; 2 = lesions covering between 25% and 50% of the root, hypocotyl or leaf surface; 3 = wilted plant with lesions covering between 50% and 75% of the root, hypocotyl or leaf surface; 4 = lesions covering more than 75% of root,

Table 3. Multinucleate and binucleate *Rhizoctonia* isolates derived from Genbank included in the phylogenetic analysis for comparison.

AG/Subgroup	Isolate	Host plant	Origin	Genbank accession number	Reference
1-IA	L31-1, L66-1, L73, L59, L38, L52, L62-1	Rice	Vietnam	EF206342, EF429208, EF429211, EF429212, EF429210, EF429209, EF429207	unpublished
	RM61	Water spinach	Vietnam	EF429216	unpublished
	LB71	Water hyacinth	Vietnam	EF429215	unpublished
	DP38	Peanut	Vietnam	EF429214	unpublished
	CLV72-2	Barnyard grass	Vietnam	EF429213	unpublished
	BV71-2, BV61-2, BV50-1	Cotton	Vietnam	EF429206, EF429205, EF206341	unpublished
	CC72	Bermuda grass	Vietnam	EF429204	unpublished
	B34-1	Corn	Vietnam	EF429203	unpublished
1-IG	RMPG28	Chickpea	India	JF701750	[29]
1-ID	BV62-1, BV61-6, BV61-5, BV61-4, BV61-1	Cotton	Vietnam	EF197803, EF197804, EF197802, EF197801, EF197800	unpublished
	SR61, SR650	Durian	Vietnam	EF197798, EF197797	[30]
	B61-1	Corn	Vietnam	EF197796	unpublished
	CCD61-1	Sugar beet	Vietnam	EF197799	unpublished
4-HGI	XL4	Cauliflower	Vietnam	EF203247	unpublished
	CB63, CB34-2	Cabbage	Vietnam	EF203251, EF203245	unpublished
	CP50-2	Coffee	Vietnam	EF203250	unpublished
	BV68-1, BV68-2	Cotton	Vietnam	EF203249, EF203248	unpublished
	KT63-1	Potato	Vietnam	EF203246	unpublished
Fc	BS-YT-06-5-14, YT, BS-J-06-6-3, DL-jiang-06-2-4, DL-YT-06-4-10, DL-YT-06-4-9, DL-YT-06-3-4	Taro, Ginger	China	HM623619, HM623631, HM623615, HM623622, HM623625, HM623624, HM623623	unpublished

AG/Subgroup	Isolate	Disease Index						
		White cabbage	Chinese cabbage	Pak choi	Mustard cabbage	Chinese flowering cabbage	Rice	Water spinach
Control		0.00 a	0.00 a	0.00 a	0.00 a	0.00 a	0.00 a	0.00 a
1-IA	STMX04-2	4.00 d	2.63 d	1.79 d	1.79 c	2.17 e	3.92 d	3.21 f
1-IB	LDDL05-3	3.83 c	0.00 a	4.00 e	3.92 e	3.96 g	1.96 c	2.83 ef
1-ID	DNBH05-4	4.00 d	0.00 a	0.42 b	1.54 c	1.67 d	2.38 c	3.54 f
1-IG	DNBH05-1-2	4.00 d	1.00 b	0.63 bc	1.04 b	1.04 c	1.04 b	2.29 de
2-2	HNDA01-1	3.83 c	0.00 a	0.33 b	0.08 a	0.04 ab	1.29 b	1.75 cd
4-HGI	STST01-1	3.88 cd	1.54 c	2.13 d	3.88 e	3.71 f	2.33 c	2.67 e
7	HNDA02-1	3.13 b	0.00 a	0.00 a	0.13 a	0.17 b	2.29 c	0.46 b
A	DNBH04-1	0.00 a	0.00 a	0.00 a	0.00 a	0.00 a	0.00 a	0.00 a
Fc	LDDL02-1	3.38 b	0.71 b	1.17 c	2.58 d	1.17 c	0.00 a	1.38 c

Figure 2. Aggressiveness of *Rhizoctonia* isolates towards detached leaves of white cabbage, Chinese cabbage, pak choi, mustard cabbage, Chinese flower cabbage, rice and water spinach. Leaves were scored using a scale ranging from 0 (no disease symptoms) to 4 (lesions covered more than 75% of leaf surface or dead leaf). For rapid visual evaluation of the data, a coloring scale with green ($0<DI\le1$), yellow ($1<DI\le2$), orange ($2<DI\le3$) and red ($3<DI\le4$) was used. The experiment was conducted twice and each treatment consisted of 12 leaves or leaf pieces. The data of the two experiments were pooled before Mann-Whitney comparisons were applied at $p=0.05$. Within columns, disease severities followed by the same letter are not significantly different.

AG/Subgroup	Isolate	Disease Index			
		Root		Leaf	
		White cabbage	Chinese cabbage	White cabbage	Chinese cabbage
Control		0.00 a	0.00 a	0.00 a	0.00 a
1-IA	STMX04-2	3.63 d	3.08 d	3.42 d	2.71 cd
1-IB	LDDL05-3	4.00 e	3.75 ef	4.00 e	3.54 d
1-ID	DNBH05-4	1.88 bc	1.04 b	2.17 c	0.63 b
1-IG	DNBH05-1-2	4.00 e	4.00 g	4.00 e	4.00 e
2-2	HNDA01-1	4.00 e	3.96 fg	4.00 e	3.21 d
4-HGI	STST01-1	4.00 e	4.00 g	4.00 e	4.00 e
7	HNDA02-1	2.50 c	1.83 c	2.29 c	1.58 c
A	DNBH04-1	1.17 b	1.42 bc	0.29 b	0.00 a
Fc	LDDL02-1	3.75 de	3.42 de	3.83 de	3.50 d

Figure 3. Pathogenic potential of *Rhizoctonia* isolates on seedlings of white cabbage and Chinese cabbage in *in vitro* bio-assays. Disease severity was assessed on a scale ranging from 0 (no symptoms) to 4 (lesions covering more than 75% of root, hypocotyl or leaf surface or dead plant). For rapid visual evaluation of the data, a coloring scale with green (0<DI≤1), yellow (1<DI≤2), orange (2<DI≤3) and red (3<DI≤4) was used. The experiment was done twice with 12 seedlings maintained in two square Petri plates for one treatment. The data of the two experiments were pooled before Mann-Whitney comparisons were applied at p = 0.05. Within columns, disease severities followed by the same letter are not significantly different.

hypocotyl or leaf surface or dead plant. In this test, a complete randomized design was applied with two Petri dishes (six seedlings each) per treatment and this experiment was done twice.

In vivo pathogenic potential of *Rhizoctonia* spp. on roots and hypocotyls of white cabbage and Chinese cabbage

Surface-sterilized seeds of white cabbage and Chinese cabbage were germinated on wet filter paper in Petri dishes at 22°C one day before sowing into 600 g of soil substrate. Four days after sowing, each perforated plastic box (22×15×6 cm) with six seedlings was inoculated by placing a row of 12 *Rhizoctonia*-colonized wheat kernels in the middle of the box. The kernels used for inoculation had equivalent sizes and were produced as described previously. Control seedlings were similarly treated with sterile wheat kernels. All plants were incubated at 22°C. Disease severity on root and hypocotyl was evaluated 14 days after inoculation using the same disease scale described for the *in vitro* experiment. A completely randomized design was used with 12 seedlings cultivated in two experimental boxes per treatment and this test was performed twice with the same nine isolates that were used in the previous tests.

AG/Subgroup	Isolate	Disease Index	
		White cabbage	Chinese cabbage
Control		0.00 a	0.00 a
1-IA	STMX04-2	0.27 b	0.21 ab
1-IB	LDDL05-3	1.57 c	0.29 bc
1-ID	DNBH05-4	0.00 a	0.08 ab
1-IG	DNBH05-1-2	0.00 a	0.13 ab
2-2	HNDA01-1	3.50 e	1.57 d
4-HGI	STST01-1	3.63 e	3.08 e
7	HNDA02-1	0.26 b	0.79 c
A	DNBH04-1	0.00 a	0.00 a
Fc	LDDL02-1	2.33 d	2.96 e

Figure 4. Pathogenic potential of *Rhizoctonia* isolates on roots and hypocotyls of white cabbage and Chinese cabbage in *in vivo* experiment. Disease severity on roots was assessed on a scale ranging from 0 (no symptoms) to 4 (seedling dead). For rapid visual evaluation of the data, a coloring scale with green (0<DI≤1), yellow (1<DI≤2), orange (2<DI≤3) and red (3<DI≤4) was used. The experiment was performed twice with 12 seedlings cultivated in two plastic boxes per treatment. The data of the two experiments were pooled before Mann-Whitney comparisons were applied at p = 0.05. Within columns, disease severities followed by the same letter are not significantly different.

Statistical analysis

The severity of *Rhizoctonia* diseases on roots and leaves of *Brassica* seedlings are presented in Figures 2–4 and Tables S3–S5 in File S1 as Disease index (DI). DI was calculated using the following formula:

$$DI = \frac{\sum(\text{Disease class x number of plants within that class})}{\text{Total number of plants within treatment}}$$

Pathogenicity data for the two experiments with nine isolates of nine different AGs were always very similar and no significant interaction was found between the experiments. Therefore, statistical analysis was done on pooled data for the different repeats. The non-parametric Kruskal-Wallis test for k independent samples was used, after which pair-wise comparisons were performed for all treatments using Mann-Whitney tests at a confidence level of p = 0.05.

To determine the correlation between the distributions of AGs and sampling locations and between the distributions of AGs and sampled *Brassica* spp., contingency tables were constructed using Excel. Then, potential significant differences between the variables were revealed with Fisher's Exact Tests. All statistical analyses were conducted in SPSS 22.0 (SPSSinc, Illinois, USA).

Table 4. Pairwise sequence similarities of unknown isolates LDDL02-1 and DNBH05-1-2 to all known AGs from the curated database in Table S2 in File S2.

	LDDL02-1	DNBH05-1-2
AG 1-IA	0.88	0.92
AG 1-IB	0.84–0.87	0.84–0.87
AG 1-IC	0.89–0.90	0.89–0.90
AG 1-ID	0.85–0.86	0.85
AG 1-IE	0.89	0.94
AG 1-IF	0.84	0.85
AG 2-1	0.87–0.90	0.82–0.85
AG 2-2	0.86–0.87	0.84–0.85
AG 2-3	0.89–0.90	0.85
AG 3	0.89–0.90	0.84–0.85
AG 4	0.86–0.88	0.85–0.88
AG 5	0.91	0.86–0.87
AG 6	0.89–0.94	0.88–0.90
AG 7	0.91	0.89
AG 8	0.92–0.93	0.89–0.90
AG 9	0.9	0.86
AG 10	0.89–0.90	0.84–0.85
AG 11	0.88–0.89	0.84
AG 12	0.89–0.90	0.87–0.88
AG 13	0.90	0.90
AG 2-BI	0.85–0.86	0.81–0.82
AG-A	0.84	0.84
AG-K	0.84	0.84
AG-Bb	0.79–0.80	0.78–0.79
AG-Q	0.80	0.79
AG-Bo	0.83	0.83
AG-Ba	0.83–0.84	0.83–0.84
AG-C	0.80–0.82	0.82
AG-H	0.81–0.82	0.81
AG-I	0.81	0.80–0.81
AG-D	0.78–0.81	0.77–0.79
AG-G	0.85	0.85
AG-L	0.85	0.85–0.86
AG-O	0.86	0.86
AG-Fb	0.93	0.89
AG-P	0.86–0.89	0.86–0.90
AG-R	0.88	0.85
AG-S	0.88	0.88
AG-Fa	0.91	0.91
AG-E	0.90–0.91	0.89
UNR1	0.82–0.83	0.84
UNR2	0.80	0.82
AG-N	0.67	0.66
W. circinata	0.66–0.67	0.64–0.66

LDDL02-1 shows most similarity to AG 6 and AG-Fb. DNBH05-1-2 shows highest pairwise sequence similarity to AG 1-IA and AG 1-IE.

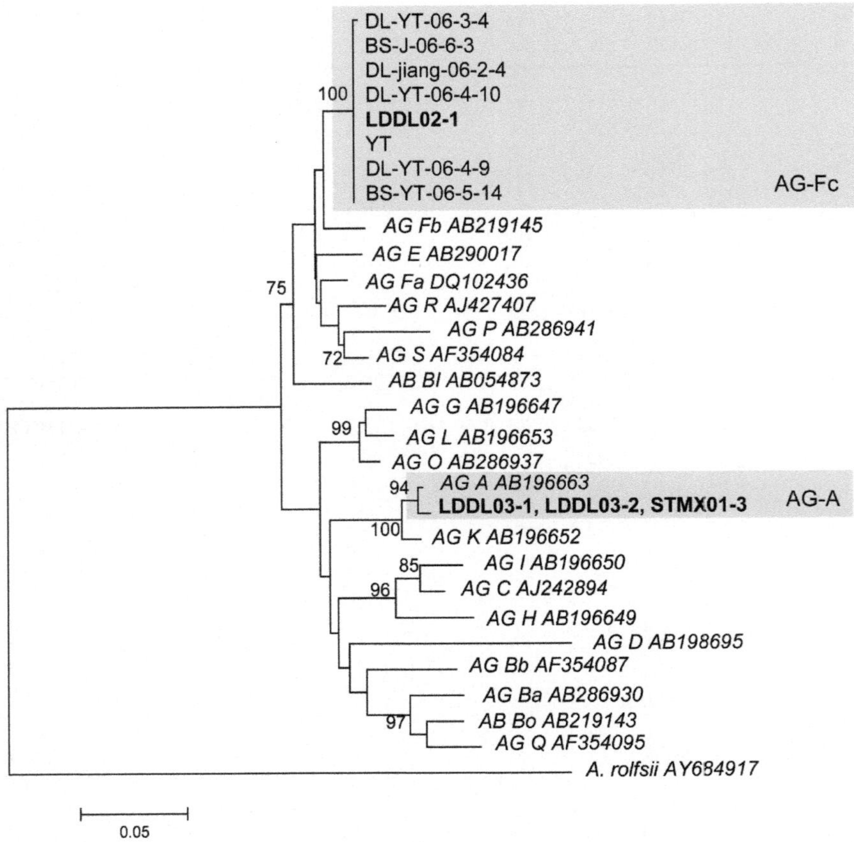

Figure 5. rDNA-ITS phylogeny of binucleate *Rhizoctonia* spp. sampled from *Brassica* spp. in Vietnam. Neighbour joining tree derived from the alignment of 31 binucleate *Rhizoctonia* isolates and the outgroup *Athelia rolfsii* (AY684917). Isolates in bold are the 4 isolates derived from *Brassica* spp. in Vietnam during this study. For each known binucleate *Rhizoctonia* AG, a representative isolate (in italics) from the curated database (Table S2 in File S2) is included. Bootstraps are only given for those branches with bootstrap support higher than 70. The tree was made using only isolates with unique sequences. Isolates with identical sequences were added afterwards on the same line.

Results

Molecular characterization and phylogenetic analysis of *Rhizoctonia* isolates

A total of 142 *Brassica* plants with *Rhizoctonia*-like symptoms were sampled in various important vegetable producing regions in Vietnam (see Figure 1). Of the 97 *Rhizoctonia* isolates recovered, only four were binucleate with two nuclei per hyphal cell. The other 93 isolates had multinucleate cells (data not shown).

Analysis of the rDNA-ITS region using the BLASTn tool (against Genbank and against the curated database) revealed that three binucleate *Rhizoctonia* isolates belonged to AG-A while the fourth isolate (LDDL02-1) could not be assigned to any known AG.

Pairwise sequence similarity scores of isolate LDDL02-1 against all isolates in the curated database (representing all known AGs) were determined. Highest pairwise sequence similarities were found with isolates of multinucleate AG 6 (94%) and binucleate AG-Fb (93%) (Table 4). When blasting to Genbank, several isolates from taro and ginger from Yunnan Province in China were found to be nearly identical to isolate LDDL02-1.

From the binucleate phylogenetic tree (Figure 5), it is clear that the unknown isolate LDDL02-1 clusters together with the isolates from Yunnan province, forming a clade with high bootstrap support that is different from any known binucleate AG, with the closest related AG being AG-Fb. So far, nothing has been published about the Chinese isolates. Pairwise sequence similarity within AG-F is 90–100% [2], and the closest related AGs are AG-Fb (93%) and AG-Fa (91%). Therefore, we propose to assign these isolates as a new AG-F subclade, namely AG-Fc.

In the multinucleate tree, all our Vietnamese isolates form clades with high bootstrap supports together with representative isolates from the curated database (Figure 6). The largest group belonged to AG 1-IA (44 isolates), followed by AG 1-ID (17 isolates), AG 1-IB (13 isolates) and AG 4-HGI (12 isolates). Five isolates belonged to AG 2-2. AG 2-2 is further divided into four groups: AG 2-2 IV, AG 2-2 LP, AG 2-2 IIIB and AG 2-2 WB. The similarity scores between our isolates and AG 2-2 IIIB and AG 2-2 LP were comparable (between 97 and 98%). Also, our isolates show 99% similarity with isolate Barranca (DQ452119), which is assigned as AG 2-2 WB by Godoy-Lutz et al. [25]. However, with other isolates from AG 2-2 WB (DQ452111-114), pairwise sequence similarity is lower (between 96 and 97%). Therefore, we decided that it is impossible at this point to assign our isolates to any of the AG 2-2 phylogenetic subgroups. One isolate belonged to AG 7 and the last isolate (DNBH05-1-2) could not be assigned to any of the known multinucleate AG groups in our reference database. The isolate showed the highest pairwise sequence similarity with AG 1-IE (94%, see Table 4). A BLASTn search on Genbank identified an isolate from chickpea in India (RMPG28, [29]) assigned to AG 2-3 to be>99% similar (1 bp substitution) to DNBH05-1-2. However, when comparing these

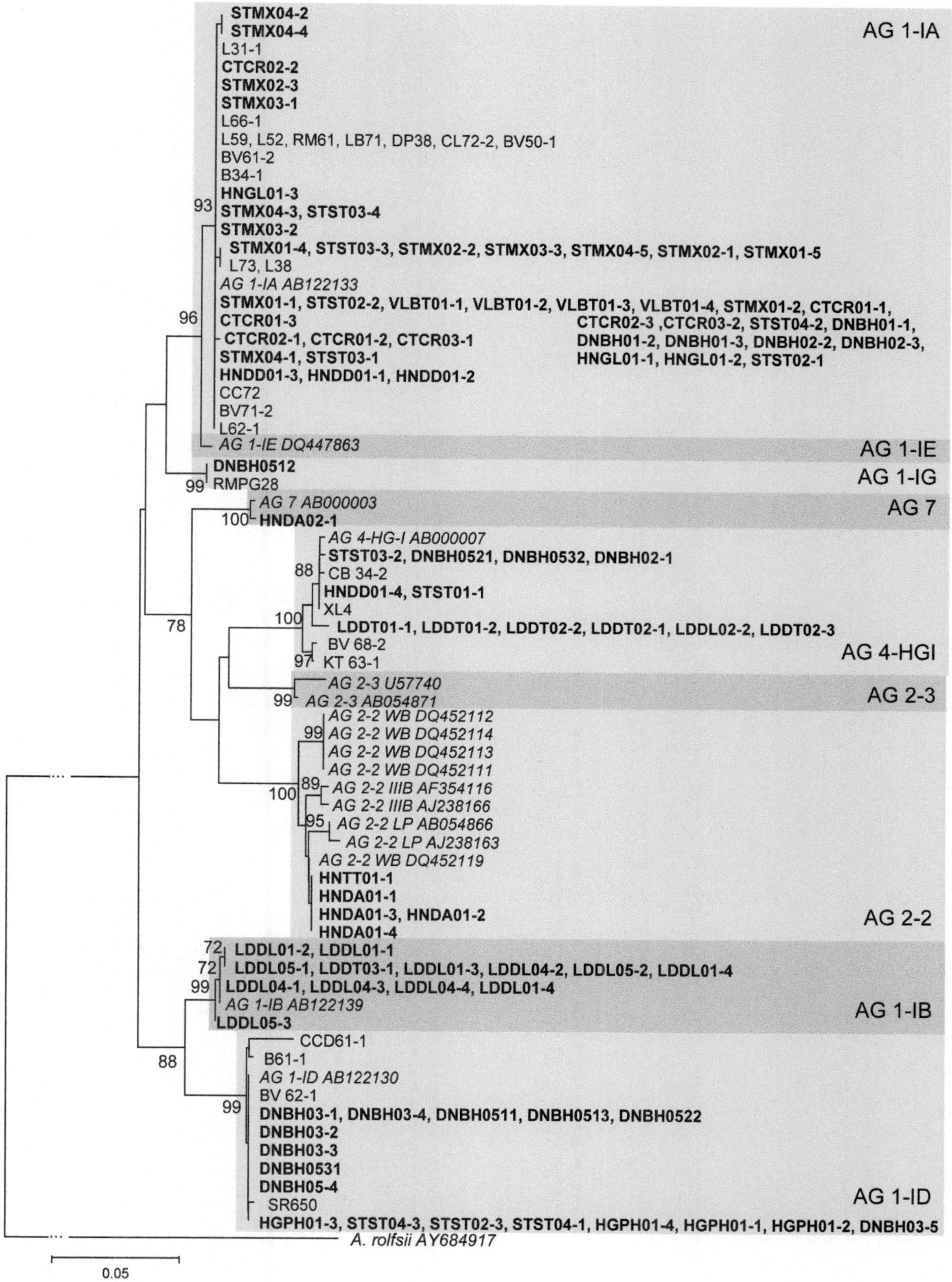

Figure 6. rDNA-ITS phylogeny of multinucleate *Rhizoctonia* spp. sampled from *Brassica* spp. in Vietnam. Neighbour joining tree derived from the alignment of 128 multinucleate *Rhizoctonia* isolates and the outgroup *Athelia rolfsii* (AY684917). Isolates in bold are the isolates derived from *Brassica* spp. in Vietnam during this study. For each of the multinucleate *Rhizoctonia* AG subgroups present in our sampling, representative isolates (in italics) from the curated database (Table S2 in File S2) are included. Bootstraps are only given for those branches with bootstrap support higher than 70. The tree was made using only isolates with unique sequences. Isolates with identical sequences were added afterwards on the same line. Only half of the length of the outgroup branch is shown to increase clarity.

Table 5. Contingency table with observed and expected frequencies of anastomosis groups (AGs) of *Rhizoctonia* spp. obtained from *Brassica* fields in different provinces of Vietnam.

Province	AG/Subset									Total/province
	Multinucleate *Rhizoctonia*							Binucleate *Rhizoctonia*		
	1-IA	1-IB	1-ID	2-2	1-IG	4-HGI	7	A	Fc	
Ha Noi	6 (5.90)	0 (1.74)	0 (2.28)	5 (0.67)	0 (0.13)	1 (1.61)	1 (0.13)	0 (0.40)	0 (0.13)	13
Lam Dong	0 (9.98)*	13 (2.95)*	0 (3.86)	0 (1.13)	0 (0.45)	6 (2.72)*	0 (0.23)	2 (0.68)	1 (0.23)	22
Dong Nai	5 (8.62)	0 (2.55)	10 (3.33)*	0 (0.98)	1 (0.20)	3 (2.35)	0 (0.20)	0 (0.59)	0 (0.20)	19
Vinh Long	4 (1.81)	0 (0.54)	0 (0.70)	0 (0.21)	0 (0.04)	0 (0.49)	0 (0.04)	0 (0.12)	0 (0.04)	4
Can Tho	8 (3.63)	0 (1.07)	0 (1.40)	0 (0.41)	0 (0.08)	0 (0.99)	0 (0.08)	0 (0.25)	0 (0.08)	8
Hau Giang	0 (1.81)	0 (0.54)	4 (0.70)*	0 (0.21)	0 (0.04)	0 (0.49)	0 (0.04)	0 (0.12)	0 (0.04)	4
Soc Trang	21 (12.25)*	0 (3.62)	3 (4.73)	0 (1.39)	0 (0.28)	2 (3.34)	0 (0.28)	1 (0.84)	0 (0.28)	27
Total/AG	44	13	17	5	1	12	1	3	1	97

Data show actual numbers of isolates collected among the different provinces. According to Fisher's exact test ($p = 0.05$), the AGs found are related to the sampling locations. An asterisk* indicates significant differences between observed and expected numbers. Values in parentheses represent the expected numbers.

Table 6. Contingency table with observed and expected frequencies of anastomosis groups (AGs) of *Rhizoctonia* spp. obtained from field-grown *Brassica* crops in Vietnam.

Host plant	AG/Subset[a]									Total/host plant
	Multinucleate *Rhizoctonia*							Binucleate *Rhizoctonia*		
	1-IA	1-IB	1-ID	2-2	1-IG	4-HGI	7	A	Fc	
Mustard cabbage	14 (9.07)	0 (2.68)	5 (3.51)	0 (1.03)	0 (0.21)	1 (2.47)	0 (0.21)	1 (0.62)	0 (0.21)	20
White cabbage	8 (9.53)	8 (2.81)*	0 (3.68)	0 (1.08)	0 (0.22)	3 (2.60)	0 (0.22)	2 (0.65)	0 (0.22)	21
Chinese flowering cabbage	9 (12.25)	0 (3.62)	12 (4.73)*	1 (1.39)	1 (0.28)	4 (3.34)	0 (0.28)	0 (0.84)	0 (0.28)	27
Chinese cabbage	0 (2.27)	1 (0.67)	0 (0.88)	0 (0.26)	0 (0.05)	3 (0.62)	0 (0.05)	0 (0.15)	1 (0.05)	5
Pak choi	10 (4.99)	0 (1.47)	0 (1.93)	0 (0.57)	0 (0.11)	0 (1.36)	0 (0.11)	1 (0.34)	0 (0.11)	11
Broccoli	0 (1.81)	4 (0.54)*	0 (0.70)	0 (0.21)	0 (0.04)	0 (0.49)	0 (0.04)	0 (0.12)	0 (0.04)	4
Turnip cabbage	3 (4.08)	0 (1.21)	0 (1.58)	4 (0.46)*	0 (0.09)	1 (1.11)	1 (0.09)	0 (0.28)	0 (0.09)	9
Total/AG	44	13	17	5	1	12	1	3	1	97

Data show actual numbers of isolates collected among the different *Brassica* crops. According to Fisher's exact test ($p = 0.05$), the AGs found are related to the crops. An asterisk* indicates significant differences between observed and expected numbers. Values in parentheses represent the expected numbers.

isolates to the AG 2–3 isolates in the curated database, we found low pairwise sequence similarity (85%) (Table 4). Also in the phylogenetic tree (see Figure 6), isolate DNBH05-1-2 clusters together with isolate RMPG28, but not with the representative AG 2–3 isolates from the curated database, indicating that isolate RMPG28 is wrongly designated as AG 2–3 in Genbank. Hence, these isolates represent a new subgroup of AG 1 and we propose the name AG 1-IG.

For most of the AGs we detected in Vietnam there was very little variation in ITS sequence, except for AG 4-HGI where six variable positions were found between the isolates from Lam Dong and the isolates from the other three provinces where this AG was detected (Soc Trang, Dong Nai and Ha Noi).

The AG 1-IA isolates showed identical or nearly identical (max. 4 SNPs) sequences to AG 1-IA isolates previously isolated from rice, water spinach, water hyacinth and other crops in Vietnam (Table 3). The AG 1-ID isolates showed high similarity (identical sequence or one SNP) to isolates detected on coffee, cotton, durian [30], corn and sugar beet (Table 3) in Vietnam.

Relationship between the AGs found and sampling locations and between AGs found and *Brassica* species

There seems to be a relationship between the AGs present and the sampling areas (Figure 1 and Table 5). Thirty three out of 44 isolates belonging to AG 1-IA were recovered from samples collected from Soc Trang, Can Tho and Vinh Long in the Mekong River delta, the main rice production region of Vietnam. On the other hand, no AG 1-IA isolates were detected in Lam Dong although the expected frequency of this group in this province was high (9.98). The observed numbers were also significantly higher than the expected numbers for AG 1-IB and AG 4-HGI in Lam Dong as well as for AG 1-ID in Hau Giang and Dong Nai.

The relationship between the occurrence of different AGs and the host plants is presented in Table 6. Significant AG-host correlations were identified in the following combinations: AG 1-IB and white cabbage, AG 1-IB and broccoli, AG 1-ID and Chinese flowering cabbage, and AG 2-2 and turnip cabbage.

Aggressiveness of *Rhizoctonia* isolates towards detached leaves

Nine *Rhizoctonia* isolates, randomly selected from each identified AG, were tested for their pathogenicity on detached leaves. Due to practical reasons, experiments with *Brassica* crops were conducted in a growth chamber specifically built for *Brassica* spp. (22°C, RH = 60%, 12 h photoperiod) and experiments with rice and water spinach were done in a growth chamber specifically built for rice (28°C, RH = 60%, 16 h photoperiod). Although the temperature and humidity in these chambers are slightly lower than those of Vietnam, they are still suitable for the growth of host plants and *Rhizoctonia* isolates involved in our study.

As shown in Figure 2, leaves of white cabbage, Chinese cabbage, pak choi, mustard cabbage, Chinese flowering cabbage, rice, and water spinach responded differently to the infection of *Rhizoctonia* spp. although each host was affected by at least four AGs. For each plant species, a wide variation in symptom severity induced by different AGs was observed and the only AG that could not cause disease on any of the plants tested was AG-A. White cabbage leaves were most severely infected by all AGs (except AG-A) with a DI varying from 3.13 to 4.00. Compared to other hosts, Chinese cabbage appeared to be most resistant to *Rhizoctonia* isolates. No disease symptoms were observed on Chinese cabbage leaves challenged with AG 1-IB, AG 1-ID, AG 2-2 and AG 7, while other AGs were weak to moderately aggressive (DI varied from 0.71 in AG-Fc to 2.63 in AG 1-IA). Inoculation of pak choi with AG 1-IB resulted in a complete decay of all leaf discs (DI = 4.00). For mustard cabbage and Chinese flowering cabbage, a strong disease pattern was obtained on leaves confronted with AG 1-IB (DI≥3.92) and AG 4-HGI (DI≥3.71). Rice leaves were very susceptible to AG 1-IA and they were severely destroyed within four days of inoculation (DI = 3.92). Disease induced by other *R. solani* isolates on rice was also significantly different from the control although the two binucleate *Rhizoctonia* isolates tested were not pathogenic on rice. For water spinach, a wide variation in aggressiveness of *R. solani* isolates was observed, resulting in DI values ranging from 0.46 to 3.54. Large lesions (DI≥3.21) developed on water spinach leaves inoculated with AG 1-IA and AG 1-ID isolates. The difference in pathogenicity of these AGs towards the different *Brassica* species was confirmed when additional isolates from AGs 1-IA, 1-IB, 1-ID, 2-2, 4-HGI and A were tested in a detached leaf assay on the same series of plants (see supplementary information, Table S3 in File S1).

In vitro pathogenic potential of *Rhizoctonia* spp. isolates on seedlings of white cabbage and Chinese cabbage

The results of the detached leaf assay revealed that white cabbage leaves were remarkably more susceptible than other cabbage species, while only some AGs could attack the leaves of Chinese cabbage. Therefore, these two plant species were selected as hosts for *in vitro* and *in vivo* pathogenicity tests using the same nine *Rhizoctonia* isolates as mentioned above. Under *in vitro* conditions (Figure 3), all AGs could cause disease and symptoms were observed on both roots and leaves. Isolates of AG 1-IA, AG 1-IB, AG 2-2, AG 1-IG, AG 4-HGI and AG-Fc were most aggressive towards these hosts (for white cabbage: DI on roots ≥ 3.63 and DI on leaves ≥3.42; for Chinese cabbage: DI on roots ≥ 3.08 and DI on leaves ≥2.71). Fewer symptoms were recorded on seedlings inoculated with isolates belonging to AG 1-ID, AG 7 and AG-A. The virulence of AG 1-IA, AG 1-IB, AG 2-2 and AG 4-HGI towards white cabbage and Chinese cabbage was confirmed when *in vitro* assays were repeated with additional isolates of these AGs (Table S4 in File S1). Towards white cabbage, DI on roots varied from 2.17 to 4.00 and DI on leaves fluctuated between 2.33 and 4.00. DI on roots and leaves of Chinese cabbage ranged from 2.08 to 4.00 and from 1.92 to 4.00, respectively.

In vivo pathogenic potential of *Rhizoctonia* spp. isolates on roots and hypocotyls of white cabbage and Chinese cabbage

The results displayed in Figure 4 demonstrate that severe *Rhizoctonia*-induced damage on white cabbage roots could only be seen for AG 2-2 (DI = 3.50) and AG 4-HGI (DI = 3.63). Moderate infection (DI = 2.33) was observed in response to AG-Fc. Towards Chinese cabbage, severe disease symptoms were incited by AG 4-HGI (DI = 3.08) and AG-Fc (DI = 2.96). No symptoms were detected on roots and hypocotyls of seedlings challenged with AG-A. In addition, the inoculation of isolates belonging to AG 1-IA, AG 1-IB, AG 1-ID, AG 1-IG and AG 7 did not result in the formation of large lesions on roots of white cabbage and Chinese cabbage seedlings (DI≤1.57). Data obtained from the tests conducted with additional *Rhizoctonia* isolates belonging to AGs 1-IA, 1-IB, 1-ID, 2-2, 4-HGI and AG-A are presented in Table S5 in File S1. These data confirm the high aggressiveness of AG 2-2 on white cabbage (DI = 4.00) and that of AG 4-HGI on both white cabbage and Chinese cabbage (DI≥3.50).

Discussion

Rhizoctonia is an important fungal 'form genus' occurring worldwide and including many important plant pathogenic strains as well as mycorrhizal fungi and hypovirulent or avirulent strains among which there are strains that are capable to protect plants against pathogenic *Rhizoctonia* and other pathogens as well as increase plant growth. Plants can be infected by different *Rhizoctonia* AGs from the time of sowing resulting in the development of both foliar and root diseases. This is the first time *Rhizoctonia* species that attack field-grown *Brassica* crops in Vietnam were isolated and characterized. Ninety seven isolates of *Rhizoctonia* were recovered from symptomatic plant tissues of seven brassicaceous hosts (mustard cabbage, white cabbage, Chinese flowering cabbage, pak choi, turnip cabbage, Chinese cabbage and broccoli). Of all the isolates collected, 4% were binucleate *Rhizoctonia* and 96% were multinucleate *Rhizoctonia*. Molecular characterization by sequencing of the rDNA-ITS region showed that the binucleate isolates found belonged to AG-A (3 isolates) and an unknown AG introduced here as AG-Fc (1 isolate). Comparison to a curated sequence database of all known *Rhizoctonia* multinucleate, binucleate and uninucleate AGs did not reveal high homology of this isolate to a known AG. Pairwise sequence similarities to all known multinucleate, binucleate and uninucleate *Rhizoctonia* AGs showed highest similarity to AG-Fb, but also to AG 6, which is a multinucleate AG. Close relationships of some binucleate groups with multinucleate groups has been previously noted by Sharon et al. [2] who did a combined ITS sequence analysis of multinucleate, binucleate and uninucleate groups. These authors stated that the clustering of binucleate and uninucleate groups close to certain multinucleate clusters may indicate a possible evolutionary bridge between multinucleate and binucleate groups and our results support this hypothesis.

We conducted pathogenicity assays on detached leaves, *in vitro* seedlings and plants grown *in vivo*. In general, pathogenicity was highest on roots and leaves under *in vitro* conditions, which is probably due to the high humidity and the young age of the plants in this system. It should be mentioned that we only checked *in vivo* pathogenicity towards roots and hypocotyls, but it should be borne in mind that some of our isolates are mainly leaf pathogens.

Although AG-A isolates were obtained from symptomatic plants, they were unable to induce disease on detached leaves or on seedlings *in vivo*, and were only slightly pathogenic on cabbage seedlings under *in vitro* conditions. This low virulence may be due to the differences in humidity between our experimental conditions and the field situation in Vietnam. Alternatively, AG-A isolates might be avirulent, but since they grow very fast *in vitro*, they may have masked the presence of slower growing, virulent AGs. In contrast to AG-A, AG-Fc appeared to be moderately to highly virulent in all experiments. This also concurs with previous studies that some binucleate *Rhizoctonia* species are highly virulent [31–33], whereas others are weakly virulent or avirulent [34–36].

Among the multinucleate AGs, AG 1-IA was the dominant group, followed by AG 1-ID, AG 1-IB, AG 4-HGI, AG 2-2, AG 7 and an unknown AG that we introduced here as AG 1-IG. The occurrence of AG 1-IB and AG 4 on *Brassica* spp. is well known from previous studies. According to Pannecoucque et al. [22], AG 1-IB is one of the causal agents of wirestem in Belgian cauliflower fields. AG 1-IB isolates are also highly virulent on lettuce in Belgium [37]. In our study, AG 1-IB isolates were only recovered from Lam Dong, a province in the cool Central Highlands of Vietnam and the main lettuce production region of Vietnam, suggesting that the presence of this AG is associated with cool climates. The presence of AG 4 has also been reported on *Brassica oleracea* [21] and *B. rapa* subsp. *chinensis* [38] in China, *B. oleracea* in the UK [23], *B. napus* L. and *B. campestris* L. in Canada [39], and *B. oleracea* [18] and *B. napus* L. in the US [40]. AG 4-HGI occurred in most regions sampled. This AG has a wide host range due to its ability to adapt to temperature variation and cropping patterns [41]. AG 4 isolates are able to induce disease on all plant parts and in our pathogenicity trials, the highest disease ratings on cabbage in all bioassays were shown for the AG 4-HGI isolates. For white cabbage, severe disease symptoms were observed on seedlings inoculated with AG 2-2 and AG 4-HGI isolates *in vivo* or on detached leaves challenged with isolates of all AGs. Although *R. solani* AG 2-1 is considered the most dominant and damaging anastomosis group attacking *Brassica* spp. [22,23,42], isolates belonging to this group were not detected in our survey.

The predominance of *R. solani* isolates of AG 1-IA and also the presence of isolates belonging to AG 1-ID in our collection were not anticipated because these AGs have not been described on *Brassica* crops before. The presence of unusual AGs in our sampling is probably due to three aspects: (i) alternative hosts of *Rhizoctonia* present in the sampling locations; (ii) poor cultural practices and (iii) high temperature. AG 1-IA isolates were mainly recovered from samples collected in provinces of the Mekong delta including Vinh Long, Can Tho and Soc Trang. This is a low-lying coastal region of Vietnam, characterized by high temperature and humidity and prone to flooding every rainy season. Due to water availability and soil type, the agricultural production in this area is dominated by rice [43,44]. As previously reported, sheath blight, caused by *R. solani* AG 1-IA, is a major disease of rice cultivated in intensive production systems [45–47]. With the ability to float and to survive in water [48], sclerotia of AG 1-IA isolates may easily spread from the rice paddy fields to vegetable fields through irrigation or flood water. The recovery of AG 1-IA in the hot region located in the South of Vietnam is consistent with the findings of Harikrishnan and Yang [41] that temperature can influence growth rate, sclerotia production of *Rhizoctonia* spp. and the distribution of *Rhizoctonia* isolates belonging to different anastomosis groups. AG 1 is a high temperature group [8] and its vegetative growth as well as sclerotia production and survival are inhibited at low temperatures. The occurrence of AG 1-IA on *Brassica* crops is probably even increased due to farmers' lack of knowledge about crop protection. AG 1 (not specific) has been considered as one of the causal agents of foliar diseases on water hyacinth (*Eichhornia crassipes*) [49], water lettuce (*Pistia stratioites*) and anchoring hyacinth (*E. azurea*) [50]. In this study, water spinach (*Ipomoea aquatic*) was found to be susceptible to the AG 1-IA isolates that we collected from *Brassica* spp. Vietnamese farmers commonly use these aquatic plants as cover materials in vegetable production and use irrigation water from sources where these plants are present, thus bringing the fungus from these alternative hosts to *Brassica* spp. (Figure 7). Isolates of AG 1-IA appeared to be very pathogenic towards leaves of cabbages but they could not induce severe disease on roots, especially under *in vivo* conditions. This finding is in agreement with results reported previously by Yang and Li [51] that AG 1-IA isolates have a tendency to attack aerial parts of plants. These data also support our hypothesis about the spread of AG 1-IA isolates from rice and water spinach to vegetables. In other words, rice and water spinach that are infected by *R. solani* could be an important source of inoculum that may contribute to the disease caused by AG 1-IA on *Brassica* spp.

In our assays AG 1-ID was mainly pathogenic on leaves of white cabbage. *Rhizoctonia* AG 1-ID was previously found to be

Figure 7. *Rhizoctonia*-infected water hyacinth is introduced as cover material to a white cabbage field in Vietnam. *Rhizoctonia*-infected water hyacinth is taken from a nearby water ditch and used as cover material on the white cabbage field. Via this common practice, Vietnamese farmers unintentionally introduce the *Rhizoctonia* fungus to their crops.

pathogenic on durian [30] and coffee [52]. Interestingly, ten out of 17 AG 1-ID isolates were collected from Dong Nai, a province where durian, coffee and cotton are widely cultivated [53], again suggesting a correlation between cropping patterns and *Rhizoctonia* distribution in Vietnam.

Collectively, it seems that the distribution of *Rhizoctonia* AGs in Vietnam is correlated with the cropping patterns and climatic conditions. However, it is difficult to draw a strong conclusion about the influence of cropping patterns and climatic conditions on the occurrence of *Rhizoctonia* spp. in Vietnam because the sampling regime for *R. solani* isolates comprises two variable parameters, namely the plant species infected by *R. solani* and the sampling site. Due to the variation in soil and climatic conditions, different *Brassica* species are grown in different geographic regions and there is a possibility that particular plant species might be more susceptible to infections by specific *R. solani* AGs than others. Therefore, another field survey with a systematic sampling regime needs to be conducted to confirm our hypothesis.

Our research also points towards the need to have good extension programs to improve the farmers' knowledge about crop protection. Additionally, knowing which AGs are responsible for *Rhizoctonia* diseases on *Brassica* spp. in Vietnam is an essential prerequisite for developing successful disease management strategies in this country.

Supporting Information

File S1 Contains the following files: Table S1. GPS co-ordinates of the wards in each province of Vietnam where *Rhizoctonia*-infected *Brassica* crops were sampled. Table S3. Aggressiveness of *Rhizoctonia* isolates towards white cabbage, Chinese cabbage, pak choi, mustard cabbage, Chinese flower cabbage, rice and water spinach in detached leaf bio-assays. Leaves were scored using a scale ranging from 0 (no disease symptoms) to 4 (lesions covered more than 75% of leaf surface or dead leaf). For rapid visual evaluation of the data, a coloring scale with green ($0<DI\leq1$), yellow ($1<DI\leq2$), orange ($2<DI\leq3$) and red ($3<DI\leq4$) was used. The test was done once with 12 leaves or leaf pieces per treatment. All data were statistically analyzed and within columns, disease severities followed by the same letter are not significantly different. Table S4. Aggressiveness of *Rhizoctonia* isolates towards roots and leaves of white cabbage and Chinese cabbage seedlings in *in vitro* bio-assays. Disease severity on roots or leaves was assessed on a scale ranging from 0 (no symptoms) to 4 (lesions covering more than 75% of root, hypocotyl or leaf surface or dead plant). For rapid visual evaluation of the data, a coloring scale with green ($0<DI\leq1$), yellow ($1<DI\leq2$), orange ($2<DI\leq3$) and red ($3<DI\leq4$) was used. Experiment was conducted once with 12 seedlings maintained in two square Petri plates for one treatment. All data were statistically analyzed and within columns, disease severities followed by the same letter are not significantly different. Table S5. Aggressiveness of *Rhizoctonia* isolates towards roots of white cabbage and Chinese cabbage seedlings in in planta experiment. Disease severity on roots was assessed on a scale ranging from 0 (no symptoms) to 4 (seedling dead). For rapid visual evaluation of the data, a coloring scale with green ($0<DI\leq1$), yellow ($1<DI\leq2$), orange ($2<DI\leq3$) and red ($3<DI\leq4$) was used. Experiment was performed once. Each treatment consisted of 12 seedlings cultivated in two plastic boxes. Data were statistically analyzed and within columns, disease severities followed by the same letter are not significantly different.

File S2 Table S2. Curated database of sequences containing representative rDNA-ITS sequences of all known uninucleate, binucleate and multinucleate *Rhizoctonia* AG and subgroups.

Acknowledgments

The authors wish to thank Dr. Tran Thi Thu Thuy and Dr. Nguyen Thi Thu Nga (Can Tho University) and Dr. Ha Viet Cuong (Ha Noi University of Agriculture) for helpful advice and Dr. Michal Sharon for providing the rDNA-ITS sequence alignment of UNR, BNR and MNR reference isolates.

Author Contributions

Conceived and designed the experiments: GKHH MH. Performed the experiments: GKHH SS. Analyzed the data: GKHH LB MH. Wrote the paper: GKHH LB MH.

References

1. Vietnam trade promotion agency (2008) Report on Vietnamese vegetable and fruit sector. Available: http://www.aseankorea.org/aseanZone/downloadFile2.asp?boa_filenum=1575. Accessed 2014 Jan 24.
2. Sharon M, Kuninaga S, Hyakumachi M, Naito S, Sneh B (2008) Classification of *Rhizoctonia* spp. using rDNA-ITS sequence analysis supports the genetic basis of the classical anastomosis grouping. Mycoscience 49: 93–114.
3. Sharon M, Kuninaga S, Hyakumachi M, Sneh B (2006) The advancing identification and classification of *Rhizoctonia* spp. using molecular and biotechnological methods compared with the classical anastomosis grouping. Mycoscience 47: 299–316.
4. Carling DE, Baird RE, Gitaitis RD, Brainard KA, Kuninaga S (2002) Characterization of AG-13, a newly reported anastomosis group of *Rhizoctonia solani*. Phytopathology 92: 893–899.
5. Carling DE, Pope EJ, Brainard KA, Carter DA (1999) Characterization of mycorrhizal isolates of *Rhizoctonia solani* from an orchid, including AG-12, a new anastomosis group. Phytopathology 89: 942–946.
6. Carling DE, Kuninaga S, Brainard KA (2002) Hyphal anastomosis reactions, rDNA-internal transcribed spacer sequences, and virulence levels among subsets of *Rhizoctonia solani* anastomosis group-2 (AG-2) and AG-BI. Phytopathology 92: 43–50.

7. Hyakumachi M, Priyatmojo A, Kubota M, Fukui H (2005) New anastomosis groups, AG-T and AG-U, of binucleate *Rhizoctonia* spp. causing root and stem rot of cut-flower and miniature roses. Phytopathology 95: 784–792.
8. Sneh B, Burpee L, Ogoshi A (1991). Identification of *Rhizoctonia* species. St. Paul Minnesota: APS Press. 133 p.
9. Schwartz HF, Gent DH, Franc GD, Harveson RM (2007) Dry bean-Rhizoctonia root rot. Available: http://www.scarab.msu.montana.edu/HpIPMSearch/AuthorSearch.exe?Complete=true&sort=asc. Accessed 2014 Jan 24.
10. Wharton P, Kirk W, Berry D, Snapp S (2007) Rhizoctonia stem canker and black scurf of potato. Available: http://www.potatodiseases.org/pdf/rhizoctonia-bulletin.pdf. Accessed 2014 Jan 24.
11. Keijer J (1996) The initial steps of the infection process in *Rhizoctonia solani*. In: Sneh B, Jabaji-Hare S, Neate S, Dijst G, editors. *Rhizoctonia* species: Taxonomy, molecular biology, ecology, pathology and disease control. Dordrecht: Springer Netherlands. pp. 149–162.
12. Verma PR (1996) Biology and control of *Rhizoctonia solani* on rapeseed: A review. Phytoprotection 77: 99–111.
13. Yang J, Kharbanda PD, Wang H (1996) Characterization, virulence, and genetic variation of *Rhizoctonia solani* AG-9 in Alberta. Plant Dis 80: 513–518.
14. Khangura RK, Barbetti MJ, Sweetingham MW (1999) Characterization and pathogenicity of *Rhizoctonia* species on canola. Plant Dis 83: 714–721.
15. Sayama A (2000) Occurrence of damping-off disease caused by *Rhizoctonia solani* AG 2-2-IIIB on cabbage plug seedlings. Annu Rep Soc Plant Protect N Jpn 51: 54–57.
16. Homma Y, Yamashita Y, Ishii M (1983) A new anastomosis group AG-7 of *Rhizoctonia solani* Kuhn from Japanese radish fields. Ann Phytopathol Soc Jpn 49: 184–190.
17. Keinath AP, Farnham MW (1997) Differential cultivars and criteria for evaluating resistance to *Rhizoctonia solani* in seedling *Brassica oleracea*. Plant Dis 81: 946–952.
18. Rollins PA, Keinath AP, Farnham MW (1999) Effect of inoculum type and anastomosis group of *Rhizoctonia solani* causing wirestem of cabbage seedlings in a controlled environment. Can J Plant Pathol 21: 119–124.
19. Paulitz TC, Okubara PA, Schillinger WF (2006) First report of damping-off of canola caused by *Rhizoctonia solani* AG 2-1 in Washington State. Plant Dis 90: 829–829.
20. Kuramae E, Buzeto A, Ciampi M, Souza N (2003) Identification of *Rhizoctonia solani* AG 1-IB in lettuce, AG 4 HG-I in tomato and melon, and AG 4 HG-III in broccoli and spinach, in Brazil. Eur J Plant Pathol 109: 391–395.
21. Yang GH, Chen JY, Pu WQ (2007) First report of head rot of cabbage and web blight of snap bean caused by *Rhizoctonia solani* AG-4 HGI. Plant Pathol 56: 351–351.
22. Pannecoucque J, Van Beneden S, Höfte M (2008) Characterization and pathogenicity of *Rhizoctonia* isolates associated with cauliflower in Belgium. Plant Pathol 57: 737–746.
23. Budge GE, Shaw MW, Lambourne C, Jennings P, Clayburn R, et al. (2009) Characterization and origin of infection of *Rhizoctonia solani* associated with *Brassica oleracea* crops in the UK. Plant Pathol 58: 1059–1070.
24. White TJ, Bruns TD, Lee S, Taylor J (1990) Amplification and direct sequencing of fungal ribosomal RNA genes for phylogenetics. In: Innis MA, Gelfland DH, Sninsky JJ, White TJ, editors. PCR protocols: A guide to methods and applications. New York: Academic Press. pp. 315–322.
25. Godoy-Lutz G, Kuninaga S, Steadman JR, Powers K (2008) Phylogenetic analysis of *Rhizoctonia solani* subgroups associated with web blight symptoms on common bean based on ITS-5.8S rDNA. J Gen Plant Pathol 74: 32–40.
26. Tamura K, Stecher G, Peterson D, Filipski A, Kumar S (2013) MEGA6: Molecular evolutionary genetics analysis version 6.0. Mol Biol Evol 30: 2725–2729.
27. Scholten OE, Panella LW, De Bock TSM, Lange W (2001) A greenhouse test for screening sugar beet (*Beta vulgaris*) for resistance to *Rhizoctonia solani*. Eur J Plant Pathol 107: 161–166.
28. Keijer J, Korsman MG, Dullemans AM, Houterman PM, De Bree J, et al. (1997) *In vitro* analysis of host plant specificity in *Rhizoctonia solani*. Plant Pathol 46: 659–669.
29. Dubey SC, Tripathi A, Upadhyay BK (2012) Molecular diversity analysis of *Rhizoctonia solani* isolates infecting various pulse crops in different agro-ecological regions of India. Folia Microbiol (Praha) 57: 513–524.
30. Thuan TTM, Tho N, Tuyen BC (2008) First report of *Rhizoctonia solani* subgroup AG 1-ID causing leaf blight on durian in Vietnam. Plant Dis 92: 648–648.
31. Martin B (1988) Identification, isolation, frequency, and pathogenicity of anastomosis groups of binucleate *Rhizoctonia* spp. from strawberry roots. Phytopathology 78: 379–384.
32. Demirci E, Eken C, Zengin H (2002) First report of *Rhizoctonia solani* and binucleate *Rhizoctonia* from Johnsongrass in Turkey. Plant Pathol 51: 391–391.
33. Babiker EM, Hulbert SH, Schroeder KL, Paulitz TC (2013) Evaluation of *Brassica* species for resistance to *Rhizoctonia solani* and binucleate *Rhizoctonia* (*Ceratobasidum* spp.) under controlled environment conditions. Eur J Plant Pathol 136: 763–772.
34. Herr LJ (1995) Biological control of *Rhizoctonia solani* by binucleate *Rhizoctonia* spp. and hypovirulent *R. solani* agents. Crop Prot 14: 179–186.
35. Poromarto SH, Nelson BD, Freeman TP (1998) Association of binucleate *Rhizoctonia* with soybean and mechanism of biocontrol of *Rhizoctonia solani*. Phytopathology 88: 1056–1067.
36. Ross RE, Keinath AP, Cubeta MA (1998) Biological control of wirestem on cabbage using binucleate *Rhizoctonia* spp. Crop Prot 17: 99–104.
37. Van Beneden S, Pannecoucque J, Debode J, De Backer G, Höfte M (2008) Characterisation of fungal pathogens causing basal rot of lettuce in Belgian greenhouses. Eur J Plant Pathol 124: 9–19.
38. Yang GH, Chen XQ, Chen HR, Naito S, Ogoshi A, et al. (2004) First report of foliar blight in *Brassica rapa* subsp. *chinensis* caused by *Rhizoctonia solani* AG-4. Plant Pathol 53: 260–260.
39. Yitbarek SM, Verma PR, Morrall RAA (1987) Anastomosis groups, pathogenicity, and specificity of *Rhizoctonia solani* isolates from seedling and adult rapeseed/canola plants and soils in Saskatchewan. Can J Plant Pathol 9: 6–13.
40. Baird RE (1996) First report of *Rhizoctonia solani* AG-4 on canola in Georgia. Plant Dis 80: 104–104.
41. Harikrishnan R, Yang XB (2004) Recovery of anastomosis groups of *Rhizoctonia solani* from different latitudinal positions and influence of temperatures on their growth and survival. Plant Dis 88: 817–823.
42. Ohkura M, Abawi GS, Smart CD, Hodge KT (2009) Diversity and aggressiveness of *Rhizoctonia solani* and *Rhizoctonia*-like fungi on vegetables in New York. Plant Dis 93: 615–624.
43. Ministry of Natural Resources and Environment Sub-Institute of Hydrometeorology and Environment of South Vietnam (2010) Climate change in Mekong delta: Climate scenario's, sea level rise, other effects. Available: http://wptest.partnersvoorwater.nl/wp.../CLIMATE-CHANGE-final-draft.pdf. Accessed 2014 Jan 24.
44. International Federation of Red Cross and Red Crescent Societies (2013) Emergency appeal final report Viet Nam: Mekong delta floods. Available: http://reliefweb.int/sites/reliefweb.int/files/resources/MDRVN009fr.pdf. Accessed 2014 Jan 24.
45. Lee FN, Rush MC (1983) Rice sheath blight: A major rice disease. Plant Dis 67: 829–832.
46. Ogoshi A (1987) Ecology and pathogenicity of anastomosis and intraspecific groups of *Rhizoctonia solani* Kuhn. Annu Rev Phytopathol 25: 125–143.
47. Taheri P, Gnanamanickam S, Höfte M (2007) Characterization, genetic structure, and pathogenicity of *Rhizoctonia* spp. associated with rice sheath diseases in India. Phytopathology 97: 373–383.
48. Hashiba T, Yamaguchi T, Mogi S (1972) Biological and ecological studies on the sclerotium of *Pellicularia sasakii* (Shirai) S. Ito. I. Floating on the water surface of sclerotium. Ann Phytopathol Soc Jpn 38: 414–425.
49. Freeman TE, Charudattan R, Cullen RE (1982) Rhizoctonia blight on waterhyacinth in the United States. Plant Dis 66: 861–862.
50. Zettler FW, Freeman TE (1972) Plant pathogens as biocontrols of aquatic weeds. Annu Rev Phytopathol 10: 455–470.
51. Yang G, Li C (2012) General description of *Rhizoctonia* species complex. In: Cumagun CJR, editor. Plant Pathology. Croatia: InTech. pp. 41–52.
52. Priyatmojo A, Escopalao VE, Tangonan NG, Pascual CB, Suga H, et al. (2001) Characterization of a new subgroup of *Rhizoctonia solani* anastomosis group 1 (AG-1-ID), causal agent of a necrotic leaf spot on coffee. Phytopathology 91: 1054–1061.
53. People's committee of Dong Nai province (2011) Program to develop main crops, domestic animals and build up trade name of agricultural products, duration 2011–2015. Available: http://laws.dongnai.gov.vn/2011_to_2020/2011/201109/201109260003/lawdocument_view. Accessed 2014 Jan 24.
54. Commerce (2014) Can Tho province. Available: http://www.thuongmai.vn/thuong-mai/du-an-keu-goi-dau-tu/54004-can-tho.html. Accessed 2014 Jan 24.
55. Department of Culture, Sport and Tourism of Soc Trang (2014) Overview about Soc Trang province. Available: http://www.sovhttdl.soctrang.gov.vn/wps/portal/!ut/p/c4/04_SB8K8xLLM9MSSzPy8xBz9CP0os3gLR1dvZ09LYwOL4GAnA08TRwsfvxBDI4swc_2CbEdFAM7mSnc!/. Accessed 2014 Jan 24.
56. Department of Planning and Investment of Vinh Long (2014) Overview about Vinh Long province. Available: http://www.skhdt.vinhlong.gov.vn/Default.aspx?tabid=36. Accessed 2014 Jan 24.
57. Department of Propaganda and Training of Hau Giang (2014) Overview about Hau Giang province. Available: http://www.haugiang.gov.vn/Portal/Default.aspx?pageindex=2&pageid=2979&siteid=59#. Accessed 2014 Jan 24.
58. Integrated Water Resources Management (2014) Lam Dong province. Available: http://www.iwrm.vn/index.php?page=2&id=21. Accessed 24 January 2014.
59. Viet's life (2014) Ha Noi capital. Available: http://www.cuocsongviet.com.vn/index.asp?act=dp&id_tinh=27. Accessed 2014 Jan 24.

8

Azoarcus sp. CIB, an Anaerobic Biodegrader of Aromatic Compounds Shows an Endophytic Lifestyle

Helga Fernández[1], Nicolás Prandoni[1], Mercedes Fernández-Pascual[2], Susana Fajardo[2], César Morcillo[2], Eduardo Díaz[1], Manuel Carmona[1]*

1 Department of Environmental Biology, Centro de Investigaciones Biológicas-CSIC, Madrid, Spain, **2** Plant Protection Department, Instituto de Ciencias Agrarias-CSIC, Madrid, Spain

Abstract

Background: Endophytic bacteria that have plant growth promoting traits are of great interest in green biotechnology. The previous thought that the *Azoarcus* genus comprises bacteria that fit into one of two major eco-physiological groups, either free-living anaerobic biodegraders of aromatic compounds or obligate endophytes unable to degrade aromatics under anaerobic conditions, is revisited here.

Methodology/Principal Findings: Light, confocal and electron microscopy reveal that *Azoarcus* sp. CIB, a facultative anaerobe β-proteobacterium able to degrade aromatic hydrocarbons under anoxic conditions, is also able to colonize the intercellular spaces of the rice roots. In addition, the strain CIB displays plant growth promoting traits such nitrogen fixation, uptake of insoluble phosphorus and production of indoleacetic acid. Therefore, this work demonstrates by the first time that a free-living bacterium able to degrade aromatic compounds under aerobic and anoxic conditions can share also an endophytic lifestyle. The phylogenetic analyses based on the 16S rDNA and *nifH* genes confirmed that obligate endophytes of the *Azoarcus* genus and facultative endophytes, such as *Azoarcus* sp. CIB, locate into different evolutionary branches.

Conclusions/Significance: This is the first report of a bacterium, *Azoarcus* sp. CIB, able to degrade anaerobically a significant number of aromatic compounds, some of them of great environmental concern, and to colonize the rice as a facultative endophyte. Thus, *Azoarcus* sp. CIB becomes a suitable candidate for a more sustainable agricultural practice and phytoremediation technology.

Editor: Ya-Wen He, Shanghai Jiao Tong University, China

Funding: This work was supported by Ministry of Economy and Competitiveness of Spain Grants BIO2009-10438, CSD2007-00005, BIO2012-3950, AGL2009-10371, Castilla-La Mancha Autonomous Community Grant POII10-0211-5015, Madrid Autonomous Community Grant S2009/AMB-1511 and European Union FP7 Grant 311815. The funders had no role in study design, data collection and analysis, decision to publish, or preparation of the manuscript.

Competing Interests: The authors have declared that no competing interests exist.

* Email: mcarmona@cib.csic.es

Introduction

Rice (*Oryza sativa* L.) is probably the most important cereal crop in the world, feeding more than 50% of worlds population [1]. To feed the increasing population in a sustainable manner without the utilization of chemical fertilizers or pesticides it will be necessary the application of green biotechnologies. Thus, endophytic bacteria that have plant growth promoting characteristics are crucial in this endeavour. The so-called "true endophytes" spend most of their life cycle inside plant tissues without causing symptoms of plant damage [2], but some endophytes are able to live outside of the plant without losing the capacity to colonize disinfected seedlings [3]. Some endophytic bacteria exhibit beneficial effects on the host plant, such as plant growth promotion, the induction of increased resistance to pathogens, as well as the supply of fixed nitrogen to the host plant [4]. In addition some endophytes contribute to enhanced biodegradation of environmental soil pollutants [5,6], and it has been described that endophytic bacteria equipped with the appropriate degradation pathways improve *in planta* degradation of aromatic hydrocarbons [7,8].

Various kinds of endophytic bacteria have been found inside rice plants [4], including nitrogen-fixing grass-associated diazotrophs bacteria from the *Azoarcus* genus [9]. Thus, *Azoarcus communis* strain SWub3 [10], *Azoarcus indigens* strain VB32 [10] or the well-characterized *Azoarcus* sp. strain BH72 invade roots of Kallar grass and rice, living as endophytic bacteria [11]. However, the ecological distribution of the *Azoarcus* genus is more widespread, and many strains are free-living bacteria that participate in the biogeochemical cycling of large number of metabolites, both organic and inorganic, such as arsenic acid, hydrogen or aromatic compounds [12–15]. A phylogenetic analysis of the 16S rDNA sequences from *Azoarcus* species known so far shows a tree with two main branches (Figure S1), [9,13,16]. One of the branches includes the free-living bacteria that usually are inhabitants of waters and soils; many strains of this group have been described and/or isolated by their ability to degrade aromatic compounds at anoxic conditions, e.g., *Azoarcus*/"*Aromatoleum*" strain EbN1 [17] and *Azoarcus evansii* KB740 [18,19]. In the other branch of the phylogenetic tree are located the endophytes such as *Azoarcus* sp. strain BH72 [11]. Interestingly, the group of free-living *Azoarcus* strains that are anaerobic biodegraders was suggested to be unable to interact with plants [9,20], and they only received particular attention for their degradation and biotransformation abilities [21–23].

Azoarcus sp. CIB is a previously described bacterium with the ability to degrade under aerobic and/or anaerobic conditions a high number of aromatic compounds, including some toxic hydrocarbons such as toluene and *m*-xylene [14,21,24]. The results presented on this work demonstrate that *Azoarcus* sp. CIB has also the ability to grow in association with plants, colonizing the intercellular spaces of the exodermis of rice roots. In addition, the strain CIB displays plant growth promoting (PGP) traits. Therefore, this work demonstrates by the first time that a free-living bacterium able to degrade aromatic compounds in aerobic and anoxic conditions can share also an endophytic lifestyle.

Materials and Methods

Bacterial strains, seeds, and plasmids used

Azoarcus sp. strain CIB [14] is deposited in the Spanish Type Culture Collection (CECT#5669). *Azoarcus communis* strain SWub3 [25] was obtained from the German Collection of Microorganisms and Cell Cultures (DMSZ#12120). *Pseudomonas syringae* pv. *syringae* [26] was a kindly gift from M. Espinosa (EEZ-CSIC). The *Escherichia coli* strains DH5α [27], S17-1λpir [28], and CC118 [29], as well as *Rhizobium leguminosarum* bv. *trifolii* TT-7C [30], were also used. Plasmid pSEVA23PlexAGFP-tir (pSEVA23GFP) that harbors a *gfp* gene under the control of the *PlexA* promoter and confers kanamycin resistance [31] was kindly provided by E. Martínez (CNB-CSIC, Spain). Seeds from *Oryza sativa* L. GLEVA and *Nicotiana tabacum* cv. Xanthi [32] plants were kindly provided by Castells Seeds Co. and F. Tenllado (CIB-CSIC), respectively.

Media and growth conditions

MC medium was prepared as previously described [14]. The nitrogen-free MNF medium used was a modification of a previously described medium [33]: A Solution (g/900 ml): 0.8 g K_2HPO_4, 0.2 g KH_2PO_4, 0.1 g NaCl, 28 mg $Na_2FeEDTA$, 25 mg $Na_2MoO_4 \times 2H_2O$, 100 mg yeast extract, pH 7.0; B Solution (g/100 ml): 0.2 g $MgSO_4 \times 7H_2O$ and 0.06 g $CaCl_2 \times 2H_2O$. The two solutions were autoclaved separately and then the suitable carbon source, vitamins and trace elements solutions were added. For checking the utilization of insoluble phosphate, the MALP medium (MA modified medium with low phosphate) was used: (per liter) 0.2 g KH_2PO_4, 0.2 g Na_2HPO_4, 0.1 g $MgSO_4 \times 7H_2O$, 0.1 g NH_4Cl, 5 g calcium phosphate [34]. Once autoclaved, the $CaCl_2$, vitamins and trace elements were added at the same concentrations than were used in the MC medium. VM-ethanol rich medium was prepared as previously described [35].

E. coli cells and *P. syringae* were grown at 37°C in lysogeny broth (LB) medium [36]. When required, *E. coli* was grown in VM-ethanol medium [37]. *Azoarcus* sp. CIB, *Azoarcus communis* and their derivatives were grown at 30°C either in MC medium, as previously described [14], or in VM-ethanol medium. *R. leguminosarum* was grown at 30°C in MC medium supplemented with 0.4% (w/v) glucose. When appropriate, kanamycin (50 μg ml^{-1}) was added to the medium.

Nitrogen fixation assays

Nitrogenase activity was measured by using the acetylene reduction activity assay for free-living bacteria [38]. *Azoarcus* sp. CIB cells were anaerobically grown in MC medium with 3 mM benzoate as sole source of carbon and energy until mid-exponential growth phase, and then they were pelleted by centrifugation at 5000×g for 15 min and resuspended in MNF medium, either in the presence or absence of ammonia. Cultures were grown microaerobically in closed batch cultures until they reached an A_{600} of 0.4. Then, 30 ml of the *Azoarcus* sp. CIB cultures were enclosed in 100 ml tubes sealed with rubber septa. Ten ml of air were removed and the same amount of acetylene was added (10% v/v). Gas samples were taken after 15 min, 1 h, 4 h, 24 h and 48 h of acetylene exposition at 25°C. Gas samples were analyzed for ethylene and acetylene content in a Perkin-Elmer 8310 gas chromatograph using nitrogen as the carrier gas and with a flow rate of 50 ml min^{-1} as described before [39].

Inoculation of rice with *gfp*-expressing bacteria

Dehulled seeds of *O. sativa* GLEVA were surface sterilized by shaking at 25°C for 30 min in 30 ml of 1% (v/v) sodium hypochlorite. After three washes for 10 min in sterile water, the seeds were incubated in VM-ethanol for 48 hours, and only the uncontaminated seedlings were selected for inoculation. The germination continued on sterilized moist water filter paper for 24 h prior to inoculation with *gfp*-expressing bacterial cells. To obtain the *gfp*-expressing bacteria, we transferred by conjugation the pSEVA23GFP plasmid, that express the green fluorescent protein (GFP) under the *PlexA* promoter [40], from *E. coli* S17-1*λpir* (pSEVA23GFP) to *Azoarcus* sp. CIB and *A. communis* strains using protocols previously described [14]. The selection of exconjugants was made as described before [14], and the expression of GFP was monitored by observation of bright green fluorescence cells under UV light. To inoculate the rice seedlings with the *gfp*-expressing bacteria, *Azoarcus* sp. CIB (pSEVA23GFP), *Azoarcus communis* strain SWub3 (pSEVA23GFP) and *E. coli* S17-1λpir (pSEVA23GFP) were grown in VM-ethanol medium at 30°C until they reached an A_{600} of 0.5. Cells were then

Table 1. Determination of *Azoarcus* sp. CIB nitrogenase activity.

	24 h		48 h	
	μmol/cells (Abs₆₀₀ 0.4)	*μmol/g fresh culture*	*μmol/cells (Abs₆₀₀ 0.4)*	*μmol/g fresh culture*
–N (30 ml)	1,37±0.4	13,69±2.1	2,41±0.3	24,03±4.7
+N (30 ml)	0,24±0.07	1,51±0.4	0,26±0.09	1,68±0.55

Total (μmol C_2H_4/30 ml of culture at Abs_{600} of 0.4) and specific (μmol C_2H_4 g^{-1}bacteria fresh weight 24 h^{-1} or 48 h^{-1}) nitrogenase activity of *Azoarcus* CIB grown in MNF medium with 0.37 g/l NH_4Cl (+N) or without ammonia (–N) as nitrogen source. The values are the average from three independent experiments +/– S. D.

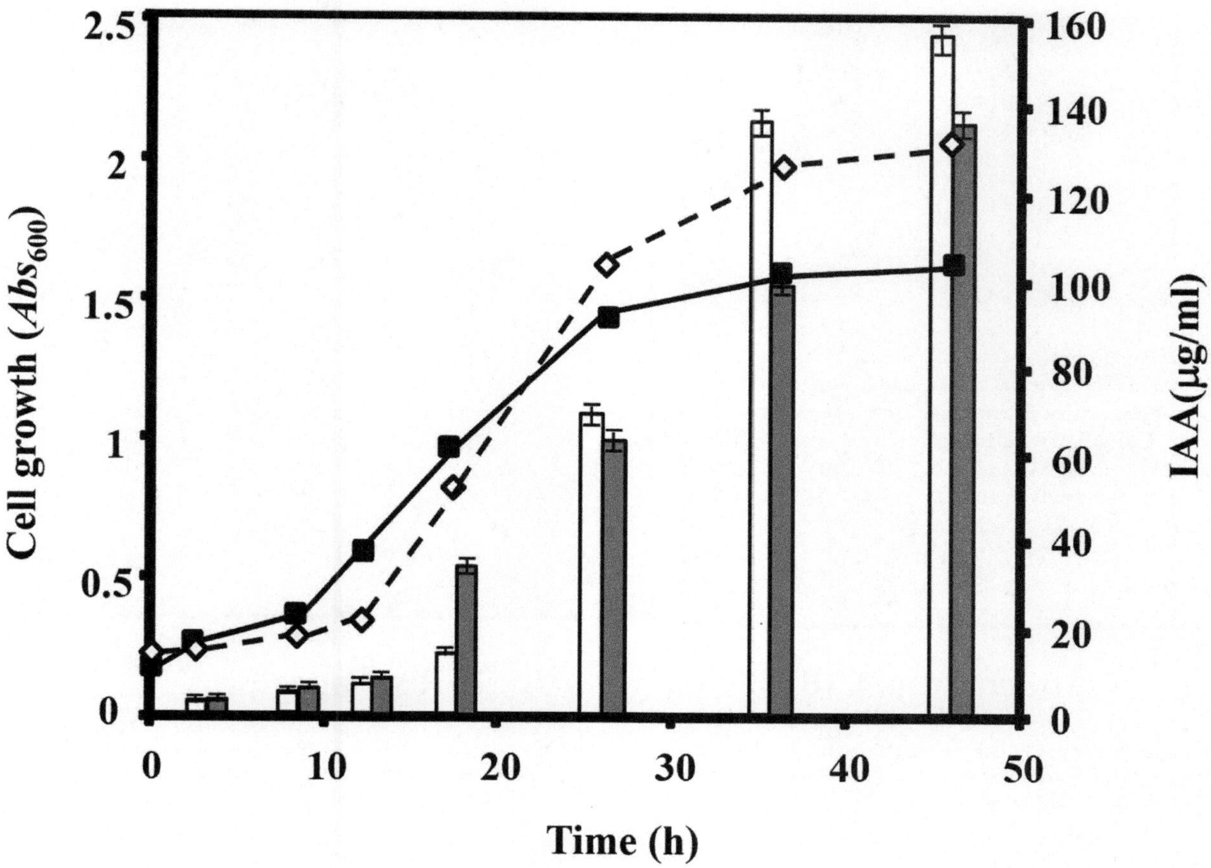

Figure 1. ***Azoarcus*** **sp. CIB produces IAA.** Production of IAA by *Azoarcus* sp. CIB (white bars) and *R. leguminosarum* bv. *trifolii* TT-7C (grey bars) along the growth curve. Growth of *Azoarcus* sp. CIB (black squares) and *R. leguminosarum* bv. *trifolii* TT-7C (grey rhombus) is also represented. IAA was quantified as detailed in Materials and Methods. Graphed values of are the average from three independent experiments +/− S. D.

harvested by centrifugation, washed with sterile 0.9% NaCl (w/v) solution, resuspended in sterile distilled water, and inoculated independently by pipetting 1 ml of the cell suspension onto the surface of the seedling in aseptic conditions. After the inoculation of the seedlings, plants were grown at 25°C with the natural daily light period (approximately 10 h of light and 14 of darkness) for 5–10 days.

Quantification of *gfp*-expressing bacteria inside rice roots

Plants were sampled at 5 days after the inoculation of seedlings with the *gfp*-expressing bacteria. Loosely attached bacteria were removed by exhaustively washing the roots with sterile water. Then, the roots were surface sterilized by immersion for 3 min in 1% sodium hypochlorite solution. After three washes with sterile distilled water, the roots were homogenized using a sterile pestle and mortar, and the extracts diluted in 1 ml sterile saline solution (0.9% NaCl) as described before [1]. Serial dilutions of the extracts were plated onto VM-ethanol solid medium in the presence of kanamycin to determine the number of CFU containing the pSEVA23GFP plasmid. Moreover, the expression of GFP in the kanamycin-resistant recovered bacteria was confirmed by epi-fluorescence microscopy.

Fluorescence microscopy studies

The root samples were collected 5 or 10 days after inoculation, washed with water, hand cut, mounted on a microscope slide and immediately examined under an inverted microscope Leica DMI6000B to visualize the *gfp*-labelled bacteria, or a Leica TCS-SP5-AOBS confocal laser microscope, on green channel, with excitation of 488 nm (laser beam) and emission of 500–547 nm to see the bacteria inside the plant tissues.

Light microscopy and immunocytochemistry studies

Small samples of rice roots (0.5–1 cm) inoculated with *Azoarcus* sp. CIB for 7 days were cut with a razor blade and fixed in 2.5% glutaraldehyde in 0.05 M Na-cacodylate buffer (pH 7.4) containing 25 mg per ml sucrose [41], and vacuum-infiltrated to enhance penetration of the fixative. Dehydration was performed using an ethanol series and nodules pieces were infiltrated and finally embedded in LR White resin (London Resin Corporation), using gelatine capsules. The post-fixation with osmium tetroxide was not carried out in order to overcome the masking of antigenic sites with osmium. Two kinds of polymerization were performed: by heat at 60°C for 24 h (for light microscopy) or by ultraviolet light at −20°C for 48 h to avoid protein denaturalization (for immunocytochemistry). Semi-thin (1 μm) and ultrathin (70–80 nm) sections of each sample were cut with a Reicher Ultracut S Ultramicrotome (Vienna, Austria) fitted with a diamond knife. Semithin sections for light microscopy (LM) were collected on glass sides and stained with 1% (w/v) toluidine blue in aqueous sodium borate [42] and directly observed with a Zeiss Axiophot photomicroscope.

Immunolocalization was carried out using an antibody against the NifH subunit of the nitrogenase enzyme, kindly provided by

Figure 2. ***Azoarcus*** **sp. CIB is a non-pathogenic bacterium.** Pathogenicity test on tobacco leaves. Bacteria were infiltrated into the intercellular spaces of a tobacco leaf, which was inspected for a visible necrosis after 2 days. Necrosis is visible as light spots (arrows) on leaves infiltrated with *Pseudomonas syringae pv. syringae* (right) and not visible on leaves infiltrated with *Azoarcus* sp. CIB (left).

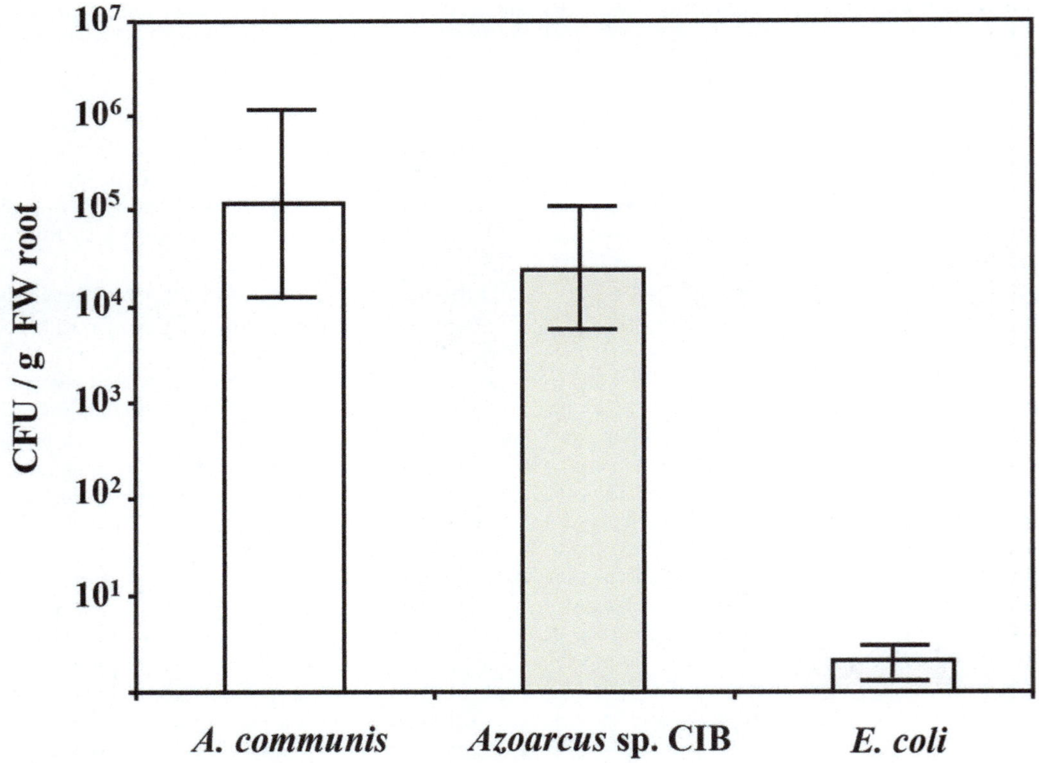

Figure 3. Quantification of the bacterial endophytic population within inoculated rice roots. Rice seedlings were inoculated with *Azoarcus communis* Swub3 (*A. communis*), *Azoarcus* sp. CIB (*Azoarcus* sp. CIB), and *E. coli* S17-1λ*pir* (*E. coli*) containing plasmid pSEVA23GFP) that expresses the *gfp* gene as indicated in Materials and Methods. Plants were grown at 25°C for 5 days, and the kanamycin-resistant bacteria within the root tissue were determined as detailed in Materials and Methods. Graphed values of CFU per g of fresh root (FW) are the average from three independent experiments +/− S. D.

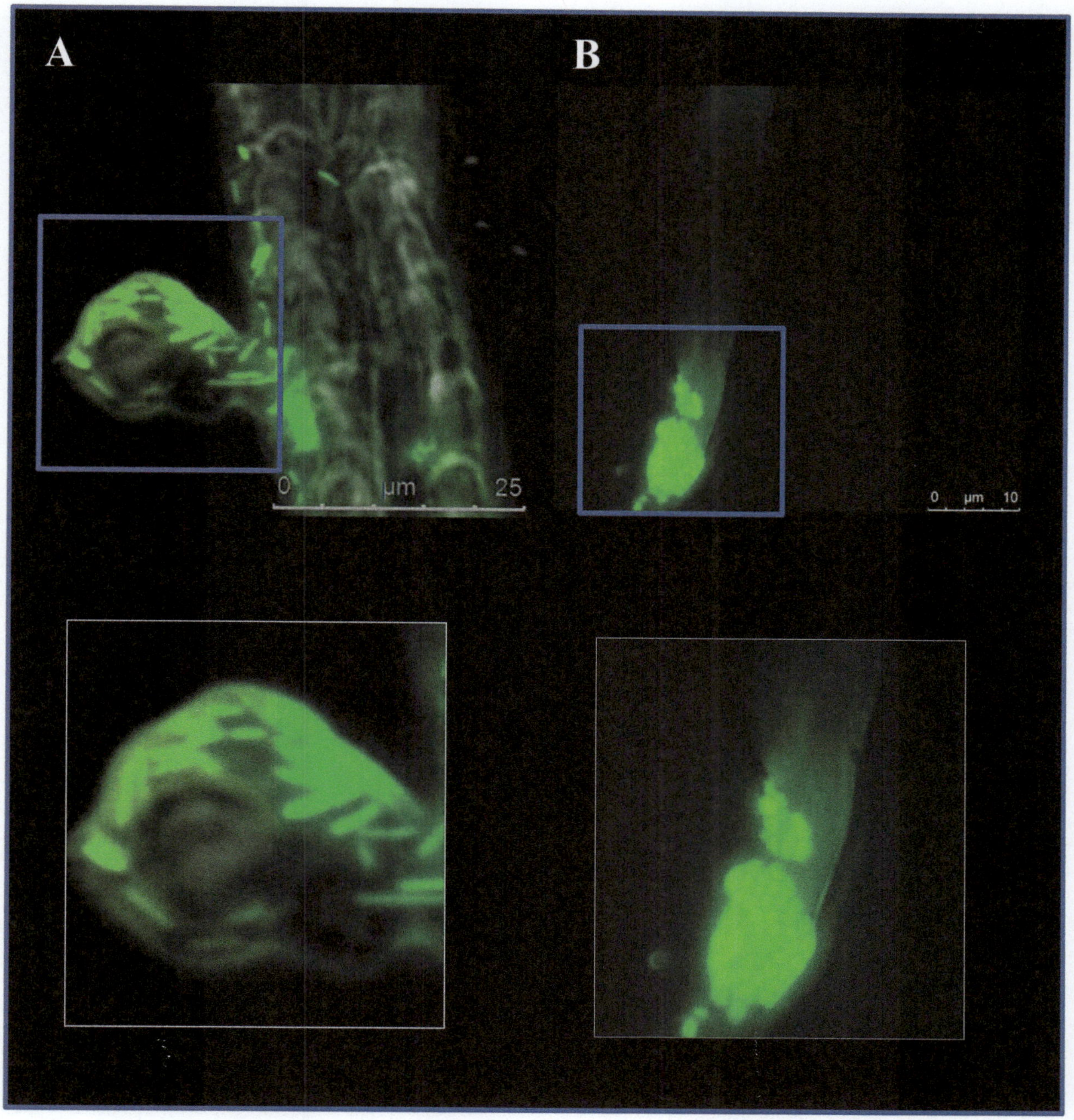

Figure 4. Images of rice roots inoculated with *Azoarcus* sp. CIB. Epifluorescence microscopy images of rice roots after 5 days (A) or 10 days (B) of inoculation with *Azoarcus* sp. CIB (pSEVA23GFP). A detail of each image is included (lower panels) to illustrate the morphological change of the cells.

Dr. Imperial (Centre for Plant Biotechnology and Genomics, UPM-INIA). Immunogold localisation was performed on 70 nm sections picked on pioloform-coated nickel grids as previously described [43], with some modifications. Grids were floated on TBS (10 mM Tris-HC1; 150 mM NaC1; 0.1% (v/v) Tween-20, pH 7.4), containing 20 mg of BSA per ml, for 1 h at 37°C. Sections were then incubated in primary antibody, anti NifH, diluted 1:500 in the same buffer (TBS containing 20 mg of BSA per ml) for 2 h at 37°C. Grids were then washed (five times, 3 min) in TBS containing 2 mg of BSA per ml. Incubation in goat anti-rabbit IgG-gold conjugate (GAR 15 nm; Bio-Cell) diluted 1:40 in TBS with 2 mg of BSA per ml was performed for 1 h at 37°C. Sections were then rinsed in TBS containing 20 mg of BSA per ml (five times, 3 min), in the same buffer containing 0.05% Triton X-100 (three times, 3 min), and finally in distilled water (3 min). Controls were performed without anti NifH antibody. Counter-staining of sections was obtained with 2% aqueous uranyl acetate and lead citrate [44] for 1 min. After rinsing and air-drying, the sections were observed with a STEM-LEO 910 microscope at 80 kV.

Indoleacetic acid (IAA) production assay

Production of IAA was estimated by growing the bacterial strains at 30°C for 48 h in MC minimal medium supplemented

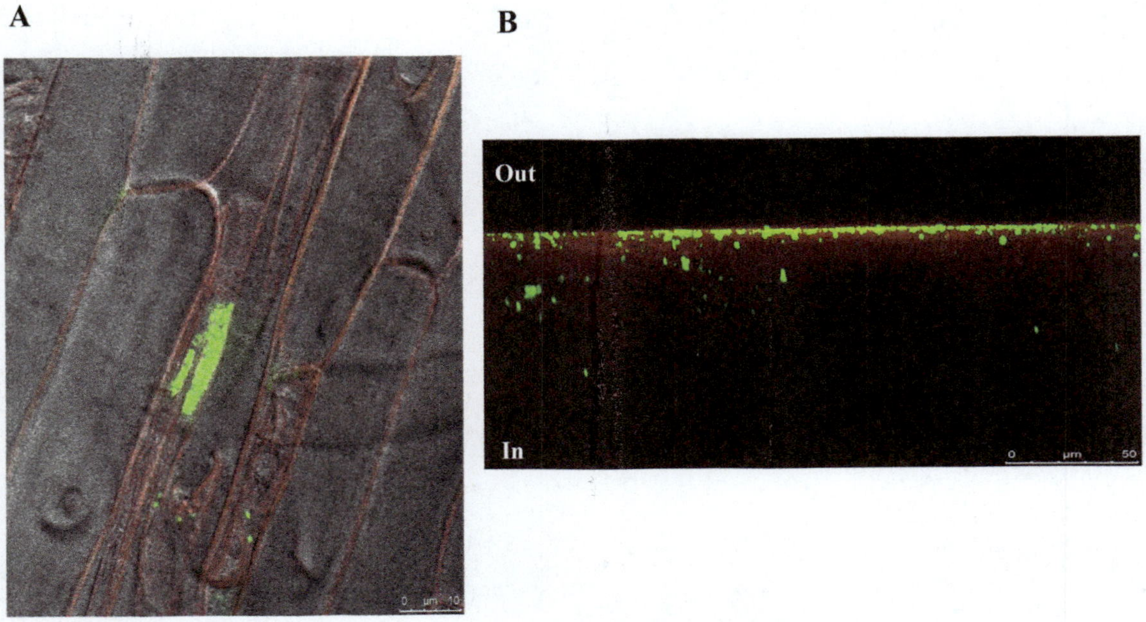

Figure 5. Colonization of rice roots by *Azoarcus* sp. CIB observed by confocal microscopy. Confocal images of rice roots after seven days colonization with *Azoarcus* sp. CIB (pSEVA23GFP) cells. (A) Bacteria are observed as single cells or as clusters of cells attached to the inner intercellular space of the exodermis. (B) XYZ projection of bacterial cells distributed on the inner part (In) or the surface (Out) of the rice root.

with 2.5 mM *L*-tryptophan which acts as a precursor for IAA synthesis. IAA produced per milliliter of culture was estimated by mixing 14 ml of Salkowski reagent (12 g/l $FeCl_3$ dissolved in 7.9 M H_2SO_4) [45] with 1 ml culture supernatants, followed by measuring absorbance at 530 nm wavelength after 30 min of incubation at 25°C [46]. A calibration curve for the estimation of the amount of IAA was made by using IAA as standard [47].

Insoluble phosphate solubilisation assay

Bacterial cells were grown on MC medium until they reached mid exponential growth phase, then they were collected by centrifugation and washed three times with sterile saline solution. The cell pellet was resuspended in 0.5 ml of sterile MALP medium. A drop of 10 µl of the resuspended cell solution was deposited in a MALP solid media. The halo around the colonies was visualized after 7 days of incubation of plates at 30°C. The ability of the bacteria to solubilise insoluble phosphate was described by the halo formation as described before [48].

Plant pathogenicity test

Azoarcus sp. CIB and *Pseudomonas syringae* pv. *syringae*, a plant pathogen used as control, were infiltrated into the intercellular spaces of an old leaf of an intact plant of *Nicotiana tabacum* at a dose of 10^7 colony forming units (CFU) per leaf area as described previously [49].

Sequence data analyses

The nucleotide sequence of the *nifH* gene from *Azoarcus* sp. CIB has been submitted to GenBank with Accession number KJ814970. Phylogenetic analysis was carried out by using the Phylogeny.fr program [50].

Results and Discussion

Azoarcus sp. CIB is a nitrogen fixing bacterium

As indicated in Introduction, *Azoarcus* sp. CIB strain is a free-living bacterium able to degrade a high number of aromatic compounds either in the presence or in the absence of oxygen [14,21,24]. The recent sequencing of the genome of *Azoarcus* sp. CIB revealed that its size (5.25 Mb) is significantly bigger than that of other well-characterized *Azoarcus* strains, such as the rice endophyte *Azoarcus* sp. BH72 (4.37 Mb) or the free-living aromatic biodegrader "*A. aromaticum*" EbN1 strain (4.72 Mb), indicating that the strain CIB contains a more complete genetic repertoire that might be used for the adaptation to different habitats. Since free-living *Azoarcus* strains have never been isolated from inside of living plants [34] and the plant-associated *Azoarcus* strains could not be isolated from root-free soil behaving as obligate endophytes [13], we explored the possibility that CIB strain may behave as a facultative endophyte, i.e., a bacterium that has a stage in its life cycle in which it exists outside host plants as a free-living bacterium [51]. Giving that certain plant-associated bacteria have the ability to fix nitrogen, solubilize minerals, and produce phytohormones [52–54], we tested whether *Azoarcus* sp. CIB showed some of these traits.

To investigate the nitrogen fixation ability of *Azoarcus* sp. CIB, we firstly cultivated bacterial cells in nitrogen-free MNF medium, and we observed that the CIB strain was able to grow, suggesting that it utilizes atmospheric nitrogen. In addition, nitrogenase activities of cells growing in MNF medium were one order of magnitude higher than those of cells grown in MNF medium supplemented with NH_4Cl as nitrogen source (Table 1). Therefore, these data indicate that *Azoarcus* sp. CIB is a nitrogen fixing bacterium.

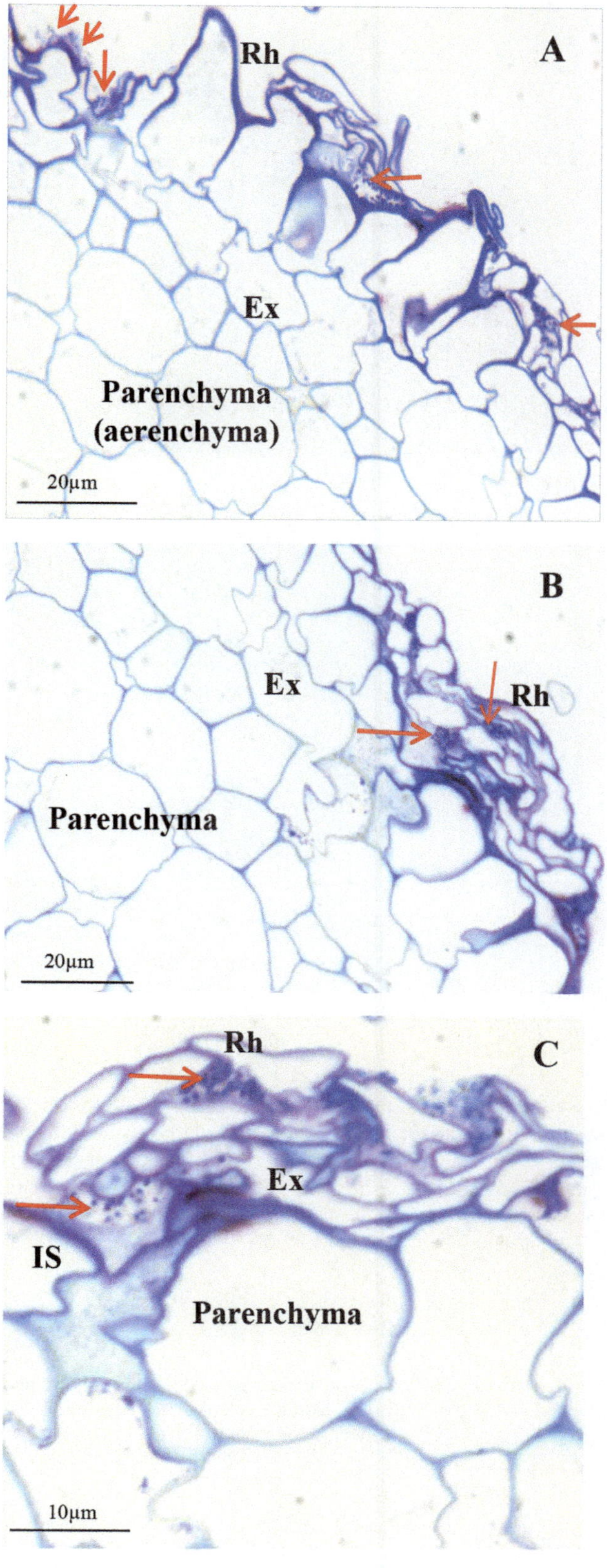

Figure 6. Light micrographs of transversal sections of rice roots inoculated with *Azoarcus* sp. CIB (pSEVA23GFP) cells and incubated for 7 days. Colonization of root surface and intercellular spaces under rhizodermis and the first layer of exodermis (A); intercellular colonization of the second and third layers of the exodermis (B); intercellular colonization of the deeper layer of the exodermis, just in contact with the parenchyma (C). IS, intercellular space; Ex, exodermis, Rh, rhizodermis. Red arrows indicate the localization of the bacterial cells.

In silico analysis of the unpublished draft of *Azoarcus* sp. CIB genome sequence identified a *nifH* gene whose product shows 76.4% amino acid sequence identity with the NifH subunit of the nitrogenase protein involved in nitrogen fixation in *Azoarcus* sp. strain BH72 [55]. A phylogenetic analysis of the *nifH* gene product from strain CIB revealed that it branched with that of other soil *Azoarcus* strains and some other β subclass of Proteobacteria (Figure S2). In contrast, NifH from *Azoarcus* sp. BH72 and other endophytic *Azoarcus* strains (aligned with the orthologs in *Azotobacter vinelandii*, *Pseudomonas stutzerii* and *Klebsiella oxytoca* of the γ subclass of Proteobacteria (Figure S2). Phylogenetic distances of NifH proteins indicate a possible lateral gene transfer of *nif* genes to *Azoarcus* from a common donor of the α Proteobacteria at the time of species diversification, with more recent and independent transfer events in some plant-associated *Azoarcus* species, such as the BH72 strain, forming a monophyletic unit with those of γ *Proteobacteria* [56]. The apparent divergence of the NifH protein in the *Azoarcus* genus was described previously where it was postulated the existence of two groups of *nifH* sequences that correspond to the soil and plant-associated *Azoarcus* strains, respectively [56].

Azoarcus sp. CIB solubilises insoluble inorganic phosphate

Several reports showed the ability of different bacteria to solubilise insoluble inorganic phosphate compounds [57]. In fact, since most of the soils under cultivation contain insoluble phosphates, the ability of the bacteria associated with plants to solubilise precipitated phosphates or enhance phosphate availability to the plant represents a mechanism of plant growth promotion (PGP) under field conditions [58]. To test whether *Azoarcus* sp. CIB is able to grow in minimal medium using insoluble phosphate as the major phosphorous source, we used a plate assay that provides a semiquantitative estimation of the phosphate solubilisation ability. According to this method, clearing zones around the microbial colonies in media containing insoluble mineral phosphates (mostly tricalcium phosphate or hydroxyapatite) as single phosphorous source can be visualized [53]. After 14 days incubation, *Azoarcus* sp. CIB showed an apparent cell growth and a clear halo around the colonies (Figure S3), indicating that the CIB strain is able to use insoluble phosphorous.

Azoarcus sp. CIB produces IAA

The ability to produce the plant hormone IAA, a naturally occurring auxin, is widespread among microorganisms that are commonly associated with plant surfaces [59]. Microbes from the rhizosphere of different crops appear to have a greater potential to synthesize and release IAA as secondary metabolite because of the relatively rich supply of tryptophan [60]. IAA has been implicated in regulating a variety of developmental and cellular processes in plants such as cell extension, cell division, vascular differentiation, root formation, apical dominance, and tropism [61]. To analyze the ability of *Azoarcus* sp. CIB to produce IAA, cells were grown on MC medium supplemented with 2.5 mM tryptophan. Relative levels of IAA along the growth curve were estimated by a colorimetric assay (see details in Material and Methods). Maximal production of IAA was achieved after 48 h of growth (Figure 1). When tryptophan was not added to the medium, no colorimetric change was detected (data not shown). *Rhizobium leguminosarum* bv. *trifolii* TT-7C, a strain that is able to produce IAA [60], was

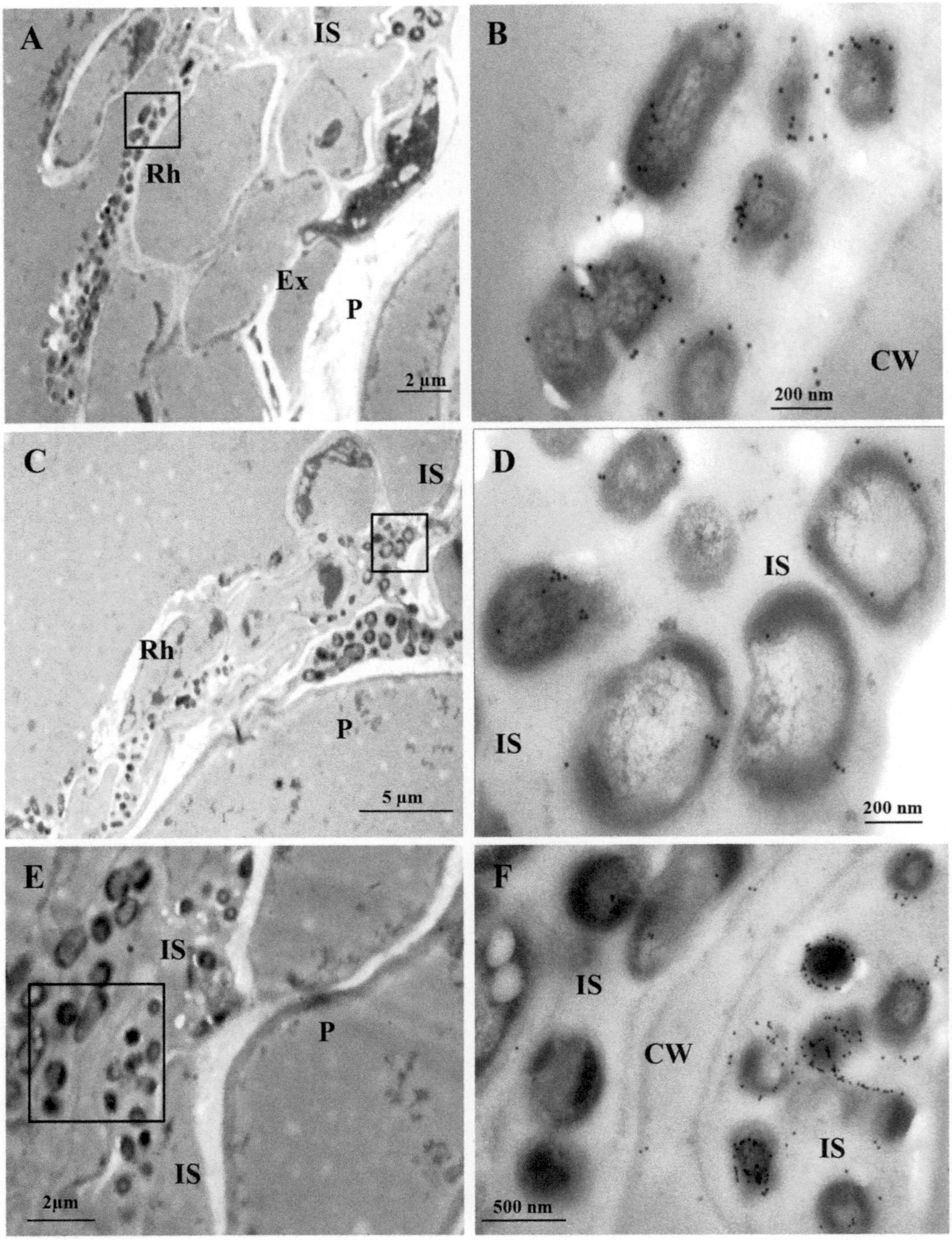

Figure 7. Electron microscopy observation of rice roots colonized by *Azoarcus* sp. CIB. Electron micrographs showing immunogold localization of NifH epitopes in rice roots inoculated with *Azoarcus* sp. CIB (pSEVA23GFP) cells for 7 days. Surface colonization (A), intercellular colonization of the second and third layers of the exodermis (C), intercellular colonization of the deeper layer of the exodermis, just in contact with the parenchyma (E). CW, cell wall; Ex, exodermis; IS intercellular space; Rh, rhizodermis; P, parenchyma. In order to distinguish gold particles, the framed areas in A, C and E are magnified in panels B, D and F, respectively.

used as a positive control and it showed levels of IAA production similar to those of *Azoarcus* sp. CIB, with a maximal production also after 48 h of growth (Figure 1). It is interesting to mention that there is an important increase of IAA production at the end of the exponential growth-phase (Figure 1), suggesting a phase-dependent production of IAA as it has been already reported in other bacteria and fungi strains [60,62,63]. Although that as far as we know, this is the first report of an *Azoarcus* strain able to produce IAA to understand the role of IAA production on the putative relationship between *Azoarcus* sp. CIB and plants, more experiments have to be done.

Azoarcus sp. CIB does not cause plant pathogenic effects

The results showed above reveal that *Azorcus* sp. CIB elicits properties of plant-associated bacteria such as the ability to fix nitrogen, produce IAA and solubilise insoluble phosphate, suggesting that the CIB strain could interact with plants. To check whether *Azoarcus* sp. CIB could cause defense reactions in a plant, a commonly used pathogenicity test [49] was done. Whereas the inoculation of *Azoarcus* sp. CIB did not elicit a hypersensitive response leading to visible necrosis in tobacco leaves within 10 days, leaves treated with the plant pathogen *P. syringae* pv. *syringae* developed this phenotype in two days (Figure 2). Moreover, no macroscopically visible disease symptoms were observed on rice seedlings infected with *Azoarcus* sp. CIB cells (data not shown). These results suggest that *Azoarcus* sp. CIB is not a plant phatogen, which is in agreement with a previous observation that demonstrated the non-pathogenic effect of some endophytic *Azoarcus* strains on tobacco and rice [64].

All these results showed that *Azorcus* sp. CIB is not a pathogen of plants that elicit some of the properties of the plant-associated bacteria such ability to fix nitrogen, to produce IAA and to solubilise insoluble phosphate. On behalf of that we next explored the ability of *Azoarcus* sp. CIB to associate with plants.

Azoarcus sp. CIB interacts with rice roots as an endophyte

Since *Azoarcus* sp. CIB shows typical traits of plant-associated bacteria (see above) and it is well-known that some *Azoarcus* strains are rice endophytes [2], it was tempting to speculate that the CIB strain could share also an endophytic lifestyle. To check whether *Azoarcus* sp. CIB is also an endophyte, we monitored the presence of CIB cells in the interior of rice roots inoculated with this bacterium. Since the *gfp* marker gene has been proved to be very useful in colonisation studies and to visualise plant endophytes [65,66], we generated bacterial cells containing plasmid pSEVA23GFP that expresses the *gfp* reporter gene. Inoculation of rice seedlings with *Azoarcus* sp. CIB (pSEVA23GFP) or with *Azoarcus communis* SWub3 (pSEVA23GFP), a previously reported rice endophyte bacterium [25], led to the detection of about 6×10^4 CFU.gr^{-1} root and 1×10^5 CFU.gr^{-1} root, respectively, in the interior of the roots 5 days after inoculation (Figure 3). These values are similar to those previously reported for re-isolations of plant endophytes [67]. Moreover, following rice inoculation with *Azoarcus* sp. CIB (pSEVA23GFP), most of the kanamycin-resistant CFUs (>90%) detected were fluorescent under UV light in an epifluorescence stereomicroscope (data not shown), confirming that most of the endophytic bacteria isolated from the inner tissues of the roots corresponded to the inoculated *Azoarcus* strains. In contrast, an insignificant number of CFUs (<10) of *E. coli* S17-1λpir (pSEVA23GFP) was isolated from inoculated rice seedlings. In summary, these results suggest that *Azoarcus* sp. CIB behaves as rice endophytic bacterium. Nevertheless, it is known that bacteria closely attached to crevices and/or embedded in mucilage might escape chemical surface sterilization of roots and behave as false endophytes. Therefore, to confirm that *Azoarcus* sp. CIB behaves as a true rice endophyte an *in planta* visualization of the bacterium should be accomplished by microscopy studies [2].

In planta visualization of the endophytic lifestyle of *Azoarcus* sp. CIB

By using epifluorescence microscopy, *Azoarcus* sp. CIB (pSEVA23GFP) cells expressing GFP were visible inside sterilised roots obtained 5 days (Figure 4A) and 10 days (Figure 4B) after the inoculation of the corresponding seedlings. The *gfp*-expressing CIB cells were particularly abundant just beneath the epidermal surface at the area of growing hairs (Figure 4A). Interestingly, there was a noticeable change in the cellular morphology of the bacteria at long colonization times. Hence, *Azoarcus* sp. CIB cells appeared to be shorter and more spherical after 10 days (Figure 4B) than after 5 days (Figure 4A) of inoculation. These changes in cellular shape are dependent on environmental conditions and have been reported previously in other plant-associated bacteria [68,69]. In fact, it has been proposed that bacterial cells are better nourished upon successful colonisation, and cell shape is related to the growth rates within a particular environment [66].

Internal colonization of rice roots by *Azoarcus* sp. CIB was confirmed by using confocal microscopy after 7 days of colonization (Figure 5). Internal distribution of cells showed a diverse pattern since bacteria could be either as single cells or cell clusters. *Azoarcus* sp. CIB (pSEVA23GFP) cells were mainly detected in the intercellular spaces of the exodermis (Figure 5A). A xyz projection of the image obtained from roots rice inoculated with *Azoarcus* sp. CIB (pSEVA23GFP) showed that bacteria distribute also in the inner parts of the exodermis (Figure 5B).

To see in more detail how the *Azoarcus* sp. CIB cells locate in the internal part of the rice root, we prepared resin-embedded roots of rice seedlings grown axenically with *Azoarcus* sp. CIB (pSEVA23GFP) cells. Light and electron microscopy examination of theses samples confirmed that bacteria colonize the exodermis of roots. Thus, bacteria moved from the surface to intercellular spaces of rhizodermis and the second and third layer of the exodermis (Figure 6A and 6B), until they reached the deeper layer of the exodermis, just in contact with the parenchyma of the roots (Figure 6C). No bacteria were observed in rice roots that were not inoculated with *Azoarcus* sp. CIB (pSEVA23GFP) (data not shown), which confirms that the observed bacteria correspond to *Azoarcus* sp. CIB cells. Immunogold labelling of the ultraviolet ligh-polymerized samples with antiserum raised against the NifH protein and electron transmission microscopy confirmed the superficial (Figure 7A and 7B) and intercellular spaces colonization of all the layers of exodermis (Figure 7C to 7F). Labelling of NifH protein was specifically confined to the *Azoarcus* sp. CIB (pSEVA23) bacteria, although unspecific and very scarce gold particles were observed in the matrix of the intercellular spaces and in the cell wall of exodermis and parenchyma cells of rice roots. Gold particles were never observed in the cytoplasm or organelles of plant cells. Although gold particles can be visualized in *Azoarcus* sp. CIB (pSEVA23GFP) cells colonizing surface and first and second layer of exodermis, a higher amount of gold particles were detected in bacterial cells that colonize the deepest layers of exodermis (Figure 7E to 7F). This observation may reflect the fact that nitrogen fixation is an anaerobic process and bacteria that occupy the deeper layers of the exodermis have lower oxygen availability, hence being able to express abundantly the nitrogenase protein.

In summary, all the three different microscopy techniques used, i.e., laser confocal, optical and electron microscopy, allow us to conclude that the CIB bacterial strain is a new rice endophyte from the *Azoarcus* genus.

Conclusions

This is the first report of a bacterium, *Azoarcus* sp. CIB, that was shown to degrade anaerobically a high number of aromatic compounds [14,21,24] and displays the ability to colonize two different ecological niches, i.e., soil and water as free-living bacteria and the inner tissues of the rice roots as a facultative endophyte. The previous thought that the *Azoarcus* genus comprises bacteria that fit into one of two major eco-physiological groups, i.e., either the free-living anaerobic biodegraders of aromatic compounds or the obligate endophytes unable to degrade aromatics under anaerobic conditions [9,20], should be now revisited. *Azoarcus* sp. CIB may represent the prototype of a subgroup of *Azoarcus* strains that share the anaerobic biodegradation of aromatic hydrocarbons with an endophytic lifestyle. On the other hand, there is another subgroup of *Azoarcus*, whose prototype is the EbN1 strain [20], that degrade aromatics under anaerobic conditions but are unable to interact with plants. Phylogenetic analyses based on the 16S rDNA and *nifH* genes confirmed that obligate endophytes such as *Azoarcus* sp. BH72, *A. communis*, and *A. indigens*, and facultative endophytes, such as *Azoarcus* sp. CIB, belong indeed to different evolutionary branches.

Azoarcus sp. CIB shows also some traits of PGP bacteria such the ability to uptake insoluble phosphorous, production of IAA or nitrogen fixation, which makes this endophyte as a potential candidate for a more sustainable agricultural practice [70]. In addition, since *Azoarcus* sp. CIB is able to degrade, both aerobically and anaerobically, toxic aromatic compounds [14], the use of this bacterium in association with plants could offer an efficient, economic and sustainable phytoremediation technology [71].

Supporting Information

Figure S1 Neighbor-joining phylogenetic tree of *Azoarcus* bacteria based on the comparison of their 16SrRNA genes. The branch length is proportional to the number of substitutions per site. The sequence of the 16SrRNA genes from *Azoarcus* sp. strain EC1-7 (EU708505), *Azoarcus* sp. strain EC2-7 (EU708500), *Azoarcus* sp. strain HxN1 (AF331975), *Azoarcus buckelii* strain U120 (AJ315676), *Azoarcus* sp. strain EbN1 (X83531), *Azoarcus anaerobius* strain DSM12081 (Y14701), *Azoarcus* sp. strain GYP_24 (JX981924), *Azoarcus taiwanensis* strain NSC3 (GQ389714), *Azoarcus* sp. strain C3b_A2 (JX575077), *Azoarcus tolulyticus* strain Tol-4 (NR_037058), *Azoarcus denitrificans* (U82665), *Azoarcus toluclasticus* strain MF63 (NR_024970), *Azoarcus* sp. strain KH32C (NC_020516), *Azoarcus* sp. strain T (AF129465), *Azoarcus* sp. strain CIB (AF515816), *Azoarcus* sp. strain DAO1 (DQ336177), *Azoarcus evansii* strain KB740 (NR_029266), *Azoarcus toluvorans* strain Td-21 (NR_025915), *Azoarcus* sp. strain S5b2 (L15532), *Azoarcus communis* strain SWub3 (NR_024850), *Azoarcus indigens* strain. VB32 (NR_024851), and *Azoarcus* sp. strain BH72 (NR_074801) were included in the analysis. The previously described as free-living *Azoarcus* are dashed in orange color and the plant-associated *Azoarcus* are dashed in green.

Figure S2 Neighbor-joining phylogenetic tree based on the NifH protein sequence. The bacteria included in the analysis are: *Azotobacter vinelandii* (YP_00297378.1), *Pseudomonas stutzeri* (YP_001171863.1), *Klebsiella oxytoca* (YP_005020938.1), *Azoarcus* sp. BH72 (YP_932042), *Oscillatoria* sp. (WP_007356926.1), *Azospirillum brasilense* Sp245 (YP_005030951.1), *Rhodopseudomonas palustris* CGA009 (NP_949954.1), *Rhodobacter capsulatus* (AAA26140.1), *Magnetospirillum magneticum* AMB-1 (YP_420937.1), *Rhizobium leguminosarum* (WP_0035927442.1), *Rhizobium etli* (NP_659736.1), *Mesorhizobium* sp. STM4661 (WP_006331760), *Sinorhizobium meliloti* (WP_018097454.1), *Leptospirillum ferrooxidans* (AF097517.1), *Azoarcus* sp. KH32C (YP_00550106.1), *Azoarcus* sp. CIB (KJ814970), *Azoarcus toluclasticus* MF63 (WP_018989049.1), *Bradyrhizobium japonicum* (WP_018319598.1), *Xanthobacter autotrophicus* Py2 (YP_001415004.1), *Azorhizobium caulinodans* ORS571 (YP_001526359.1), *Burkolderia xenovorans* LB400 (YP_553849.1), *Cupriavidus* sp. WS (WP_020202091.1) and *Herbaspirillum frisingense* (WP_006463090.1). The branch length is proportional to the number of substitutions per site. The position of *Azoarcus* sp. CIB is indicated by a red arrow.

Figure S3 Solubilisation of the low soluble inorganic phosphate by *Azoarcus* sp. CIB. The halos were observed after 7 days of incubation on solid MALP medium at 30°C.

Acknowledgments

The technical work of A. Valencia is greatly appreciated. We are gratefully acknowledged to E. Martínez, J. Imperial and "Semillas Castells" for providing plasmid pSEVA23GFP, NifH antibody and rice seeds, respectively.

Author Contributions

Conceived and designed the experiments: MC ED MF-P. Performed the experiments: HF NP SF CM. Analyzed the data: MC HF MF-P ED. Contributed to the writing of the manuscript: MC MF-P ED.

References

1. Gyaneshwar P, James EK, Mathan N, Reddy PM, Reinhold-Hurek B, et al. (2001) Endophytic colonization of rice by diazotrophic strain of *Serratia marcescens*. J Bacteriol 183: 2634–2645.
2. Reinhold-Hurek B, Hurek T (1998) Life in the grasses: diazotrophic endophytes. Trends Microbiol 6: 13 9–144.
3. Rosenblueth M, Martínez-Romero E (2006) Bacterial endophytes and their interactions with hosts. Mol Plant Microbe Interact 19: 827–837.
4. Mano H, Morisaki H (2008) Endophytic bacteria in the rice plant. Microbes Environ 23: 109–117.
5. Bacon C, White J (2000) Microbial endophytes. New York, NY (USA): Marcel Dekker, Inc.
6. Sessitsch A, Reiter B, Pfeifer U, Wilhelm E (2002) Cultivation-independent population analysis of bacterial endophytes in three potato varieties based on eubacterial and Actinomycetes-specific PCR of 16S rRNA genes. FEMS Microbiol Ecol 39: 23–32.
7. Barac T, Taghavi S, Borremans B, Provoost A, Oeyen L, et al. (2004) Engineerd endophytic bacteria improve phytoremediation of water-soluble, volatile, organic compounds. Nat Biotech 5: 583–588.
8. Singh OV, Jain RK (2003) Phytoremediation of toxic aromatic pollutants from soil. Appl Microbiol Biotechnol 63: 128–135.
9. Hurek T, Reinhold-Hurek B (1995) Identification of grass-associated and toluene-degrading diazotrophs, *Azoarcus* spp., by analyses of partial 16S ribosomal DNA sequences. Appl Environ Microbiol 61: 2257–2261.
10. Hurek T, Wagner B, Reinhold-Hurek B (1997a) Identification of N_2-fixing plant- and fungus-associated *Azoarcus* species by PCR-based genomic fingerprints. Appl Environ Microbiol 63: 4331–4339.
11. Reinhold B, Hurek T, Niemann EG, Frendrik I (1986) Close association of *Azospirillum* and diazotrophic roods with different roots zones of Kallar grass. Appl Environ Microbiol 52: 520–526.
12. Rabus R, Widdel F (1995) Anaerobic degradation of ethylbenzene and other aromatic hydrocarbons by a new denitrifying bacteria. Arch Microbiol 163: 96–103.
13. Hurek T, Reinhold-Hurek B (2003) *Azoarcus* sp. strain BH72 as a model for nitrogen-fixing grass endophytes. J Biotechnol 106: 169–178.
14. López-Barragán MJ, Carmona M, Zamarro MT, Thiele M, Boll M, et al. (2004) The *bzd* gene cluster, coding for anaerobic benzoate catabolism, in *Azoarcus* sp. strain CIB, J Bacteriol 186: 5762–5774.
15. Rhine ED, Phelps CD, Young LY (2006) Anaerobic arsenite oxidation by novel denitrifying isolates. Environ Microbiol 8: 899–908.

16. Harms G, Zengler K, Rabus R, Aeckersberg F, Minz D, et al. (1999) Anaerobic oxidation of *o*-xylene, *m*-xylene, and homologous alkylbenzenes by new types of sulfate-reducing bacteria. Appl Environ Microbiol 65: 999–1004.
17. Rabus R, Trautwein K, Wöhlbrand L (2014) Towards habitat-oriented systems biology of "*Aromatoleum aromaticum*" EbN1: Chemical sensing, catabolic network modulation and growth control in anaerobic aromatic compound degradation. Appl Microbiol Biotechnol 98: 3371–3388.
18. Anders HJ, Kaetzke A, Kämpfer P, Ludwig W, Fuch G (1995) Taxonomic position of aromatic-degrading denitrifying pseudomonad strain K 172 and KB 740 and their description as new members of the genera *Thauera*, as *Thauera aromatica* sp. nov., and *Azoarcus*, as *Azoarcus evansii* sp. nov., respectively, members of the beta subclass of the Proteobacteria. Int J Syst Bacteriol 45: 327–333.
19. Ebenau-Jehle C, Thomas M, Scharf G, Kockelkorn D, Knapp B, et al. (2012) Anaerobic metabolism of indoleacetate. J Bacteriol 194: 2894–2903.
20. Rabus R (2005) Functional genomics of an anaerobic aromatic-degrading denitrifying bacterium, strain EbN1. Appl Microbiol Biotechnol 68: 580–587.
21. Carmona M, Zamarro MT, Blázquez B, Durante-Rodríguez G, Juárez JF, et al. (2009) Anaerobic catabolism of aromatic compounds: a genetic and genomic view. Microbiol. Mol Biol Rev 73: 71–133.
22. Porter AW, Young LY (2014) Benzoyl-CoA, a universal biomarker for anaerobic degradation of aromatic compounds. Adv Appl Microbiol 88: 167–203.
23. Boll M, Löffler C, Morris BE, Kung JW (2014) Anaerobic degradation of homocyclic aromatic compounds via arylcarboxyl-coenzyme A esters: organisms, strategies and key enzymes. Environ Microbiol 16: 612–627.
24. Juárez JF, Zamarro MT, Eberlein C, Boll M, Carmona M, et al. (2013) Characterization of the *mbd* cluster encoding the anaerobic 3-methylbenzoyl-CoA central pathway. Environ Microbiol 15: 148–166.
25. Reinhold-Hurek B, Hurek T, Gillis M, Hoste B, Vancanneyt M, et al. (1993) *Azoarcus* gen, *nov.*, nitrogen-fixing Proteobacteria associated with roots of Kallar grass *(Leptochoa fusca* (L.) Kunth), and description of two species, *Azoarcus indigens* sp. nov. and *Azoarcus communis* sp. nov. Int J Syst Bact 3: 574–584.
26. Anzai Y, Kim H, Park JY, Wakabayashi H, Oyaizu H (2000) Phylogenetic affiliation of the pseudomonads based on 16S rRNA sequence. Int J Sys Evol Microbiol 50: 1563–1589.
27. Sambrook J, Rusell DW (2001) Molecular cloning: a laboratory manual. Sambrook and Russell 3rd Ed. New York. Cold Spring Harbor Laboratory Press.
28. De Lorenzo V, Timmis KN (1994) Analysis and construction of stable phenotypes in gram-negative bacteria with Tn*5*- and Tn*10*-derived minitransposons. Methods Enzymol 235: 386–405.
29. Herrero M, de Lorenzo V, Timmis KN (1990) Transposon vectors containing non-antibiotic resistance selection markers for cloning and stable chromosomal insertion of foreign genes in gram-negative bacteria. J Bacteriol 172: 6557–6567.
30. Ruiz-Díez B, Quiñones MA, Fajardo S, López MA, Higueras P, et al. (2012) Mercury-resistant rhizobial bacteria isolated from nodules of leguminous plants growing in high Hg-contaminated soils. App Microbiol Biotechnol 96: 543–554.
31. Silva-Rocha R, Martínez-García E, Calles B, Chavarría M, Arce-Rodríguez A, et al. (2013) The standard european vector architecture (SEVA): a coherent platform for the analysis and development of complex prokaryotic phenotypes. Nucl Ac Res 41: D666–D675.
32. Lewis RS, Nicholson JS (2007) Aspects of the evolution of *Nicotiana tabacum* L. and the status of the United States *Nicotiana* Germplasm Collection. Genet Resour Crop Evol 54: 727–740.
33. Elbeltagy A, Nishioka K, Sato T, Suzuki H, Ye B, et al. (2001) Endophytic colonization and *in planta* nitrogen fixation by a *Herbaspirillum* sp. isolated from wild rice species. Appl Environ Microbiol 67: 5285–5293.
34. Chen YP, Rekha PD, Arun AB, Shen FT, Lai WA, et al. (2006) Phosphate solubilizing bacteria from subtropical soil and their tricalcium phosphate solubilizing abilities. App Soil Ecol 34: 33–41.
35. Reinhold-Hurek B, Hurek T (2000) Reassessment of the taxonomic structure of the diazotrophic genus *Azoarcus* sensu-lato and description of three new genera and species, *Azovibrio restrictus* gen. nov., sp. nov., *Azospira oryzae* gen. nov. sp. nov., and *Azonexus fungusphilus* gen. nov., sp. nov. Int J Syst Evol Microbiol 50: 649–659.
36. Miller JH (1972) Experiments in molecular genetics. Cold Spring Harbor Laboratory Press, Cold Spring Harbor, N.Y.
37. Karg T, Reinhold-Hurek B (1996) Global changes in protein composition of N_2-fixing *Azoarcus* sp. strain BH72 upon diazosome formation. J Bacteriol 178: 5748–5754.
38. Bedmar EJ, Olivares J (1979) Nitrogen fixation (acetylene reduction) by free-living *Rhizobium meliloti*. Current Microbiol 2: 1–13.
39. Fernández-Pascual M, de Felipe MR, Serra MT, Pozuelo JM (1988) Effects of cianazine and linuron on chloroplast development, nodule activity and protein metabolism in *Lupinus albus* L. J Plant Physiol 133: 288–294.
40. Brent R, Ptashne M (1980) The *lexA* gene product represses its own promoter. Proc Natl Acad Sci USA 77: 1932–1936.
41. Fedorova E, Redondo FJ, Koshiba T, de Felipe MR, Pueyo JJ, et al. (2005) Aldehyde oxidase (AO) in the root nodules of *Lupinus albus* and *Medicago truncatula:* Identification of AO in meristematic and infection zones. Mol. Plant Microbe Interact 18: 405–413.
42. De María N, de Felipe MR, Fernández-Pascual M (2005) Alterations induced by glyphosate on lupin photosynthetic apparatus and nodule ultrastructure and some oxygen diffusion related proteins. Plant Physiol Biochem 43: 985–996.
43. De Lorenzo CA, Fernández-Pascual M, de Felipe MR (1998) Subcellular localization of glycoprotein epitopes during the development of lupin root nodules. Protoplasma 201: 71–84.
44. Reynolds ES (1963) The use of the lead citrate at high pH as an electron opaque stain in electron microscopy. J Cell Biol 17: 208–213.
45. Salkowski E (1885) Ueber das verhalten der skatokarbonsäure in organismus. Z Physiol Chem 9: 23–33.
46. Gordon SA, Weber RP (1951) Colorimetric estimation of indolacetic acid. Plant Physiol 26: 192–195.
47. Glickmann E, Dessaux Y (1995) A critical examination of the specificity of the Salkowski reagent for indolic compounds produced by phytopathogenic bacteria. Appl Environ Microbiol 61: 793–796.
48. Rodríguez H, González T, Selman G (2000) Expression of a mineral phosphate solubilizing gene from *Erwinia herbicola* in two rhizobacterial strain. J Biotech 84: 155–161.
49. Klement Z (1963) Rapid detection of the pathogenicity of phytopathogenic pseudomonads. Nature 199: 299–300.
50. Dereeper A, Guignon V, Blanc G, Audic S, Buffet S, et al. (2008) Phylogeny.fr: robust phylogenetic analysis for the non-specialist. Nucleic Acids Res 36: W465–469.
51. Rajkumar M, Ae N, Freitas H (2009) Endophytic bacteria and their potential to enhance heavy metal phytoextraction. Chemosphere 77: 153–160.
52. Glick BR (1995) The enhancement of plant growth by free-living bacteria. Can J Microbiol 41: 109–117.
53. Rodríguez H, Fraga R (1999) Phosphate solubilizing bacteria and their role in plant growth promotion. Biotechnol Adv 17: 319–339.
54. Mitter B, Petric A, Shin MW, Chain PSG, Hauberg-Lotte L, et al. (2013) Comparative genome analysis of *Burkholderia phytofirmans* PsJN reveals a wide spectrum of endophytic lifestyle based on interaction strategies with host plants. Trends in Plant Sci 4: 1–15.
55. Hurek T, van Montagu M, Kellenberger E, Reinhold-Hurek B (1995) Induction of complex intracytoplasmic membranes related to nitrogen fixation in *Azoarcus* sp. BH72. Mol Microbiol 18: 225–236.
56. Hurek T, Egener T, Reinhold-Hurek B (1997b) Divergence in Nitrogenases of *Azoarcus* spp., *Proteobacteria* of the β Subclass. J Bacteriol 179: 4172–4178.
57. Goldstein AH (1986) Bacterial solubilisation of mineral phosphates: historical perspective and future prospects. Am J Altern Agri 1: 51–57.
58. Verma SC, Ladha JK, Tripathi AK (2001) Evaluation of plant growth promoting and colonization ability of endophytic diazotrophs from deep water rice. J Biotech 91: 127–141.
59. Wichner S, Libbert E (1968) Interactions between plants and epiphytic bacteria regardling their auxin metabolism. I. Detection of IAA-producing epiphytic bacteria and their role in long duration experiments on tryptophan metabolism in plant homogenates. Physiol Plant 21: 227–241.
60. Camerini S, Senatore B, Lonardo E, Imperlini E, Blanco C, et al. (2008) Introduction of a novel parthway for IAA biosynthesis to rhizobia alters vetch root nodule development. Arch Microbiol 190: 67–77.
61. Patten CL, Blakney AJC, Coulson JD (2013) Activity, distribution and function of indole-3-acetic acid byosinthetic pathways in bacteria. Crit Rev Microbiol 39: 395–415.
62. Quyet-Tien P, Park YM, Ryu CM, Park SH, Ghim SY (2008) Functional identification and expression of indole-3-pyruvate decarboxylase from *Paenibacillus polymyxa* E681. J Microbiol Biotechnol 18: 1235–1244.
63. Tsavkelova E, Oeser B, Oren-Young L, Israeli M, Sasson Y, et al. (2012) Identification and functional characterization of indole-3-acetamide-mediated IAA biosynthesis in plant-associated *Fusarium* species. Fungal Gen Biol 49: 48–57.
64. Hurek T, Reinhold-Hurek B, van Montagu M, Kellenberger E (1994) Root colonization and systemic spreading of *Azoarcus* sp. strain BH72 in grasses. J Bacteriol 176: 1913–1923.
65. Egener T, Hurek T, Reinhold-Hurek B (1998) Use of green fluorescent protein to detect expression of *nif* genes of *Azoarcus* sp. BH72, a grass-associated diazotroph, on rice roots. Mol Plant Microb Interaction 11: 71–75.
66. Germaine K, Keogh E, García-Cabellos G, Borremans B, van der Lelie D, et al. (2004) Colonisation of poplar trees by *gfp* expressing bacterial endophytes. FEMS Microbiol Ecol 48: 109–118.
67. Bulgarelli D, Schlaeppi K, Spaepen S, van Thermaat EVL, Schulze-Lefert P (2013) Structure and functions of the bacterial microbiota of plants. Ann Rev Plant Biol 64: 807–838.
68. Li L, Li Y, Meng Lim T, Pan SQ (1999) Gfp-aided confocal laser scanning microscopy can monitor *Agrobacterium tumefaciens* cell morphology and gene expression associated with infection. FEMS Microbiol Lett 179: 141–146.
69. Ramos C, Molbak L, Molin S (2000) Bacterial activity in the rizosphere analysed at the single cell level by monitoring ribosome content and synthesis rates. Appl Environ Microbiol 66: 801–809.
70. Engelhard M, Hurek T, Reinhold-Hurek B (2000) Preferential occurrence of diazotrophic endophytes, *Azoarcus* spp., in wild rice species and land races of *Oryza sativa* in comparison with modern races. Environ Microbiol 2: 131–141.
71. Chappell J (1998) Phytoremediation of TCE in groundwater using *Populus*. US Environmental Protection Agency, 1998. Available: http://clu-in.org/products/intern/phytotce.htm. Accessed 2014 September 29.

9

A Preliminary Study on *Oxya fuscovittata* (Marschall) as an Alternative Nutrient Supplement in the Diets of *Poecillia sphenops* (Valenciennes)

Arijit Ganguly[1], Ranita Chakravorty[1], Angshuman Sarkar[2], Dipak K. Mandal[1], Parimalendu Haldar[1], Julieta Ramos-Elorduy[3], Jose Manuel Pino Moreno[3]*

1 Department of Zoology, Visva-Bharati University, Santiniketan, West Bengal, India, **2** Department of Statistics, Visva-Bharati University, Santiniketan, West Bengal, India, **3** Departamento de Zoología, Universidad Nacional Autonoma de Mexico, Mexico City, Distrito Federal, México

Abstract

Growth of the ornamental fish industry is being hindered by the scarcity of low cost feed; hence alternative protein supplements should be explored. In this context the present study aims to evaluate whether the grasshopper *Oxya fuscovittata* could be used as a supplement for fish meal in the diets of *Poecillia sphenops*, which is one of the most common ornamental fishes worldwide. The present work is divided into three phases: In the first phase proximate composition of the grasshopper is obtained and five diets are prepared where fish meal is gradually replaced by *Oxya* meal and named as control, D1, D2, D3 and D4. All the diets are formulated on iso-nitrogenous basis where the protein percentage is fixed at 400 g/kg. The second phase deals with feeding trial and in the third phase all the data of the feeding trial are subjected to a linear model. The feeding trial shows that the control, D1 and D2 fed fishes have almost similar results. The linear model proves that the variation in the indices are mainly due to replacement of fish meal by *Oxya* meal, not due to the variations of rice husk and mustard oil cake that are also used to formulate the diets of the present study. From the results two *Oxya* supplemented diets, i.e. D1 and D2 are proved to be almost equivalent to the control diet. Hence it is concluded that *Oxya* meal is able to replace 25% to 50% of fish meal from the diets of *P. sphenops*.

Editor: Alexander Chong Shu-Chien, Universiti Sains Malaysia, Malaysia

Funding: These authors have no support or funding to report.

Competing Interests: The authors have declared that no competing interests exist.

* Email: jpino@ib.unam.mx

Introduction

Fish farmers in general are facing difficulty in raising their fishes because of an alarming increase in the cost of feed which represents 60–70% of the total production cost [1]; and this is also true for ornamental fish industry. In concert with the high rising production cost, the market price of the ornamental fishes is also gaining height [2]. Moreover the conventional protein sources for fish industry i.e. fish meal, soy meal and other grains are being competed for by the human population as well [3]. Recently, with a global population boom the increasing demand and uncertain availability of these traditional protein sources resulted in the quest of other protein rich alternatives for inclusion into the fish diets [4,5]. In this context, insects being nutritionally rich could be a possible alternative [1]. Among the edible insects, acridids (i.e. short-horn grasshoppers of family Acrididae of the order Orthoptera) might have a great future to be established as a sustainable mini-livestock. Acridids are suitable insects for consumption because according to the report of Xiaoming et al. acridids have a high nutritional value and can be used to formulate good quality feed for livestock [6]. But for a constant supply to the feed developing companies there is a need to build acridid farms, for which a suitable species should be found out. In this context Anand et al. reports an estimation of probable annual biomass that could be produced in acridid farms [7]. According to their conclusions *Oxya fuscovittata* (Marschall) is most suitable because being tetra-voltine, they can complete four life cycles annually and could produce a huge biomass. Keeping this in mind the present study aims to utilize *O. fuscovittata* as a supplement of fish meal in the diets of *Poecilia sphenops* (Valenciennes).

Materials and Methods

The present work is divided into three phases; in the first phase analysis of nutrient composition of the ingredients are carried out, which is followed by feed formulation and diet quality assessment; in the second phase the feeding trial is conducted with the selected ornamental fish. During diet formulation fish meal is gradually replaced by *Oxya* meal; but to formulate diets on iso-nitrogenous basis, two other ingredients are needed to vary. Consequently the amounts of those ingredients differ from diet to diet. In this context it is hard to infer that the significant differences in the results of feeding trial is due to replacement of fish meal by *Oxya* meal, or due to the variations in the amounts of the other major ingredients. Hence, in the third phase all the data of the feeding trial are fitted to a linear model, to conclude which ingredient has greater effect on the variations of the observed indices during the feeding trial.

Table 1. Proximate composition of main ingredients.

Ingredients	CP(g/Kg±SD)	EE(g/Kg±SD)	C(g/Kg±SD)	TA(g/Kg±SD)
Fish Meal	483.1±07.2c	92.8±19.0b	381.7±10.1b	42.4±17.5a
Oxya Meal	640.7±02.3d	64.8±03.1a	241.8±04.7a	52.7±0.9a
Rice Husk	54.42±01.5a	49.0±08.0a	742.3±11.8d	154.3±11.1b
Mustard Oil Cake	306.3±12.8b	52.7±12.5a	606.5±12.2c	34.5±11.9a

Note: CP = crude protein, EE = ether extract (crude fat), C = carbohydrate, TA = total ash. Values with different letters within a column are significantly different ($P<0.001$) using Tukey's range test.

The present work has been conducted according to the rules of "Institutional Animal Ethics Committee" of Visva-Bharati University, and approved by the committee according to the Indian law.

Selection of model ornamental fish

To accomplish the present work we have chosen *P. sphenops* (more commonly "black molly") which could be easily reared, could complete life cycle within a short period and easy to breed. Moreover, it is one of the most common ornamental fishes to be seen in the aquariums.

Diet formulation

For the formulation of various supplementary diets *Oxya* meal, fish meal (dried *Harpadon nehereus* F. Hamilton 1822), mustard oil cake (MOC) and rice husk are used as major ingredients. Carboxymethyl cellulose (CMC) is used as a binder for the diets and calcium, table salt (NaCl) and vitamin-mineral mixture are used as additive. First of all proximate composition of the major ingredients are carried out according to AOAC methods [8]. Nitrogen is estimated using Kjeldahl method and crude protein is calculated by multiplying the amount of nitrogen with the factor 6.25 (988.05). Crude fat is estimated with soxhlet apparatus (920.39). Ash contents are determined by subjecting the samples to muffle furnace at 550°C for 6 hrs (942.05). The amount of carbohydrate content is calculated by difference method (100–sum of protein, fat and ash).

Five diets are prepared gradually replacing fish meal by *Oxya* meal with the ratio of 100:0, 75:25, 50:50, 25:75 and 0:100. The diet having "fish meal: *Oxya* meal" with the ratio of 100:0 is considered as control because there is no supplement of *Oxya* meal. The other diets are named D1, D2, D3 and D4 respectively. The amount of ingredients of all of the formulated diets are mixed in such a way so that each of them contain about 400 g/kg crude protein, because according to literature mollies need this much protein in their diets for optimal growth [9]. Maintaining the

Table 2. Ingredient composition and quality of formulated diets.

Ingredients	Control	D1	D2	D3	D4
Fish meal	600.00 g	450.00 g	300.00 g	150.00 g	0.00 g
Acridid meal	0.00 g	150.0 g	300.0 g	450.0 g	600.0 g
Mustard oil cake	339.00 g	267.80 g	173.70 g	80.10 g	12.70 g
Rice husk	11.00 g	82.20 g	176.30 g	269.90 g	337.30 g
Carboxy methyl cellulose	30.00 g	30.00 g	30.00 g	30.00 g	30.00 g
Vitamin-Mineral mixture[a]	7.00 g	7.00 g	7.00 g	7.00 g	7.00 g
NaCl	8.00 g	8.00 g	8.00 g	8.00 g	8.00 g
Calcium additive	5.00 g	5.00 g	5.00 g	5.00 g	5.00 g
Total	1 Kg	1 Kg	1 Kg	1 Kg	1 Kg
Formulated diet quality					
CP (g/Kg ± SD)	399.4±6.2a	400.1±5.3a	401.1±6.7a	399.9±8.1a	400.6±7.6a
EE (g/Kg ± SD)	63.1±11.3a	57.3±9.8a	52.7±12.3a	61.2±11.8a	52.3±8.7a
C (g/Kg ± SD)	498.4±7.7a	517.1±4.8b	522.1±10.1b	523.1±8.6b	523.3±4.4b
TA (g/Kg ± SD)	44.1±4.6b	27.5±3.1a	25.2±4.8a	21.8±3.6a	23.8±2.7a
Energy (kJ/g ± SD)	18.94±0.06a	19.15±0.04b	19.15±0.07b	19.39±0.06c	19.19±0.05b
P/E(mg protein per kJ ± SD)	20.82±0.19a	20.89±0.13a	21.14±0.11b	20.61±0.08a	21.18±0.19b
Striking time (in seconds)	6.50±1.22a	6.72±0.68a	7.05±1.57a	7.24±0.28a	6.36±0.16a

Note: CP = crude protein, EE = ether extract (crude fat), C = carbohydrate, TA = total ash, P/E = protein to energy ratio. Values with different letters within a column are significantly different ($P<0.001$) using Tukey's range test.

[a]Provided *per* Kg of diet: Vitamin A (as Acetate) 69000 IU, Vitamin D3 6900 IU, Vitamin E Acetate 72.5 mg, Ascorbic Acid 1035 mg, Thiamine Mononitrate 69 mg, Riboflavine 69 mg, Pyridoxine HCl 20.7 mg, Cyanocobalamine 103.5 mcg, Niacinamide 690 mg, Vitamin K 1.7 mg, Calcium Pentothenate 112.4 mg, Calcium Phosphate 890.1 mg, Magnesium Oxide 414 mg, Ferrous Fumerate 221.1 mg, Zinc (from zinc sulphate) 15.2 mg, Copper (from copper sulphate) 23.4 mg, Manganese (from manganese sulphate) 14 mg, Sodium Molybdate 1.7 mg, Sodium Borate 6.1 mg.

Table 3. Water quality of the experimental tanks.

Parameters	Range	Mean ± SD
Temperature (°C)	21.40–30.50	27.18±1.25
pH	7.34–7.71	7.52±0.48
Dissolve Oxygen (mg/L)	3.14–3.82	3.61±0.52
Hardness (mg/L)	162.40–167.60	165.67±2.20

amount of protein by adjusting the proportion of ingredients is done by the Pearson's square method. After formulation, diets are first made to dough forms that are subjected to a pelletizer (RJK-FFF-PM-52, Rajkumar Agro Engineers Pvt. Ltd, India), to make floating type pellets.

Assessment of diet quality

Crude protein, crude fat (ether extract), carbohydrate and total ash (TA) are estimated according to AOAC as already mentioned [8]. Energy contents of the diets are estimated by oxygen bomb calorimeter and expressed in kJ/g. Protein to energy (P/E) ratio is an important measure for the quality assessment of the formulated diets and it is expressed as "mg of protein/kJ of energy". Feed acceptability test is assessed as the time taken by the first fish to strike the first particle after the diet dropped into the tanks [10]. The time is recorded in seconds using a stop watch.

Water quality

Analysis of water of the experimental tanks has been done fortnightly in terms of temperature, pH, dissolved oxygen (DO) and hardness. Temperature and pH is measured using a combined pH meter and thermometer (ZPHI-9100), Zico India Ltd., but DO and hardness is measured by titration according to APHA [11].

Collection of experimental species

Same variety of twenty females and ten males of *P. sphenops* are bought from the local market of Bhubandanga, Santiniketan, West Bengal, India. They are reared in the laboratory using a *Spirulina* based diet till the females give birth to sufficient amount of hatchlings. One hundred and fifty (150) fingerlings of similar initial weight (about 0.01 g) are selected to start the feeding experiment, and are kept for 7 days of acclimation period prior to feeding.

Experimental setup

150 fingerlings are divided into 5 sets with 3 replications having 10 individuals in each group. They are placed in 15 experimental tanks of 26 L capacities. Half of the total volume of water is changed daily from each tank. The tanks are completely drained and thoroughly scrubbed once a week to prevent algal growth. Adult body weight and length are measured at the end of experiment after 176 days.

Feeding the fishes

The experimental fishes are fed with known amount of oven dried feed twice daily at 6.30 and 18.30 hrs for five minutes. The uneaten diets are siphoned off and oven dried. The amount of dried feed consumed is calculated by subtracting the left over dried feed from the known amount of dried feed offered.

Estimation of consumption, utilization, survival, growth and reproductive potential of *P. sphenops* when fed with the formulated diets

Growth is measured in terms of specific growth rate (SGR) and condition factor (K) according to De Silva & Anderson, and Moyle & Cech using the following equations [12,13].

$$SGR = [(\log W_{t2} - \log W_{t1}) \times 100]/T$$

$$K = 100 \times (\text{weight}/\text{length}^3)$$

Food consumption and utilization is measured in terms of protein efficiency ratio (PER), feed conversion ratio (FCR), and feed conversion efficiency (FCE) as suggested by De Silva and Anderson using the following equations [12].

Table 4. Results of Survival, SGR, PER, FCR, FCE, K and reproductive potential of *P. sphenops* fed with various formulated diets.

Response indices	Control	D1	D2	D3	D4
% Survival	84.19±2.36a	83.78±2.11a	83.93±1.82a	82.97±2.32a	83.31±1.97a
SGR	1.52±0.10c	1.52±0.08c	1.5±0.17c	1.26±0.06b	1.04±0.14a
PER	14.97±0.28d	15.97±0.24e	12.18±0.40c	3.16±0.26b	1.13±0.05a
FCR	1.16±0.03a	1.14±0.02a	1.18±0.05a	1.86±0.08b	2.21±0.11c
FCE	3.40±0.23d	3.62±0.14d	2.76±0.14c	0.97±0.13b	0.45±0.09a
K	1.66±0.0187c	1.55±0.192c	1.94±0.206c	1.01±0.097b	0.55±0.016a
Reproductive potential/female	14.33±1.53a	14.67±1.15a	15.67±0.58a	14.33±1.15a	15.33±0.58a

Note: SGR = Specific growth rate, PER = Protein efficiency ratio, FCR = Feed conversion ratio, FCE = Feed conversion efficiency, K = Condition factor. Values are means ± SD. Values with different letters within a row are significantly different (P<0.001) using Tukey's range test.

Table 5. Estimate of the regression coefficients of the effects of different factors used in the linear model on each response variable.

Response indices	Estimate of the regression coefficients			
	Intercept [P value]	Replacement [P value]	Rice husk [P value]	Mustard oil cake [P value]
SGR	0.0003 [0.4999]	0.0080 [0.4970]	0.0055 [0.4979]	0.0047 [0.4982]
PER	0.0004 [0.4998]	0.0074 [0.4972]	0.0107 [0.4960]	0.0044 [0.4983]
FCR	0.0002 [0.4999]	0.0089 [0.4967]	0.0016 [0.4994]	0.0051 [0.4981]
FCE	0.0004 [0.4999]	0.0070 [0.4974]	0.0090 [0.4966]	0.0042 [0.4984]
K	0.0003 [0.4999]	0.0077 [0.4971]	0.0067 [0.4975]	0.0045 [0.4983]

Note: SGR = Specific growth rate, PER = Protein efficiency ratio, FCR = Feed conversion ratio, FCE = Feed conversion efficiency, K = Condition factor, Intercept = constant effect, Replacement = *Oxya* meal that replaced fish meal, P value = Area under the t distribution above the observed t value of the student t-test that was conducted to know whether their corresponding factors had significant effect on the response.

$$PER = WWt/DWp$$

$$FCR = F/WWt$$

$$FCE = WWt/F$$

[Where W_{t1} = initial weight, W_{t2} = final weight, T = number of days, F = average dry weight of food ingested per individual, WWt = wet weight gained and DWp = dry weight of protein in feed].

Survival is measured in terms of percentage of individuals alive up to sexual maturity, whereas reproductive potential is measured in terms of total number of offspring born per female in the experimental sets.

Statistical analysis phase 1 and 2

All the data are presented as means ± SD. Each variable is first subjected to one way analysis of variance (ANOVA). Thereafter separation of mean values according to significance is done within each variable by Tuky's range test. All the statistical analyses are carried out using S Plus version 4.0 and PAST version 3.02.

Statistical analysis phase 3

To make sure whether the responses obtained in the feeding trial are due to the replacement of fish meal by *Oxya* meal or due to the variation of rice husk and mustard oil cake, the data of the response indices are fitted to the following linear model using MATLAB program (MATLAB 7.1)–

$$y = \beta_0 + \beta_1 x + \beta_2 (60 - x) + \beta_3 Z_1 + \beta_4 Z_2 + e$$

[Where, y = response indices, β_0 = effect of some constant factors that are identical for all the formulated diet fed sets, β_1 = effect due to x, x = *Oxya* meal, β_2 = effect due to (60-x), (60-x) = fish meal, β_3 = effect due to Z_1, Z_1 = rice husk, β_4 = effect due to Z_2, Z_2 = mustard oil cake, e = error].

$$\text{Or, } y = \beta_0 + 60\beta_2 + (\beta_1 - \beta_2)x + \beta_3 Z_1 + \beta_4 Z_2 + e$$

This equation could be simplified as–

$$y = \beta_0^* + \beta_1^* x + \beta_3 Z_1 + \beta_4 Z_2 + e$$

[Where, $\beta_0^* = (\beta_0 + 60\beta_2)$, and $\beta_1^* = (\beta_1 - \beta_2)$. In other words according to this model, β_0^*, β_1^*, β_3, and β_4 are the "estimate of the regression coefficients" of the effects due to their corresponding factors (i.e. intercept or the constant effect, *Oxya* meal that replaced fish meal, rice husk and mustard oil cake respectively) on the response indices].

In the next step, "estimate of the regression coefficients" are individually subjected to student t-test to know whether their corresponding factors have significant effect on the response indices. If the null hypothesis is accepted, one could conclude that the corresponding factor has no significant effect and if the null hypothesis is rejected, one could conclude that the corresponding factor has significant effect on the response.

Table 6. Comparison between the sum of squares of the effects due to Group1 and Group2 on the response indices.

Response Variables	Sum of squares due to Group1 [$\times 10^3$]	Sum of squares due to Group2 [$\times 10^3$]	P-value of the ratio of sum of squares
SGR	215.9000000	0.7788000	<0.001
PER	0.9156000	0.2053000	0.0883
FCR	543.9200000	26.7680000	<0.01
FCE	21.7300000	3.3322000	<0.01
K	135.9900000	3.4697000	<0.01

Note: SGR = Specific growth rate, PER = Protein efficiency ratio, FCR = Feed conversion ratio, FCE = Feed conversion efficiency, K = Condition factor, Group1 = replacement of fish meal by *Oxya* meal, Group2 = rice husk and mustard oil cake in combination.

After this test the four major factors of the formulated diets (i.e. *Oxya* meal, fish meal, rice husk and mustard oil cake) are divided into two groups–

replacement of fish meal by *Oxya* meal = Group1

rice husk + mustard oil cake = Group2

Then the effect of Group1 on the response indices are statistically compared with the same of Group2, to conclude which group has greater effect on the variation of the response indices. This is done by comparing the sum squares (SS) of the effect of Group1 and Group 2 because SS represents the sum of squared difference from the mean, and this difference gives rise to variance. This part is carried out using the same model already programmed in MATLAB.

Results

Diet formulation

Proximate compositions of the major ingredients i.e. rice husk, *Oxya* meal, MOC and fish meal are tabulated in table 1. It reveals that crude protein level is highest in *Oxya* meal (more than 640 g/kg) and crude fat is highest in fish meal (more than 90 g/kg). Contents of carbohydrate and total ash show higher values for rice husk. After getting the proximate composition of the major ingredients, their required proportions in the formulated diets are calculated. The final compositions of the diets are depicted in table 2.

Assessment of diet quality

Quality of the formulated diets is presented in table 2. Almost 400 g/kg of crude protein level is obtained in each of the diets with no significant difference ($p>0.05$, Tuky's range test). Similarly the values of crude fat (ether extract) again show statistically insignificant variation. The values of carbohydrate contents are also very similar; however, in the control diet it is slightly lower than the others ($p<0.05$, Tuky's range test). Total ash has a result which is reciprocal to the values of carbohydrate where the control diet shows statistically higher ($p<0.001$, Tuky's range test) results. Energy is quite similar, but the control and D3 has significantly lowest and highest mean values respectively ($p<0.001$, Tuky's range test). Protein to energy ratios (P/E) are almost the same (nearly 21 mg/kJ) for all the diets. Results of the acceptability test reveals that the time taken by the first fish to strike the first particle of the formulated diets is within about 6–7 seconds for each case. Hence no significant variation is obtained by Tuky's range test.

Assessment of water quality of the experimental tanks

The temperature, pH, dissolved oxygen and hardness of water during the feeding period are tabulated in table 3. Temperature of all the experimental tanks range from 21.4 to 30.5°C with a mean value of 27.18°C. The water is found to be slightly alkaline as the pH range falls between 7.34–7.71 with a mean value of 7.52. Dissolved oxygen has an average value of 3.61 mg/L and it ranges between 3.14–3.82 mg/L in all the experimental tanks. Assessment of hardness reveals that the water of the experimental tanks is hard with a mean value of 165.67 mg/L that ranges between 162.4–167.6 mg/L.

Consumption, utilization, survival, growth and reproductive potential of *P. sphenops* when fed with the formulated diets

The results of survival, SGR, PER, FCR, FCE, K and reproductive potential of *P. sphenops* are summarized in table 4. More than 80% survival is obtained for all the diet fed sets and they do not vary significantly. But for the other indices as a whole there is a tendency of the control diet, along with D1 and D2 fed *P. sphenops* to have similar results that are significantly better than D3 and D4 fed groups. In case of specific growth rate (%SGR) the values vary insignificantly ($p>0.05$, Tuky's range test) between the control, D1 and D2 fed sets and these values are higher than D3 and D4 fed ones. The values of feed conversion ratio (FCR) also show better results for the control, D1 and D2 diets as the results are significantly lower ($p<0.05$, Tuky's range test) than the groups fed with D3 and D4. On the contrary unlike SGR and FCR, the results of protein efficiency ratio (PER) and feed conversion efficiency (FCE) statistically vary between the control, D1 and D2 fed sets. For PER among these three diets D1 fed groups show highest values ($p<0.05$, Tuky's range test) followed by control and D2 fed groups. For FCE, though the values are insignificant between control and D1 fed groups, the sets fed with D2 shows a slightly lower value ($p<0.05$, Tuky's range test). Condition factor (K) also shows a similar trend, values vary insignificantly between control, D1 and D2 fed mollies and then it gradually decreases in D3 and D4 fed sets. Each female of all the experimental tanks are found to give birth to 13–16 offspring. On an average the values are insignificant ($p>0.05$, ANOVA, Tuky's range test) when data are compared between the sets fed on different formulated diets.

Linear model of response variables

The estimate of the regression coefficients of the effects of different factors used in the linear model on each response index are depicted in table 5. Survival and reproductive potential are excluded from this analysis as they vary insignificantly from diet to diet ($p>0.05$). Values of the student t-test that are conducted to know whether their corresponding factors have significant effect on the response indices are also presented in the same table. The insignificant p values ($p>0.05$, Tuky's range test) show that the null hypothesis is rejected for all. This means all the factors in question (i.e. replacement of fish meal by *Oxya* meal, rice husk and mustard oil cake) has certain amount of effect on the variation of the response data that are obtained during the feeding trial. Now, as it has been evident that all the factors in question are playing some role on the variation of the response data, in the next step it is essential to find out which factor is playing greater role. For this purpose the sum of squares (SS) of the effects due to replacement of fish meal by *Oxya* meal (Group1) is compared with the sum of squares (SS) of the effects due to rice husk and mustard oil cake in combination (Group 2). Results of the sum of squares along with their corresponding p values have been tabulated as table 6. This table has revealed that only excluding PER, for rest of the cases, Group1 has much greater effect on the variations in response data that are obtained during the feeding trial. This proves that the effect of replacement of fish meal by *Oxya* meal has greater influence on the variation of response than the effect of rice husk and mustard oil cake in combination.

Discussion

It is already known that mollies prefer hard water with a pH range of 7.5–8.2, and 18°C−28°C is the suitable temperature for them [14]. In the present study the result of water quality reveals that the water in the experimental tanks is hard with a temperature and pH within the optimum range. However, dissolved oxygen is a little lower than the optimum need (i.e. above 5 mg/L) for most of the fishes [15,16]. The main cause behind this lower DO may be because no aerator is used in the tanks. However, mollies have been reported to get adapted to hypoxic conditions as they could tolerate a level of DO as low as 1 mg/L [17]. In this context it could be stated that the water condition of the experimental tanks are ideal for rearing mollies.

Protein is one of the most important constituents in feed for fish growth. Before designing any experiment on the effect of formulated diets on growth of fishes, the dietary percentage of protein needed for their optimal growth should be obtained first [18]. Shim and Chua worked on poeciliids like *Poecilia reticulata* Peters 1859 and according to them 300–400 g/kg of protein in feed is optimum for their growth [19]. Parker also opines that 400 g/kg dietary protein is optimum for mollies [9]. Hence 400 g/kg of crude protein is fixed for all the formulated diets in the present study. According to some workers, 100–200 g/kg of lipid in fish diets gives optimum growth rates without producing an excessive fatty carcass [20]. In this context it may seem that the diets of the present study contain a little lower amount of fat (near about 50–60 g/kg). However, report of Fernando et al. states that poeciliids are fed with diets having 30–70 g/kg of fat in Singapore [21]. This report supports the view that the formulated diets of the present study contain sufficient amount of fat. It is well known that fish apparently do not have carbohydrate requirement, but it could be metabolized by many fishes and usually herbivorous fishes can metabolize carbohydrates better than carnivorous species [22,23]. According to the report of National Research Council warm water fish (such as mollies) are able to use high carbohydrate levels as an energy source [24]. This phenomenon is advantageous because carbohydrate sources are inexpensive as seen in the present case where all of the formulated diets contain a high amount of carbohydrate. Protein to energy (P/E) ratio is another important measure that gives us a clear picture whether the balance between protein and energy contents is maintained in the diets in proper amount. According to some authors for achieving maximum growth, the P/E ratio in fish diet should range from 19–22 mg/kJ [25–27]. These reports are encouraging because in the present case all of the formulated diets have P/E value of nearly 21 mg/kJ.

Before feeding trial it is necessary to be sure whether the experimental fish recognizes the diets as a palatable one. The results of low feed striking time with insignificant variation positively indicate about the palatability of all the formulated diets. The feeding trial reveals that the results of growth, food consumption and utilization are very similar when fed on control, D1 and D2. On the contrary the values when fed on D3 and D4 are of a lower grade. Among the very little information in literature on using insects as alternative protein supplement for fish, satisfactory results for *Heterobranchus longifilis* Valenciennes 1840 has been reported, where 50% fish meal is replaced by termite meal [28]. Martinez et al. also report satisfactory results when 30% cockroach meal is added to formulate diets for Japanese carp [29]. Along with food consumption, utilization and growth, the measurement of condition factor (K) is also important as it could express the final health of the adult individuals in terms of length and weight at the end of the feeding trial. Results of the condition factor follow a similar trend as observed in case of the other indices already mentioned, because here also the results of control, D1 and D2 diets show ideal results (more than 1.5 [30,31]). Unsatisfactory results are obtained in case of D3 and D4 fed groups might be because of mollies cannot tolerate a composition where more than 50% of fish meal is replaced by the *Oxya* meal. Or this could be an effect of variations in the amount of rice husk and mustard oil cake.

To clear the doubt whether the difference in response variables of feeding trial are due to replacement of fish meal by *Oxya* meal, or due to the effect of change in the amount of rice husk and mustard oil cake, a statistical explanation is important. In this regard results of the linear model quite clearly reveal that only excluding PER, variations of all the indices under consideration occur mainly due to the replacement of fish meal by *Oxya* meal. Here it should be kept in mind that PER is actually a measure of weight gain with respect to the amount of protein consumed through food. The statistical model shows that *Oxya* meal, fish meal, rice husk and mustard oil cake, all have certain effect on the response variables; it is quite clear that all these four ingredients contribute in the protein part of the formulated diets. In this context it could be explained that may be a significant amount of protein contribution by all the four major ingredients leads to an insignificant result when the effect of replacement of fish meal by *Oxya* meal is compared with the combined effect of rice husk and mustard oil cake in case of PER. Results of the present work is encouraging because it is quite clear from this study that at least up to 50% of fish meal could be effectively replaced by *Oxya* meal. This amount of replacement will have no negative effect on the growth and reproduction of *P. sphenops*. Moreover, it is also clear that the variations on the response indices are mainly due to this replacement and not due to the varied amount of rice husk and mustard oil cake.

The results of the present study as a whole support the view of rearing *O. fuscovittata* in "acridid farms" that will easily and constantly provide nutritionally rich animal protein supplement to the aqua-feed developers to formulate supplementary diets for ornamental fishes making the industry more viable. Moreover, if the concept of "acridids as alternative food source" is popularized it could lower the rate of overexploitation of fish meal. Consequently the "demand: supply" ratio of the fish meal could come down, resulting a lower market price of the fish meal also. However, similar studies with various proportions of *Oxya* supplement in the diets of other ornamental fishes are necessary in future. These works also should emphasize on obtaining the digestibility of the formulated diets, which is lacking in the present work. Then only one could be more assured of the suitability of *Oxya* meal as an effective alternative protein source.

Acknowledgments

The authors are thankful to the head of the Department of Zoology, Visva-Bharati University for providing necessary laboratory facilities. Mrs. Mousumi Das is specially acknowledged for her kind help regarding crude protein estimation.

Author Contributions

Conceived and designed the experiments: AG RC DM PH. Performed the experiments: AG RC AS DM PH JR JP. Analyzed the data: AS. Contributed reagents/materials/analysis tools: AG RC DM PH. Contributed to the writing of the manuscript: AG RC As DM PH JR JP.

References

1. Van Huis A, Itterbeeck JV, Klunder H, Mertens E, Halloran A, et al. (2013) Edible insects: Future prospects for food and feed security. FAO Forestry paper 171. Rome: Food and Agriculture Organization of the United Nations. 201 p.
2. Mohanta KN, Subramanian S (2011) Nutrient of common fresh water ornamental fishes. Technical Bulletin No. 27. Goa: ICAR Research Complex for Goa. 55 p.
3. Hossain MH, Ahammad MU, Howlider MAR (2003) Replacement of fish meal by broiler offal in broiler diet. Int J Poult Sci 2: 159–163.
4. Naylor RL, Goldburg RJ, Primavera JH, Kautsky N, Beveridge MCM, et al. (2000) Effect of aquaculture on world fish supplies. Nature 405: 1017–1024.

5. Nyirenda J, Mwabumba M, Kaunda E, Sales J (2000) Effect of substituting animal protein source with soybean meal in diets of *Oreochromis karongae* (Trewavas 1941). Naga, ICLARM q 23: 13–15.
6. Xiaoming C, Ying F, Hong Z, Zhiyong C (2010) Review of the nuritive value of edible insects. In: Durst PB, Johnson DV, Leslie RL, Shono K, editors. Forest insects as food: humans bite back. Proceedings of a workshop on Asia-Pacific resources and their potential for development. Bangkok: FAO Regional Office for Asia and the Pacific. 85–92.
7. Anand H, Das S, Ganguly A, Haldar P (2008) Biomass production of acridids as possible animal feed supplement. **J Environ &** Sociobiol 5: 181–190.
8. AOAC (Association of official analytical chemists) (1990) Official methods of analysis, 15 th edition, volume 1. Virginia: Association of Official Analytical Chemists. 737 p.
9. Parker R (2011) Aquaculture Science. Stamford, CT: Cengage Learning. 672 p.
10. Sogbesan OA, Ugwumba AAA (2006) Bionomics evaluation of garden snail (*Limicolaria aurora*, Jay, 1937; Gastropoda: Limicolaria) meat meal in the diet of *Clarias gariepinus* fingerlings (Burchell, 1822). Nig J Fisheries 2: 358–371.
11. APHA (American Public Health Association) (1985) Standard methods for the examination of water and waste water, 16 th edition. Washington DC: American Public Health Association. 1268 p.
12. De Silva SS, Anderson A (1998) Fish nutrition in aquaculture. London: Chapman and Hall. 319 p.
13. Moyle PB, Cech JJ (1996) Fishes: An introduction to ichthyology. Englewood Cliffs, NJ: Prentice Hall Inc. 590 p.
14. Bailey M, Sandford G (2001) The ultimate aquarium. London: Anness publishing Ltd. 256 p.
15. Lawrence C (2007) The husbandry of zebrafish (*Danio rerio*): A review. Aquaculture 269: 1–20.
16. Bahnasawy MH, El-Ghobashy AE, Abdel-Hakim NF (2009) Culture of the Nile tilapia (*Oreochromis niloticus*) in a recirculating water system using different protein levels. Egypt Aquat Biol Fish 13: 1–15.
17. Timmerman CM, Chapman LJ (2004) Hypoxia and interdemic variation in *Poecilia latipinna*. J Fish Biol 65: 635–650.
18. Guillaume J (1997) Protein and amino acids. In: D'Abramo LR, Conclin DE, Akiyama DM, editors. Crustacean nutrition, Advances in world aquaculture. Louisiana: World Aquaculture Society. 26–41.
19. Shim KF, Chua YL (1986) Some studies on the protein requirement of the guppy *Poecilia reticulata* (Peters). J Aquariculture Aquat Sci 4: 79–84.
20. Cowey CB, Sargent JR (1979) Nutrition. In: Hoar WS, Randall J, editors. Fish physiology. New York: Academic Press. 1–69.
21. Fernando AA, Phang VPE, Chan SY (1991) Diets and feeding regimes of poeciliid fishes in Singapore. Asian Fish Sci 4: 99–107.
22. Furuichi M, Yone Y (1981) Availability of carbohydrate in nutrition of carp and red sea breem. B Japan Soc Sci Fish 48: 945–948.
23. Kalita P, Mukhopadhyay PK, Mukherjee AK (2007) Evaluation of the nutritional quality of four unexplored aquatic weeds from North East India for the formulation of cost-effective fish feeds. Food Chem 103: 204–209.
24. National Research Council (NRC) (1983) Nutrient requirements of warm water fishes and shellfishes. Washington: National Academy Press.102 p.
25. Hasan MR, Moniruzzaman M, Farooque AMO (1990) Evaluation of leucaena and water hyacinth leaf meal as dietary protein sources for the fry of Indian major carp, Labeo rohita (Hamilton). In: Hirano R, Hanyu I, editors. Second Asian Fisheries Forum. Manila: Asian Fisheries Society. 275–278.
26. Akand AM, Hassan MR, Habib MAB (1991) Utilization of carbohydrate and lipid as dietary energy sources by stinging catfish H. fossilis (Bloch). In: DeSilva S, editor. Fish nutrition research in Asia. Proceedings of the fourth Asian fish nutrition workshop. Manila: Asian Fisheries Society. 93–100.
27. Arockiaraj AJ, Muruganandam M, Marimuthu K, Haniffa MA (1999) Utilization of carbohydrates as a dietary energy source by stripped murrel *Channa striatus* (Bloch) fingerlings. Acta Zool Taiwanica 10: 103–111.
28. Sogbesan O, Ugwumba AAA (2008) Nutritional evaluation of termite (*Macrotermes subhyalinus*) meal as animal protein supplements in the diets of *Heterobranchus longifilis* (Valenciennes, 1840) Fingerlings. Turk J Fish Aquat Sc 8: 149–157.
29. Martínez HH, Ramos-Elorduy J, Pino-Moreno JM, Acosta-Castaneda C (2008) Evaluation of diets including nymphs of *Periplaneta americana*. Cienc Pesq 16: 25–30. (In Spanish with English abstract).
30. Lagler KF (1956) Freshwater Fishery Biology. Iowa: W.C. Brown Company. 421 p.
31. Wade JW (1992) The relationship between temperature, food intake and growth of brown trout, *Salmon trutta* (L.) fed natural and artificial pelleted diet in earth ponds. J Aquat Sci 7: 59–71.

10

Divergence of Gene Body DNA Methylation and Evolution of Plant Duplicate Genes

Jun Wang, Nicholas C. Marowsky, Chuanzhu Fan*

Department of Biological Sciences, Wayne State University, Detroit, Michigan, United States of America

Abstract

It has been shown that gene body DNA methylation is associated with gene expression. However, whether and how deviation of gene body DNA methylation between duplicate genes can influence their divergence remains largely unexplored. Here, we aim to elucidate the potential role of gene body DNA methylation in the fate of duplicate genes. We identified paralogous gene pairs from *Arabidopsis* and rice (*Oryza sativa* ssp. *japonica*) genomes and reprocessed their single-base resolution methylome data. We show that methylation in paralogous genes nonlinearly correlates with several gene properties including exon number/gene length, expression level and mutation rate. Further, we demonstrated that divergence of methylation level and pattern in paralogs indeed positively correlate with their sequence and expression divergences. This result held even after controlling for other confounding factors known to influence the divergence of paralogs. We observed that methylation level divergence might be more relevant to the expression divergence of paralogs than methylation pattern divergence. Finally, we explored the mechanisms that might give rise to the divergence of gene body methylation in paralogs. We found that exonic methylation divergence more closely correlates with expression divergence than intronic methylation divergence. We show that genomic environments (e.g., flanked by transposable elements and repetitive sequences) of paralogs generated by various duplication mechanisms are associated with the methylation divergence of paralogs. Overall, our results suggest that the changes in gene body DNA methylation could provide another avenue for duplicate genes to develop differential expression patterns and undergo different evolutionary fates in plant genomes.

Editor: Shin-Han Shiu, Michigan State University, United States of America

Funding: CF was supported by start-up fund from Wayne State University. The funders had no role in study design, data collection and analysis, decision to publish, or preparation of the manuscript.

Competing Interests: The authors have declared that no competing interests exist.

* Email: cfan@wayne.edu

Introduction

Epigenetic modifications of DNA and histones can inheritably regulate the access to the genetic information encoded by DNA [1–4]. DNA methylation, defined as adding a methyl group to the cytosine base of DNA to form 5-methylcytosine, is one of such epigenetic modifications [5]. In plants, cytosine can be methylated in three sequence contexts, CG, CHG and CHH (H = A, C, or T) [6]. DNA methylation can be established *de novo* and maintained by various methyltransferases, e.g. DRM2, MET1, and CMT3 in plants [7–11]. It can also be removed either passively by malfunction of methylation maintenance pathway [12], or actively by demethylation enzymes, e.g. DME and ROS1 in plants [13,14]. DNA methylation is involved in various important biological processes such as repressing the expression of TEs and repetitive elements, and participating in early embryogenesis, stem cell differentiation, X chromosome inactivation, genomic imprinting, neuronal and cancer development [15–22].

DNA methylation can occur in the promoter and gene body regions of genes, which influence gene expression differently. DNA methylation in promoter regions is usually negatively associated with gene expression [23,24]. Whereas, in gene body extreme low or high DNA methylation level is associated with lower gene expression, while modest DNA methylation levels is related to higher expression [25–28]. However, the underlying mechanisms of gene body methylation in regulating gene expression have not been well understood. The causal relationships between gene body methylation and gene expression have been found in preventing transcriptional initiation from alternative promoters within genes, ensuring the accuracy of alternative-splicing and hindering transcriptional elongation [26,29,30]. Although an alternative hypothesis claimed that gene body methylation could be the by-product of expression [31,32], the conservation of gene body methylation among orthologs in evolutionarily distant species and higher proportion of body-methylated genes with phenotypic effects supported that gene body methylation may play certain functional roles [33,34].

Gene duplication plays a critical role in the origination of functional novelties in organisms. How duplicate genes evolve and become fixed in a genome is one of the central questions in molecular evolution [35]. Duplicate genes can be fixed through genetic drift or subfunctionalization/neofunctionalization driven by natural selection [35–45]. Previous studies revealed that the expression and function divergence of duplicate genes can be achieved through various mechanisms, such as nucleotide substitution, *cis*-regulation, post-translational regulation, promoter epigenetic marks and so on [37,40,46,47]. A recent study demonstrated that the DNA methylation in promoter may play a significant role for functional divergence of duplicated genes in human [48].

In previous studies of methylation in duplicate genes, Wildman et al (2009) was the first to investigate the methylation pattern conservation between duplicate genes in *Arabidopsis* [49]. Chang and Liao (2012) discovered that DNA methylation in upstream flanking regions of genes repressed and thus "rebalance" the overall expression dosage of paralogs [46]. Wang et al (2013) found that ~20% of the lineage-specific new duplicate genes had methylation pattern significantly more divergent from their parental genes in gene body than the methylation conservation of all the paralogs in *Arabidopsis* [50]. Wang et al. (2013) examined the divergence of gene body methylation levels of duplicate genes generated from different mechanisms, and the relationship between gene body methylation patterns and synonymous substitutions of paralogs [51]. However, it remains to be explored whether and how DNA methylation, particularly in gene body, influences the fate of duplicate genes in plants [52–54].

Here, we aim to determine how divergence of DNA methylation level and pattern in gene body is associated with expression divergence and evolution of duplicate genes in plant genomes. We chose rice (*Oryza sativa* ssp. *japonica*) and *Arabidopsis thaliana* as the model systems, which have high abundance of gene duplication and modest DNA methylation level in the genomes (in CG context, the methylation levels of rice and *Arabidopsis* are 23.27% and 38.21%, respectively, see Table S1 and [27,55]). Our results show that gene body methylation divergence is associated with divergence of duplicate genes, suggesting that gene body methylation might play an important role in the expression divergence and evolutionary fate of duplicate genes in plant genomes.

Materials and Methods

Plant species selected and genome sequences, bisulfite-sequencing (BS-seq), expression and small RNA data sets

We selected two species, *A. thaliana*, and *O. sativa* ssp. *japonica* for our analyses. We downloaded their genome, CDS, protein sequences, and ".gff" files from Phytozome v8.0 http://www.phytozome.net/with *A. thaliana* 167 (TAIR10) for *A. thaliana*, and http://rice.plantbiology.msu.edu/downloads_gad.shtml (MSU7) for *O. sativa* ssp. *japonica*. We downloaded the BS-seq raw data from the NCBI Short Read Archive (SRA, http://www.ncbi.nlm.nih.gov/sra/) with accession numbers SRA000284 (immature floral tissue) for *Arabidopsis* [55] and SRA012190 (panicle) for *O. sativa* ssp. *japonica* [27], and from the NCBI Gene Expression Omnibus database (http://www.ncbi.nlm.nih.gov/geo/) with accession numbers GSM560562 (embryo), GSM560563 (endosperm), GSM560564 (seedling roots), and GSM560565 (seedling shoots) for rice [28,56]. We downloaded the RNA-seq raw data from SRA with accession number SRA000286 (immature floral tissue) for *Arabidopsis* [55], and digital gene expression (DGE) raw data with accession number GSE20871 (panicle) from the NCBI Gene Expression Omnibus (GEO, http://www.ncbi.nlm.nih.gov/geo) for *O. sativa* ssp. *japonica* [27]. We downloaded *Arabidopsis* Affymetrix expression data of 56 microarray tissues/conditions that were not based on genetic mutants or overlapping tissues from http://www.ebi.ac.uk/arrayexpress/experiments/with accession number E-AFMX-9 [57]. We downloaded the RNA-seq data of endogenous small RNAs in panicle of rice from GSE32973 (GSM816731) of GEO [58], and that of small RNAs in wild-type immature flower of *Arabidopsis* from GSM277608 of GEO [55], which are the same types of tissues, where the methylation data were measured for the two species.

Process methylation data, expression data and small RNA data

We re-analyzed the BS-seq data for *Arabidopsis* and *O. sativa* ssp. *japonica* with Bismark [59]. The intermediate steps included (1) running quality control and trimming the low quality bases with trim_galore (Version 0.2.5) (http://www.bioinformatics.babraham.ac.uk/projects/trim_galore/), (2) mapping the reads with Bismark v0.7.7, (3) removing the duplicates generated by PCR (deduplicate_bismark_alignment_output.pl), (4) generating cytosine methylation reports with Bismark v0.7.7. The plant chloroplast genome has no methylation activity [60], so any methylation reads detected in chloroplast genome should be accounted for the error. To test whether a cytosine is methylated, we conducted binomial test for each cytosine site based on the number of methylated reads, non-methylated reads and the error rate estimated from the chloroplast genome. To correct for multiple comparison problem, we computed the false discovery rate (FDR) '*q*' value for each binomial test '*p*' value [61–63]. The '*q*' value≤0.05 was taken as the criterion of the methylated cytosine [27,55]. We only considered the cytosine mapped with ≥ 5 BS-reads.

We reprocessed the raw RNA-seq data of *Arabidopsis* and DGE data of *O. sativa* ssp. *japonica*. For RNA-seq data, we trimmed the low quality data with trim_galore (Version 0.2.5), mapped the RNA-seq reads to *Arabidopsis* genome with Bowtie-0.12.8 [64], removed duplications due to PCR with picard-tools-1.79 (http://picard.sourceforge.net/), and finally we used cufflinks (v2.0.2) [65] to calculate gene-level relative abundance of reads in Fragments Per Kilobase per Million mapped fragments (FPKM) format. For reprocessing DGE data, we chose the longest transcript for each gene in *O. sativa* ssp. *japonica* genome to build the reference sequences, then mapped DGE reads to the reference sequences with Bowtie-0.12.8, finally counted the number of DGE reads for each gene and normalized the number of DGE reads with the gene length.

We annotated Affymatrix microarray expression data to *Arabidopsis* genes with customized CDF file downloaded from http://brainarray.mbni.med.umich.edu/Brainarray/Database/CustomCDF/17.1.0/tairg.asp. We processed the expression data with RMA package of R. The expression specificity index of each gene was computed according to the approach by Yanai et al [66]. Namely, we computed the tissue specificity index, τ.

$$\tau = \frac{\sum_{i=1}^{N}(1-\chi_i)}{N-1},$$

where N is the number of tissues, and χ_i is the expression component (EC) in the *i*th tissue normalized by the maximal EC value among all the tissues. For the detailed steps to compute χ_i and τ, please refer to Yanai et al's paper [66]. The τ value of each gene ranges from 0 to 1. The higher the τ value, the more specifically expressed the gene is [66].

We collected the processed counts of each small RNA in the genomes. We then mapped the small RNA sequences to the genomes of corresponding species with BWA (bwa-12-17-2013-git) [67]. Based on the above two information plus the coordinates of gene annotations, we estimated the number of small RNAs mapped in each paralogous gene.

Identification and classification of paralogous gene pairs in the two species

We used all the peptide sequences of the genome to Blat [68] against all the peptides for the two species, respectively. We discarded the self-hit pairs and only kept the gene pairs whose reciprocal best hits were each other (namely, we only kept the gene pair of A and B, when the best hit of A is B and the best hit of B is A in the genome) with ≥70% sequence coverage and ≥50% sequence identity at the protein level. The Blat identity score was calculated following the formula in http://genome.ucsc.edu/FAQ/FAQblat.html#blat4. We classified the duplicate gene pairs according to four duplication mechanisms: dispersed duplication, retrotransposition, tandem duplication and segmental duplication. If a group (≥2) of genes, where any adjacent ones are within at most 10-gene distance, and their corresponding paralogs can also be clustered together with any adjacent ones within at most 10-gene distance, we defined them as generated by segmental duplication. If query and target genes of a paralogous pair were adjacent to each other, we defined them as generated by tandem duplication. If one duplicate gene had one exon, and the other copy had multiple exons, and intron losses in the CDS regions of the two genes were identified, we defined them as generated by RNA-based retrotransposition. The rest of paralogous gene pairs, both of which have single or multiple exons and are not belong to segmental, tandem duplication or retrotransposition, were defined as dispersed duplication. If a paralogous pair was classified into more than one duplication mechanism, we didn't count this pair in the analyses of duplication mechanisms. We used Codeml of PAML4.7 [69] to compute Ka, Ks, and K for all the paralogs.

Defining the methylation level/pattern divergence

For methylation level comparison, we only considered the paralogous pairs, where both genes have ≥50% of the cytosines mapped by BS-seq. We conducted Fisher test using 2×2 contingency table to compare numbers of methylated cytosines and non-methylated cytosines (in CG, CHG, CHH context, respectively) in the gene bodies of two paralogous genes. We used FDR (the '*q*' value of the Fisher test '*p*') to correct for multiple test problems. With the '*q*' value of the Fisher test '*p*' value≤0.05, we classified the paralogous pairs as the ones with divergent methylation level, otherwise as the ones with conserved methylation level. And we used the Fisher test '*p*' value as the proxy of methylation level conservation to perform other analyses.

To define the paralogs with conserved methylation level in the promoter regions, we first chose different cutoff length of the regions at 5′ upstream transcription starting sites for two species respectively, and tested the correlation between methylation level and gene expression level in these regions. We identified that 200 bp in rice and 500 bp in *Arabidopsis* have the strongest correlation. Thus, we extracted 200 bp (for rice) and 500 bp (for *Arabidopsis*) upstream regions of 5′ end transcription starting sites as the potential promoter regions. We conducted Fisher test with 2×2 contingency table to compare the methylated cytosines and non-methylated cytosines in the promoter regions of the two paralogous genes. And we chose the paralogs with Fisher test '*p*' value and FDR '*q*' value>0.05 as the ones with conserved methylation level in the promoter regions.

For methylation pattern comparison, we first aligned the CDS of the two paralogous genes with MAFFT [70]. We only considered the paralogous pairs with ≥50% of the aligned cytosines mapped by BS-seq. We classified the aligned cytosines in the same genomic context (both in CG, CHG, or CHH context) as conserved (both methylated or neither methylated), or non-conserved (one methylated and the other un-methylated). For each cytosine context, we computed the percentage of cytosines with conserved status, e.g., the number of cytosines with conserved status/(the number of cytosines with conserved +non-conserved status), as the proxy of methylation pattern conservation for other analyses. We then computed the average percentage of methylation pattern conservation for all the paralogs. We conducted a binominal test with this average percentage, the number of cytosines with conserved status, and the number of cytosine with non-conserved status. We computed the FDR '*q*' value associated with the '*p*' value of binominal test. We defined the paralogs with pattern conservation percentage significantly (FDR '*q*' ≤0.05) lower than the average value as the ones with non-conserved methylation patterns, and the rest as those with conserved methylation patterns.

We further divided paralogous gene pairs into conserved highly methylated, conserved lowly methylated, and conserved non-methylated pairs. If the methylation of both paralogous copies was zero, we classified the pair into the conserved non-methylated group. If the methylation levels of both copies of a pair were higher than the median methylation level of all paralogs with conserved methylation level/pattern, we classified this pair into the conserved highly methylated group. And if they were lower than the median, we classified the pair into the conserved lowly methylated group.

Population genetics analysis and mutation frequency estimation

We obtained *A. thaliana* single nucleotide polymorphism (SNP) data generated from a complete re-sequencing of 80 strains using next-generation sequencing technology [71], downloading from http://1001genomes.org/data/MPI/MPICao2010/releases/current/. We then computed the θ_w values for duplicate genes in *Arabidopsis*. θ_w was defined as the number of segregation sites, which can be estimated using SNP data, divided by a_n where $a_n = \sum_{i=1}^{n-1} \frac{1}{i}$ [72].

To estimate the G/C to A/T mutation frequency of *A. thaliana* genes, we compared *A. thaliana* gene to its *A. lyrata* orthologous sequences. The extracted genomic sequences of *A. thaliana* genes were Blat against *A. lyrata* genome. The genomic sequences of *A. thaliana* genes and their *A. lyrata* orthologous sequences were identified and then aligned. We collected the SNPs of the aligned regions of these *A. thaliana* genes. For each SNP, the derived allele was distinguished from ancestral using the *A. lyrata* orthologous sequence as ougroup. We only considered the G/C→A/T mutations in the sites of the *A. thaliana* gene. The mutation frequency of G/C→A/T was calculated from the quotient of the total number of G/C→A/T mutations and the total number of G/C bases in the aligned region of *A. thaliana* genes.

Analysis of TE distribution of paralogs

We downloaded the repeat library repeatmaskerlibraries-20130422 from http://www.girinst.org/, and mapped the repeat sequences to the genome with RepeatMasker4.0.1 [73]. Based on the mapping result, we divided paralog pairs into three categories: 1) both copies contain TEs/repeat sequences within 1 kb upstream or downstream region; 2) neither copy has TEs/repeat sequences within 1 kb upstream or downstream region; 3) only one of the two copies has TEs/repeat sequences within 1 kb upstream or downstream region. We conducted further analyses based on this classification. Additionally, we used different cutoff length of paralog flanking region to search for the nearby TEs and examined the reliability of our analyses.

Statistical analyses

Correlations were measured with Spearman's rank correlation method. We used 'ppcor' package in R to compute partial correlation [51,74,75]. 'ppcor' can calculate the pairwise partial correlations for each pair of variables while controlling a third or more other variables with three correlation methods (i.g. Pearson, Kendall, and Spearman). Each variable has mathematical expressions and variances to compute the partial correlation with the response variable [76]. It provides the 'p' value as well as statistic for each pair of variables.

The goodness of fit was measured with R-squared (coefficient of determination, R^2), and coefficient significance of linear regression with 'lm' function in R. R^2 is defined as the percentage of the response variable variation that can be explained by a statistical model, so $R^2 =$ explained variation/total variation, and specifically $R^2 = 1 - \frac{\sum_i (y_i - f_i)^2}{\sum_i (y_i - \bar{y})^2}$. The binomial test of the significance of paralog pairs overlapped in the same categories of the two species was conducted using 'binom.test' function in R. The expected probability to overlap is P = (conserved speciesA%*conserved speciesB%+non-conserved speciesA%*non-conserved speciesB%). The number of paralog pairs with data available in both species A and species B is 'N', and the observed number of paralog pairs with same categories in the two species is 'X'.

All the intermediate steps were carried out with Perl and R scripts.

Results

Plant duplicate genes identification and gene body DNA methylation measurement

We identified total 3459 and 2911 paralogous gene pairs in *Arabidopsis* and rice genomes, respectively. We first reprocessed the BS-seq data generated from the immature floral tissue of *Arabidopsis* and the panicle of rice. Our reprocessed results are consistent with previous reports (Table S1) [27,55]. We assessed the DNA methylation divergence between the two paralogous genes from two aspects: (1) methylation level, which estimates the overall methylation level difference in the entire gene body; (2) methylation pattern, which estimates the effect of methylation change in specific cytosine sites. The methylation level was defined as the percentage of methylated cytosines over the total cytosines mapped by BS-seq, for each of the cytosine contexts respectively. Methylation pattern was measured by the percentage of cytosines with conserved methylation status (both methylated/neither methylated) among all mapped aligned cytosines in the coding sequences (CDS) of paralogs, for each of the cytosine contexts respectively. Gene body is typically only methylated in the CG context, so methylation on cytosine in CG context is most abundant in gene body [55,77] and CG methylation in genic region is associated with up-regulation of gene expression [78]. Whereas, CHG and CHH methylation is often associated with RNA-directed DNA methylation and represses gene transcription [5]. Therefore, we focused on CG methylation in the following analyses. Our results showed that the methylation levels of all duplicate genes are bi-modally distributed for both species (Figure 1).

Further, to examine the consistency of gene body methylation level of duplicate genes across different tissues, we reprocessed the raw BS-seq data from embryos, endosperms, seedling roots, and seedling shoots in rice. We pair-wisely plotted the gene body methylation levels of rice duplicate genes among the four tissues with the contour plots. The methylation levels between any pair of the tissues are highly correlated (Spearman's rank correlation coefficients of all pairs of tissues >0.96, p<2.2e-16, Figure 2). This result suggested that the majority of duplicate genes have similar methylation levels across different tissues with few genes with tissue-specific methylation pattern, which is consistent with previous results for all the genes [34,51]. Therefore, it is rational to use the methylation level from one tissue (e.g. the immature floral tissue of *Arabidopsis* and the panicle of rice) as the representative to study the general patterns of methylation for a large set of genes in the following analyses.

Gene body DNA methylation correlates with gene properties

To investigate whether the methylation divergence influences the divergence of duplicate genes, we first measured the relationship between DNA methylation and the properties of duplicate genes including gene structure, transcription and mutation rate.

DNA methylation and gene structure. Previous studies revealed inconsistent relationships between gene length/exon number and methylation. For example, in honeybee and silkworm, lowly methylated genes were significantly longer than highly methylated ones; whereas in the sea squirt and sea anemone, highly methylated genes were significantly longer than lowly methylated ones [79]. And in *Arabidopsis* body-methylated genes are longer and have more exons [33,80]. Using smoothing spline function in R with different degrees of freedom, we plotted the regression line between the methylation level and gene length/exon number for duplicate genes in *Arabidopsis* and rice, respectively [51]. The methylation level 0.6 was present as the inflection point dividing the positive and negative correlation with gene length/exon number (Figure 3). Thus we investigated this relationship by separating the duplicate genes into two groups with methylation level <0.6 and ≥0.6. We found that methylation level positively and negatively correlates to gene length/exon number for the two groups of genes, respectively. Total number of duplicate genes in *Arabidopsis* and rice are 6498 and 4992, respectively. Spearman's rank correlation p<2.2e-16 and R^2 of linear regression model in the two methylation intervals are in the range of 0.14–0.38. The linear coefficients are statistically significant with p value<2e-16 (Table S2). However, the data size of duplicate genes with methylation level ≥0.6 was small with 247 in *Arabidopsis* and 631 in rice, and 26%–43% of them fell into the range of methylation level = 1 (Figure 3), which is corresponding to the methylation level distribution of duplicate genes (Figure 1). Therefore, given the biased sampling, it needs caution to interpret the correlation in this interval. Further, to explore the relative association of gene length and exon numbers with methylation level and their interactions, we examined the partial correlation among the methylation level, gene length and exon number with the "ppcor" package in R. We found gene length and exon number are highly correlated with Spearman's rank correlation coefficient around 0.7 and $p \approx 0$. Gene length is more closely correlated to methylation level than exon number.

Our results show that the gene length/exon number does not linearly correlate to methylation level and intermediately methylated genes are longer and contain more exons in plants. The underlying mechanisms of gene body methylation remain to be explored. However, our analyses imply that gene body methylation may prevent transcriptional initiation from cryptic promoters within genes. Therefore, longer genes with more exons require higher methylation level to ensure their precise expression. Thus methylation level is positively correlated with gene length/exon

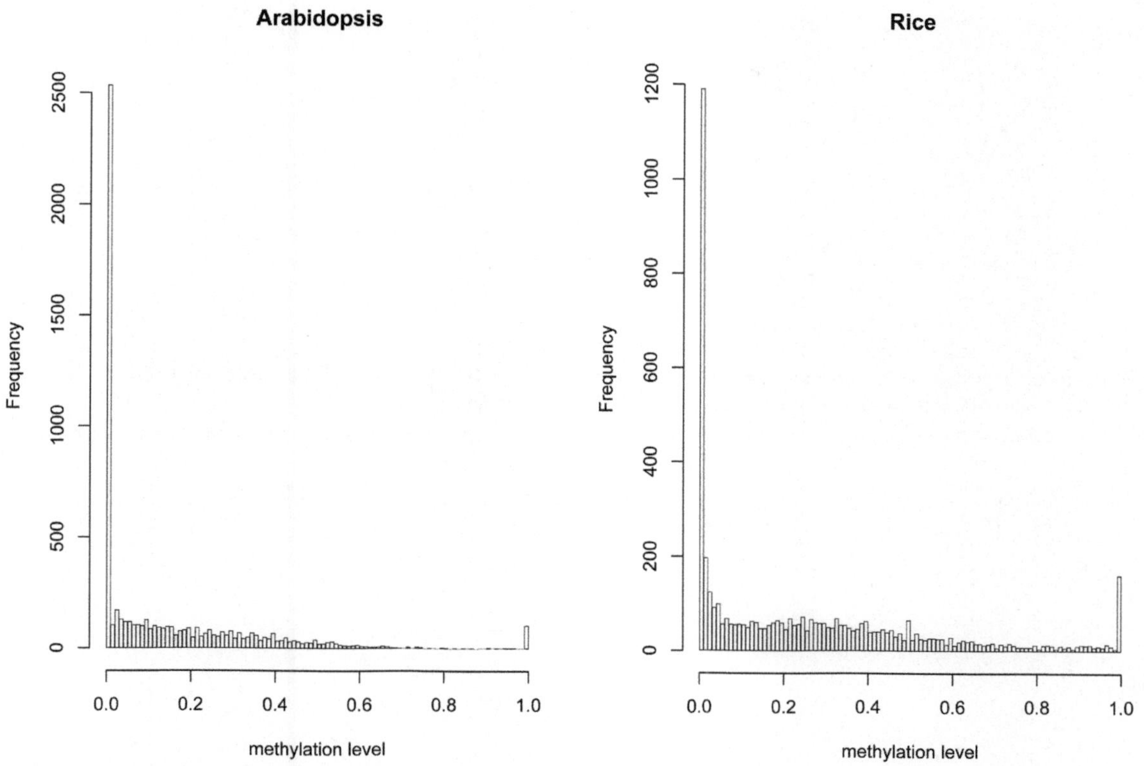

Figure 1. Histograms of methylation level of all duplicate genes in *Arabidopsis* and rice.

number below certain threshold. Meanwhile, extreme high methylation level could hinder transcriptional elongation, so shorter genes can be associated with higher methylation level as well. Thus methylation level is negative correlated with gene length/exon number after gene body methylation level passes certain threshold.

DNA methylation and gene expression. We measured the correlation between gene body methylation and gene expression in terms of expression level and specificity in *Arabidopsis*. Expression level was measured with RNA-seq in the immature floral tissue where the methylation was measured. Whereas, the expression specificity was computed based on 56 tissues/conditions. Previous studies showed that moderately methylated genes had higher expression level, and extremely high or low methylation accompanied lower gene expression [27]. We plotted correlation between methylation level and gene expression level/specificity. Methylation level 0.5 appears as the turning point of the curves, which was determined based on smoothing spline regression (Figure 4). Therefore, we separated the duplicate genes into two groups with methylation level <0.5 and ≥0.5. For the ones with methylation level <0.5, expression level positively and expression specificity negatively correlate with methylation level; and for the ones with methylation level ≥0.5, expression level negatively and expression specificity positively correlate with methylation level. Spearman's rank correlation $p<0.05$ and R^2 of linear regression model in the two methylation intervals are in the range of 0.0079–0.2873 (Table S3). The linear coefficients are statistically significant with p value<1e-5 (Table S3). Additionally, the expression level and specificity strongly negatively correlate with each other (Spearman's rank correlation coefficient: −0.6845 p< 2.2e-16).

The small R^2 between gene body methylation and expression may be due to regulation of gene expression by multiple factors, so the contribution of gene body methylation to expression is relatively small and/or mirrored by other confounding factors. Thus, we examined the general correlation between expression levels/specificity and other factors including small RNA abundance (we focused on the 21nt and 24nt small RNA here, see explanation in the later section), promoter methylation, gene length, exon number, and mutation rate. Mutation rate was measured by theta (θ_w) estimated from SNPs [81]. We found that expression level was significantly negatively associated with the 21nt and 24nt small RNA abundance, promoter methylation level, and mutation rate (p<2.2e-16), but was positively associated with gene length and exon number (p<2.2e-16, Table S4). Expression specificity was significantly negatively associated with gene length and exon number (p<2.2e-16), but was positively associated with 21nt and 24nt small RNA abundance, promoter methylation level, and mutation rate (p<1e-7, Table S4).

To control for the effects of the above factors and test whether gene body methylation exclusively correlates to gene expression, we computed the partial correlation between methylation level and expression level/specificity, considering all the above factors simultaneously using the "ppcor" package of R [74,75]. Our analyses show that the partial correlations between methylation and expression are similar but weaker than the above general correlations (Spearman's rank correlation coefficients. Methylation level <0.5: *vs.* expression level: 0.1375, p = 6.760e-16; *vs.* expression specificity: −0.1694, p = 3.9333e-17. Methylation level ≥0.5: *vs.* expression level: −0.0898, p = 0.1102; *vs.* expression specificity: 0.0934, p = 0.1858). These results suggest that gene body methylation is indeed associated with gene expression, and the associations for duplicate genes are consistent with the overall association for the non-TE genes revealed by the previous studies [25–28,33,77].

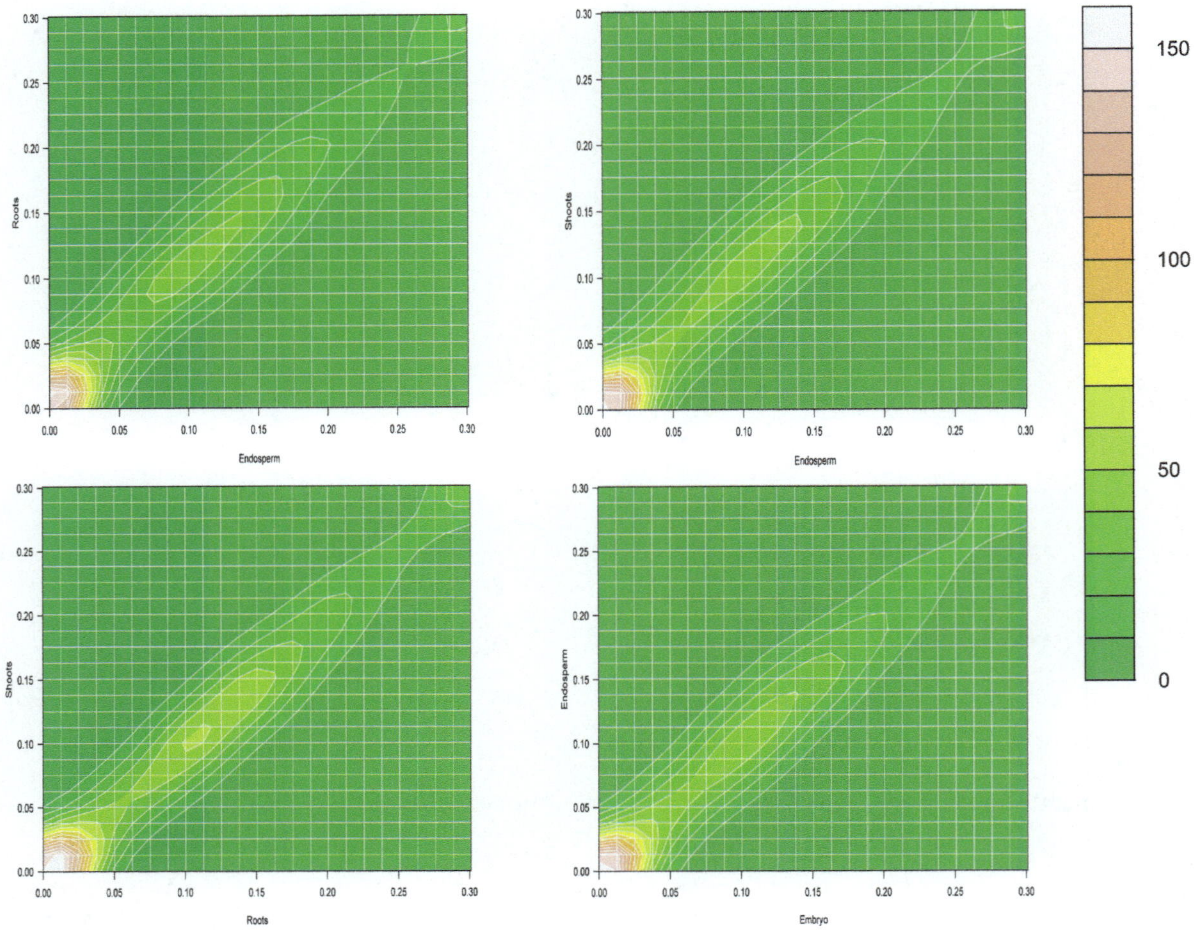

Figure 2. Contour plots of gene body methylation level of rice duplicate genes among embryos, endosperms, seedling roots, and seedling shoots. The 'x' axis and 'y' axis are methylation level in two tissues, respectively. We take log(methylation level +1, base = 10) as the measured values, since a large portion of genes with methylation level = 0.

DNA methylation and gene mutation rate. To investigate the association between methylation level and sequence changes of duplicate genes, we looked into the correlation between methylation level and DNA sequence mutation rate, θ_w, in *Arabidopsis* as the SNP data corresponding to the current reference genome was only available in *Arabidopsis*. Given the inflection point of the regression line between θ_w and methylation level as 0.5 (Figure 5), we also split the duplicate genes into two groups with methylation level <0.5 and ≥0.5. Methylation level negatively and positively correlates with θ_w for the two groups of genes, respectively. Spearman's rank correlation $p<1e\text{-}7$ and R^2 of linear regression model in the two methylation intervals are in the range of 0.0070–0.4304. The linear coefficients are statistically significant, p value< 1e-7 (Table S5). However, the correlation of θ_w and other factors including exon number, gene length, expression, small RNAs and promoter methylation level could be the cofounding factors between methylation level and θ_w (Table S6). After controlled all these factors simultaneously, we still observed the consistent partial correlations between θ_w and methylation level (Spearman's rank correlation coefficients: Methylation level <0.5: 0.0004, p = 9.7994e-01. Methylation level ≥0.5: 0.2241, p = 4.6118e-05). This result suggests that highly methylated genes have higher mutation rates and intermediately methylated genes have lower mutation rates. This is conceivable, since highly methylated genes may experience higher mutation rate due to the mutagenic property of the methylated cytosines, which frequently mutate to thymine [82–84]. To further test this speculation, we computed frequency of G/C→A/T mutation as the percentage of G/C sites with G/C→A/T mutation among all the G/C sites for *Arabidopsis* genes based on *Arabidopsis* SNP data and *A. lyrata* genome sequence (see Method and Materials). We found that when methylation level <0.5, G/C→A/T mutation frequency is negatively correlated with methylation level (Spearman rank's correlation coefficient = −0.1853, p<2.2e-16). Whereas, G/C→A/T mutation frequency is positively correlated with methylation level when methylation level ≥0.5 (Spearman rank's correlation coefficient = 0.1100, p = 0.01869). Thus, this result supported the above conjecture, suggesting methylation and DNA mutation may be able to interact with each other.

Methylation divergence positively correlates with genetic divergence of paralogs

We estimated how divergence of methylation level/pattern was associated with genetic divergence of paralogs, measured by synonymous (Ks) and non-synonymous (Ka) substitution rate. Previous studies suggested that orthologous genes with high methylation levels in both species evolved more slowly than those with low methylation levels in both species, or those with high methylation level in one species but low methylation level in the other species [33,79]. Wang et al (2013) found that the

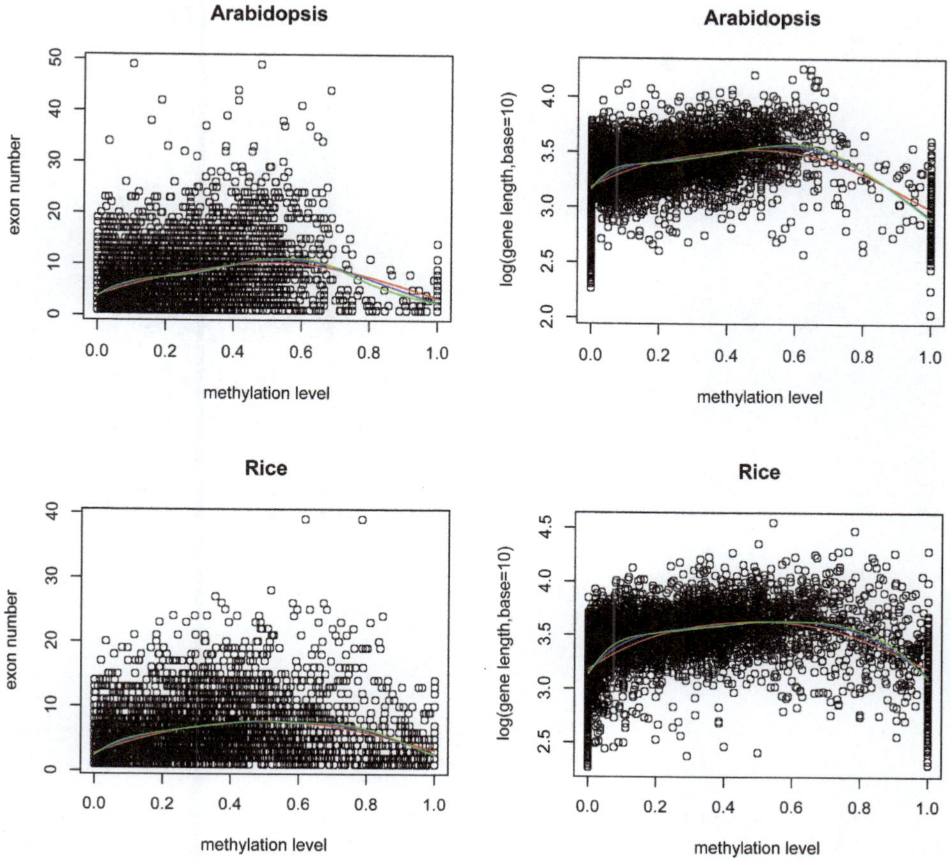

Figure 3. Plots of the methylation level *vs.* gene length/exon number for duplicate genes in *Arabidopsis* and rice. The red, blue, and green lines are corresponded to smoothing spline regression from degree of freedom (*df*) as 4, 6 and 8, respectively. The different *df* values were taken to avoid the over fitting in smoothing spline regression. 0.6 appears as the inflection point of the curves.

methylation level divergence of paralogs was positively correlated with Ks given the gene physical distance (measured by the number of genes between two paralogs) and duplication mechanisms [51]. Consistent to previous study [51], we observed that the physical distance of paralogs negatively correlated with the conservation of methylation level and pattern (Spearman's rank correlation coefficients. *Arabidopsis* methylation level: −0.0582, $p = 4.375e\text{-}03$ and methylation pattern: −0.1068, $p = 1.937e\text{-}07$. Rice methylation level: −0.0217, $p = 0.3082$ and methylation pattern: −0.0848, $p = 7.708e\text{-}05$). Further, we found that the physical distance positively correlated with Ks or Ka (Spearman's rank correlation coefficients. *Arabidopsis* Ks: 0.4639, $p < 2.2e\text{-}16$ and Ka: 0.0586, $p = 4.076e\text{-}4$. Rice Ks: 0.2799, $p < 2.2e\text{-}16$ and Ka: −0.0033, $p = 0.8775$).

Physical distance could be a confounding factor for the relationship between Ks or Ka and methylation conservation. Therefore, we computed the partial correlation between Ks or Ka and methylation level/pattern conservation given physical distance. We found that methylation level conservation negatively correlates with both Ks and Ka in both genomes, but methylation pattern conservation only negatively correlates with Ks and Ka in rice (Spearman's rank correlation coefficients, $p < 0.05$, Table 1; also see Figure S1). Theoretically, Ks implies the divergent time or neutral DNA evolutionary rate, and Ka suggests the amino acid changes. Thus, these results suggest that methylation level divergence of paralogs may increase along with evolutionary time and correlate with the DNA and protein sequence divergence. For example, genetic variables could give rise to methylation divergence or *vice versa* as we showed the correlation between methylation level and mutation rate. Further, no significant correlations between methylation pattern divergence and Ks in *Arabidopsis* could be explained by the process of recurrent methylation and demethylation at specific sites/regions in short period [85].

Methylation divergence is associated with the expression divergence of paralogs

As shown by the above analyses that gene body methylation indeed correlated with gene expression, we examined whether divergence of gene body methylation contributes to expression divergence of paralogs. We divided paralogs into four categories based on the conservation of methylation level/pattern in gene body: conserved highly methylated (CHM), conserved lowly methylated (CLM), conserved non-methylated (CNM), and non-conserved methylated (NCM) (see Materials and Methods). The numbers of paralogous pairs in the four categories for the two species were listed in Table 2.

We further test whether the patterns of methylation divergence of duplicate genes were consistent across different tissues in rice. Interestingly, for methylation level divergence, most duplicate gene pairs (85.16%–89.50%) were classified into the same categories (either conserved or non-conserved) for any pair of the tissues (e.g. embryos, endosperms, seedling roots, and seedling shoots). For methylation pattern, similarly, the majority of duplicate gene pairs (88.22%–92.63%) were in the same categories

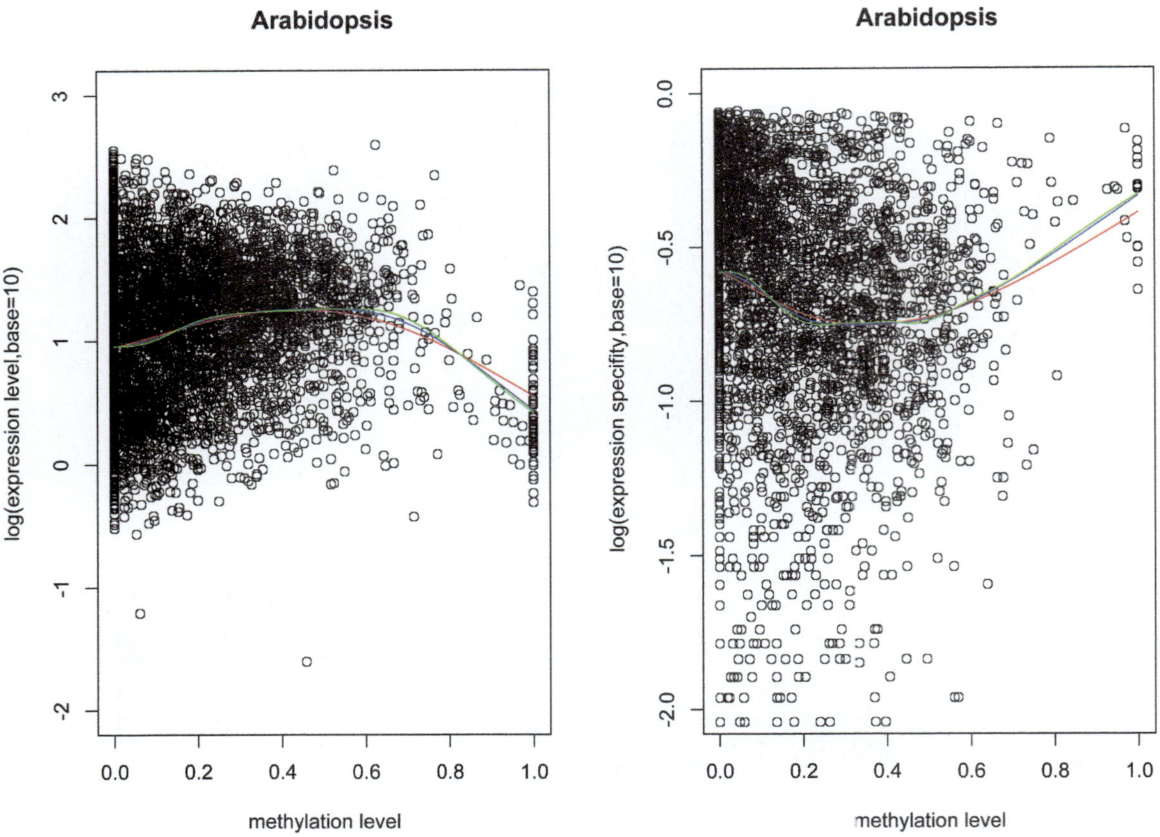

Figure 4. Plots of methylation level *vs.* expression level/specificity for duplicate genes in *Arabidopsis*. The red, blue, and green lines are corresponded to smoothing spline regression from degree of freedom as 4, 6 and 8, respectively. 0.5 appears as the inflection point of the curves.

for any pair of the tissues. Binomial tests indicate that the proportion of these observed gene pairs in either methylation level or pattern are significantly larger than random expectation (p value<2.2e-16). This result suggested that duplicate genes in general maintained the same patterns of methylation divergence across different tissues, which is consistent with the observation in human tissues [48] and was first reported in plants here. It also implies that it is feasible to use the methylation divergence of duplicate genes in one tissue to study the general relationship between methylation divergence and expression divergence.

We computed the expression level fold change and expression specificity difference between the two paralogous genes as the two proxies of their expression divergence, and compared the two proxies among the four categories. The expression level fold change was determined by the absolute quotient of expression level of two paralogous genes and the paralog with higher expression level was always assigned as numerator. We found the expression level and specificity changes of CHM paralogs were significantly lower than those of CLM, CNM and NCM paralogs for the two species in both methylation level and pattern (Figure 6. Wilcoxon rank sum test, $p<0.05$, the detailed p values see Table S7 and sample size see Table 2).

Cofounding factors that associated with methylation and expression divergence of paralogs

As shown previously, expression level/specificity could be affected by other factors, e.g. nucleotide substitution, difference in gene structure (gene length and exon number), *cis*-regulatory motifs, promoter methylation, small RNA abundance, and physical distance (Table S4). Thus expression divergence of paralogs could also result from these factors. We controlled these factors one by one and finally altogether, to determine whether and how the methylation divergence was associated with expression divergence of duplicate genes. Due to the similar behavior of CNM and CLM (Figure 6) and CNM paralogs may be not regulated by methylations, we only performed the comparison among CHM, CLM, and NCM in the following analyses.

The effect of gene structure. To control for the effect of gene structure difference, we choose the duplicate genes with the same ranges of exon number and gene length, which are not significantly different among the three categories in *Arabidopsis* and rice. We found the expression level changes of the CHM paralogs are significantly lower than those of the CLM and NCM ones for methylation level and pattern in both genomes (Wilcoxon rank sum test, $p<0.05$, Table S8).

The effect of nucleotide substitutions. To control for the effects of nucleotide substitutions, we calculated the nucleotide substitution rate (K) in CDS regions, for each of the paralogs, in *Arabidopsis* and rice. We choose the paralogs with the same range of K values (e.g. in the range of 0.05 to 0.2), which were not significantly different among the three categories (Table S9). We found the expression level differences/specificity differences of the CHM duplicate genes are significantly lower than those of the CLM and NCM ones in methylation level and pattern for both species (Figure 7). Wilcoxon rank sum test shows $p<0.05$ except the comparison of expression specificity differences between CHM

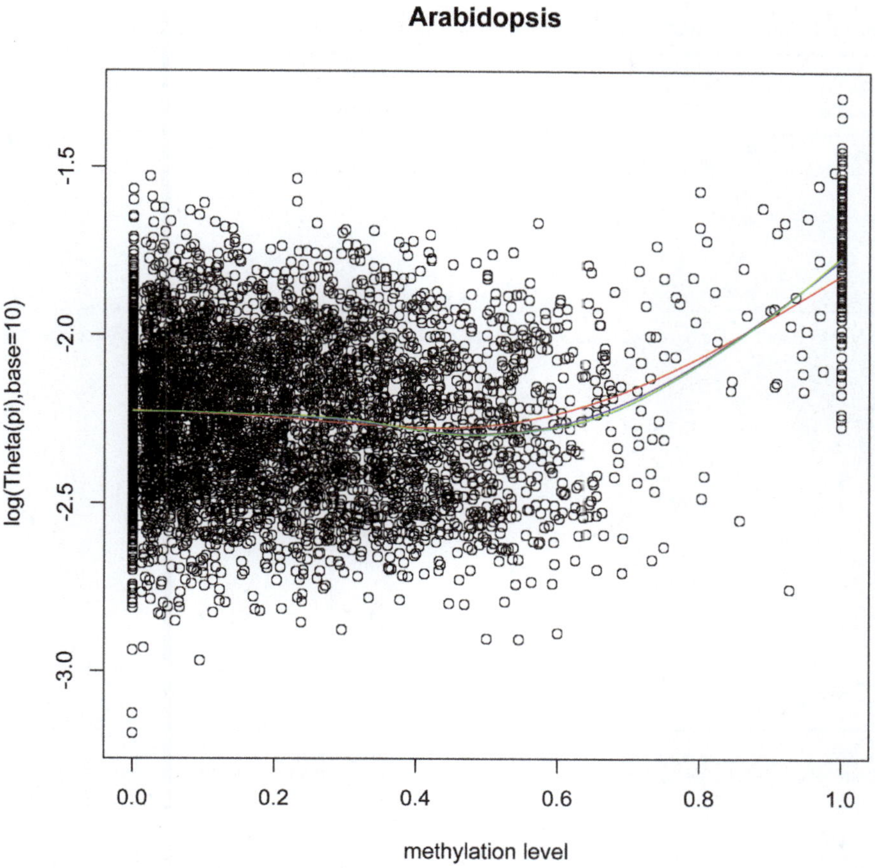

Figure 5. The plot of methylation level *vs.* mutation rate θ_w for duplicate genes in *Arabidopsis*. The red, blue, and green lines are corresponded to smoothing spline regression from degree of freedom as 4, 6 and 8, respectively.

and CLM for methylation level ($p = 0.05418$), and between CHM and CLM for methylation pattern ($p = 0.05397$) (Table S9).

The effect of cis-regulatory difference. To control for the effect of *cis*-regulatory difference, we extracted the *cis*-regulatory binding sites of *Arabidopsis* duplicate genes, based on all the *cis*-regulatory binding sites in *Arabidopsis* genome downloaded from AGRIS (http://arabidopsis.med.ohio-state.edu/downloads.html). We picked the paralogous pairs whose two copies have the same types of *cis*-regulatory binding sites, we still observed that the expression level divergence of the CHM paralogs is significantly lower than that of the NCM ones in methylation level (Wilcoxon rank sum test, $p = 0.0565$ with 101 CHM paralogs and 103 NCM paralogs). However, because the false positive rate of *cis*-regulatory binding site mapping is high, this analysis might not be able to completely and precisely control the effect of *cis*-regulatory difference.

The effect of promoter methylation difference. To control for the effect of promoter methylation difference, we choose the paralogous pairs with conserved methylation level (using Fisher test, see Materials and Methods) in their promoter regions in *Arabidopsis* and rice. Among them, we found that the expression level/specificity changes of the gene body CHM paralogs were significantly lower than those of the CLM and NCM ones in methylation level and pattern for both genomes

Table 1. The partial Spearman's rank correlations between methylation level/pattern conservation and substitution rates in *Arabidopsis* and rice.

Rice	Ks~level	Ka~level	Ks~pattern	Ka~pattern
Coefficient	−0.2218	−0.2175	−0.1175	−0.0702
p value	1.242e-26	1.241e-25	3.753e-08	0.0011
Arabidopsis	**Ks~level**	**Ka~level**	**Ks~pattern**	**Ka~pattern**
Coefficient	−0.1644	−0.1108	−0.0230	0.0212
p value	3.386e-16	4.872e-08	0.2634	0.3022

Note: Correlation coefficient and *p* value were computed with Spearman's rank sum correlation.

Table 2. The number of paralogous genes with different gene body methylation conservation.

Paralogs	For methylation level		For methylation pattern	
	Arabidopsis	Rice	*Arabidopsis*	Rice
CHM	633	403	565	463
CLM	627	402	558	462
CNM	758	159	737	156
NCM	797	729	561	456

Note: CHM-Conserved Highly Methylated; CLM-Conserved Lowly Methylated; CNM- Conserved Non-Methylated; NCM- Non-Conserved Methylated.

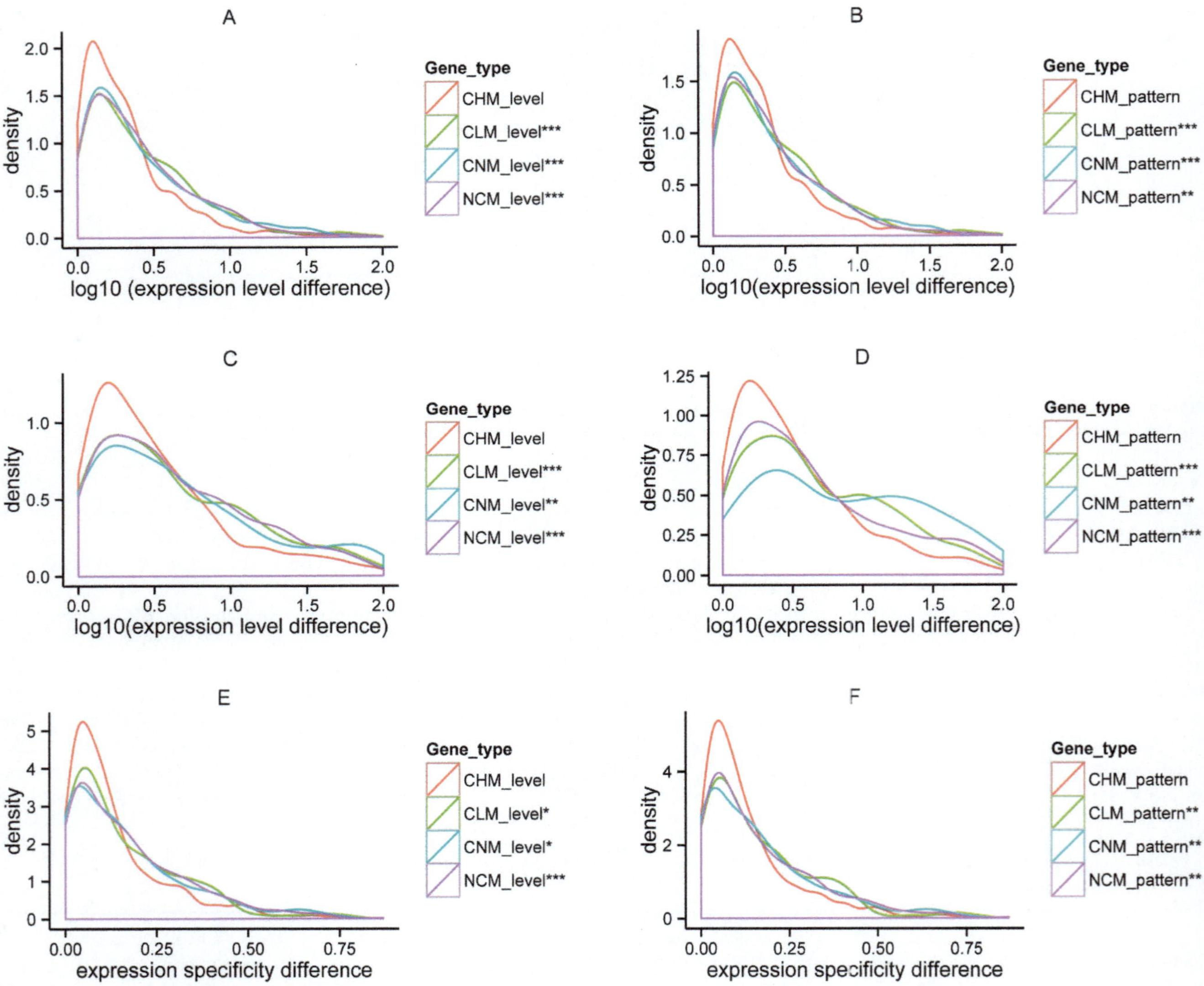

Figure 6. Relationship between expression and methylation divergence of paralogs in *Arabidopsis* and rice. A: Distribution of expression **fold** changes of *Arabidopsis* paralogs in the three methylation **level** conservation categories. B: Distribution of expression **fold** changes of *Arabidopsis* paralogs in the three methylation **pattern** conservation categories. C: Distribution of expression **fold** changes of rice paralogs in the three methylation **level** conservation categories. D: Distribution of expression **fold** changes of rice paralogs in the three methylation **pattern** conservation categories. E: Distribution of expression **specificity** changes of *Arabidopsis* paralogs in the three methylation **level** conservation categories. F: Distribution of expression **specificity** changes of *Arabidopsis* paralogs in the three methylation **pattern** conservation categories. CHM: conserved highly methylated paralogs. CLM: conserved lowly methylated paralogs. CNM: conserved non-methylated paralogs. NCM: non-conserved methylated paralogs. Significant levels (*) are assigned to CLM, CNM, or NCM If their expression changes are significantly higher than those of CHM by Wilcoxon rank sum test. Significant levels were shown as "*": $p<0.05$, "**": $p<0.01$, and "***": $p<0.001$.

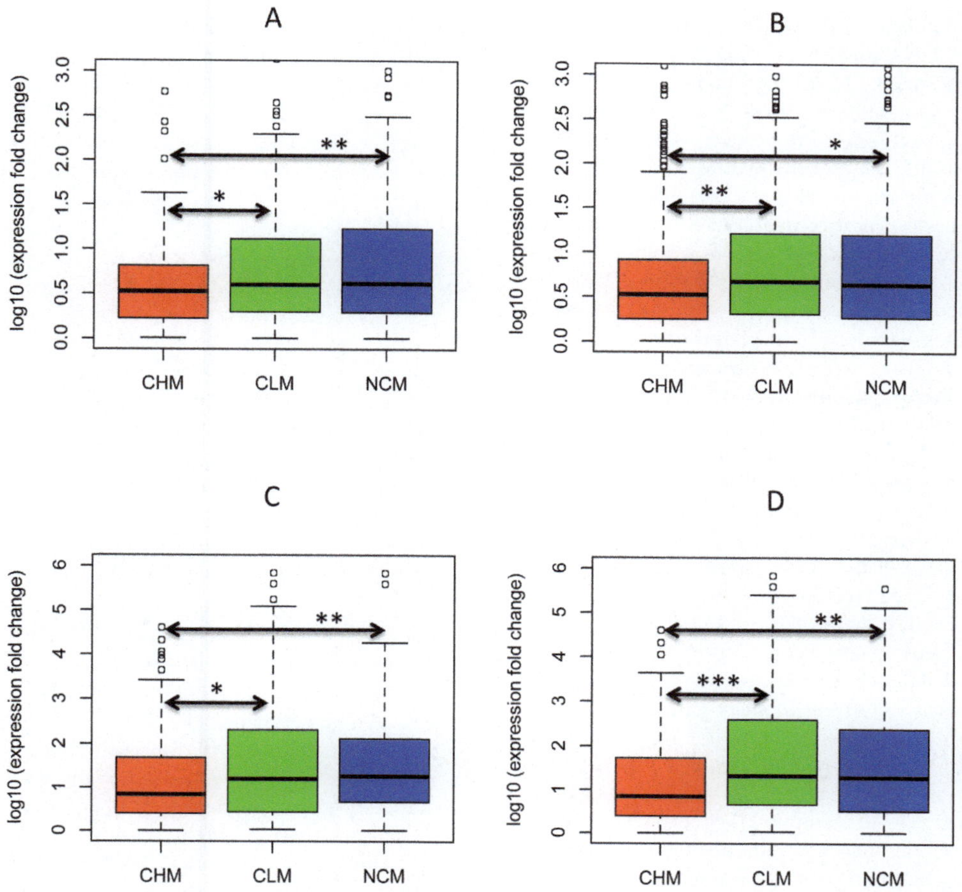

Figure 7. Relationship between methylation level and expression divergence of paralogs in *Arabidopsis* and rice genomes controlling DNA substitution rates (K) in the certain range, which are not significant different among the three methylation conservation categories. A: boxplot of expression **fold** changes of *Arabidopsis* paralogs in the three methylation **level** conservation categories with K in the range of 0.05–0.2. B: boxplot of expression **fold** changes of *Arabidopsis* paralogs in the three methylation **pattern** conservation categories with K in the range of 0.01–0.3. C: boxplot of expression **fold** changes of rice paralogs in the three methylation **level** conservation categories with K in the range of 0.05–0.25. D: boxplot of expression **fold** changes of rice paralogs in the three methylation **pattern** conservation categories with K in the range of 0.05–0.25. Significant levels of Wilcoxon rand sum test were shown as "*": $p<0.05$, "**": $p<0.01$, and "***": $p<0.001$.

(Wilcoxon rank sum test, $p<0.05$, except the comparison of expression specificity changes between CHM and CLM for methylation level $p=0.1246$. Table 3).

The effect of post-transcription regulation. To control for the effect of post-transcriptional regulation (e.g. regulation of the small RNAs), we mapped small RNA sequences to *Arabidopsis* and rice genomes and estimated the numbers of the small RNAs mapped to each duplicate gene normalized by gene length. Because small RNAs with different length may regulate gene expression differently [86]. To simplify and standardize the experiment, we focused on 21nt and 24nt small RNAs, which are the most abundant small RNA species in *Arabidopsis* and rice

Table 3. Comparison of expression divergences (*p* value) in three gene body methylation conservation categories after controlling promoter methylation divergence.

Species	Expression level divergence		Expression specificity divergence	
Arabidopsis	CHM<CLM	CHM<NCM	CHM<CLM	CHM<NCM
Methylation level	3.432e-06	7.388e-07	0 1246	0.02
Methylation pattern	4.929e-04	0.02239	0 005676	0.005267
Rice	CHM<CLM	CHM<NCM		
Methylation level	1.831e-05	1.458e-05		
Methylation pattern	2.067e-07	8.958e-04		

Note: *p* value was computed with Wilcoxon rank sum test.

[55,58,87] and can influence gene expression through microRNA and siRNA induced DNA methylation and microRNA mediated mRNA cleavage [8,86,88]. Then, we choose gene pairs with similar proportions (the number of small RNAs/the gene length) of 21nt and 24nt small RNAs respectively between the two paralogous copies, and found that the expression level changes of the CHM paralogs were still significantly lower than those of the CLM and NCM ones for most comparisons in methylation level and pattern for the two genomes (Wilcoxon rank sum test, $p<0.05$, although some comparison had insignificant p values (Table S10). However, the results should be interpreted with the caution since small RNAs data used here were limited and our operation may not completely control the effect of post-transcription regulation.

Controlling all factors simultaneously. Finally, to control for nucleotide substitution, differences in exon number/gene length, promoter methylation and small RNA abundance, and physical distance simultaneously, we pooled the CHM and NCM duplicate genes together to compute the partial correlation between expression divergence and methylation conservation while controlling for all of the above factors in *Arabidopsis* and rice. We excluded CLM paralogs in this analysis, because, mechanistically, the CLM paralogs are both lowly methylated, thus their genetic and expression divergences are less likely controlled by methylation. In contrast, expression of CHM paralogs would have a stronger effect imposed by DNA methylation. Indeed, we observed that CLM paralogs have different expression divergence pattern (significant higher expression divergence) from CHM paralogs. If we pooled CLM and CHM paralogs together to compute the correlation between methylation conservation and expression divergence, the correlation would contain noises and could not stand out. We found that the partial correlations are significantly negative in both genomes (Spearman's rank correlation $p<0.05$, Table 4 and also see Figure S2). Further, the two proxies of expression divergence, e.g. expression level and specificity divergence, positively correlated with each other (Spearman's rank correlation coefficient 0.3551, $p<2.2e\text{-}16$), and methylation conservation has a stronger negative correlation with expression level change than specificity difference (Table 4). Overall, these results indicated that methylation divergence does associate with expression divergence of duplicate genes.

Methylation level divergence is overall more important for expression divergence of paralogs than methylation pattern divergence

To determine whether divergence of methylation level or pattern is more relevant to expression divergence of paralogs, we classified the paralogs into the following four categories: 1) Conserved high methylation Level and Pattern (CLCP); 2) Conserved high methylation Level and Non-conserved methylation Pattern (CLNP); 3) Non-conserved methylation Level and Conserved high methylation Pattern (NLCP); and 4) Non-conserved methylation Level and Pattern (NLNP). We found that in *Arabidopsis* there was no significant difference between the expression level changes of CLCP and CLNP paralogs, as well as between those of NLCP and NLNP ones (Figure 8). However, the expression changes of NLCP and NLNP paralogs are significant larger than those of CLCP and CLNP ones, respectively. Furthermore, the expression changes of NLCP ones are significantly larger than those of CLNP ones (Figure 8). This observation suggests that methylation level divergence overall is more relevant to expression change of paralogs than methylation pattern divergence in *Arabidopsis*.

Exonic regions contribute more to gene body methylation divergence than intronic regions

To estimate whether the methylation level divergence in exonic and intronic regions contribute equally to the gene body methylation divergence, we divided the paralogs into the following categories using previous approaches: 1) conserved methylation level (CM) in exonic regions; 2) non-conserved methylation level (NCM) in exonic regions; 3) CM in intronic regions; 4) NCM in intronic regions. With Fisher test, we found that the proportion of exonic NCM among exonic NCM+CM was significantly higher than that of intronic regions in *Arabidopsis* and rice (Table 5, Fisher test, *Arabidopsis*: $p<2.2e\text{-}16$; rice: $p = 2.212e\text{-}10$), suggesting exonic methylation divergence may contribute more to gene body methylation divergence than intronic methylation. Further, we found exonic methylation divergence is more closely correlated with expression divergence than intronic methylation divergence (Table S11), suggesting the exonic methylation divergence may play a more important role in the association with gene expression compared to intronic methylation divergence, for example, through regulating the alternative-splicing of transcripts.

Dispersed duplicate genes tend to have more divergent DNA methylation level/pattern

At the end, we explored the potential mechanisms contributing to the differential methylation between paralogs. Previous studies showed that DNA-based dispersed duplications in *Arabidopsis* tend to differ in DNA methylation in promoter regions than other gene duplication mechanisms [89]. In rice, different duplication mechanisms generate paralogs with different physical distance; which are associated with different body methylation level conservation [51]. Our analyses indicated that DNA-based dispersed gene duplication (Fisher test: methylation level: $p = 2.69e\text{-}05$; pattern: $p = 0.0074$) and RNA-based retrotransposition (Fisher test: methylation level: $p = 0.06467$) generated a higher proportion of paralogs with divergent methylation level and/or pattern than tandem and segmental duplication in rice. We also found the methylation level (Wilcoxon rank sum test, $p = 3.248e\text{-}06$) and pattern (Wilcoxon rank sum test, $p = 1.689e\text{-}9$) of segmental duplication is more conserved than those of the duplicate genes generated by single events, such as tandem duplication, dispersed duplication in rice. However, we did not observe such pattern in *Arabidopsis*, which might be due to the intrinsic methylation property of *Arabidopsis* genome. And the underlying mechanism remains further to be explored.

Previous studies have demonstrated that various mechanisms contributed to the methylation variation in trans-generational epialleles, and indicated that DNA methylation could spread from highly methylated transposable elements (TEs) and repetitive sequences to the nearby genes [3,4,77,85,90,91]. Given our discovery that dispersed duplications generate a higher proportion of divergently methylated paralogs in rice, we asked whether the dispersed paralogs are more frequently associated with asymmetric TE/repeat sequence distribution, namely one duplicate copy flanked with TE/repeat sequence but not the other in rice. This could lead to the one flanked by TE/repeat sequence being more likely to be methylated, while the other without flanking TE/repeat sequence was not. Interestingly, we found that DNA-level dispersed duplication and retrotransposition paralogs were more frequently associated with asymmetric TE/repeat sequence distribution than tandem and segmental duplication ones in rice (Fisher test, $p<0.05$). The same trend was observed when we used different cutoff length of paralog flanking regions to search for the nearby TEs (Table S12).

Table 4. Spearman correlations between methylation conservation and expression divergence with other factors simultaneously controlled.

Species	Expression level divergence			
Arabidopsis	Partial coef	*p* value	General coef	*p* value
Methylation level	−0.1405	2.003e-06	−0.1431	4.1e-05
Methylation pattern	−0.0955	0.0038	−0.0815	0.0355
Arabidopsis	**Expression specificity divergence**			
	Partial coef	*p* value	General coef	*p* value
Methylation level	−0.0928	0.0081	−0.1135	0.0012
Methylation pattern	−0.0922	0.0177	−0.1007	0.0093
	Expression level divergence			
Rice	Partial coef	*p* value	General coef	*p* value
Methylation level	−0.1141	0.0014	−0.1524	2.048e-05
Methylation pattern	−0.1333	9.135e-04	−0.1410	4.451e-04

Note: Correlation coefficient and p value was computed with spearman correlation.

Discussion and Conclusions

In human, a exclusive relationship of gene body DNA methylation with evolutionary and expression divergence of paralogs could not be revealed [48]. Our analysis in plants added a different aspect regarding with the property and significant role of gene body DNA methylation in gene duplication. Gene body methylation is a pervasive phenomenon in various species and their pattern is highly conserved in orthologs from highly diverged species [23,28,34,79,92,93]. Furthermore, as demonstrated previously, gene body DNA methylation in plant genomes has been considered as the main mode of DNA methylation and links with gene transcription [34]. Expectedly, the divergence of gene body methylation could play a functional role in influencing the evolution and divergence of paralogs given the interplay between gene transcription and DNA methylation [26,29,30,94]. Specifically, paralogs initially tended to maintain same expression pattern and retained similar function; and gene body DNA methylation could regulate gene expression. Therefore, the divergence of gene body DNA methylation could provide a fast trajectory to differentiate gene expression of paralogs, leading to the divergence and thus the preservation of duplicate genes [52].

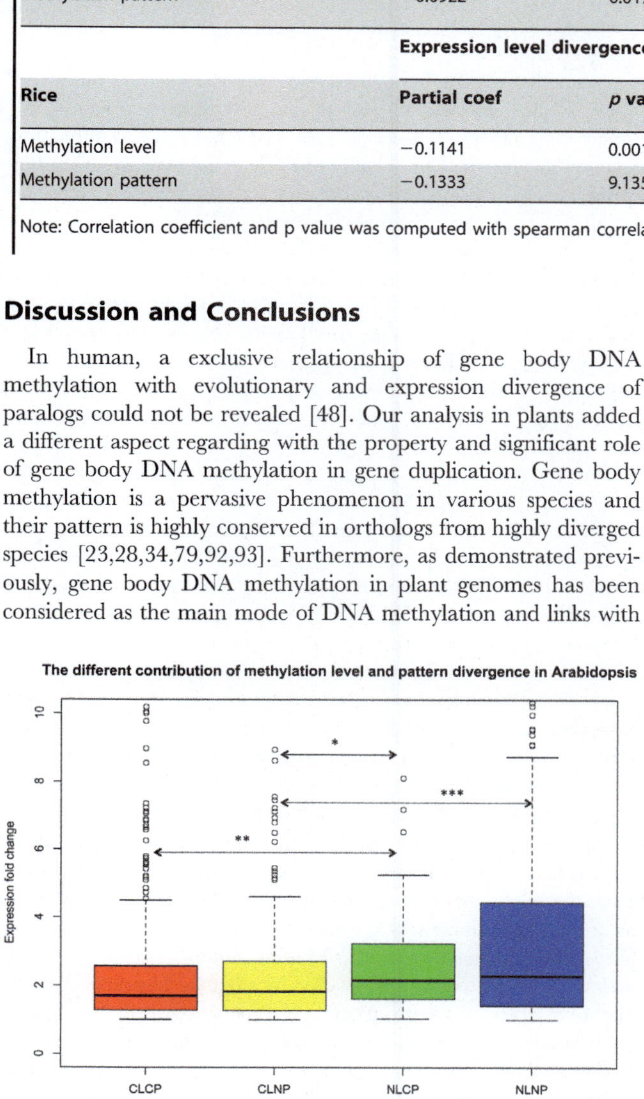

Figure 8. Differential contributions of methylation level and pattern divergence to expression level divergence of *Arabidopsis* paralogs. CLCP: paralogs with Conserved high methylation Level and Pattern. CLNP: paralogs with Conserved high methylation Level and Non-conserved methylation Pattern. NLCP: paralogs with Non-conserved methylation Level and Conserved high methylation Pattern. NLNP: paralogs with Non-conserved methylation Level and Pattern. Significant levels of Wilcoxon rank sum test were shown as "*": $p<0.05$, "**": $p<0.01$, and "***": $p<0.001$.

To validate the association between gene body DNA methylation and divergence of plant duplicate genes and the contribution of methylation to the divergence of plant duplicate genes, we analyzed a large number of reliable paralogs spanning different evolutionary ages from two distantly related plant species, *Arabidopsis* and *Oryza*, which represent dicots and monocots, respectively. We first showed that divergence of methylation level was positively correlated to the genetic evolution of paralogs in terms of Ks and Ka. This result suggested that methylation level variation accumulates along evolutionary time, and companies with neutral DNA mutation and amino acid changes. These postulations are supported by the previous discoveries that CG-SMP (single methylation polymorphism)-based phylogenetic tree is correlated to the SNP-based phylogenetic tree among multiple populations of *Arabidopsis* or closely related rice species [27,77].

Differential gene body DNA methylation covaries with gene expression divergence between duplicate genes. It has been hypothesized that epigenetic modification of paralogs could facilitate functional divergence of duplicate genes [52]. Consistently, we demonstrated that conserved highly methylated duplicate genes have lower expression level and specificity divergence than conserved lowly- or non-conserved methylated duplicate genes. Remarkably, this pattern still robustly exists even after we controlled other factors known to affect gene expression divergence. Our observation is reflecting with previous trans-generational epigenetic variation studies, which indicated that epigenetic variation could contribute to phenotypic diversity [85,90] and that *de novo* genes could originate through change

Table 5. Paralogs with CM and NCM in exonic and intronic regions.

Species	Exonic		Intronic		Exonic	Intronic	
	CM	NCM	CM	NCM	NCM/(CM+NCM)		*p* value[a]
Arabidopsis	1741	601	1165	13	25.66%	1.10%	<2.2e-16
Rice	1435	693	1214	367	32.57%	23.21%	2.212e-10

[a]*p* value was computed with Fisher test.

the DNA methylation status (Silveira, et al. 2013). However, two additional considerations should be explored futher. First, although we quantitatively controlled for other factors, we could not controll for all the factors completely. For example, the same amount of nucleotide subsititions occuring in the different gene regions might also contribute to expression divergence of paralogs, and the data available for *cis*- and post-transcription regulation are limited. Also, gene body methylation might interplay with other mechanisms together and contribute to the divergence of duplicate genes.

Methylation level measures the overall methyaltion density of genes. Whereas methylation pattern measures methylation property of specific cytocine sites of genes. Therefore, methylation level and pattern divergence could behave differentially in affecting the expression divergence. We identified the substantial roles of DNA methylation level for expression and genetic divergence of *Arabidopsis* duplicate genes. This result supports the hypothesis that gene body methylation might be maintained as a property of regions with a minimal/maximal methylation level over the whole region while experiencing site to site methylation stochasticity, driven by natural selection [34]. Alternatively, methylation changes in specific sites could also affect gene expression. For example, it was previously found that enhancer access could be modulated by a single methylation event [24].

The methylation divergence of duplicate genes may be the outcome of spontaneous methylation variation, which might be impacted by natural selection (see reviews by [95–97]). A few factors could contribute to the methylation divergence of paralogs. For example, methylation divergence from exons and introns could differently contribute to the divergence of gene body methylation in duplicate genes. We found that more methylation divergence occurs in the exonic regions than intronic regions. Previously, two studies reported conflicting results about the involvement of exons and introns in differentially methylated regions (DMR) among different *Arabidopsis* lines due to spontaneous epigenetic variation. Becker et al. (2011) showed that CG-DMRs in exons were more abundant than those in introns [85], while Schmitz et al. (2011) found the opposite pattern [90]. Previous studies also showed that methylation in exons may be associated with mRNA-splicing, thus exonic methylation divergence might be relevant to the mRNA-splicing divergence and expression divergence of paralogs [29,98].

We demonstrated that exonic methylation divergence is more closely correlated with expression divergence than intronic methylation divergence, which implies that the exonic methylation divergence might have more functional consequences than intronic methylation divergence. Conclusively, exonic methylation variation might have a higher chance to be maintained than intronic methylation variation, which can explain our observation that exonic regions bear more methylation divergence than intronic regions. However, it remains to be explored other functional roles of methylation in exonic and intronic regions. Furthermore, the genomic environments of duplicate genes can also impact the divergence of methylation among paralogs. Our analysis confirmed that the physical distance of paralogs negatively correlated with the conservation of methylation level and pattern, and that dispersed duplicate genes more tend to have divergent methylation level and pattern.

We found inconsistent patterns, for some analyses, in *Arabidopsis* and rice. This may be due to two factors. First, the limitation of availability and the quality of the data in two genomes are different. For example, the expression data of *Arabidopsis* were obtained from RNA-seq but those of rice were generated from DGE. Second, properties of methylation in the two genomes are different. For example, the methylation level in rice is at least two-fold higher than that in *Arabidopsis*.

In conclusion, the correlation between the divergence of gene body DNA methylation and expression in plant duplicate genes implicates that gene body DNA methylation could serve as another avenue for duplicate genes to develop different expression and undergo different evolutionary fates. Our study indicates that gene body DNA methylation as one type of epigenetic modifications is an important facilitator that potentially drove divergence and evolution of plant duplicate genes.

Supporting Information

Figure S1 Plots of methylation level *vs.* Ka and Ks for duplicate genes in *Arabidopsis* and rice. Red line is the linear regression fit to the data, which were generated with *lm* function in R.

Figure S2 Plots of the methylation level *vs.* the expression level for duplicate genes in *Arabidopsis* and rice. Red line is the linear regression fit to the data, which were generated with *lm* function in R.

Table S1 Reprocessed error rate in un-methylated chloroplast genome and methylation data for *Arabidopsis* and rice.

Table S2 The correlation between methylation level and gene length/exon number.

Table S3 The correlation between methylation level and expression level/specificity in *Arabidopsis*.

Table S4 The correlation between expression level and other factors.

Table S5 The correlation between methylation level and theta in *Arabidopsis*.

Table S6 The correlation between theta and other factors.

Table S7 The relationship of methylation conservation and expression divergence that is corresponded to Figure 6.

Table S8 The relationship of methylation conservation and expression divergence with gene structure controlled.

Table S9 The relationship of methylation conservation and expression divergence with nucleotide substitution rate controlled.

Table S10 The relationship of methylation conservation and expression divergence with small RNA abundance controlled.

Table S11 The correlation between exonic/intronic methylation divergence and expression level divergence.

Table S12 The p-values corresponding to the different flanking region length cutoffs in searching nearby TEs.

Acknowledgments

We thank Dr. Xin Li from Johns Hopkins University School of Medicine, Mr. Muhua Wang from University of Arizona, and Dr. Chengjun Zhang from University of Chicago for helpful discussions. We thank the grid computing service from Computing & information Technology of Wayne State University.

Author Contributions

Conceived and designed the experiments: JW CF. Performed the experiments: JW NCM. Analyzed the data: JW CF. Contributed reagents/materials/analysis tools: JW CF. Contributed to the writing of the manuscript: JW NCM CF.

References

1. Cubas P, Vincent C, Coen E (1999) An epigenetic mutation responsible for natural variation in floral symmetry. Nature 401: 157–161.
2. Manning K, Tor M, Poole M, Hong Y, Thompson AJ, et al. (2006) A naturally occurring epigenetic mutation in a gene encoding an SBP-box transcription factor inhibits tomato fruit ripening. Nat Genet 38: 948–952.
3. Martin A, Troadec C, Boualem A, Rajab M, Fernandez R, et al. (2009) A transposon-induced epigenetic change leads to sex determination in melon. Nature 461: 1135–1138.
4. Bender J, Fink GR (1995) Epigenetic control of an endogenous gene family is revealed by a novel blue fluorescent mutant of Arabidopsis. Cell 83: 725–734.
5. Law JA, Jacobsen SE (2010) Establishing, maintaining and modifying DNA methylation patterns in plants and animals. Nat Rev Genet 11: 204–220.
6. Henderson IR, Jacobsen SE (2007) Epigenetic inheritance in plants. Nature 447: 418–424.
7. Wassenegger M, Heimes S, Riedel L, Sanger HL (1994) Rna-Directed De-Novo Methylation of Genomic Sequences in Plants. Cell 76: 567–576.
8. Matzke M, Kanno T, Claxinger L, Huettel B, Matzke AJM (2009) RNA-mediated chromatin-based silencing in plants. Current Opinion in Cell Biology 21: 367–376.
9. Vongs A, Kakutani T, Martienssen RA, Richards EJ (1993) Arabidopsis-Thaliana DNA Methylation Mutants. Science 260: 1926–1928.
10. Bartee L, Malagnac F, Bender J (2001) Arabidopsis cmt3 chromomethylase mutations block non-CG methylation and silencing of an endogenous gene. Genes & Development 15: 1753–1758.
11. Lindroth AM, Cao XF, Jackson JP, Zilberman D, McCallum CM, et al. (2001) Requirement of CHROMOMETHYLASE3 for maintenance of CpXpG methylation. Science 292: 2077–2080.
12. Jullien PE, Mosquna A, Ingouff M, Sakata T, Ohad N, et al. (2008) Retinoblastoma and its binding partner MSI1 control imprinting in Arabidopsis. PLoS Biol 6: e194.
13. Choi YH, Gehring M, Johnson L, Hannon M, Harada JJ, et al. (2002) DEMETER, a DNA glycosylase domain protein, is required for endosperm gene imprinting and seed viability in Arabidopsis. Cell 110: 33–42.
14. Gong ZH, Morales-Ruiz T, Ariza RR, Roldan-Arjona T, David L, et al. (2002) ROS1, a repressor of transcriptional gene silencing in Arabidopsis, encodes a DNA glycosylase/lyase. Cell 111: 803–814.
15. Bird A (2002) DNA methylation patterns and epigenetic memory. Genes Dev 16: 6–21.
16. Hsieh TF, Shin J, Uzawa R, Silva P, Cohen S, et al. (2011) Regulation of imprinted gene expression in Arabidopsis endosperm. Proc Natl Acad Sci U S A 108: 1755–1762.
17. Dowen RH, Pelizzola M, Schmitz RJ, Lister R, Dowen JM, et al. (2012) Widespread dynamic DNA methylation in response to biotic stress. Proc Natl Acad Sci U S A 109: E2183–2191.
18. Lisch D (2009) Epigenetic regulation of transposable elements in plants. Annu Rev Plant Biol 60: 43–66.
19. Morgan HD, Santos F, Green K, Dean W, Reik W (2005) Epigenetic reprogramming in mammals. Hum Mol Genet 14 Spec No 1: R47–58.
20. Feng J, Fouse S, Fan G (2007) Epigenetic regulation of neural gene expression and neuronal function. Pediatr Res 61: 58R–63R.

21. Kulis M, Heath S, Bibikova M, Queiros AC, Navarro A, et al. (2012) Epigenomic analysis detects widespread gene-body DNA hypomethylation in chronic lymphocytic leukemia. Nat Genet 44: 1236–1242.
22. Diez CM, Roessler K, Gaut BS (2014) Epigenetics and plant genome evolution. Curr Opin Plant Biol 18: 1–8.
23. Suzuki MM, Bird A (2008) DNA methylation landscapes: provocative insights from epigenomics. Nat Rev Genet 9: 465–476.
24. Bell AC, Felsenfeld G (2000) Methylation of a CTCF-dependent boundary controls imprinted expression of the Igf2 gene. Nature 405: 482–485.
25. Zhang X, Yazaki J, Sundaresan A, Cokus S, Chan SW, et al. (2006) Genome-wide high-resolution mapping and functional analysis of DNA methylation in arabidopsis. Cell 126: 1189–1201.
26. Zilberman D, Gehring M, Tran RK, Ballinger T, Henikoff S (2007) Genome-wide analysis of Arabidopsis thaliana DNA methylation uncovers an interdependence between methylation and transcription. Nat Genet 39: 61–69.
27. Li X, Zhu J, Hu F, Ge S, Ye M, et al. (2012) Single-base resolution maps of cultivated and wild rice methylomes and regulatory roles of DNA methylation in plant gene expression. BMC Genomics 13: 300.
28. Zemach A, McDaniel IE, Silva P, Zilberman D (2010) Genome-wide evolutionary analysis of eukaryotic DNA methylation. Science 328: 916–919.
29. Shukla S, Kavak E, Gregory M, Imashimizu M, Shutinoski B, et al. (2011) CTCF-promoted RNA polymerase II pausing links DNA methylation to splicing. Nature 479: 74–79.
30. Maunakea AK, Nagarajan RP, Bilenky M, Ballinger TJ, D'Souza C, et al. (2010) Conserved role of intragenic DNA methylation in regulating alternative promoters. Nature 466: 253–257.
31. Teixeira FK, Colot V (2009) Gene body DNA methylation in plants: a means to an end or an end to a means? Embo Journal 28: 997–998.
32. Roudier F, Teixeira FK, Colot V (2009) Chromatin indexing in Arabidopsis: an epigenomic tale of tails and more. Trends in Genetics 25: 511–517.
33. Takuno S, Gaut BS (2012) Body-Methylated Genes in Arabidopsis thaliana Are Functionally Important and Evolve Slowly. Mol Biol Evol 29: 219–227.
34. Takuno S, Gaut BS (2013) Gene body methylation is conserved between plant orthologs and is of evolutionary consequence. Proc Natl Acad Sci U S A 110: 1797–1802.
35. Ohno S (1970) Evolution by gene duplication. Berlin, New York,: Springer-Verlag. xv, 160 p.
36. Ohta T (1973) Slightly deleterious mutant substitutions in evolution. Nature 246: 96–98.
37. Harada E, Nakagawa J, Asano T, Taoka M, Sorimachi H, et al. (2012) Functional evolution of duplicated odorant-binding protein genes, Obp57d and Obp57e, in Drosophila. PLoS One 7: e29710.
38. Piatigorsky J, Wistow G (1991) The recruitment of crystallins: new functions precede gene duplication. Science 252: 1078–1079.
39. Ferris SD, Whitt GS (1979) Evolution of the differential regulation of duplicate genes after polyploidization. J Mol Evol 12: 267–317.
40. Hittinger CT, Carroll SB (2007) Gene duplication and the adaptive evolution of a classic genetic switch. Nature 449: 677–681.
41. Force A, Lynch M, Pickett FB, Amores A, Yan YL, et al. (1999) Preservation of duplicate genes by complementary, degenerative mutations. Genetics 151: 1531–1545.
42. Lynch M, Conery JS (2003) The origins of genome complexity. Science 302: 1401–1404.
43. Lynch M, O'Hely M, Walsh B, Force A (2001) The probability of preservation of a newly arisen gene duplicate. Genetics 159: 1789–1804.
44. Shiu SH, Byrnes JK, Pan R, Zhang P, Li WH (2006) Role of positive selection in the retention of duplicate genes in mammalian genomes. Proc Natl Acad Sci U S A 103: 2232–2236.
45. Hughes AL (1994) The evolution of functionally novel proteins after gene duplication. Proc Biol Sci 256: 119–124.
46. Chang AY, Liao BY (2012) DNA methylation rebalances gene dosage after mammalian gene duplications. Mol Biol Evol 29: 133–144.
47. Li J, Musso G, Zhang Z (2008) Preferential regulation of duplicated genes by microRNAs in mammals. Genome Biol 9: R132.
48. Keller TE, Yi SV (2014) DNA methylation and evolution of duplicate genes. Proc Natl Acad Sci U S A 111: 5932–5937.
49. Widman N, Jacobsen SE, Pellegrini M (2009) Determining the conservation of DNA methylation in Arabidopsis. Epigenetics 4: 119–124.
50. Wang J, Marowsky NC, Fan C (2013) Divergent evolutionary and expression patterns between lineage specific new duplicate genes and their parental paralogs in Arabidopsis thaliana. PLoS One 8: e72362.
51. Wang Y, Wang X, Lee TH, Mansoor S, Paterson AH (2013) Gene body methylation shows distinct patterns associated with different gene origins and duplication modes and has a heterogeneous relationship with gene expression in Oryza sativa (rice). New Phytol 198: 274–283.
52. Rodin SN, Riggs AD (2003) Epigenetic silencing may aid evolution by gene duplication. J Mol Evol 56: 718–729.
53. Rapp RA, Wendel JF (2005) Epigenetics and plant evolution. New Phytol 168: 81–91.
54. Chen ZJ, Ni Z (2006) Mechanisms of genomic rearrangements and gene expression changes in plant polyploids. Bioessays 28: 240–252.
55. Lister R, O'Malley RC, Tonti-Filippini J, Gregory BD, Berry CC, et al. (2008) Highly integrated single-base resolution maps of the epigenome in Arabidopsis. Cell 133: 523–536.
56. Zemach A, Kim MY, Silva P, Rodrigues JA, Dotson B, et al. (2010) Local DNA hypomethylation activates genes in rice endosperm. Proc Natl Acad Sci U S A 107: 18729–18734.
57. Schmid M, Davison TS, Henz SR, Pape UJ, Demar M, et al. (2005) A gene expression map of Arabidopsis thaliana development. Nature Genetics 37: 501–506.
58. Jeong DH, Park S, Zhai JX, Gurazada SGR, De Paoli E, et al. (2011) Massive Analysis of Rice Small RNAs: Mechanistic Implications of Regulated MicroRNAs and Variants for Differential Target RNA Cleavage. Plant Cell 23: 4185–4207.
59. Krueger F, Andrews SR (2011) Bismark: a flexible aligner and methylation caller for Bisulfite-Seq applications. Bioinformatics 27: 1571–1572.
60. Fojtova M, Kovarik A, Matyasek R (2001) Cytosine methylation of plastid genome in higher plants. Fact or artefact? Plant Science 160: 585–593.
61. Storey JD (2003) The positive false discovery rate: A Bayesian interpretation and the q-value. Annals of Statistics 31: 2013–2035.
62. Storey JD, Tibshirani R (2003) Statistical significance for genomewide studies. Proceedings of the National Academy of Sciences of the United States of America 100: 9440–9445.
63. Storey JD (2002) A direct approach to false discovery rates. Journal of the Royal Statistical Society Series B-Statistical Methodology 64: 479–498.
64. Langmead B, Trapnell C, Pop M, Salzberg SL (2009) Ultrafast and memory-efficient alignment of short DNA sequences to the human genome. Genome Biol 10: R25.
65. Trapnell C, Williams BA, Pertea G, Mortazavi A, Kwan G, et al. (2010) Transcript assembly and quantification by RNA-Seq reveals unannotated transcripts and isoform switching during cell differentiation. Nat Biotechnol 28: 511–U174.
66. Yanai I, Benjamin H, Shmoish M, Chalifa-Caspi V, Shklar M, et al. (2005) Genome-wide midrange transcription profiles reveal expression level relationships in human tissue specification. Bioinformatics 21: 650–659.
67. Li H, Durbin R (2009) Fast and accurate short read alignment with Burrows-Wheeler transform. Bioinformatics 25: 1754–1760.
68. Kent WJ (2002) BLAT–the BLAST-like alignment tool. Genome Res 12: 656–664.
69. Yang Z (2007) PAML 4: phylogenetic analysis by maximum likelihood. Mol Biol Evol 24: 1586–1591.
70. Katoh K, Kuma K, Toh H, Miyata T (2005) MAFFT version 5: improvement in accuracy of multiple sequence alignment. Nucleic Acids Res 33: 511–518.
71. Cao J, Schneeberger K, Ossowski S, Gunther T, Bender S, et al. (2011) Whole-genome sequencing of multiple Arabidopsis thaliana populations. Nat Genet 43: 956–963.
72. Nei M, Li WH (1979) Mathematical model for studying genetic variation in terms of restriction endonucleases. Proc Natl Acad Sci U S A 76: 5269–5273.
73. Smit AFA, Hubley R, Green P (1996–2010) RepeatMasker Open-3.0.
74. Kim SH, Yi SV (2007) Understanding relationship between sequence and functional evolution in yeast proteins. Genetica 131: 151–156.
75. Kim SH, Yi SV (2006) Correlated asymmetry of sequence and functional divergence between duplicate proteins of Saccharomyces cerevisiae. Mol Biol Evol 23: 1068–1075.
76. Whittaker J (1996) Graphical models in applied multivariate statistics. New York: John Wiley and Sons.
77. Schmitz RJ, Schultz MD, Urich MA, Nery JR, Pelizzola M, et al. (2013) Patterns of population epigenomic diversity. Nature 495: 193–198.
78. Hua Z, Pool JE, Schmitz RJ, Schultz MD, Shiu SH, et al. (2013) Epigenomic programming contributes to the genomic drift evolution of the F-Box protein superfamily in Arabidopsis. Proc Natl Acad Sci U S A 110: 16927–16932.
79. Sarda S, Zeng J, Hunt BG, Yi SV (2012) The evolution of invertebrate gene body methylation. Mol Biol Evol 29: 1907–1916.
80. Flores K, Wolschin F, Corneveaux JJ, Allen AN, Huentelman MJ, et al. (2012) Genome-wide association between DNA methylation and alternative splicing in an invertebrate. BMC Genomics 13: 480.
81. Watterson GA (1975) On the number of segregating sites in genetical models without recombination. Theor Popul Biol 7: 256–276.
82. Pfeifer GP (2006) Mutagenesis at methylated CpG sequences. DNA Methylation: Basic Mechanisms 301: 259–281.
83. Messeguer R, Ganal MW, Steffens JC, Tanksley SD (1991) Characterization of the Level, Target Sites and Inheritance of Cytosine Methylation in Tomato Nuclear-DNA. Plant Molecular Biology 16: 753–770.
84. Bird AP (1980) DNA methylation and the frequency of CpG in animal DNA. Nucleic Acids Res 8: 1499–1504.
85. Becker C, Hagmann J, Muller J, Koenig D, Stegle O, et al. (2011) Spontaneous epigenetic variation in the Arabidopsis thaliana methylome. Nature 480: 245–249.
86. Arikit S, Zhai JX, Meyers BC (2013) Biogenesis and function of rice small RNAs from non-coding RNA precursors. Current Opinion in Plant Biology 16: 170–179.
87. Johnson C, Kasprzewska A, Tennessen K, Fernandes J, Nan GL, et al. (2009) Clusters and superclusters of phased small RNAs in the developing inflorescence of rice. Genome Research 19: 1429–1440.
88. Wu L, Zhou HY, Zhang QQ, Zhang JG, Ni FR, et al. (2010) DNA Methylation Mediated by a MicroRNA Pathway. Molecular Cell 38: 465–475.

89. Wang Y, Wang X, Tang H, Tan X, Ficklin SP, et al. (2011) Modes of gene duplication contribute differently to genetic novelty and redundancy, but show parallels across divergent angiosperms. PLoS One 6: e28150.
90. Schmitz RJ, Schultz MD, Lewsey MG, O'Malley RC, Urich MA, et al. (2011) Transgenerational Epigenetic Instability Is a Source of Novel Methylation Variants. Science 334: 369–373.
91. Ahmed I, Sarazin A, Bowler C, Colot V, Quesneville H (2011) Genome-wide evidence for local DNA methylation spreading from small RNA-targeted sequences in Arabidopsis. Nucleic Acids Res 39: 6919–6931.
92. Hunt BG, Brisson JA, Yi SV, Goodisman MA (2010) Functional conservation of DNA methylation in the pea aphid and the honeybee. Genome Biol Evol 2: 719–728.
93. Zeng J, Yi SV (2010) DNA methylation and genome evolution in honeybee: gene length, expression, functional enrichment covary with the evolutionary signature of DNA methylation. Genome Biol Evol 2: 770–780.
94. Jones PA (2012) Functions of DNA methylation: islands, start sites, gene bodies and beyond. Nat Rev Genet 13: 484–492.
95. Klironomos FD, Berg J, Collins S (2013) How epigenetic mutations can affect genetic evolution: model and mechanism. Bioessays 35: 571–578.
96. Bonduriansky R (2012) Rethinking heredity, again. Trends Ecol Evol 27: 330–336.
97. Shea N, Pen I, Uller T (2011) Three epigenetic information channels and their different roles in evolution. J Evol Biol 24: 1178–1187.
98. Chodavarapu RK, Feng SH, Bernatavichute YV, Chen PY, Stroud H, et al. (2010) Relationship between nucleosome positioning and DNA methylation. Nature 466: 388–392.

11

Optimizing Hill Seeding Density for High-Yielding Hybrid Rice in a Single Rice Cropping System in South China

Danying Wang[1], Song Chen[1], Zaiman Wang[2], Chenglin Ji[1], Chunmei Xu[1], Xiufu Zhang[1]*, Bhagirath Singh Chauhan[3]

1 State Key Laboratory of Rice Biology, China National Rice Research Institute, Hangzhou, Zhejiang, China, **2** Key Laboratory of Key Technology on Agricultural Machine and Equipment, South China Agricultural University, Guangzhou, Guangdong, China, **3** Queensland Alliance for Agriculture and Food Innovation, University of Queensland, Queensland, Australia

Abstract

Mechanical hill direct seeding of hybrid rice could be the way to solve the problems of high seeding rates and uneven plant establishment now faced in direct seeded rice; however, it is not clear what the optimum hill seeding density should be for high-yielding hybrid rice in the single-season rice production system. Experiments were conducted in 2010 and 2011 to determine the effects of hill seeding density (25 cm×15 cm, 25 cm×17 cm, 25 cm×19 cm, 25 cm×21 cm, and 25 cm×23 cm; three to five seeds per hill) on plant growth and grain yield of a hybrid variety, Nei2you6, in two fields with different fertility (soil fertility 1 and 2). In addition, in 2012 and 2013, comparisons among mechanical hill seeding, broadcasting, and transplanting were conducted with three hybrid varieties to evaluate the optimum seeding density. With increases in seeding spacing from 25 cm×15 cm to 25 cm×23 cm, productive tillers per hill increased by 34.2% and 50.0% in soil fertility 1 and 2. Panicles per m^2 declined with increases in seeding spacing in soil fertility 1. In soil fertility 2, no difference in panicles per m^2 was found at spacing ranging from 25 cm×17 cm to 25 cm×23 cm, while decreases in the area of the top three leaves and aboveground dry weight per shoot at flowering were observed. Grain yield was the maximum at 25 cm×17 cm spacing in both soil fertility fields. Our results suggest that a seeding density of 25 cm×17 cm was suitable for high-yielding hybrid rice. These results were verified through on-farm demonstration experiments, in which mechanical hill-seeded rice at this density had equal or higher grain yield than transplanted rice.

Editor: Jose Luis Gonzalez-Andujar, Instituto de Agricultura Sostenible (CSIC), Spain

Funding: The authors are grateful for grants from the National Rice Industry Technology System (CARS-01-04A), the 3 Nong 6 Fang Program, and the Five-twelfth National Science and Technology Support Program (2011BAD16B14). The authors state that the funders had no role in study design, data collection and analysis, decision to publish, or preparation of the manuscript.

Competing Interests: The authors have declared that no competing interests exist.

* Email: chais.zju@gmail.com

Introduction

Although transplanting has been a major traditional method of rice establishment in most Asian countries, the rising labor cost and developments in rice production technology have improved the desirability of direct-seeded rice [1]. Direct seeding of rice is a low-cost establishment technology. It avoids nursery raising, seedling uprooting, and transplanting, and has the benefits of saving labor, facilitating timely establishment of rice, and earlier crop maturity (by 7–10 days) [2–4]. Direct seeding helps in solving the labor scarcity problem, which is now very critical in China's agricultural development. Simultaneously, the chemical weed control method has made such a switch technically viable. In 2008, for example, 8.3% of the rice-growing area in China was direct-seeded; in some provinces of south China, it was more than 30% [5].

In the south China irrigated rice ecosystem, broadcasting in wet conditions (wet seeding) is the principal method of rice establishment. In this method, pregerminated seeds are sown onto wet (saturated) puddled soils. But, some difficulties are faced in wet seeding. First, no specific varieties have been developed in China for wet seeding. The existing varieties used for transplanting do not appear to be well-adapted for seedling growth in an initially oxygen-depleted micro environment. Second, the extent of laser leveling in China is currently extremely smaller compared with that in other countries (e.g., 50–80% of the rice land in Australia is laser-leveled) [4]. Due to lack of uniform water distribution associated with unevenness of the land, the problem of excess- or no-water-caused yield variability within a field is common, which leads to poor establishment of direct-seeded rice [3,6]. Third, surface-sown seeds are damaged by birds and rodents. As a result, farmers often resort to the costly practice of increasing the seeding rate for direct-seeded rice, even by two to three times [7]. High seeding rates can result in large yield losses due to excessive number of tillers, increased proportion of ineffective tillers, higher spikelet sterility, and fewer grains per panicle [8]. Moreover, a dense canopy and less ventilation around the plants at high seeding rates can create favorable conditions for diseases (e.g., sheath blight) [9,10] and insects (e.g., brown planthoppers) and can make plants more prone to lodging [11]. High seeding rates using broadcasting restrict individual plant growth, which is not suitable for the production of high-yielding hybrid rice.

High-yielding hybrid rice, which is crucial to meet expanding food demand, had large panicles and high aboveground biomass [12,13]. Hybrid rice varieties grow well with proper planting density, but their lodging resistance decreases dramatically under dense seeding [14]. The use of high seeding rate in a broadcast culture may increase the gap between farmers' potential and actual yield.

Mechanical hill seeding of rice is one of the options used to solve the problem of high seeding rate and uneven plant establishment. Here, rice seeds are sown uniformly in the field at designated spacing and seed number per hill. Combined with a harrow or furrower opener, many hill seeders have land-preparation functions to improve soil evenness. Some hill seeders enable seeding and puddling to be performed simultaneously [15]; others can make furrows and plant the crop at the same time [16]; still others can effectively incorporate seeds into puddled soils at depths of 5–20 mm, which helps decrease losses caused by birds [15]. Even soil conditions and uniform seeding result in an even plant establishment. A well-ventilated micro-environment not only reduces pest and disease incidence but also enhances photosynthesis [17]. In previous studies, lodging resistance of hill-seeded rice was found to be higher than mechanically transplanted rice, row-seeded rice, and broadcast-seeded rice [15,18].

However, the practical use of mechanical hill seeders had not been seen in China until the 2000s. The adoption of this implement was hampered by the belief that broadcasting is not a suitable method for hybrid rice varieties and both government and farmers still have misgivings about the use of mechanical hill seeding for this purpose. Research is thus imperative to shed light on this technology. There is limited information on hill seeding density for high-yielding hybrid rice varieties, and results for optimum hill seeding density in a single-season rice production system varied from 20 cm×14 cm to 30 cm×20 cm spacing [17,19–21]. It is difficult to draw a concrete conclusion about what hill seeding density is optimum for high-yielding hybrid rice varieties.

The aim of this study was to evaluate the optimum hill seeding density to support the development of mechanical hill seeders for hybrid rice. In this study, 2-year field experiments and 2-year on-farm demonstration experiments were conducted in Zhejiang, China. In the field experiments, a high-yielding hybrid rice variety (Nei2you6) was hill-seeded manually at five seeding densities in two fields with different fertility profiles. The objectives were to evaluate the effects of hill seeding density on plant growth and grain yield, the interrelation between seeding density and soil fertility, and the optimum hill seeding density for high-yielding hybrid rice. In the on-farm demonstrations, three high-yielding hybrid rice varieties were planted using three methods: mechanical hill-seeded with a hill seeder at optimum seeding density, broadcasting, and transplanting following standard farmers' practice. The objective was to verify the optimum seeding density in the field by comparing the yields of hill-seeded, broadcast, and transplanted rice.

Materials and Methods

Field experiments and on-farm demonstrations were conducted; field experiments were done in 2010 and 2011 and the on-farm demonstration experiments were carried out in 2012 and 2013. Field experiments were conducted at the experimental farm of the China Rice Research Institute and on-farm demonstration experiments were conducted at the agricultural demonstration base of the China Rice Research Institute in Fuyang and Yuhang. No permission was needed to use fields at these facilities. Authors also confirmed that the study did not involve endangered or protected species.

Field experiments

The 2010 and 2011 experiments were conducted in two paddy fields with different fertility levels (soil fertility 1 and 2) at the China National Rice Research Institute (CNRRI) (120.2′ E, 30.3′ N, 11 m above sea level), Zhejiang Province, China. The area has a subtropical monsoon climate with an annual mean temperature of 13–20°C (range from 2°C in January to 39°C in July). Mean annual precipitation is 1200–1600 mm, with about 80% rainfall occurring between April and September. Soil at the experimental site is classified as ferric-accumulic stagnic anthrosol [22]. The cropping system of soil fertility 1 was monoculture rice-fallow without rice straw incorporation, and the cropping system of soil fertility 2 was rice-potato (*Solanum tuberosum*) with rice and potato straw incorporation. After 10 years of continuous farming, the soil properties in the two fields differed and this was noted at the start of the experiment in 2010. The soil in field 1 had 2.46 g total N kg^{-1}, 0.53 g total P kg^{-1}, 0.25 g available K kg^{-1}, 10.73 C mol kg^{-1} cation exchange capacity (CEC), and pH 5.78 (soil fertility 1). The corresponding soil properties in field 2 were 2.96 g total N kg^{-1}, 0.72 g total P kg^{-1}, 0.31 g available K kg^{-1}, 10.80 C mol kg^{-1} CEC, and pH 5.15 (soil fertility 2). The soil test was based on samples taken from the upper 10 cm of the soil. Soil chemical properties at the 10–20 cm depth were also different [23].

Pregerminated seeds of a high-yielding hybrid rice variety, Nei2you6, were hill-seeded manually in the two fields with different fertility levels using five spacing schemes (between and within rows: 25 cm×15 cm, 25 cm×17 cm, 25 cm×19 cm, 25 cm×21 cm, and 25 cm×23 cm). Each hill was planted with three to five seeds. The experiment was replicated four times and subplot size was 30 m^2. In both years, the fields were puddled in ponded conditions and drained 2 d before seeding. Pre-germinated seeds were sown on the surface of puddled soil on June 2, 2010 and May 30, 2011. Before sowing, seeds were soaked for 24 h and incubated for 24 h for proper germination, and seeds dressed with chemical pesticides to avoid the attack of soil insects, birds, and rodents.

To the crop, 225 kg N ha^{-1} (as urea) was applied in three splits in a ratio of 5:3:2 at basal, tillering, and earing stage, respectively. Potassium at 180 kg K_2O ha^{-1} was applied in two splits: 50% as basal dressing and 50% as topdressing at earing in the form of KCl. Phosphorus at 75 kg P_2O_5 ha^{-1} in the form of calcium superphosphate was applied as basal. Water management followed standard farmers' practice, in which the field was drained 2 d before seeding. There was no standing water on the soil surface from sowing to the three-leaf stage of rice, and then the field was reflooded and a water depth of 1 to 3 cm was maintained until the end of the tillering period. After this, water was drained for 7–10 days to control unproductive tillers and later, a water depth of 3–5 cm was maintained until the grain-filling stage. Weed and insects were intensively controlled by chemicals to avoid biomass and yield loss.

In 2010, the number of tillers from 30 hills of rice was counted at 1-week intervals, starting from 20 d after sowing (DAS) to panicle initiation stage. Tillers with at least one visible leaf were included. In 2011, 30 hills of rice plants were sampled at heading and maturity to calculate average panicle number per hill. In both years, six representative hills of rice plants from each plot were separately sampled at heading and divided into leaf blades, stem plus sheath, and panicles. Green leaves were grouped into the top

Table 1. Grain yield of a high-yielding hybrid rice variety Nei2you6 grown at five seeding spacings in soil fertility 1 and soil fertility 2 at the China National Rice Research Institute, Zhejiang Province, China, in 2010 and 2011.

Year	Field	Seeding spacing	Grain yield (kg ha^{-1})	Grain yield (g $hill^{-1}$)	Grain yield (g $shoot^{-1}$)
2010	Soil fertility 1	25 cm×15 cm	7804 c	29.37 d	3.09 d
		25 cm×17 cm	8348 a	36.01 c	3.49 b
		25 cm×19 cm	8152 ab	38.05 bc	3.41 bc
		25 cm×21 cm	7892 bc	41.65 b	3.36 c
		25 cm×23 cm	7938 b	46.05 a	3.68 a
	Soil fertility 2	25 cm×15 cm	9133 b	34.69 d	3.64 b
		25 cm×17 cm	9433 a	40.63 c	3.84 a
		25 cm×19 cm	8801 c	42.08 bc	3.68 b
		25 cm×21 cm	8437 d	45.09 b	3.50 c
		25 cm×23 cm	8311 d	48.35 a	3.38 d
2011	Soil fertility 1	25 cm×15 cm	7842 c	29.79 e	3.14 d
		25 cm×17 cm	8573 a	36.80 d	3.56 b
		25 cm×19 cm	8218 b	39.61 c	3.55 b
		25 cm×21 cm	8121 b	43.24 b	3.48 c
		25 cm×23 cm	8140 b	47.19 a	3.70 a
	Soil fertility 2	25 cm×15 cm	9380 b	35.39 d	3.71 bc
		25 cm×17 cm	9611 a	41.13 c	3.88 a
		25 cm×19 cm	8942 c	42.91 bc	3.76 b
		25 cm×21 cm	8831 c	47.14 ab	3.66 c
		25 cm×23 cm	8582 d	49.94 a	3.49 d
Analysis of variance					
Seeding density (A)			*	*	*
Soil fertility (B)			*	*	*
Year (C)			NS	NS	NS
A×B			*	*	*
A×C			NS	NS	NS
B×C			NS	NS	NS
A×B×C			*	*	*

Within a column for each year, means followed by the same letters are not significantly different according to the LSD test (0.05).
*Significance at the 0.05 level based on analysis of variance.
NS denotes nonsignificance based on analysis of variance.

three green leaves and others, and their leaf area was measured with a leaf area meter (LI-3000A, LICOR, Lincoln, NE, USA). LAI was calculated as the total green leaf area of the plant, divided by the corresponding ground sampling area. All samples were oven-dried at 80°C until constant weight for determination of aboveground dry weight at flowering. At maturity, 30 hills were sampled diagonally from each subplot, panicle numbers were counted for each hill to determine the number of panicles per m^2. Six representative hills of rice plants were selected to determine aboveground total dry weight and harvest index. Grain yield per hill was measured from the sampled 30 hills and grain yield per shoot was calculated as the ratio of yield per hill to panicle numbers per hill. In each subplot, grain yield was determined from a harvesting area of 30 m^2 and converted to kg ha^{-1} at 14% moisture content. Panicles were hand-threshed and filled spikelets were separated from unfilled spikelets by submerging them in tap water. Dry weights of straw, rachis, and filled and unfilled spikelets were determined after oven drying at 80°C to constant weight. Three subsamples of 20 g of filled spikelets were taken and counted to determine grain weight (20 divided by the counted filled spikelets number); and the number of filled spikelets was calculated as total dry weight of filled spikelets divided by grain weight. All unfilled spikelets were counted and the number of total spikelets (filled and unfilled) was calculated. Aboveground total dry weight at maturity included the total dry matter of straw, rachis, and filled and unfilled spikelets. Dry weight accumulation during the grain-filling phase was calculated as the difference in total aboveground dry weight between flowering and maturity stages. Spikelets per panicle (ratio of total spikelets number to panicles number), grain-filling percentage (ratio of filled spikelets number to total spikelets number), and harvest index (ratio of dry grain yield to total aboveground biomass dry weight) were calculated.

On-farm demonstration experiments

On-farm demonstrations were conducted in 2012 and 2013 in Fuyang (120.0′ E, 30.1′ N; 14 m above sea level) and Yuhang (120.3′ E, 30.4′ N; 40 m above sea level), Zhejiang Province. Soil at both sites is classified as ferric-accumulic stagnic anthrosol [22]. The soil properties in Fuyang were 2.48 g total N kg-1, 0.56 g total P kg-1, 0.26 g available K kg-1, and pH 5.63. The soil

Table 2. Yield components of a high-yielding hybrid rice variety Nei2you6 grown at five seeding spacings in soil fertility 1 and soil fertility 2, China National Rice Research Institute, Zhejiang Province, China, 2010 and 2011.

Year	Field	Seeding spacing	Spikelets panicle^{-1} (no.)	Grain filling (%)	Grain weight (mg)	Panicles m^{-2} (no.)	Panicles hill^{-1} (no.)
2010	Soil fertility 1	25 cm×15 cm	115.52 b	76.82 e	30.30 a	240.97 a	9.03 c
		25 cm×17 cm	125.92 a	77.74 d	30.24 a	228.08 b	9.71 bc
		25 cm×19 cm	126.70 a	78.85 c	30.46 a	223.75 b	10.60 b
		25 cm×21 cm	125.42 a	79.99 b	30.72 a	226.41 b	11.91 a
		25 cm×23 cm	127.55 a	81.51 a	30.52 a	210.76 c	12.11 a
	Soil fertility 2	25 cm×15 cm	135.01 a	81.63 b	30.73 a	247.01 a	9.25 cd
		25 cm×17 cm	135.78 a	83.15 a	30.44 a	238.86 b	10.17 c
		25 cm×19 cm	130.83 a	81.44 b	30.22 a	229.02 b	10.86 c
		25 cm×21 cm	135.80 a	78.06 c	30.56 a	234.84 b	12.36 b
		25 cm×23 cm	133.09 a	74.11 d	30.33 a	236.59 b	13.59 a
2011	Soil fertility 1	25 cm×15 cm	127.68 b	84.90 c	33.48 a	266.33 a	9.98 c
		25 cm×17 cm	142.00 a	86.66 b	34.10 a	257.20 ab	10.95 bc
		25 cm×19 cm	140.04 a	87.15 b	33.66 a	247.31 b	11.72 b
		25 cm×21 cm	135.88 a	86.65 b	33.28 a	245.27 b	12.91 a
		25 cm×23 cm	140.97 a	90.09 a	33.74 a	232.94 c	13.39 a
	Soil fertility 2	25 cm×15 cm	143.37 a	86.67 b	32.63 a	262.29 a	9.83 e
		25 cm×17 cm	147.10 a	90.07 a	32.98 a	258.76 a	11.01 d
		25 cm×19 cm	144.61 a	90.02 a	33.40 a	253.12 a	12.00 c
		25 cm×21 cm	147.12 a	84.56 c	33.10 a	254.42 a	13.40 b
		25 cm×23 cm	147.09 a	81.91 d	33.53 a	261.49 a	15.03 a

Within a column for each site, means followed by the same letters are not significantly different according to the LSD test (0.05).

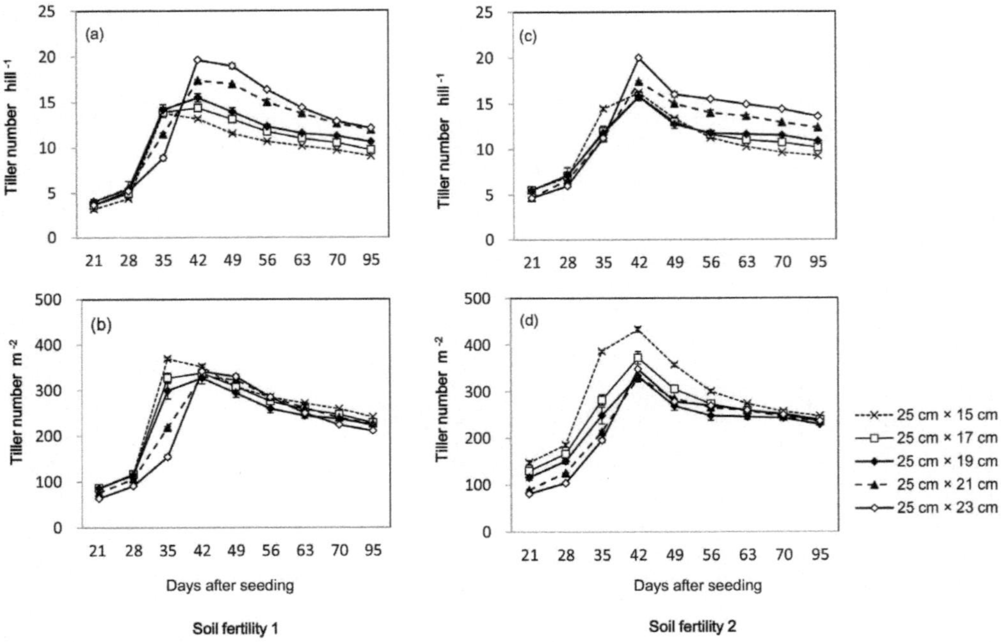

Figure 1. Tillers per hill (a, c) and tillers per m² (b, d) of a high-yielding hybrid rice variety Nei2you6 grown at five plant spacings in soil fertility 1 (a, b) and soil fertility 2 (c, d). China National Rice Research Institute, Zhejiang Province, China, 2010.

properties in Yuhang were 2.66 g total N kg-1, 0.62 g total P kg-1, 0.32 g available K kg-1, and pH 5.75. The hybrid rice variety used in Fuyang was Nei2you6. In Yuhang, high-yielding hybrid rice varieties Yongyou9 and Yongyou12 were grown. These high-yielding hybrid rice varieties are used widely in south China under a single rice cropping system. Three treatments were included at each site: mechanical hill seeding, broadcasting, and transplanting. Hill seeding was done by using a newly developed hill-seeder ZB-10 (Wang et al 2010) at a spacing of 25 cm×17 cm, three to five seeds per hill, and the seed rate was 21.0 to 25.5 kg ha^{-1}. This spacing was chosen because the highest grain yield was noted in this treatment in the 2010 and 2011 experiments. Broadcasting and transplanting were standard farmers' practices. In the broadcasting treatment, sprouted seeds were broadcast at a seeding rate of 33.8 kg ha^{-1}. In the transplanting treatment, seedlings with four to five leaves were transplanted at a spacing of 30 cm×20 cm; two to three seedlings per hill. In the mechanical hill seeding and broadcasting treatments, seed treatment was the same as used in field experiments in 2010 and 2011; all varieties were seeded on the same day (June 2) in both years. Because direct seeding could facilitate timely establishment of rice and earlier crop maturity by 7–10 d, the nursery for the transplanting treatment was raised on May 25 and transplanted on June 18 in both years. The planting area of each treatment was 3000 m^2, a total of 9000 m^2 for each variety. Fertilizer and water management in the three treatments were the same as that described for the field experiments in 2010 and 2011. Management operations such as weeding, irrigation, and plant protection measures were done as needed.

At maturity stage, a 5-m^2 area was harvested diagonally from six places in each treatment for yield determination, and grain yield was calculated at 14% moisture content. Ten hills were sampled diagonally from each 5-m^2 harvest area to determine yield components in the mechanical hill seeding and transplanting treatments, and a 0.5-m^2 area from each 5-m^2 harvest area in the broadcasting treatment. Panicles were hand-threshed and filled and unfilled grains were separated. Dry weights of filled and unfilled spikelets were determined after oven drying samples at 80°C to constant weight. Three subsamples of 20 g of filled spikelets were taken and counted to determine weight and total number of filled spikelets. Unfilled spikelets were counted and grain-filling percentage was calculated.

Statistical analyses

In the field experiments, normal distribution of data was tested for grain yield, yield components, productive tiller percentage, aboveground dry weight, and harvest index by Q-Q plots using SPSS 11.5 software, and no data transformation was needed to meet the assumptions of normality and constant variance. Data were analyzed following analysis of variance (ANOVA), and differences among treatments were compared based on the least significant difference (LSD) test at the 5% level of probability. The statistical model used in the analysis of grain yield per hectare, per hill, and per shoot included the sources of variation due to replication, seeding density, soil fertility, year, and the interactions of seeding density×soil fertility, seeding density×year, soil fertility×year, and seeding density×soil fertility×year. The bivariate correlations between grain yield and yield components, including grain weight, panicles per hectare and per hill, spikelets per panicle, and grain-filling percentage, were conducted using the Pearson correlation coefficient (two-tailed).

In the on-farm demonstration experiments, ANOVA and mean comparisons were based on the LSD test at the 0.05 probability level for each variety and year (SPSS 11.5).

Results

Field experiments

In both years, a significant grain yield response to seeding density was observed (Table 1). Grain yield per hill increased consistently with decreases in seeding density. The plants grown at 25 cm× 23 cm spacing had the highest grain yield per hill and

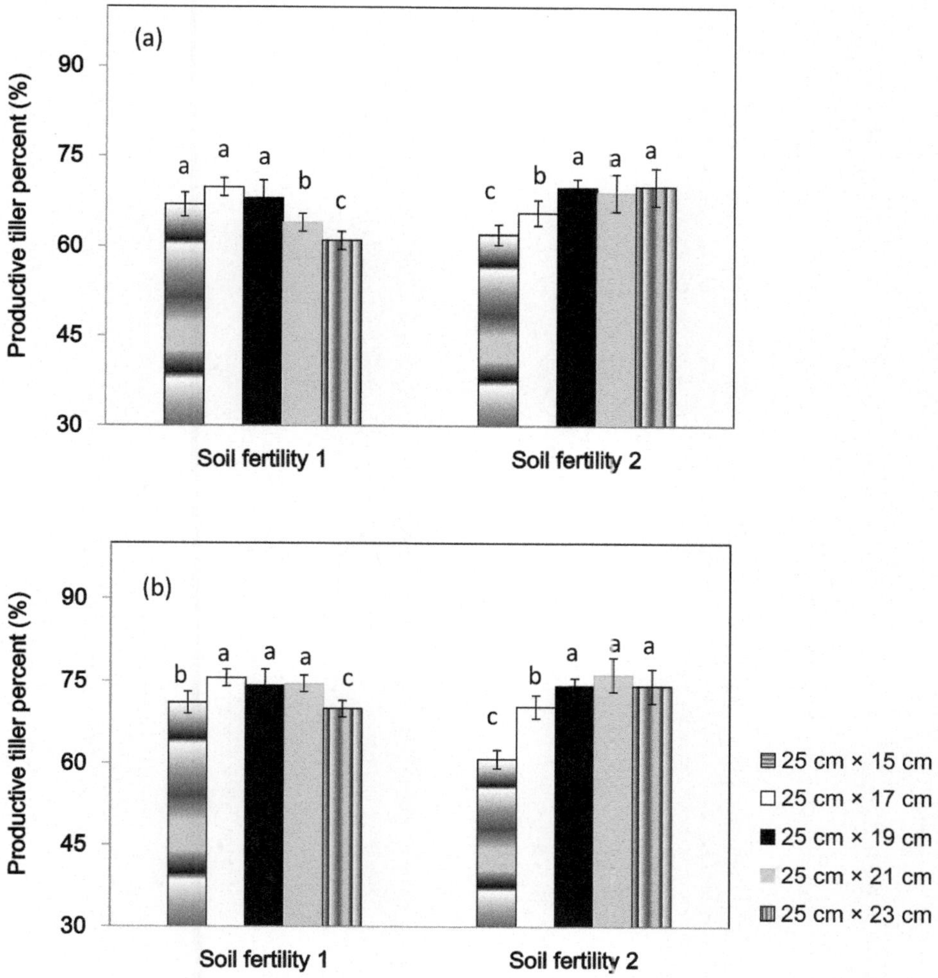

Figure 2. Productive tiller percentage of a high-yielding hybrid rice variety Nei2you6 grown at five seeding spacings in soil fertility 1 and soil fertility 2. China National Rice Research Institute, Zhejiang Province, China, 2010 (a) and 2011 (b).

plants grown at 25 cm×15 cm spacing had the lowest. The rate of change in grain yield per hectare with seeding spacing, however, was different between soil fertility 1 and 2. In soil fertility 1, 25 cm×17 cm spacing had the highest grain yield per ha, while 25 cm×15 cm spacing had the lowest value. In soil fertility 2, highest grain yield per ha was obtained at 25 cm×17 cm spacing, but, beyond this, grain yield decreased with an increase in seeding spacing and rice plants grown at 25 cm× 23 cm produced the lowest grain yield per ha. There were large differences in grain yield per shoot among the five seeding spacings in both soil fertility fields. In soil fertility 1, the spacing of 25 cm× 23 cm produced the maximum grain yield per shoot and the lowest value was observed at 25 cm×15 cm spacing. In soil fertility 2, the response of grain yield per shoot to seeding density was the same as that for grain yield per ha; 25 cm×17 cm had the maximum and 25 cm× 23 cm had the minimum grain yield per shoot. The difference in grain yield between the two years was nonsignificant (Table 1).

In both years, there was no significant difference in grain weight among different spacing treatments in both fields (Table 2). The number of panicles per hill consistently increased with increases in seeding spacing in both years and both fields. Across years, with increases in seeding spacing from 25 cm×15 cm to 25 cm×23 cm, panicles per hill increased by 34.2% in soil fetility 1 and by 50.0% in soil fertility 2. In both years, the number of panicles per m^2 was not significantly different at spacing ranging from 25 cm×17 cm to 25 cm× 21 cm in soil fertility 1; the maximum and minimum number of panicles per m^2 was observed at 25 cm×15 cm and 25 cm× 23 cm spacing, respectively. In soil fertility 2, 25 cm×15 cm spacing had higher number of panicles per m^2 in 2010, while no difference was observed among other treatments in this year and among all spacings in 2011. In both years, 25 cm×15 cm spacing had the minimum number of spikelets per panicle in soil fertility 1, while no difference was found among treatments in soil fertility 2. Grain-filling percentage in 2011 was higher than that in 2010 in both soil fertility fields. The plants grown at 25 cm× 23 cm spacing had the highest and 25 cm×15 cm spacing had the lowest grain-filling percentage in soil fertility 1 Meanwhile, in soil fertility 2, 25 cm× 17 cm spacing had the highest grain filling percentage and 25 cm× 23 cm spacing had the lowest value in both years.

In 2010, maximum tiller number per hill was observed at 25 cm×15 cm spacing at 35 DAS in soil fertility 1, while at 42 DAS at other seeding spacings in soil fertility 1 and all treatments in soil fertility 2 (Figure 1). A decreasing trend of tiller number per hill was observed after the reproductive stage, irrespective of treatment. Maximum tiller numbers per hill and per m^2 were affected by seeding density. In both soil fertility fields, maximum tiller number per hill increased with increases in seeding spacing;

Table 3. Leaf area index (LAI), leaf area per hill, and leaf area per shoot at flowering of a high-yielding hybrid rice variety Nei2you6 grown at five seeding spacings in soil fertility 1 and soil fertility 2, China National Rice Research Institute, Zhejiang Province, China, 2010 and 2011.

Year	Field	Seeding spacing	LAI	Leaf area (m^2 per hill)			Leaf area ($\times 10^{-3}$ m^2 per shoot)		
				Total	Top 3 leaves	Others	Total	Top 3 leaves	Others
2010	Soil fertility 1	25 cm×15 cm	5.72 a	0.21 e	0.12 d	0.11 c	26.87 c	14.16 c	12.71 b
		25 cm×17 cm	5.39 c	0.23 d	0.13 d	0.11 c	27.76 c	15.18 c	12.58 b
		25 cm×19 cm	5.42 bc	0.26 c	0.14 c	0.12 b	28.78 b	16.15 b	12.63 b
		25 cm×21 cm	5.49 b	0.29 b	0.16 b	0.13 a	29.08 b	16.46 b	12.62 b
		25 cm×23 cm	5.53 b	0.32 a	0.18 a	0.13 a	31.18 a	17.80 a	13.35 a
	Soil fertility 2	25 cm×15 cm	7.03 a	0.27 e	0.16 c	0.10 d	35.66 a	21.02 a	13.64 c
		25 cm×17 cm	7.07 a	0.30 d	0.17 bc	0.13 c	35.55 a	20.97 a	14.57 b
		25 cm×19 cm	6.68 b	0.32 c	0.18 b	0.13 c	34.67 b	19.59 b	15.07 b
		25 cm×21 cm	6.66 b	0.35 b	0.18 b	0.16 b	34.06 c	17.89 c	16.17 a
		25 cm×23 cm	6.46 c	0.37 a	0.19 a	0.17 a	32.43 d	17.05 c	15.37 b
2011	Soil fertility 1	25 cm×15 cm	6.20 a	0.23 d	0.12 d	0.11 d	29.11 c	15.34 c	13.77 ab
		25 cm×17 cm	5.61 d	0.23 d	0.13 d	0.11 d	28.90 c	15.80 c	13.10 b
		25 cm×19 cm	5.88 c	0.28 c	0.16 c	0.12 c	31.18 b	17.49 b	13.69 ab
		25 cm×21 cm	5.83 c	0.31 b	0.18 b	0.13 b	31.88 b	17.48 b	13.40 b
		25 cm×23 cm	5.99 b	0.34 a	0.20 a	0.15 a	33.78 a	19.28 a	14.47 a
	Soil fertility 2	25 cm×15 cm	7.77 a	0.29 e	0.18 c	0.12 e	38.30 a	23.24 a	15.08 c
		25 cm×17 cm	7.51 b	0.32 d	0.19 b	0.13 d	37.75 a	22.27 ab	15.47 c
		25 cm×19 cm	7.24 c	0.34 c	0.20 ab	0.15 c	37.55 a	21.23 b	16.33 b
		25 cm×21 cm	7.01 d	0.37 b	0.19 b	0.17 b	35.81 b	18.81 c	17.00 a
		25 cm×23 cm	6.86 d	0.39 a	0.21 a	0.19 a	34.43 b	18.11 c	16.33 b

Green leaves per hill were subdivided into top three green leaves and others. Leaf area per shoot was calculated as the ratio of leaf area per hill to productive tiller number per hill.
Within a column for each site, means followed by the same letters are not significantly different according to the LSD test (0.05).

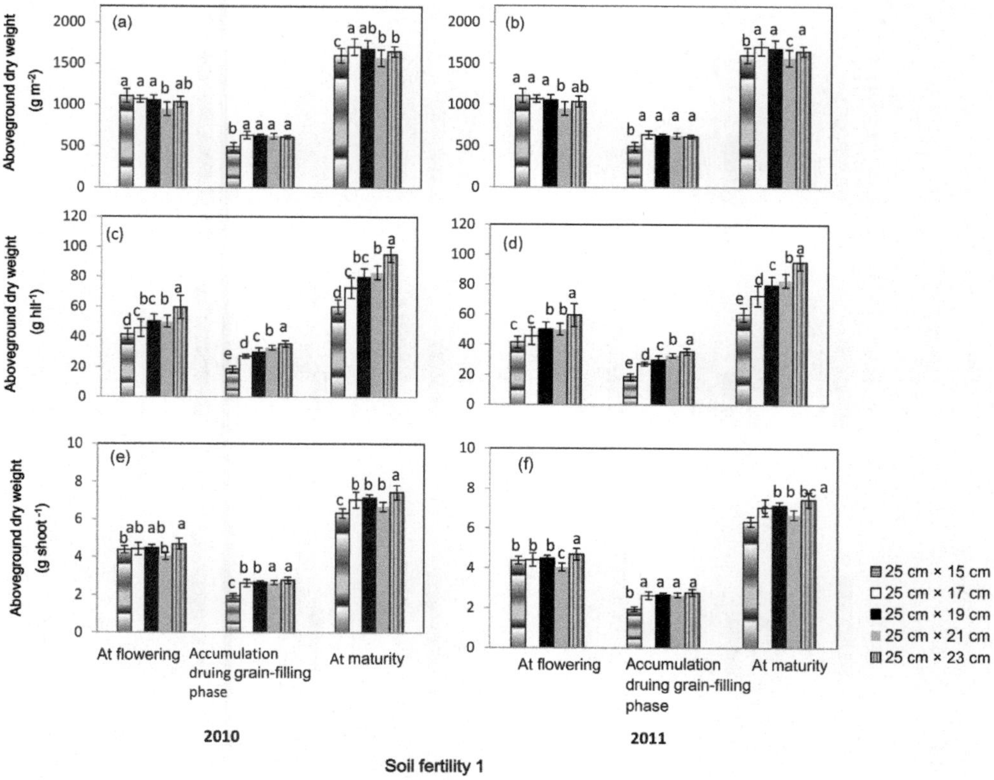

Figure 3. Aboveground dry weight at flowering and maturity, and dry weight accumulation during grain-filling phase of a high-yielding hybrid rice variety Nei2you6 grown at five seeding spacings in soil fertility 1 at the China National Rice Research Institute, Zhejiang Province, China, in 2010 and 2011. Aboveground dry weight was calculated as dry weight per m^2 (a, b), dry weight per hill (c, d), and dry weight per shoot (e, f). Dry weight per shoot was the ratio of dry weight per hill to productive tiller number per hill.

25 cm× 23 cm spacing had the highest tiller number per hill and 25 cm×15 cm spacing had the lowest. However, maximum tiller number per m^2 in both fields decreased with increases in seeding spacing; it was highest at 25 cm×15 cm spacing. From the maximum tillering to flowering stage, the difference in tiller number per m^2 among treatments gradually decreased with the reduction in unproductive tillers.

The effects of seeding density on the percentage of productive tillers were also different between the two soil fertility fields (Figure 2). In soil fertility 1, 25 cm× 23 cm spacing had the lowest productive tiller percentage in both years, and no significant difference was observed among spacings ranging from 25 cm×15 cm to 25 cm×19 cm in 2010 and from 25 cm×17 cm to 25 cm×21 cm in 2011. In both years, in soil fertility 2, 25 cm×15 cm spacing had the lowest productive tiller percent, followed by 25 cm×17 cm spacing, and both these spacings had significantly lower productive tillers than the other ones.

Although leaf area per hill at flowering consistently increased in both years with increases in seeding spacing in both soil fertility fields, the change in leaf area per shoot was different between the two fields (Table 3). In both years, total leaf area per shoot and top three leaf blades per shoot increased with increases in seeding spacing in soil fertility 1, while they decreased in soil fertility 2. The highest leaf area index (LAI) in soil fertility 1 was obtained at 25 cm×15 cm spacing, followed by that at 25 cm×23 cm spacing. The highest LAI in soil fertility 2 was also observed at 25 cm×15 cm spacing, while it gradually decreased with increases in seeding spacing; 25 cm×23 cm spacing had the lowest LAI in both years.

Aboveground dry weight per hill at flowering and maturity, and dry weight accumulation per hill during grain-filling phase increased with increases in seeding spacing; 25 cm× 23 cm spacing had the highest values and 25 cm×15 cm spacing had the lowest in both soil fertility fields in both years (Figures 3 and 4). In terms of aboveground dry weight per shoot, a difference existed between the two soil fertility fields. In both years, the highest aboveground dry weight per shoot at flowering and maturity was observed at 25 cm× 23 cm spacing in soil fertility 1; rice plants grown at 25 cm×15 cm spacing had the lowest dry weight accumulation per shoot during the grain-filling phase and aboveground dry weight per shoot at maturity. In soil fertility 2, no significant difference in dry weight accumulation during the grain-filling phase was observed in spacing ranging from 25 cm×17 cm to 25 cm× 23 cm in both years; and aboveground dry weight per shoot at flowering and maturity decreased with increases in spacing from 25 cm×17 cm to 25 cm× 23 cm. Significant differences also existed in aboveground dry weight per m^2 between the two soil fertility fields. In soil fertility 1, no significant difference was found for aboveground dry weight per m^2 at flowering at spacings ranging from 25 cm×15 cm to 25 cm×19 cm. Rice plants grown at 25 cm×15 cm spacing had the lowest dry weight accumulation during grain filling, and no significant difference was observed at spacings ranging from 25 cm×17 cm to 25 cm×23 cm. Aboveground dry weight at maturity was highest at 25 cm×17 cm spacing in both years and lowest at 25 cm×15 cm spacing in 2010 and at 25 cm×15 cm spacing in 2011. In soil fertility 2, no significant difference in dry weight accumulation during grain-filling phase was observed among the treatments; aboveground dry weights per m^2 at

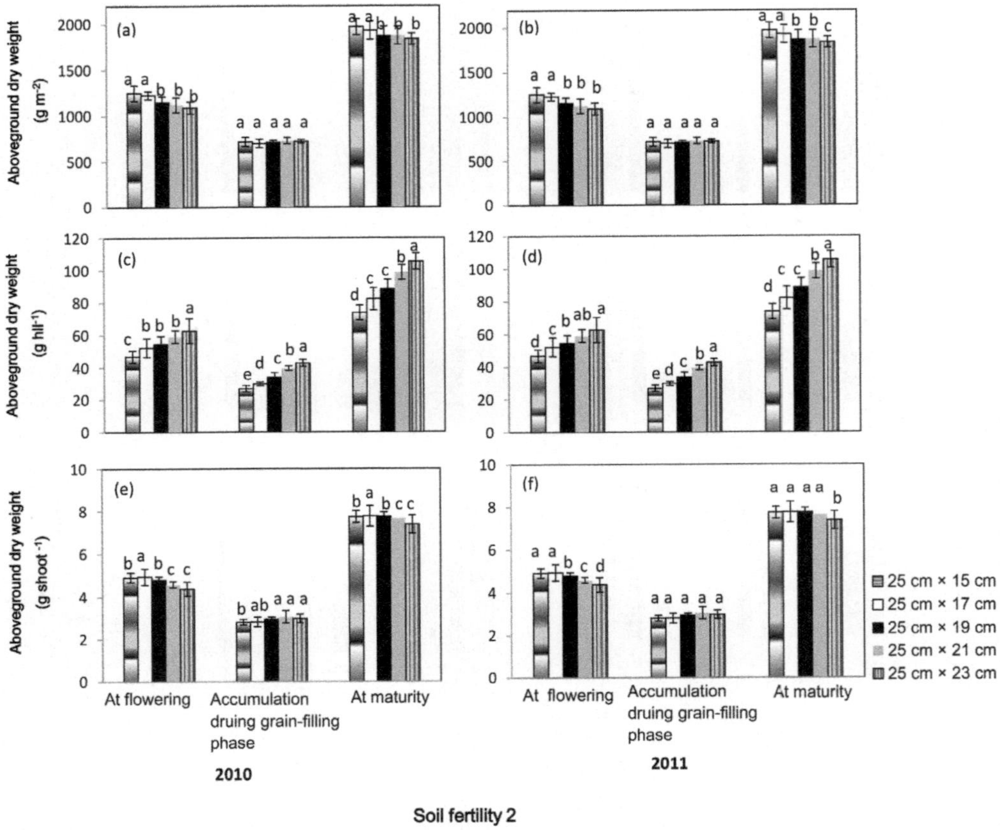

Figure 4. Aboveground dry weight at flowering and maturity, and dry weight accumulation during grain-filling phase of a high-yielding hybrid rice variety Nei2you6 grown at five seeding spacings in soil fertility 2 at the China National Rice Research Institute, Zhejiang Province, China, in 2010 and 2011. Aboveground dry weight was calculated as dry weight per m^2 (a, b), dry weight per hill (c, d), and dry weight per shoot (e, f). Dry weight per shoot was the ratio of dry weight per hill to productive tiller number per hill.

flowering and maturity were significantly higher at 25 cm×15 cm and 25 cm×17 cm spacing than at spacing wider than 25 cm×19 cm.

The effect of seeding density on harvest index (HI) was also different between the two soil fertility fields (Figure 5). In soil fertility 1, 25 cm×23 cm spacing and 25 cm×21 cm spacing had higher HI than that at the other three narrower spacings in both years. In soil fertility 2, plants grown at spacings ranging from 25 cm×17 cm to 25 cm×21 cm had significantly higher HI than plants grown at 25 cm×15 cm and 25 cm×15 cm spacings.

On-farm demonstration experiments

In the on-farm demonstration experiments in 2012 and 2013, mechanical hill seeding, using seeder ZB-10 at a spacing of 25 cm×17 cm, outperformed broadcasting in terms of grain yield for all three high-yielding hybrid rice varieties, increasing grain yield by 10.4% in Yongyou 9 and 12.3% in Yonyou 12 across the 2 years. In 2012, variety Nei2you6 produced 26.6% higher grain yield in mechanical hill seeding compared with broadcasting because more than 30% of the plants lodged in the broadcast culture due to heavy rain during the later grain-filling phase (Figure 6). In both years, no significant difference in grain yield between mechanical hill-seeded rice and transplanted rice was observed for Nei2you 6 and Yonyou 9. Yongyou 12 produced more than 1000 kg ha^{-1} grain yield in the mechanical hill seeding treatment across 2 years, increasing grain yield by 9.1% compared with that of transplanted rice.

Mechanical hill-seeded rice had more spikelets per panicle and fewer panicles per m^2 compared with broadcasting, but fewer spikelets per panicle and more panicles per unit area than transplanted rice (Table 4). Across the years, mechanical hill-seeded hybrid rice varieties Nei2you 6, Yongyou 9, and Yonyou 12 had 36.0%, 28.3%, and 33.1% higher number of spikelets per panicle, and 17.4%, 16.7%, and 14.3% lower number of panicles per m^2, respectively, compared with the broadcast culture. Compared with the transplanted treatment, 7.3%, 15.5%, and 13.4% decline in spikelets per panicle, and 11.2%, 20.7%, and 28.5% increase in panicle number per m^2 were observed in mechanical hill-seeded hybrid rice varieties Nei2you 6, Yongyou 9, and Yonyou 12, respectively.

Discussion

Our study showed that high-yielding hybrid rice variety Nei2you6 produced maximum grain yield at a seeding spacing of 25 cm×17 cm in two soil fertility fields in both 2011 and 2012. Optimal hill seeding density for hybrid rice has also been reported in previous studies: Peilianyou 986 had the highest grain yield at a spacing of 25 cm×18 cm [20]; Jinyou 207 had the highest grain yield when hill-seeded at 30 cm×20 cm spacing [17]; and hill-seeded Peizataifen and Yuyouxiangzhang [19, 20] gave the highest grain yield at 25 cm×14 cm. However, these studies were conducted at one location in a single season and used only one hybrid rice variety with one to three seeding rates. It is thus difficult to conclude which seeding density can fullfill the yield

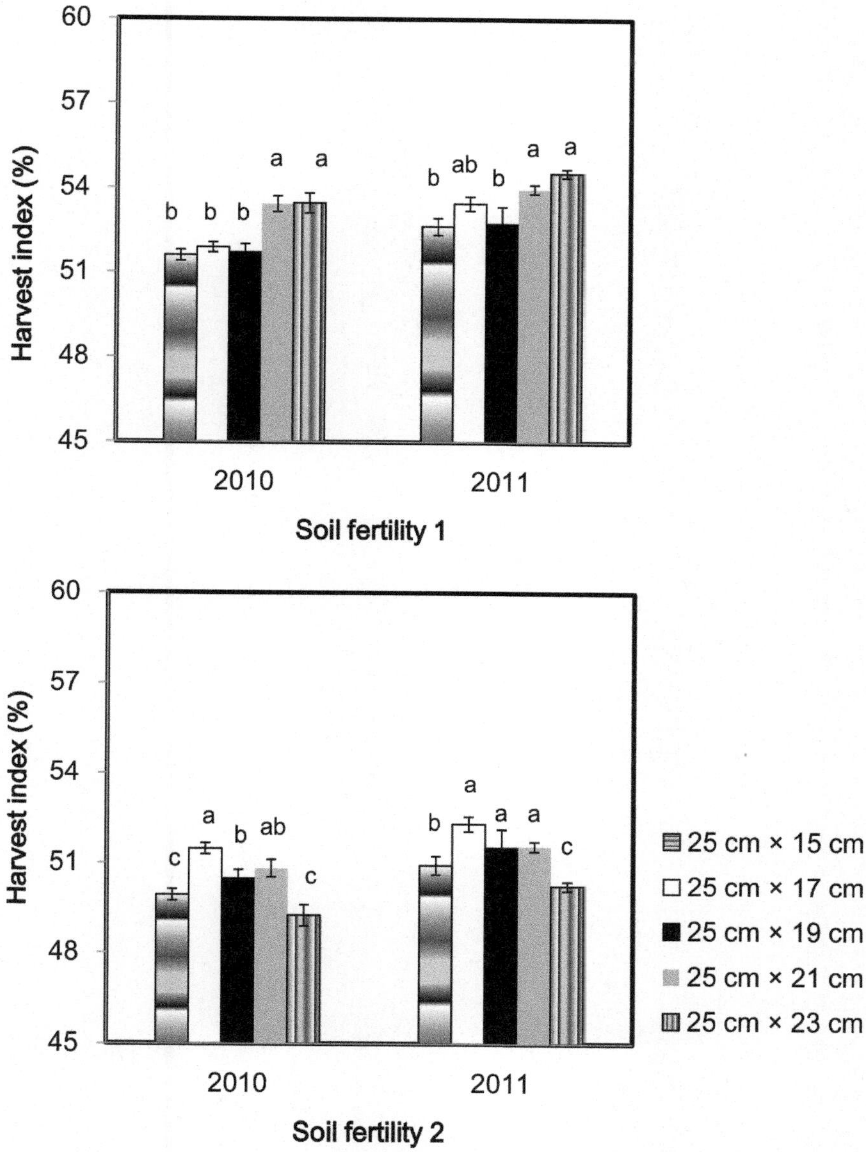

Figure 5. Harvest index of a high-yielding hybrid rice variety Nei2you6 grown at five seeding spacings in soil fertility 1 and soil fertility 2 at the China National Rice Research Institute, Zhejiang Province, China, in 2010 and 2011. Dry weight per shoot was calculated as the ratio of dry weight per hill to productive tiller number per hill.

potential of most high-yielding hybrid rice varieties. In our study, we compared plant yield and dry weight at five hill seeding spacings in two fields of different fertilities during a 2-year period. Besides that, three high-yielding hybrid rice varieties were mechanically seeded with a hill seeder at 25 cm×17 cm spacing in the on-farm demonstration experiments and compared with broadcasting and transplanting. Mechanical hill seeding outperformed broadcasting in terms of grain yield for all three high-yielding hybrid rice varieties; they also had grain yield equal to or higher than that of transplanted rice. We conclude that, for the south China region, a hill seeding spacing of 25 cm×17 cm, with three to five seeds per hill, is the appropriate density to be used for single cropping of high-yielding hybrid rice.

There have been extensive studies on the relationships between yield and plant density under nonstressed conditions [24–29]. For decades, rice reseachers in China have emphasized the importance of regulating the compensatory and competitive relationships among plants to increase grain yield by modifying planting density. In the 1960s, dense planting to increase panicle number per unit area was recommended in inbred rice production [30,31]. In the 1980s, with the development of inbred and hybrid rice varieties with large panicles, the advice on rice planting density changed. It was reported that planting at wide spacing decreased plant number per unit area, which improved the morphological traits of the rice plants such as plant height, tiller angle, and leaf angle [32]. It has been widely accepted at present in high-yielding hybrid rice production systems that increasing panicle size, rather than panicle number per unit area, increases yield [32–34]. In considering best management practices for rice, the basic concept applied is the use of young seedlings at optimal spacing to stimulate individual plant growth [35]. Recently, on-farm surveys conducted in Zhejiang, China, showed that some farmers transplanted hybrid rice using extremely wide spacing because of rising labor cost. They opined that the reduced plant number

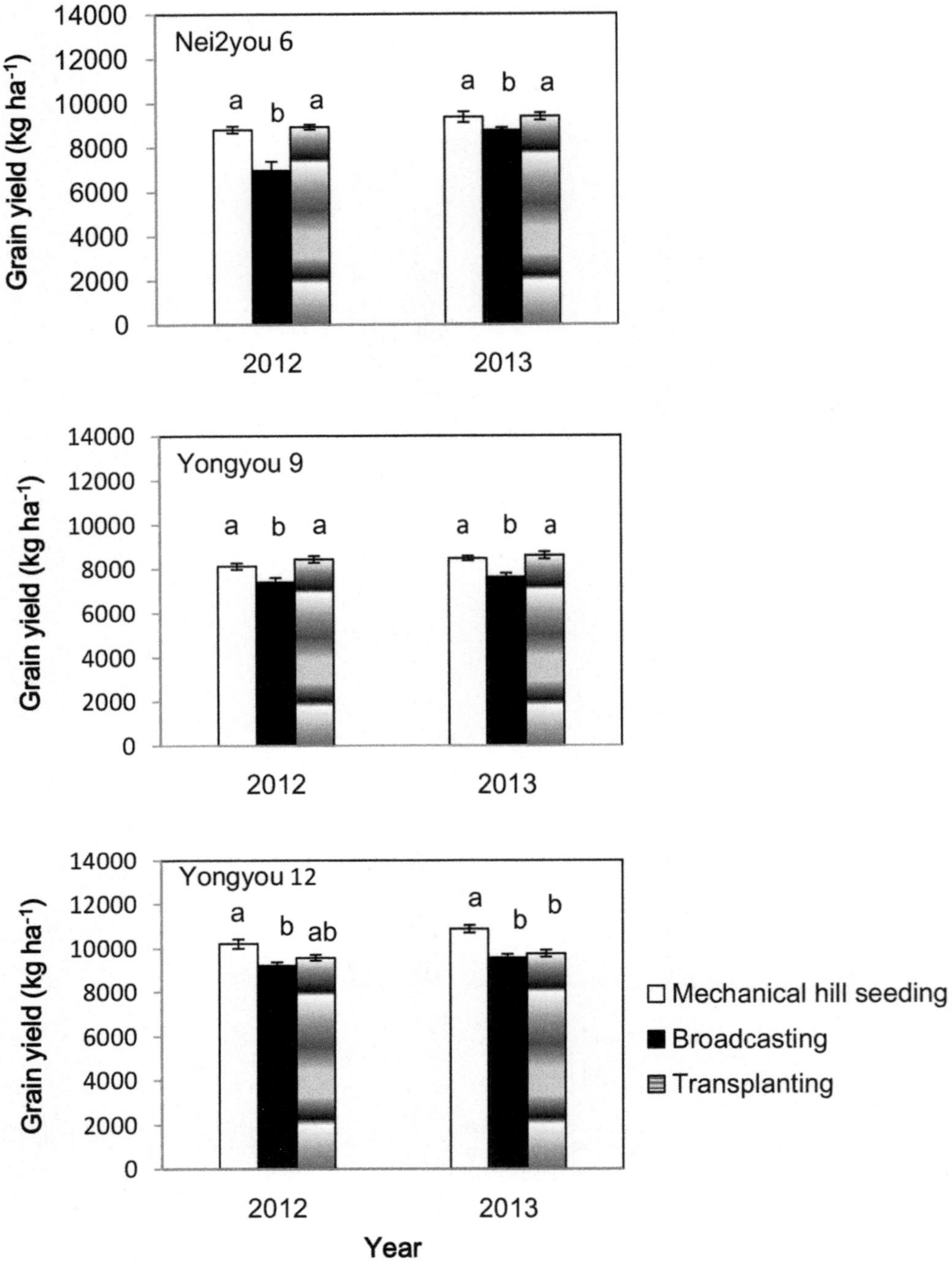

Figure 6. Grain yield of three high-yielding hybrid rice varieties under three plant establishment methods in Fuyang and Yuhang, Zhejiang Province, in 2012 and 2013. Plant spacing for mechanical hill seeding was 25 cm×17 cm; broadcasting and transplanting done according to standard farmer's practice.

caused by low planting density could be compensated for by the improvement in growth of the individual plants [36]. In our study, although grain yield per hill across two years consistently increased with increases in seeding spacing, grain yield per hill of hybrid rice variety Nei2you6 at 25 cm×23 cm spacing was 57.6% and 40.3% higher than at 25 cm×15 cm spacing in soil fertility 1 and 2, respectively. Grain yield per area across two years was highest at 25 cm×17 cm spacing, which was 5.3% and 12.7% higher than at 25 cm×23 cm spacing in soil fertility 1 and 2, respectively. These results suggest that yield per unit area cannot be fully compensated for by the yield increase of individual plants. Therefore, enough seedlings are necessary to achieve high yield. The on-farm surveys in Zhejiang Province also showed that the lowest grain yield was obtained with the widest planting spacing because of the lack of adequate plant population [36].

Rice grain yield per area or per hill is ultimately determined by the number of panicles per area or per hill and the grain yield per panicle. It was reported that yield was positively related to panicle number per unit area [36–37]. In our study, there was a significantly positive correlation between yield per hill and panicles per hill ($r = 0.98^{**}$), and a significantly negative correlation between grain yield per ha and panicles per hill ($r = -0.89^{*}$)

Table 4. Grain yield components of three high-yielding hybrid varieties under three plant establishment methods in Fuyang and Yuhang, Zhejiang Province, China, in 2012 and 2013.

Variety	Year	Treatment	Spikelets panicle^{-1} (no.)	Grain filling (%)	Grain weight (mg)	Panicles m^{-2} (no.)
Nei2you 6	2012	Mechanical hill seeding	176.03 b	83.25 a	26.76 b	235.56 b
		Broadcasting	130.82 c	80.32 b	26.72 b	283.65 a
		Transplanting	193.86 a	84.41 a	27.18 a	213.09 c
	2013	Mechanical hill seeding	174.03 b	85.75 a	27.42 a	250.60 b
		Broadcasting	126.57 c	82.40 b	26.84 b	305.00 a
		Transplanting	183.58 a	86.52 a	27.41 a	224.30 b
Yongyou 9	2012	Mechanical hill seeding	169.87 b	85.16 b	21.88 a	251.52 b
		Broadcasting	133.13 c	83.37 c	21.77 a	303.61 a
		Transplanting	203.08 a	87.63 a	22.17 a	210.60 c
	2013	Mechanical hill seeding	166.79 b	87.55 ab	22.79 a	262.00 b
		Broadcasting	129.37 c	85.49 b	22.45 a	313.00 a
		Transplanting	195.33 a	89.61 a	22.62 a	214.90 c
Yongyou 12	2012	Mechanical hill seeding	388.02 b	82.16 a	19.46 b	171.93 b
		Broadcasting	294.69 c	80.24 b	19.79 ab	202.85 a
		Transplanting	457.15 a	84.45 a	20.21 a	136.71 c
	2013	Mechanical hill seeding	389.09 b	84.46 ab	20.71 a	182.90 b
		Broadcasting	289.35 c	82.40 b	20.62 a	211.30 a
		Transplanting	439.71 a	86.52 a	20.62 a	139.50 c

Plant spacing for mechanical hill seeding was 25 cm×17 cm; broadcasting and transplanting done according to standard farmer's practice.

(Table 5). Panicle number is largely determined by the number of tillers that develop during the vegetative stage [38] and there is a large variation in tillering capacity among rice varieties [39,40]. Environmental factors and agronomic practices also alter tiller production and survival [41–43]. Plant density and fertilization are important factors influencing tiller production in rice [39,44–45]. LAI probably affected tillering by attenuation of light intensity and/or by influencing light quality at the base of the canopy where tiller buds and young tillers are located [46]. Furthermore, there are evidences showing that plant N status and LAI depend on each other in determining tillering [42]. A higher plant N concentration was needed to prevent tillers from dying when LAI was high [42]. In our study, maximum tiller number per hill increased with increases in seeding spacing, suggesting that plant tillering capacity increased with a decrease in seeding density. Compared with soil fertility 1, 17.2% and 10.1% increase in maximun tiller number per hill was obtained at 25 cm×15 cm and 25 cm×17 cm spacing, respectively, in soil fertility 2. However, no difference in maximum tiller number per hill was found between the two soil fertility fields at spacings ranging from 25 cm×19 cm to 25 cm×23 cm, suggesting that the negative effect of dense seeding on plant tillering capacity could be partly compensated for by an increase in nutrition supply in soil fertility 2. However, the rate of change in productive tiller percentage against seeding spacing was different between the two soil fertility fields. The lower productive tiller percentage at 25 cm×15 cm in soil fertility 2 offset its advantage of maximun tiller number per hill, and no significant difference in productive tillers per hill was observed at this spacing between the two fields. A 7% higher productive tillers at 25 cm×23 cm spacing in soil fertility 2 resulted in significantly higher panicle number in this field compared with soil fertility 1. The number of surviving tillers depends mainly on the extent of competition among tillers for carbohydrates produced and N absorbed from the maximum tillering stage to heading [40]. The differences in productive tiller percentage among seeding densities and between two soil fertility fields indicate that nutrition supply could increase panicle number at wide seeding spacing, but it cannot compensate for the negative effect of high seeding density on panicle number per hill because of the reduced productive tiller percentage.

There are compensatory and competitive relationships among tillers and their yield components over a wide range of plant densities [39]. In soil fertility 2, when seeding spacing was increased from 25 cm×17 cm to 25 cm×23 cm, the number of panicles per hill increased by 33.6% in 2010, and by 22.3% in 2011, while grain yield per shoot declined by 12.0% and by 10.0%, respectively. Grain filling percentage declined with increases in seeding spacing in both years in soil fertility 2, and no significant differences in spikelets per panicle and grain weight were found among treatments, suggesting that grain filling percentage decreased with the increase in productive tillers per hill, which resulted in the decline in yield per shoot. However, there was an interaction between seeding density and soil fertility. Less change in grain yield per shoot was in soil fertility 1 than soil fertility 2 at seeding spacings ranging from 25 cm×17 cm to 25 cm×21 cm. Rice plants grown at 25 cm×23 cm spacing had the highest grain yield per shoot in soil fertility 1, but had the the lowest grain yield per shoot in soil fertility 2. It was reported that high-yielding varieties usually have high biomass accumulation during the reproductive phase [47–51]. In our study, although aboveground dry weight per hill at flowering and maturity increased with increases in seeding spacing, aboveground dry weight per shoot was different between the two fertility fields. Rice plants seeded at 25 cm×23 cm spacing had the lowest aboveground dry weight per shoot both at flowering and maturity in soil fertility 2, while they had the highest aboveground dry weight per

Table 5. Correlation coefficients between grain yield and yield components of a high-yielding hybrid rice variety Nei2you6 grown at five seeding spacings at the China National Rice Research Institute, Zhejiang Province, China, in 2010 and 2011.

Yield and yield components	Spikelets per panicle	Grain filling percentage	Grain weight	Panicles per unit area	Panicles per hill	Grain yield per ha	Grain yield per hill
Grain filling percentage	−0.218						
Grain weight	−0.001	−0.802					
Panicles per unit area	0.207	−0.105	−0.449				
Panicles per hill	0.229	−0.842	0.973**	−0.378			
Grain yield per ha	0.159	0.828	−0.947*	0.409	−0.891*		
Grain yield per hill	0.291	−0.729	0.942*	−0.502	0.980**	−0.824	
Grain yield per shoot	0.041	0.965**	−0.791	−0.098	−0.774	0.868	−0.635

*Significant at $p<0.05$.
**Significant at $p<0.01$.

shoot in soil fertility 1. A different response of leaf area per shoot to seeding spacing was one of the factors that influence dry matter accumulation. The top three-leaf area per shoot increased with increases in seeding spacing in soil fertility 1, while it decreased in soil fertility 2. Considering that there are more panicles per hill in soil fertility 2 than in soil fertility 1 at wider seeding spacing and no extra fertilizer application, the lower area of the top three leaves and aboveground dry weight per shoot at 25 cm×23 cm spacing in soil fertility 2 may be attributed to limited N supply and/or N uptake. A recent study reported low N concentrations in shoots due to dilution as a result of more biomass production [43], which suggests that rice plants seeded at wider spacing might need more fertilizer to obtain high yield than those subjected to dense spacing. More fertilizer is needed not only to improve productive tiller percentage at wide spacing but also to increase N concentrations in plants and to improve dry weight accumulation and grain yield per shoot. Nevertheless, we think that the application of extra N to compensate for the reduced grain yield due to low planting density is not a good option because of potential environmental problems and increased cost for rice farmers.

In the on-farm demonstration experiments, mechanical hill seeding of rice at 25 cm×17 cm spacing had equal or higher grain yield than transplanted rice, indicating that this spacing was the optimum seeding density for high-yielding hybrid rice varieties. Hill-seeded rice had more panicles per unit area and less spikelets per panicle compared with transplanted rice.

Acknowledgments

The authors are grateful to Ms. Priscilla Grace Cañas of the International Rice Research Institute, for providing valuable comments on the manuscript.

Author Contributions

Conceived and designed the experiments: DYW. Performed the experiments: DYW SC ZMW CLJ XFZ. Analyzed the data: DYW CMX. Contributed reagents/materials/analysis tools: XFZ. Contributed to the writing of the manuscript: DYW BSC.

References

1. Pandey S, Velasco L (2002) Economics of direct seeding in Asia: patterns of adoption and research priorities. In: Pandey S, Mortimer M, Wade L, Tuong TP, Lopez K, editors. Seeding: research strategies and opportunities. Los Baños: International Rice Research Institute. pp. 3–14.
2. Farooq M, Siddique KHM, Rehman H, Aziz T, Lee DJ, et al. (2011) Rice direct seeding: experiences, challenges and opportunities. Soil Till Res 111: 87–98.
3. Chauhan BS, Mahajan G, Sardana V, Timsina J, Jat ML (2012) Productivity and sustainability of the rice-wheat cropping system in the Indo-Gangetic Plains of the Indian subcontinent: problems, opportunities, and strategies. Adv Agron 117: 315–369.
4. Joshi E, Kumar D, Lal B, Nepalia V, Gautam P, et al. (2013) Management of direct seeded rice for enhanced resource use efficiency. Plant Knowledge J 2: 119–134.
5. Chen F, Chen P (2011) Status and economic evaluation of direct-seeded rice production in south China-based on farmer's investigation. China Rice 17: 1–5 (in Chinese).
6. Chauhan BS (2013) Strategies to manage weedy rice in Asia. Crop Prot. 48: 51–56.
7. Wang D, Zhang X, Lu Y, Xu C, Jin B, et al. (2010) Grain yield difference investigation and development trends discussion of direct seeding rice production in Zhejiang province. China Rice 16: 22–26 (in Chinese).
8. Kabir MH, Saha A, Mollah IU, Kabir MS, Rahman F (2008) Effect of crop establishment methods and weed management practices on the productivity of boro rice in lowland ecosystem. Int J Biol Res 5: 42–51.
9. Mithrasena YJPK, Adikari WP (1986) Effect of density on sheath blight (ShB) incidence. Int Rice Res Notes 11: 20–21.
10. Sittisuang P (1995) Extent of problems and policies in wet-seeded rice production in Thailand. In: Moody K, editor. Constraints, opportunities, and innovations for wet-seeded rice. Los Baños: International Rice Research Institute, pp. 1–5.

11. Islam MF, Sarkar MAR, Islam MS, Parveen S, Hossain MS (2008) Effects of crop establishment methods on root and shoot growth, lodging behavior of Aus rice. Int J Biol Res 5: 60–64.
12. Ying J, Peng S, He Q, Yang H, Yang C, et al. (1998) Comparison of high-yield rice in tropical and subtropical environments. I. Determinants of grain and dry matter yields. Field Crops Res 57: 71–84.
13. Wu W, Zhang H, Qian Y, Cheng Y, Wu G, et al. (2008) Analysis on dry matter production characteristics of super hybrid rice. Rice Sci 15: 110–118.
14. Yang S, Xie L, Zheng S, Li J, Yuan J (2009) Effects of nitrogen rate and transplanting density on physical and chemical characteristics and lodging resistance of culms in hybrid rice. Acta Agron Sin 35: 93–103 (in Chinese, with English abstract).
15. Yoshinaga S (2005) Improved lodging resistance in rice (*Oryza sativa* L.) cultivated by submerged direct seeding using a newly developed hill seeder. Jpn Agric Res Q 39: 147–152.
16. Wang Z, Luo X, Tang X, Ma G, Zhang G, et al. (2010) Precision rice hill-direct-seeding technology and machine based on the combination of agricultural machinery and agronomic technology. J South China Agric Univ 31: 91–95 (in Chinese, with English abstract).
17. Zhu D, Wang J, Xu W, Wang Y, Xu W, et al. (2008) Rice mechanical precision hill seeded experiments. Anhui Agri Sci Bull 14: 133–135 (in Chinese).
18. Lei X, Liu L, Gou W, Ma R, Ren W (2013) Effects of planting methods on culm lodging resistance of indica hybrid rice (*Oryza sativa* L.). Acta Agron Sin 39: 1814–1825 (in Chinese with English abstract).
19. Shu S, Zheng T, Jia X, Luo X, Li G, et al. (2009) Effects of nitrogen fertilizer and density on precision hill-direct-seeding rice I – Yield formation characteristics. Chin Agric Sci Bull 25: 142–146 (in Chinese with English abstract).
20. Cheng J, Luo X, Fan Q, Zhang J, Wu J, et al. (2010) Influence of different planting methods on growth and development characteristics and yield of rice. J Huazhong Agric Univ 29: 1–5 (in Chinese with English abstract).
21. Zeng S, Huang Z, Wang Z, Luo X, Tang X (2014) Effects of different planting density on grain yield of precision hill-drop drilling rice. J Huazhong Agric Univ 33: 12–18 (in Chinese with English abstract).
22. Gong Z, Zhang G, Luo G (1999) Diversify of Anthrosols in China. Pedosphere 9: 193–204.
23. Cheng S, Zheng X, Wang D, Chen L, Xu C, et al. (2012) Effect of long-term paddy-upland yearly rotations on rice (*Oryza sativa*) yield, soil properties, and bacteria community diversity. Sci World J 279641. doi:10.1100/2012/279641.
24. Counce PA, Moldenhauer KAK, Marx DB (1989) Rice yield and plant yield variability responses to equidistant spacing. Crop Sci 29: 175–179.
25. Miller BC, Hill JE, Roberts SR (1991) Plant population effects on growth and yield in water-seeded rice. Agron J 83: 291–297.
26. Gravios KA, Helms RS (1982) Path analysis of rice yield and yield components as affected by seeding rate. Agron J 84: 1–4.
27. Baloch AW, Soomro AM, Javad MA, Ahmed M, Bughio HR, et al. (2002) Optimum plant density for high yield in rice (*Oryza sativa* L.). Asian J Plant Sci 1: 25–27.
28. Oziegbe M, Faluyi JQ (2007) Effects of seeding rate on the yield components of an enhanced rice cultivar (DTPMFe+) *Oryza sativa* Linn. Int J Bot 3: 317–320.
29. Lin X, Zhu D, Chen H, Zhang Y (2009) Effects of plant density and nitrogen application rate on grain yield and nitrogen uptake of super hybrid rice. Rice Sci 16: 138–142.
30. Ding Y (1961) China rice cultivation. Beijing: China Agriculture Press. pp. 53–56 (in Chinese).
31. Chen Y (1959) Chen Yongkang rice cultivation experience. Shanghai: Shanghai Scientific and Technical Publishers. pp. 1–54 (in Chinese).
32. Jiang P, Feng L (1987) The effect of cultivation techniques on plant type in view of the high yield result induced by TFS cultivation method in paddy rice. Chin J Rice Sci 1: 111–117.
33. Lin Q, Zhang H, Cheng G, Zhu J, Lu W (1982) Technical approaches to achieve large-area high-yield cultivation of IR 24 and discuss on the small groups, strong individual cultivation patterns. Jiangsu Agric Sci 9: 1–10 (in Chinese).
34. Ling QH (2000) Crop population quality. Shanghai. Shanghai Scientific and Technical Publishers. pp. 42–107 (in Chinese).
35. Uphoff N (2003) Higher yields with fewer external inputs? The system of rice intensification and potential contributions to agricultural sustainability. Int J Agric Sustain 1: 38–50.
36. Wang D, Zhang X, Zhou C, Zheng G, Zhang G, et al. (2010) Grain yield difference investigation and reasonable planting density analysis of rice production in Zhejiang Province. Acta Agric Zhejiang 22: 330–336 (in Chinese with English abstract).
37. Huang M, Zou Y, Jiang P, Xia B, Ibrahim M, et al. (2011) Relationship between grain yield and yield components in super hybrid rice. Agric Sci China 10: 1537–1544.
38. De Datta SK (1981) Principles and practices of rice production. New York: John Wiley & Sons. pp. 220–358.
39. Wu G, Wilson LT, McClung AM (1998) Contribution of rice tillers to dry matter accumulation and yield. Agron J 90: 317–323.
40. Nuruzzaman M, Yamamoto Y, Nitta Y, Yoshida T, Miyazaki A (2000) Varietal differences in tillering ability of fourteen japonica and indica rice varieties. Soil Sci Plant Nutr 46: 381–391.
41. Hanada K (1993) Tillers. In: Matsuo T, Hoshikawa K, editors. Science of the rice plant. Tokyo: Morphology, Food and Agricultural Policy Center. pp. 222–258.
42. Zhong X, Peng S, Sanico AL, Liu H (2003) Quantifying the interactive effect of leaf nitrogen and leaf area on tillering of rice. J Plant Nutr 26: 1203–1222.
43. Huang M, Zou Y, Jiang P, Xia B, Feng Y, et al. (2011) Yield component differences between direct-seeded and transplanted super hybrid rice. Plant Prod Sci 14: 331–338.
44. Yoshida S, Hayakawa Y (1970) Effects of mineral nutrition on tillering of rice. Soil Sci Plant Nutr 16: 186–191.
45. Fagada SO, De Datta SK (1971) Leaf area index, tillering capacity, and grain yield of tropical rice as affected by plant density and nitrogen level. Agron J 63: 503–506.
46. Zhong X, Peng S, Sheehy JE, Visperas RM, Liu H (2002) Relationship between tillering and leaf area index: quantifying critical leaf area index for tillering in rice. J Agric Sci 138: 269–279.
47. Katsura K, Maeda S, Horie T, Shiraiwa T (2007) Analysis of yield attributes and crop physiological traits of Liangyoupeijiu, a hybrid rice recently bred in China. Field Crops Res 103: 170–177.
48. Horie T, Lubis I, Takai T, Ohsumi A, Kuwasaki K, et al. (2003) Physiological traits associated with high yield potential in rice. In: Mew TW, Brar DS, Peng S, Dawe D, Hardy B, editors. Rice science: innovations and impact for livelihood. Proceedings of the International Rice Research Conference. Beijing: International Rice Research Institute. pp. 117–145.
49. Takai T, Matsuura S, Nishio T, Ohsumi A, Shiraiwa T, et al. (2006) Rice yield potential is closely related to crop growth rate during later reproductive period. Field Crops Res 96: 328–335.
50. Wu W, Zhang H, Qian Y, Cheng Y, Wu G, et al. (2008) Analysis on dry matter production characteristics of super hybrid rice. Rice Sci 15: 110–118.
51. Zhang Y, Tang Q, Zou Y, Li D, Qin J, et al. (2009) Yield potential and radiation use efficiency of "super" hybrid rice grown under subtropical conditions. Field Crops Res 114: 91–98.

12

Bt Rice Expressing Cry2Aa Does Not Harm *Cyrtorhinus lividipennis*, a Main Predator of the Nontarget Herbivore *Nilapavarta lugens*

Yu Han[1], Jiarong Meng[1], Jie Chen[1], Wanlun Cai[1], Yu Wang[1], Jing Zhao[1], Yueping He[1], Yanni Feng[2], Hongxia Hua[1]*

1 Hubei Insect Resources Utilization and Sustainable Pest Management Key Laboratory, College of Plant Science and Technology, Huazhong Agricultural University, Wuhan, P.R. China, **2** College of Life Science and Technology, Huazhong Agricultural University, P.R. China

Abstract

T2A-1 is a newly developed transgenic rice that expresses a synthesized *cry2Aa* gene driven by the maize *ubiquitin* promoter. T2A-1 exhibits high resistance against lepidopteran pests of rice. The brown planthopper, *Nilapavarta lugens* (Stål), is a main nontarget sap-sucking insect pest of rice, and *Cyrtorhinus lividipennis* (Reuter) is the major predator of the eggs and young nymphs of planthoppers. As *C. lividipennis* may expose to the Cry2Aa protein via *N. lugens*, it is therefore essential to assess the potential effects of transgenic *cry2Aa* rice on this predator. In the present study, three experiments were conducted to evaluate the ecological risk of transgenic *cry2Aa* rice to *C. lividipennis*: (1) a direct feeding experiment in which *C. lividipennis* was fed an artificial diet containing Cry2Aa at the dose of 10-time higher than that it may encounter in the realistic field condition; (2) a tritrophic experiment in which the Cry2Aa protein was delivered to *C. lividipennis* indirectly through prey eggs or nymphs; (3) a realistic field experiment in which the population dynamics of *C. lividipennis* were investigated using vacuum-suction. Both direct exposure to elevated doses of the Cry2Aa protein and prey-mediated exposure to realistic doses of the protein did not result in significant detrimental effects on the development, survival, female ratio and body weight of *C. lividipennis*. No significant differences in population density and population dynamics were observed between *C. lividipennis* in transgenic *cry2Aa* and nontransgenic rice fields. It may be concluded that transgenic *cry2Aa* rice had no detrimental effects on *C. lividipennis*. This study represents the first report of an assessment continuum for the effects of transgenic *cry2Aa* rice on *C. lividipennis*.

Editor: Guy Smagghe, Ghent University, Belgium

Funding: This research was supported by the National Genetically Modified Organisms Breeding Major Project: Technology of Environmental Risk Assessment on Transgenic Rice (2014ZX08011-001). HH received the funding. The funders had no role in study design, data collection and analysis, decision to publish, or preparation of the manuscript.

Competing Interests: The authors have declared that no competing interests exist.

* Email: huahongxia@mail.hzau.edu.cn

Introduction

Rice, *Oryza sativa* L., is the staple food of more than three billion people of Asia [1]. More than 200 species of insect pests infest rice during its growing season [2], and this causes 15% to 25% yield losses of rice [3,4]. Among these insects, lepidopteran species such as stem borers and leaffolders are particularly serious chronic pests and cause large annual yield losses [5,6]. Traditional management of lepidopteran pests is mainly dependent on the spraying of pesticides. However, excessive or continual applications of pesticide not only cause environmental contamination and the resurgence of herbivores but also reduce populations of the natural enemies of these herbivores [7,8]. Researchers have therefore been encouraged to seek more effective and environmentally friendly methods to control lepidopteran pests.

The *Bacillus thuringiensis* (*Bt*) insecticidal δ-endotoxin has been used as a biological insecticide for more than 50 years, and it is now possible to introduce different *Bt* genes into rice. Transgenic rice plants expressing Bt proteins have been shown to be effective against many lepidopteran insect pests [9] and have led to great reductions in the use of insecticides [10,11]. A series of rice lines that express various *Bt* genes (e.g., *cry1Ab*, *cry1Ab/1Ac*, *cry1C* and *cry2A*) have been developed and can effectively suppress the infestation of target lepidopteran insect pests [12–14]. Despite these successes, concerns have been raised regarding the potential impacts of transgenic *Bt* rice on nontarget herbivores and their natural enemies through tritrophic transmission. It is therefore necessary to conduct an environmental risk assessment of any novel transgenic *Bt* rice prior to its commercialization, and the environmental risks of each rice line must be evaluated on a case-by-case basis.

The brown planthopper, *Nilapavarta lugens* (Stål) (Hemiptera: Delphacidae), is the main nontarget sap-sucking insect pest of rice and causes significant annual reductions in the rice yield [15,16]. The evaluation of nontarget effects of transgenic rice on the natural enemies of *N. lugens* is an important part of the environmental risk assessment, and should be conducted before the commercialization of any novel *Bt* rice. Several previous reports have examined the effects of transgenic *Bt* rice on the predators of *N. lugens*. Although the Cry1Ab protein can be transferred from transgenic rice plants to predators via *N. lugens*, no adverse effects have been found on any of the fitness parameters (survival, developmental time, weight and fecundity) of numerous predators [*Cyrtorhinus lividipennis* (Reuter) (Hemiptera: Miridae), *Pardosa pseudoannulata* (Bösenberg et Strand) (Araneae: Lycosidae), *Pirata subpiraticus* (Bösenberg et Strand) (Araneae: Lycosidae), *Propylea japonica* (Thunberg) (Coleoptera: Coccinellidae) and *Ummeliata insecticeps* (Bösenberg et Strand) (Araneae: Linyphiidae)] that preyed on *N. lugens* reared on *Bt*-transformed rice lines in the laboratory [17–22]. Additionally, the results from field experiments have also shown that the population densities and population dynamics of *U. insecticeps*, *C. lividipennis* and *P. pseudoannulata* are not significantly different between transgenic *cry1Ab* rice and non-*Bt* rice fields [21–23]. All of these results have indicated that the predators of *N. lugens* are not affected by the Cry1Ab protein. T2A-1 is a relatively newly developed transgenic rice line expressing *cry2Aa* driven by the maize *ubiquitin* promoter, and it exhibits high resistance against lepidopteran pests of rice [12]. However, the potential effects of this rice line on the predators of *N. lugens* have received little attention. T2A-1 did not cause direct detrimental effects on the larvae of *Chrysoperla sinica* (Tjeder) (Neuroptera: Chrysopidae), a general predator of *N. lugens* [24]; that study represents the only existing report about the effects of Cry2Aa on the predators of *N. lugens*.

Cyrtorhinus lividipennis (Reuter) (Hemiptera: Miridae) is a major predator of the eggs and young nymphs of planthoppers, and it is a primary factor regulating the population density of *N. lugens* in rice fields [7,25,26]. Thus, *C. lividipennis* may expose to the Bt protein via *N. lugens*. The potential effects of transgenic *Bt* rice on *C. lividipennis* should be evaluated. Transgenic *cry1Ab* rice had no negative effects on the life-table parameters, population density and population dynamics of *C. lividipennis* [18,23]. However, no such study of transgenic *cry2Aa* rice has yet been conducted with *C. lividipennis*. In the current study, we conducted comprehensive experiments in the laboratory and in rice fields to examine the effects of T2A-1 on *C. lividipennis*. The effects of T2A-1 via prey on the life-table parameters and the functional response of *C. lividipennis* to *N. lugens*, and the direct toxicity of the Cry2Aa protein to *C. lividipennis* were evaluated in the laboratory. The effects of T2A-1 on population density and population dynamics were investigated through a 3-year experiment in rice fields. ELISA was used to determine whether the Cry2Aa protein could be transferred to *C. lividipennis* via *N. lugens*.

Materials and Methods

Ethics Statement

All necessary permits were obtained for the described field studies. Permission of small-field test of the transgenic line (T2A-1) at the suburbs of Xiaogan City and the suburbs of Suizhou City during the 2011-2013 was issued by Ministry of Agriculture of the People's Republic of China.

Plant materials

The transgenic *Bt* rice line (T2A-1) and the nontransgenic parental *indica* rice line Minghui 63 were selected for the experiments. T2A-1 expresses a synthesized *cry2Aa* gene driven by the maize *ubiquitin* promoter. T2A-1 is homozygous and exhibits high resistance against lepidopteran pests of rice [12]. Minghui 63 is an elite *indica* restorer strain, and served as nontransgenic control. Both rice lines were gifted by National Key Laboratory of Crop Genetic Improvement, Wuhan, China.

The two rice lines used for the laboratory experiments were cultured in different plastic tanks (25 cm length×20 cm width×3 cm height) in Yoshida culture solution [27]. Fifteen-day-old rice seedlings (approximately 15 cm in height) were used in the experiments. All plants were maintained at 26°C±2°C, and the relative humidity was approximately 80%.

Insects for the laboratory experiments

The original adults of *N. lugens* were collected from paddy fields in Wuhan, Hubei Province, China. Prior to the tritrophic bioassay, independent colonies of *N. lugens* were established on T2A-1 and Minghui 63 and maintained for more than ten generations. The *C. lividipennis* individuals were collected from paddy fields in Xiaogan and reared with eggs of *N. lugens* infested on Minghui 63. The colony was maintained for more than six generations before its use in the present study. The insects were cultured at 28±1°C, RH 70±5% and with a light-dark cycle of 14 h:10 h. Both susceptible strains of *Plodia interpunctella* (Hubner) (Lepidoptera: Pyralidae) and *Cnaphalocrocis medinalis* (Guenee) (Lepidoptera: Pyralidae) were used to confirm the bioactivity of the Cry protein.

Prey egg–mediated effects of transgenic *cry2Aa* rice on the life-table parameters of *C. lividipennis*

Two reproductive females of *N. lugens* were placed into a glass tube (3 cm diameter×25 cm length) containing six or seven 15-day-old rice seedlings (Minghui 63 and T2A-1) and allowed to lay eggs for 2 days. The *N. lugens* adults were then removed, and the eggs of *N. lugens* on the rice seedlings were used as the prey of *C. lividipennis*. Newly hatched nymphs of *C. lividipennis* (<24 h) were placed individually in glass tubes. All tubes were sealed with nylon mesh. During the period from the first to the third instar of *C. lividipennis*, the rice seedlings were refreshed every 2 days, and during the period from the fourth instar to adulthood, the rice seedlings were refreshed every day. Yoshida culture solution [27] was used to keep the rice seedlings fresh. The survival and molting of the *C. lividipennis* nymphs were recorded every day. After the *C. lividipennis* adults emerged, the sex and body weight of these adults were recorded. Thirty-two nymphs of *C. lividipennis* were tested for each rice line.

Prey nymph–mediated effects of transgenic *cry2Aa* rice on the life-table parameters of *C. lividipennis*

Newly molted second-instar nymphs (<24 h) of *C. lividipennis* were placed individually in glass tubes (2 cm diameter×12 cm length) covered with tampons. The bottom of each tube was filled with a piece of wetted sponge to maintain humidity. Newly hatched nymphs of *N. lugens* (24–48 h after hatching), reared on either Minghui 63 or T2A-1 rice plants, were employed as the prey of *C. lividipennis*. During the period from the second to the third instar of *C. lividipennis*, 10 nymphs of *N. lugens* were provided daily, and during the period from the fourth instar to adulthood, 20 nymphs were provided daily. The survival and molting of the *C. lividipennis* nymphs were monitored on a daily basis. After the adults of *C. lividipennis* emerged, the sex and body weight of these adults were recorded. Eighty nymphs of *C. lividipennis* were tested for each rice line.

Cry2Aa contents in rice plants, *N. lugens* and *C. lividipennis*

The sheaths of 15-day-old rice seedlings, neonates of *N. lugens* fed on T2A-1 or Minghui 63 for 2 days, *N. lugens* eggs laid by *N. lugens* adults fed on T2A-1 or Minghui 63, and third- or fourth-instar nymphs of *C. lividipennis* that preyed on eggs or nymphs of *N. lugens* fed on T2A-1 or Minghui 63 were collected. Four or five samples were collected for each treatment. Before the assay, the insect samples were washed four times with PBST (PBS/0.55% Tween-20) to remove any Cry protein from their outer surfaces. The Cry protein contents were determined using AP005 ENVIRONLOGIX kits (ENVIRONLOGIX, USA). The kits were used according to the manufacturer's instructions.

Exposure of *C. lividipennis* to Cry2Aa at high dose

Lyophilized Cry2Aa protein was purchased from the Biochemistry Department Laboratory, School of Medicine, Case Western Reserve University, USA. The Cry2Aa protoxin was expressed by *Escherichia coli*, then the protoxin inclusion bodies were solubilized and trypsinized, subsequently the toxins were purified and lyophilized. The purity of toxins is about 95–98%, and the molecular size of activated toxin is 65 kDa. Potassium arsenate (PA, KH_2AsO_4), which has been previously reported as toxic to insects [24,28], was used as a toxic model compound and was purchased from Sigma-Aldrich (St. Louis, MO).

The results of the preliminary experiments showed that the chemically defined diet for *N. lugens* as described by Fu et al. [29] was sufficient for sustaining the growth and development of *C. lividipennis* from the second instar to adulthood (The survival of *C. lividipennis* from the first instar to adulthood was low). The artificial diet for *N. lugens* was therefore used as the medium to deliver the Cry2Aa protein to the gut of *C. lividipennis*. Second-instar nymphs of *C. lividipennis* were reared on one of three different diets: i) an artificial diet (negative control); ii) an artificial diet containing 300 μg/ml of Cry2Aa (a level over ten times higher than that to which *C. lividipennis* would realistically be exposed in rice fields); iii) an artificial diet containing 40 μg/ml of PA (positive control). The diets were refreshed daily. The molting and survival of *C. lividipennis* were observed daily. When the adults emerged, their genders and body weights were recorded. Thirty-six nymphs of *C. lividipennis* were evaluated for each treatment.

To ensure the stability of the Cry2Aa protein in the artificial diets before and after 24 h of feeding exposure, the Cry2Aa proteins were extracted from the artificial diets, and the concentrations of Cry2Aa protein in the diet were determined using AP005 ENVIRONLOGIX kits.

To examine the bioactivity of this batch of Cry2Aa on lepidopteran insects, Cry2Aa was mixed with an artificial diet for *P. interpunctella*. Each bioassay included five concentrations for Cry2Aa (2, 10, 20, 30, 40 μg/g) plus a control. Forty newly hatched *P. interpunctella* larvae javascript:void(0); were introduced into the artificial diets of each concentration and five replicates were tested. Mortality of the larvae was determined after 1 week. The LC_{50} (concentration resulting in 50% *P. interpunctella* larval mortality) of this batch of Cry2A protein was measured.

To examine the bioactivity of Cry2Aa protein in the artificial diets before and after 24 h of feeding exposure, the artificial diets containing Cry2Aa were appropriately diluted and sprinkled on the leaves of corn. After 2 h of air-drying, 15 *Bt*-susceptible second-instar larvae of *C. medinalis* were distributed onto the leaves for each treatment. Four replicates were tested in each treatment. Mortality of the insects were recorded 48 h later.

Effects of transgenic *cry2Aa* rice on the functional response of female *C. lividipennis*

C. lividipennis individuals were reared on T2A-1 or Minghui 63 rice plants infested with *N. lugens* eggs for one generation. Female adults of *C. lividipennis* (2 days after eclosion) were starved for 24 h, and each female was then transferred to one glass tube (2 cm diameter×12 cm length) containing three or four 15-day-old rice seedlings infested with first-instar nymphs of *N. lugens*. The experimental densities of *N. lugens* were 10, 20, 30, 40 and 50 nymphs per tube. The number of *N. lugens* consumed by *C. lividipennis* was recorded after 24 h. The experiment was repeated five times for each density.

Field experiment design

The experiments were conducted during the growing seasons of 2011–2013 at the two different sites, where field trials of *Bt* rice were permitted, in Hubei Province, China. The first site located in the suburbs of Xiaogan City, and the second site located in the suburbs of Suizhou City. The layout of the plots in the field followed a completely randomized block design with four replications for each rice line. Each experimental plot was 150 m^2 (10 m×15 m) and was surrounded by a 1 m wide unplanted border. The entire experimental field was bordered by five rows of the nontransgenic control plants. The *Bt* and non-*Bt* control rice lines were sown in early May and transplanted 1 month after sowing. The seedlings were manually transplanted with one seedling per pot, 13.3 cm between plants within a row, and 29.9 cm between rows. The agronomic practices used for growing the rice, including fertilization and irrigation, were consistent with those followed by local farmers, except that no insecticides were applied during the whole growing season.

Field sampling with a vacuum-suction machine

Sampling of *C. lividipennis* was conducted as described by Xu et al. [30]. Arthropods at both field sites were collected using a vacuum-suction machine, constructed basing on a description by Carino et al. [31] and supplemented by a square sampling box (50 cm length×50 cm width×120 cm height) with a metal frame enclosed by Mylar film. Samples were collected every 10–15 days, starting 1 month after transplantation and continuing until the rice was ripe (as measured by grain maturity and harvest). On each sampling date, a square sampling box was placed at random along the diagonal line of each test plot at each site, with five subsamples per plot. The sample locations in each plot were marked with bamboo stakes to avoid resampling at the same location. Arthropods inside the frame enclosure were collected using the vacuum-suction machine for 5 min at each sampling location and were preserved in 75% ethanol. All samples were taken back to the laboratory and identified to the species level.

Data analysis

ELISA data and body weights were compared using the Student's *t*-tests. The Chi-square test was used for the parameters of preimaginal survival and female ratio. Nymphal developmental time was analyzed using Mann–Whitney *U*-tests, as the data did not fulfill the assumptions required for parametric analyses (normal distribution of residues and homogeneity of error variances). Survival response to the artificial diets containing Cry2Aa was analyzed using the Kaplan-Meier procedure, and the log-rank test was used in the purified toxin experiment.

In the field experiment, the population density and population dynamics determined by vacuum-suction were used to evaluate the impacts of the transgenic *Bt* rice on *C. lividipennis* populations in the fields. The population density of *C. lividipennis* was represented by seasonal means as captured by vacuum-suction. The population dynamics of the predators were measured by the means at each sampling date. Population density and population dynamics were analyzed using the Student's *t*-test.

The data from the functional response experiment were fitted to Holling's "Type II" disc equation, which estimates *Na* as follows: $Na = aTN/(1 + aThN)$, where *Na* is the number of prey attacked, *N* is prey density, and *T* is the duration of the experiment ($T = 1$ day in the present study). The parameters *a* (instantaneous search rate) and *Th* (time required to handle a prey item) were calculated via least-squares nonlinear regression based on the Gauss-Newton method.

The percentage data were arcsine–square root transformed, and all count data were square root (x+1) or $\log_{10}$ (x+1) transformed before being subjected to data analysis. The untransformed means are presented in the results. All statistical analyses were performed using the software package SPSS (version 16.0 for Windows, 2007).

Results

Prey-mediated effects of transgenic *cry2Aa* rice on the life-table parameters of *C. lividipennis*

No significant differences were observed between the developmental time, preimaginal survival, female ratio and fresh body weight of *C. lividipennis* adults reared with eggs or nymphs of *N. lugens* fed on *Bt* and non-*Bt* rice (P>0.05) (Tables 1, 2).

The Cry2Aa content in T2A-1 rice sheaths was 13.9±1.2 μg/g. The content of Cry2Aa in nymphs of *N. lugens* fed on T2A-1 was 5.3±0.8 ng/g, only 0.04% of that in the rice sheaths. The content of Cry2Aa in the eggs of *N. lugens* fed on T2A-1 was undetectable (Table 3).

When *N. lugens* nymphs and T2A-1 seedlings were provided simultaneously to *C. lividipennis*, Cry2Aa was detectable, and the concentration of Cry2Aa in *C. lividipennis* was 48.5±13.0 ng/g; this value was much lower than that in the rice sheaths (13.9±1.2 μg/g) but higher than that in the nymphs of *N. lugens* (5.3±0.8 ng/g). However, when the *N. lugens* nymphs fed on T2A-1 were removed from T2A-1 and the nymphs alone were provided to *C. lividipennis*, no Cry2Aa was detected in the predator. When *N. lugens* eggs and T2A-1 seedlings were provided to *C. lividipennis* simultaneously, Cry2Aa was detectable, and the concentration of Cry2Aa in *C. lividipennis* was 200.2±33.2 ng/g. However, when the *N. lugens* eggs were removed from the T2A-1 seedlings and the eggs alone were provided to *C. lividipennis*, Cry2Aa was not detectable in the predator. As was expected, no Cry2Aa protein was detected in the Minghui 63 rice plants (Table 3).

Purified Cry2Aa protein bioassay

Before and after exposure to *C. lividipennis* for 24 h, the concentrations of Cry2Aa in the artificial diets were 98.4±3.2 μg/g and 91.1±3.7 μg/g, respectively. After 24 h of exposure, the concentration of Cry2Aa in the diets decreased by 7.42%, and no statistical differences were detected between the values before and after exposure (Student's *t*-test, $P = 0.185$). Therefore, it could be concluded that the Cry2Aa protein in the artificial diets was stable.

The LC_{50} of this batch of Cry2Aa for *P. interpunctella* larvae was 14.9 μg/g fresh weight. Before and after exposure to *C. lividipennis* for 24 h, the artificial diets containing 300 μg/ml Cry2Aa resulted in 90% and 80% mortality of *C. medinalis* larvae, respectively; these mortalities were significantly higher than those that occurred in larvae fed with the pure artificial diet (13%). These results indicated that the Cry2Aa protein in the artificial diets for *C. lividipennis* was bioactive.

Survival of the *C. lividipennis* fed with the artificial diet containing 40 μg/ml PA was significantly decreased compared to that of the *C. lividipennis* fed with the pure artificial diet ($P <$ 0.001) (Fig. 1). This result indicated that the test system employed in the present study could detect the dietary effects of insecticidal compounds. High-dosage exposure to the Cry2Aa protein had no adverse effects on the survival response of *C. lividipennis* in comparison to the pure artificial diet (negative control) (Fig. 1). Similarly, the developmental time from the second instar to adulthood, preimaginal survival and body weight of *C. lividipennis* were unaffected by the Cry2Aa protein in comparison to the negative control (Table 4).

Table 1. Prey-mediated effects of Cry2Aa on life-table parameters of *Cyrtorhinus lividipennis* preying eggs of *Nilapavarta lugens* reared with T2A-1 or Minghui 63 rice plants.

Parameters	T2A-1	Minghui 63	Statistics
1st instar developmental time (days ± SE)[a]	2.3±0.09 (32)	2.5±0.11 (32)	$U = 415$, $P = 0.117$
2nd instar developmental time (days ± SE)[a]	1.7±0.09 (30)	1.6±0.10 (30)	$U = 400.5$, $P = 0.387$
3rd instar developmental time (days ± SE)[a]	1.5±0.09 (29)	1.5±0.10 (30)	$U = 435$, $P = 1.000$
4th instar developmental time (days ± SE)[a]	1.8±0.09 (29)	1.8±0.07 (30)	$U = 400$, $P = 0.473$
5th instar developmental time (days ± SE)[a]	2.7±0.09 (29)	2.9±0.14 (29)	$U = 364$, $P = 0.399$
Whole nymphal stage developmental time (days ± SE)[a]	10.0±0.13 (28)	10.3±0.13 (29)	$U = 313$, $P = 0.080$
Preimaginal survival (%)[b]	87.5	90.6	$\chi^2 = 0.002$, $P = 0.967$
Female ratio (%)[b]	42.9	55.2	$\chi^2 = 0.864$, $P = 0.352$
Male weight (mg ± SE)[c]	0.55±0.03	0.49±0.03	$t = -1.492$, $P = 0.179$
Female weight (mg ± SE)[c]	0.93±0.02	0.92±0.04	$t = -0.193$, $P = 0.851$

The experiment started with 32 nymphs per treatment. (n), number of individuals at each development stage.
[a]Mann–Whitney *U*-test.
[b]Chi-square test.
[c]Student's *t*-test.

Table 2. Prey-mediated effects of Cry2Aa on life-table parameters of *Cyrtorhinus lividipennis* preying nymphs of *Nilapavarta lugens* reared with T2A-1 or Minghui 63 rice plants.

Parameters	T2A-1	Minghui 63	Statistics
2nd instar developmental time (days ± SE)[a]	1.9±0.08 (76)	1.8±0.06 (77)	U=2.538E3*, P=0.102
3rd instar developmental time (days ± SE)[a]	1.8±0.07 (63)	2.0±0.08 (70)	U=1.875E3*, P=0.087
4th instar developmental time (days ± SE)[a]	2.4±0.09 (50)	2.4±0.13 (48)	U=1.180E3*, P=0.873
5th instar developmental time (days ± SE)[a]	2.8±0.17 (29)	2.7±0.15 (27)	U=367.5, P=0.654
2nd instar-adult developmental time (days ± SE)[a]	8.8±0.21 (28)	9.2±0.20 (26)	U=292, P=0.190
Preimaginal survival (%)[b]	35.0	32.5	χ^2=0.112, P=0.738
Female ratio (%)[b]	42.9	38.5	χ^2=0.108, P=0.743
Male weight (mg ± SE)[c]	0.33±0.05	0.30±0.01	t=−0.420, P=0.696
Female weight (mg ± SE)[c]	0.46±0.02	0.43±0.09	t=−0.322, P=0.764

The experiment started with 80 nymphs per treatment. (n), number of individuals at each development stage.
*E3 = 10^{-3}.
[a]Mann–Whitney U-test.
[b]Chi-square test.
[c]Student's t-test.

Effects of Cry2Aa on the functional response of *C. lividipennis*

The results indicated that the functional response of *C. lividipennis* to *N. lugens* on both rice lines was typically Type II as described by Holling [32,33] (Fig. 2). The instantaneous search rate and handling time were not significantly affected by rice line (P>0.05) (Table 5).

Effects of transgenic *cry2Aa* rice on the population density and dynamics of *C. lividipennis*

The population density of *C. lividipennis* is shown in Table 6. Compared with that on the nontransgenic Minghui 63, the population density of *C. lividipennis* was not significantly influenced by transgenic *cry2Aa* rice at any site or in any year (Student's t-test, P>0.05) (Table 6). The population dynamics (means of each sampling date) of *C. lividipennis* are shown in Fig. 3. No significant differences in the population dynamics of *C. lividipennis* were observed between Minghui 63 and transgenic *cry2Aa* rice fields at any sampling date, at any site or in any year (Student's t-test, all P>0.05). Repeated measures ANOVA analysis showed that the population dynamics were unaffected by rice line (P>0.05).

Discussion

The ecological risk assessment of an insect-resistant transgenic crop for a nontarget arthropod (NTA) should be conducted within a tiered scheme: (i) effects on the NTA at elevated doses in a replicated controlled system; (ii) effects on the NTA at realistic doses in a replicated controlled system; (iii) effects on the population of the NTA at realistic doses in a realistic agricultural system [34]. According to these criteria, three experiments were conducted to evaluate the ecological risk of transgenic *cry2Aa* rice to *C. lividipennis*, a primary predator of *N. lugens* that is the main NTA of transgenic *Bt* rice in the present study: (1) a direct feeding experiment, in which *C. lividipennis* was fed an artificial diet containing Cry2Aa at the dose of 10-time higher than that it may encounter in the realistic field condition; (2) a tritrophic experiment, in which the Cry2Aa protein was delivered to *C. lividipennis* indirectly through prey eggs or nymphs; (3) a realistic field experiment, in which the population dynamics of *C. lividipennis* were investigated by vacuum-suction. Direct exposure to elevated doses of the Cry2Aa protein and prey-mediated exposure to realistic doses of the Cry2Aa protein did not result in significant detrimental effects on the developmental time, pre-imaginal survival, female ratio and body weight of *C. lividipennis*.

Table 3. Contents of Cry2Aa protein in rice sheath tissue, *Nilapavarta lugens* and *Cyrtorhinus lividipennis*.

Treatments	T2A-1	Minghui 63
Sheath of rice plants	13.9±1.2 μg/g	Not detectable
Eggs of *N.lugens*	Not detectable	Not detectable
Nymphs of *N. lugens*	5.3±0.8 ng/g	Not detectable
C. lividipennis provided with *N. lugens* eggs with rice plants	200.2±33.2 ng/g	Not detectable
C. lividipennis provided with *N. lugens* nymphs with rice plants	48.5±13.0 ng/g	Not detectable
C. lividipennis provided with *N. lugens* eggs without rice plants	Not detectable	Not detectable
C. lividipennis provided with *N. lugens* nymphs without rice plants	Not detectable	Not detectable
C. lividipennis provided with rice plants	17.1±8.6 ng/g	Not detectable

Data are represented as mean ± SE.

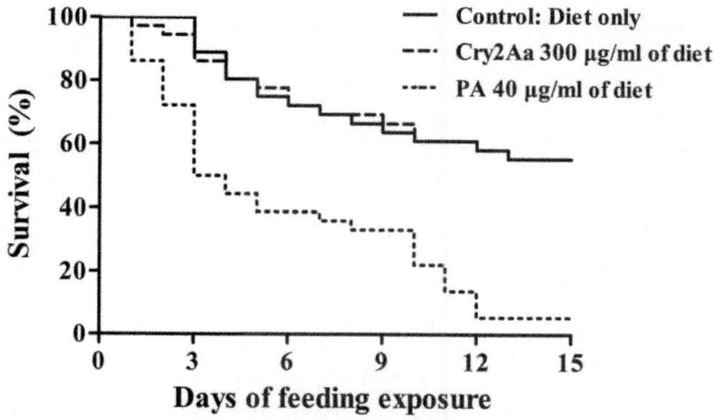

Figure 1. Survival of *Cyrtorhinus lividipennis* fed pure artificial diet or diet containing different insecticidal compounds. 300 μg Cry2Aa and 40 μg PA per ml were incorporated into artificial diets. Pure diet served as a negative control (N = 36).

No significant differences in population density and population dynamics were observed between *C. lividipennis* populations in transgenic *cry2Aa* and nontransgenic rice fields. It could be concluded that transgenic *cry2Aa* rice had no adverse effects on *C. lividipennis*. This study represents the first report of an assessment continuum for the effects of transgenic *cry2Aa* rice on *C. lividipennis*.

Planthoppers are the main nontarget herbivores in transgenic *Bt* rice fields. Determining whether the Bt protein may be transmitted to predators via planthoppers is important for the ecological risk assessment of transgenic *Bt* rice. Several reports have examined Bt protein transmission to predators via predation on planthoppers infesting transgenic *Bt* rice, but their results were inconclusive. Cry1Ab could be transferred to *P. subpiraticus* by predation on *N. lugens* fed on transgenic *cry1Ab* rice, and the content of Cry1Ab in *P. subpiraticus* was significantly higher than that in *N. lugens* fed on transgenic *cry1Ab* rice [19]. Cry1Ab could also be transferred to *U. insecticeps* via predation on *N. lugens* fed on transgenic *cry1Ab* rice, and the concentration of Cry1Ab in *U. insecticeps* was significantly lower than that in *N. lugens* [21]. However, no Cry2Aa was detected in a study of the larvae of a general predator (*C. sinica*) of planthoppers, in which *C. sinica* was provided with *Laodelphax striatellus* fed on transgenic *cry2Aa* rice [24].

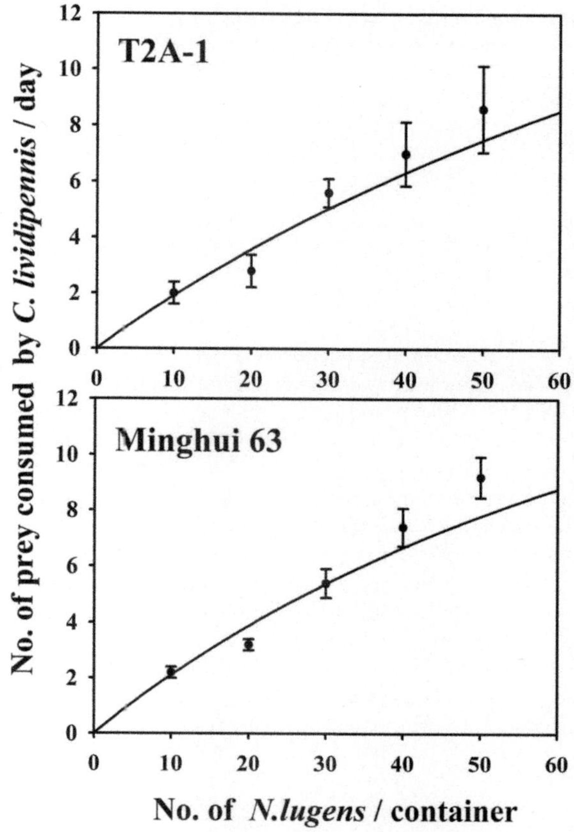

Figure 2. Functional response of *Cyrtorhinus lividipennis* collected from T2A-1 and Minghui 63.

In the present study, when *N. lugens* nymphs and T2A-1 seedlings were provided simultaneously to *C. lividipennis*, Cry2Aa was detected in the predator. However, when the *N. lugens* nymphs fed on T2A-1 were removed from the rice and provided alone to *C. lividipennis*, no Cry2Aa was detected in the predator.

Table 4. Effects of purified Cry2Aa incorporating into artificial diet on life-table parameters of *Cyrtorhinus lividipennis*.

Parameters	Treatments		
	Control	300 μg/ml Cry2Aa	40 μg/ml PA
2nd instar developmental time (days ± SE)[a]	2.1±0.15 (35)	2.3±0.13 (35)	2.4±0.21 (23)
3rd instar developmental time (days ± SE)[a]	2.5±0.10 (28)	2.6±0.13(29)	4.2±0.36 (13)**
4th instar developmental time (days ± SE)[a]	3.1±0.17 (21)	3.1±0.24 (22)	7.0±0.58 (3)**
5th instar developmental time (days ± SE)[a]	4.2±0.18 (20)	4.1±0.20 (20)	—
2nd instar-adult developmental time (days ± SE)[a]	11.8±0.40 (20)	11.7±0.40 (20)	—
Preimaginal survival (%)[b]	55.6	55.6	0.0
Male weight (mg ± SE)[c]	0.41±0.02	0.40±0.02	—
Female weight (mg ± SE)[c]	0.58±0.02	0.54±0.02	—

Nymphs of *C. lividipennis* were fed with an artificial diet containing 300 μg/ml Cry2Aa or 40 μg/ml PA (positive control). Pure diet served as a negative control (N = 36). The experiment lasted until adult eclosed. Statistical comparisons were made separately for each of the insecticidal compounds comparing with the control. Asterisks denote significant differences: $P<0.01$.

[a]Mann–Whitney *U*-test with Bonferroni correction (adjusted $\alpha = 0.025$).

[b]Chi-square test with Bonferroni correction (adjusted $\alpha = 0.025$).

[c]Student's *t*-test.

Table 5. Parameters of Type II functional response of *Cyrtorhinus lividipennis* to *Nilapavarta lugens* nymph fed on *Bt* or non-*Bt* rice.

Rice materials	*a*	*Th*	R^2 (%)
T2A-1	0.251±0.028	0.036±0.008	95.83
Minghui 63	0.231±0.025	0.037±0.007	95.20

a: instantaneous search rate (day^{-1}). *Th*: time required to handle a prey (day). Data are represented as mean ± SE.
There was no significant difference between T2A-1 and Minghui 63, based on Student's *t*-test ($P<0.05$).

Similarly, when *N. lugens* eggs and T2A-1 seedlings were provided to *C. lividipennis* simultaneously, Cry2Aa was detected in the predator. However, when the *N. lugens* eggs were removed from the T2A-1 seedlings and provided alone to *C. lividipennis*, no Cry2Aa was detected in the predator. Therefore, it may be inferred that Cry2Aa was not transmitted to *C. lividipennis* via predation on the eggs and nymphs of *N. lugens*; the protein may instead be transferred by the piercing-sucking foraging behavior of *C. lividipennis* on rice. This hypothesis was verified through the results of our supplementary experiments: when *C. lividipennis* nymphs were provided with T2A-1 seedlings alone for 1 day, Cry2Aa was detected in the predator, and the concentration of Cry2Aa in *C. lividipennis* was 17.1±8.6 ng/g (Table 3). Therefore, *C. lividipennis* may serve as a good indicator species in ecological risk assessments of transgenic *Bt* tice.

In the ecological risk assessment of an arthropod-resistant genetically engineered crop, researchers normally use "Tier-1 assays" as the initial step to determine the toxicity of the insecticidal compounds expressed by the transgenic crop on NTAs. In Tier-1 tests, insecticidal compounds are added to artificial diets for the tested NTAs, and the tested organisms are directly exposed to doses of the insecticidal compounds several times higher than those realistically present in the field. Tier-1 tests increase the likelihood that a hazard will be detected if the hazard exists, and therefore provide confidence that minimal risk is present if no adverse effect is detected [34,35]. Three important factors must be considered in Tier-1 assays: (i) the methods for the delivery of the insecticidal proteins to the test organisms; (ii) the need for and selection of the compounds used as positive controls; and (iii) the methods for monitoring the concentration, stability and bioactivity of the insecticidal proteins during the assay [35]. In

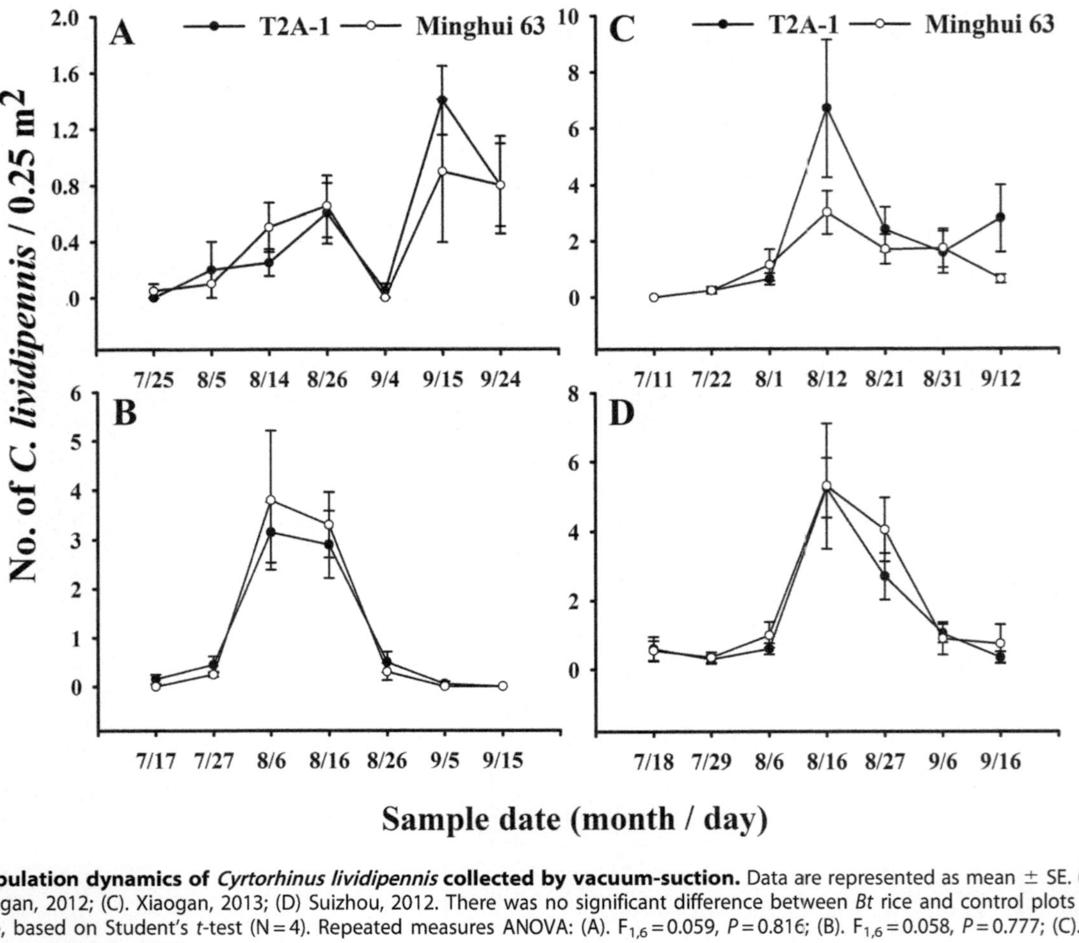

Figure 3. Population dynamics of *Cyrtorhinus lividipennis* collected by vacuum-suction. Data are represented as mean ± SE. (A). Xiaogan, 2011; (B). Xiaogan, 2012; (C). Xiaogan, 2013; (D) Suizhou, 2012. There was no significant difference between *Bt* rice and control plots at the same sampling time, based on Student's *t*-test (N=4). Repeated measures ANOVA: (A). $F_{1,6}=0.059$, $P=0.816$; (B). $F_{1,6}=0.058$, $P=0.777$; (C). $F_{1,6}=1.366$, $P=0.287$; (D). $F_{1,6}=0.964$, $P=0.364$.

Table 6. Population densities (no. of per 0.25 m^2) of *Cyrtorhinus lividipennis* collected by vacuum-suction.

Rice materials	Xiaogan			Suizhou
	2011	2012	2013	2012
T2A-1	0.47±0.10	1.03±0.09	2.06±0.68	1.55±0.06
Minghui 63	0.43±0.14	1.09±0.20	1.21±0.23	1.84±0.29

N = 4 at both sites in 2011, 2012, and 2013. Data are represented as mean ± SE. There was no significant difference between T2A-1 and Minghui 63 field, based on Student's *t*-test.

the present study, a dietary exposure experiment was conducted in which purified Cry2Aa protein was directly fed to *C. lividipennis* nymphs through its incorporation into a previously described artificial diet for *N. lugens* [29]. This artificial diet was sufficient for sustaining the growth and development of *C. lividipennis* from the second instar to adulthood. Before and after exposure to *C. lividipennis* for 24 h, the Cry2Aa protein in the artificial diets was stable and bioactive. The oral poison PA was used as a positive control to validate our dietary exposure assay. The survival of the *C. lividipennis* fed with the artificial diet containing 40 μg/ml PA significantly decreased compared to that of the *C. lividipennis* fed with the pure artificial diet ($P<0.001$) (Fig. 1). This result confirmed that the test system employed in the current study was able to detect the dietary effects of insecticidal compounds. No detrimental effects on the life-table parameters of *C. lividipennis* were observed when the insects were provided with an artificial diet containing Cry2Aa at a concentration that was nearly ten times higher than that measured in the rice sheaths. This study represents the first report of the use of a Tier-1 system to evaluate the potential effects of Cry2Aa on *C. lividipennis*.

The survival rate of *C. lividipennis* preying the nymphs of *N. lugens* was 32–35%, which was much lower than that of *C. lividipennis* preying eggs of *N. lugens* (87–91%) in the tritrophic experiment. So did body weight of adults of *C. lividipennis*. Similar results have been reported by Chua and Mikil and Chen et al., and they concluded that *N. lugens* eggs was an essential food type for *C. lividipennis*, and *N. lugens* nymphs was not an ideal food for the predator [36,37]. The different nutrient composition of prey may cause biological parameters difference of predator. *Harmonia axyridis* (Pallas) (Coleoptera: Coccinellidae) is a polyphagous species, the complete development of this predator can be accomplished using the aphid *Acyrthosiphon pisum* (Harris) (Homoptera: Aphididae) or *Ephestia kuehniella* (Zeller) (Lepidoptera: Pyralidae) eggs as substitution prey. Biochemical analyses indicated that amino acids and lipids of *E. kuehniella* eggs were richer than *A. pisum* adults, but, on the contrary, the glycogen of aphids was richer than *E. kuehniella* eggs. Some biological parameters such as larval mortality, adult weight, and fecundity, were modified according to the food eaten [38]. Whether the different survival of *C. lividipennis* in the present study is caused by different nutrient composition of prey needs to be further explored.

Conclusions

In summary, this comprehensive study, involving a Tier-1 examination system, a tritrophic bioassay, functional response experiments in the laboratory and population dynamics determinations in the field, provides the most complete information to date on the impacts of *Bt* rice expressing Cry2Aa on *C. lividipennis*, a major predator in rice ecosystems. These results indicate that *C. lividipennis* is not sensitive to the Cry2Aa protein and that *Bt* rice (T2A-1) poses a negligible risk to this nontarget organism.

Acknowledgments

We thank Prof. Yongjun Lin (National Key Laboratory of Crop Genetic Improvement at Huazhong Agricultural University) for providing the transgenic rice seeds.

Author Contributions

Conceived and designed the experiments: HH Y Han. Performed the experiments: Y Han JM JC YW. Analyzed the data: Y Han WC. Contributed reagents/materials/analysis tools: JZ Y He YF. Wrote the paper: HH Y Han.

References

1. Zeigler RS, Barclay A (2008) The relevance of rice. Rice 1: 3–10.
2. Cheng JA, He J (1996) Rice insect pests. China A gricultural Press, Beijing.
3. Kiritani K (1979) Pest management in rice. Annu Rev Entomol 24: 279–312.
4. Oerke EC, Dehne HW, Schönbeck F, Weber A (1994) Crop production and crop protection: estimated losses in major food and cash crops. Amsterdam: Elsevier.
5. Pathak MD, Khan ZR (1994) Insect Pests of Rice. Los Baños, Laguna, Philippines: International Rice Research Institute.
6. Sheng CF, Wang HT, Gao LD, Xuan JW (2003) The occurrence status, damage cost estimate and control strategies of stem borers in China. Plant Prot 29: 37–39. (in Chinese with English summary).
7. Lou YG, Zhang GR, Zhang WQ, Hu Y, Zhang J (2013) Biological control of rice insect pests in China. Biol Control 67: 8–20.
8. Matteson PC (2000) Insect pest management in tropical Asian irrigated rice. Annu Rev Entomol 45: 549–574.
9. Tu J, Zhang G, Datta K, Xu C, He Y, et al. (2000) Field performance of transgenic elite commercial hybrid rice expressing *Bacillus thuringiensis*-endotoxin. Nat Biotechnol 18: 1101–1104.
10. Huang JK, Hu RF, Rozelle S, Pray C (2005) Insect-resistant GM rice in farmers' fields: assessing productivity and health effects in China. Science 308: 688–690.
11. Shelton AM, Zhao JZ, Roush RT (2002) Economic, ecological, food safety, and social consequences of the deployment of *Bt* transgenic plants. Ann Rev Entomol 47: 845–881.
12. Chen H, Tang W, Xu CG, Li XH, Lin YJ, et al. (2005) Transgenic indica rice plants harboring a synthetic *cry2A** gene of *Bacillus thuringiensis* exhibit enhanced resistance against lepidopteran rice pests. Theor Appl Genet 111: 1330–1337.
13. Cheng X, Sardana R, Kaplan H, Altosaar I (1998) Agrobacterium-transformed rice plants expressing synthetic *cry1A(b)* and *cry1A(c)* genes are highly toxic to yellow stem borer and striped stem borer. Proc Natl Acad Sci U S A 95: 2767–2772.
14. Tang W, Chen H, Xu CG, Li XH, Lin YJ, et al. (2006) Development of insect-resistant transgenic indica rice with a synthetic *cry1C** gene. Mol Breeding 18: 1–10.
15. Heinrichs EA (1979) Control of leafhopper and planthopper vectors of rice viruses. In: Moramorosch K, Arris KF, Editors. Leafhopper vectors and planthopper disease agents. New York: Academic Press. pp. 529–558.
16. Sogawa K, Liu GJ, Shen JH (2003) A review on the hyper susceptibility of Chinese hybrid rice to insect pests. Chinese J Rice Sci 17, 23–30. (in Chinese with English summary).

17. Bai YY, Jiang MX, Cheng JA (2005) Effects of transgenic *cry1Ab* rice pollen on fitness of *Propylea japonica* (Thunberg). J Pest Sci 78: 123–128.
18. Bernal CC, Aguda RM, Cohen MB (2002) Effect of rice lines transformed with *Bacillus thuringiensis* toxin genes on the brown planthopper and its predator *Cyrtorhinus lividipennis*. Entomol Exp Appl 102: 21–28.
19. Chen M, Ye GY, Lu XM, Hu C, Peng YF, et al. (2005) Biotransfer and bioaccumulation of Cry1Ab insecticidal protein in rice plant-brown planthopper-wolf spider food chain. Acta Entomol Sinica 48: 208–213. (in Chinese with English summary).
20. Jiang YH, Fu Q, Cheng JA, Zhu ZR, Jiang MX, et al. (2004) Dynamics of Cry1Ab protein from transgenic *Bt* rice in herbivores and their predators. Acta Entomol Sinica 47: 454–460. (in Chinese with English summary).
21. Tian JC, Liu ZC, Chen M, Chen Y, Chen XX, et al. (2010) Laboratory and field assessments of prey-mediated effects of transgenic Bt rice on *Ummeliata insecticeps* (Araneida: Linyphiidae). Environ Entomol 39: 1369–1377.
22. Tian JC, Chen Y, Li ZL, Li K, Chen M, et al. (2012) Transgenic Cry1Ab rice does not impact ecological fitness and predation of a generalist spider. PLoS ONE. 7: e35164.
23. Chen M, Liu ZC, Ye GY, Shen ZC, Hu C, et al. (2007) Impacts of transgenic *cry1Ab* rice on non-target planthoppers and their main predator *Cyrtorhinus lividipennis* (Hemiptera: Miridae)-A case study of the compatibility of Bt rice with biological control. Biol Control 42: 242–250.
24. Li YH, Wang YY, Romeis J, Liu QS, Lin KJ, et al. (2013) *Bt* rice expressing Cry2Aa does not cause direct detrimental effects on larvae of *Chrysoperla sinica*. Ecotoxicology 22: 1413–1421.
25. Chen JM, Cheng JA, He JH (1992) Review on *Cyrtorrhinus livdipennis* Reuter. Entomol Knowledg 25: 370–373.
26. Sigsgaard L (2007) Early season natural control of the brown planthopper, *Nilaparvata lugens*: the contribution and interaction of two spider species and a predatory bug. Bull Entomol Res 97: 533–544.
27. Yoshida S, Forno DA, Cock JH, Gomez KA (1976) Laboratory manual for physical studies of rice, third editon. Los Baños, Laguna, Philippines: International Rice Research Institute. pp. 61–65.
28. Duan JJ, Head G, McKee MJ, Nickson TE, Martin JW, et al. (2002) Evaluation of dietary effects of transgenic corn pollen expressing Cry3Bb1 protein on a non-target ladybird beetle, *Coleomegilla maculata*. Entomol Exp Appl 104: 271–280.
29. Fu Q, Zhang ZT, Hu C, Lai FX, Sun ZX (2001) A chemically defined diet enables continuous rearing of the brown planthopper, *Nilaparvata lugens* (Stål) (Homoptera: Delphacidae). Appl Entomol Zool 36: 111–116.
30. Xu XL, Han Y, Wu G, Cai WL, Yuan BQ, et al. (2011) Field evaluation of effects of transgenic *cry1Ab/cry1Ac*, *cry1C* and *cry2A* rice on *Cnaphalocrocis medinalis* and its arthropod predators. Sci China Life Sci 54: 1019–1028.
31. Carino FO, Kenmore PE, Dyck VA (1979) The farmcop suction sampler for hoppers and predators in flooded rice fields. Int Rice Res News 4: 21–22.
32. Holling CS (1959) Some characteristics of simple types of predation and parasitism. Can Entomol 91: 385–398.
33. Holling CS (1961) Principles of insect predation. Annu Rev Entomol 6: 163–182.
34. Romeis J, Bartsch D, Bigler F, Candolfi MP, Gielkens MMC, et al. (2008) Assessment of risk of insect-resistant transgenic crops to nontarget arthropods. Nat Biotechnol 26: 203–208.
35. Li YH, Romeis J, Wu KM, Peng YF (2013) Tier-1 assays for assessing the toxicity of insecticidal proteins produced by genetically engineered plants to non-target arthropods. Insect Sci DOI: 10.1111/1744-7917.12044. Available: http://onlinelibrary.wiley.com/doi/10.1111/1744-7917.12044/abstract. Accessed 2013 Dec 6.
36. Chen JM, Cheng JA, He JH (1994) Effects of temperature and food on the development, survival and reproduction of *Cyrtorrhinus livdipennis* (Reuter). Acta Entomol Sinica 37: 63–70. (in Chinese with English summary).
37. Chua TH, Mikil E (1989) Effects of prey number and stage on the biology of *Cyrtorhinas lividipennis* (Hemiptera: Miridae): a predator of *Nilaparvata lagens* (Homoptera: Delphacidae). Environ Entomol 18: 251–255.
38. Specty O, Febvay G, Grenier S, Delobel B, Piotte C, et al. (2003) Nutritional plasticity of the predatory ladybeetle *Harmonia axyridis* (Coleoptera: Coccinellidae): comparison between natural and substitution prey. Arch Insect Biochem 52: 81–91.

13

Rice Chalky Ring Formation Caused by Temporal Reduction in Starch Biosynthesis during Osmotic Adjustment under Foehn-Induced Dry Wind

Hiroshi Wada[1]*, Chisato Masumoto-Kubo[1], Yousef Gholipour[2], Hiroshi Nonami[2], Fukuyo Tanaka[3], Rosa Erra-Balsells[4], Koichi Tsutsumi[1], Kenzo Hiraoka[5], Satoshi Morita[1]

1 Kyushu Okinawa Agricultural Research Center, National Agriculture and Food Research Organization, Chikugo, Fukuoka, Japan, **2** Department of Biomechanical Systems, Faculty of Agriculture, Ehime University, Matsuyama, Ehime, Japan, **3** Agricultural Research Center, National Agriculture and Food Research Organization, Tsukuba, Ibaraki, Japan, **4** Department of Organic Chemistry-CIHIDECAR, Faculty of Exact and Natural Sciences, University of Buenos Aires, Buenos Aires, Argentina, **5** Clean Energy Research Center, The University of Yamanashi, Kofu, Yamanashi, Japan

Abstract

Foehn-like extreme hot and dry wind conditions (34°C, >2.5 kPa vapor pressure deficit, and 7 m s^{-1}) strongly affect grain quality in rice (*Oryza sativa* L.). This is a current concern because of the increasing frequency and intensity of combined heat and water-deficit stress under climate change. Foehn-induced dry wind conditions during the grain-filling stage increase ring-shaped chalkiness as a result of spatiotemporal reduction in starch accumulation in the endosperm, but kernel growth is sometimes maintained by osmotic adjustment. Here, we assess the effects of dry wind on chalky ring formation in environmentally controlled growth chambers. Our results showed that hot and dry wind conditions that lasted for >24 h dramatically increased chalky ring formation. Hot and dry wind conditions temporarily reduced panicle water potential to −0.65 MPa; however, kernel growth was maintained by osmotic adjustment at control levels with increased transport of assimilate to the growing kernels. Dynamic tracer analysis with a nano-electrospray-ionization Orbitrap mass spectrometer and quantitative polymerase chain reaction analysis revealed that starch degradation was negligible in the short-term treatment. Overall expression of starch synthesis-related genes was found to be down-regulated at moderately low water potential. Because the events observed at low water potential preceded the packing of starch granules in cells, we concluded that reduced rates of starch biosynthesis play a central role in the events of cellular metabolism that are altered at osmotic adjustment, which leads to chalky ring formation under short-term hot and dry wind conditions.

Editor: Tai Wang, Institute of Botany, Chinese Academy of Sciences, China

Funding: This work was supported by JSPS KAKENHI Grant Number 24380013 (Grant-in-Aid Scientific Research (B)) and in part by JSPS KAKENHI Grant Number 24228004 (Grant-in-Aid Scientific Research (S)) and Grant Number 24248005 (Grant-in-Aid Scientific Research (A)). The funders had no role in study design, data collection and analysis, decision to publish, or preparation of the manuscript.

Competing Interests: The authors have declared that no competing interests exist.

* Email: hwada@affrc.go.jp

Introduction

It has been recognized that foehn-like high-speed hot and dry wind (e.g., 34°C, >2.5 kPa vapor pressure deficit [VPD], and 7 m s^{-1}) disturbs the quality of rice (*Oryza sativa* L.) grain appearance [1,2]. As the frequency and intensity of dryness are likely to increase in eastern Asia in addition to elevated global temperature under climate change [3], understanding the mechanism(s) behind rice quality under the combined stressors of heat and water deficit has become increasingly important in rice production. Hot and dry wind conditions during grain filling often impose temporary water deficit in plant shoots as a result of increasing VPD at elevated temperatures, resulting in a significant increase in ring-shaped chalky kernels, called 'milky white rice' (MWR) [4]. These kernels exhibit an annual ring-like chalky area on their transverse section that is typically composed of several cell layers in the endosperm, in which inadequate starch accumulation occurs at the midway of starch accumulation that occurs from the center towards outward in the endosperms [5,6]. Loosely packed starch granules are observed in the interior of the cells, and air spaces between starch granules [7,8] cause random light reflection [8,9] to create the appearance of a chalky ring in the endosperm.

An interesting body of evidence has emerged in recent years shows that the regulation of several α-amylase-encoded genes plays a central role in chalky ring formation under long-term high-temperature conditions [10,11,12]. Because the chalky zone occurs midway between the center and peripheral part of the endosperm in the developing kernels, the ring-shaped chalkiness could be a cell-specific event associated with the disruption of starch accumulation under dry wind conditions [4]. Given that foehn-induced dry wind conditions are a result of the simultaneous occurrence of high temperature and water deficit (i.e., temporary shoot water deficit at elevated temperature) over a relatively short time frame, starch breakdown in kernels by α-amylase may occur similarly to that which has been reported under long-term high temperature [10,12]. While it remains unknown whether starch breakdown or a reduction in starch synthesis occurs during induction of chalky formation under dry wind conditions, direct examination of the possibility of starch breakdown at the metabolic level is methodologically challenging.

In the last decade, metabolomic approaches have been widely applied in many areas of biological research due to the newest introduction of the Orbitrap mass analyzer to the family of high-resolution high-sensitivity mass spectrometers [13]. The Orbitrap-based mass spectrometer (MS) is an innovative and powerful instrument with high mass accuracy that allows useful measurements, including the correct measurement of isotopic ratios, in many analytical applications [14]. To our knowledge, this analyzer has not yet been used to determine the ratio of the carbon stable isotope (^{13}C) prefixed in plants; however, it is possible to examine the significance of starch degradation under short-term dry wind conditions by analyzing the enzymatic degradation of starch in kernels pre-fixed with ^{13}C. Here, we combined a pre-fixed ^{13}C tracer experiment with nano-electrospray ionization (nano-ESI)-Orbitrap MS to determine the isotopic ratio of ^{13}C-labeled glucose from the surface of starch granules newly synthesized in developing kernels, according to the partial starch degradation method [15].

In our previous study, we applied a dry wind treatment following a low radiation treatment to replicate foehn-associated climate conditions observed in the field [4]. Here, we sought to provide a better understanding of the mechanisms of chalky ring formation, which has often been recorded in field conditions, in developing endosperm under repeated hot and dry wind conditions (34°C, >2.5 kPa VPD, and 7 m s^{-1}) in an environmentally controlled growth chamber for 24–72 h (representing potential foehn duration). In addition to the pre-fixed ^{13}C isotope tracer experiment at the whole-plant level, plant water-status measurements–including an in situ endosperm turgor (Ψ_p) assay–were conducted under 24-h dry wind conditions. Quantitative polymerase chain reaction (qPCR) analysis was also carried out for the target genes, including starch metabolism-related genes, to determine the impact of starch breakdown and altered starch synthesis on dry wind-induced chalky ring formation.

Materials and Methods

Plant Materials

A growth-chamber experiment was conducted in 2011 and 2012 according to the previous study [4]. Simply, *Oryza sativa* L. cv. Koshihikari plants were grown outdoors in pots until the flowering stage. Ten plants per pot were prepared; the tillers were periodically removed to restrict each plant to its main culm to minimize panicle-to-panicle variations. At 5 days after heading (DAH), the plants were transferred to a growth chamber (22/22°C, 70/80% relative humidity [RH], 0.79/0.53 kPa VPD, and 750 μmol photons $m^{-2} s^{-1}$ photosynthetically active radiation [PAR]) set at the plant canopy with a photoperiod of 14 h day/10 h night. At 13 DAH, when the score of inferior kernels attached to the tertiary pedicels on the fourth to sixth secondary rachis branches (middle panicle position) reached 0.87 on average (Fig. 1A), the plants were transferred to another growth chamber (34/34°C, 50/40% RH, 2.66/3.19 kPa VPD, and 750 μmol $m^{-2} s^{-1}$ PAR) to conduct 24, 48, and 72 h dry wind treatments (referred to as '24 h W', '48 h W', and '72 h W', respectively), starting from 1200 h. The grain score ranges from 0 to 1 according to size and developmental stage, as shown in Figure 1A and 1B. Wind speed was set at approximately 7 m s^{-1} at the plant canopy. Other potted plants were kept in the same chamber in a cool and non-dry wind (control) treatment. Wind speed at the canopy in the control treatment was 0.2 m s^{-1}. After the dry wind was stopped at 1200 h each day, plants were transferred to the control chamber to grow until 33 DAH, after which the plants were placed outdoors until 40 DAH (maturing stage). Plants were supplied with water daily. For all of the following analyses (in situ Ψ_p, ^{13}C tracer, and qPCR assays), inferior spikelets attached to the tertiary pedicels on the middle panicle position were used because they showed the highest frequency of ring-shaped chalkiness under dry wind conditions at that stage of development [4].

Plant Water Status and Photosynthesis

All plant water status and photosynthesis measurements were conducted as described previously [4]. Briefly, panicle water potential (PWP), stem water potential (SWP), and leaf water potential (LWP) were determined using the pressure-chamber technique [16]. After determining PWP, the cell pressure probe technique [17] was used for a continuous in situ assay of Ψ_p of individual cells in the growing inner endosperm, from 370 to 810 μm below the epidermis [4]. The tips of glass microcapillaries were ground with a grinder to obtain a tip inner diameter between 2 and 4 μm. The Ψ_p of the endosperm cell layers, where starch granules tended to be loosely packed, was measured and the stability of Ψ_p over time was assessed according to Wada et al. [18]. In some cases, the kernel water potential of spikelets attached at the same position in a panicle was determined at 13 and 14 DAH with isopiestic thermocouple psychrometers [19] after determination of PWP. Net photosynthesis was measured concurrently with LWP using a Li-Cor LI-6400 photosynthesis meter (Li-Cor Inc., Lincoln, Nebraska, USA) between 1000 and 1200 h during the experiment.

Pre-fixed ^{13}C Isotope Tracer and Whole-Plant ^{13}C Distribution Analyses

The plants were divided into two groups for the ^{13}C feeding experiment to obtain two different treatments. The first group was labeled with $^{13}CO_2$ at 12 DAH (one day before initiation of the dry wind treatment) for the following pre-fixed ^{13}C analysis, and the second group was labeled at 13 DAH (just prior to application of the dry wind; arrows in Figure 1C–D) for the ^{13}C distribution analysis. The flag leaf in both groups was exposed to $^{13}CO_2$. $^{13}CO_2$ was applied by gently enclosing the leaf inside a 1.27-L polyester gas sampling bag (45 μm thickness, Analytic-Barrier, Ohmi Odor Air Services Inc., Tokyo, Japan) with a plastic container containing 0.5 g Ba $^{13}CO_3$ (99 atom % ^{13}C) and $^{13}CO_2$ was then generated by injecting 2 mL lactic acid from the outside of the bag. The leaf was allowed to assimilate ^{13}C under the light conditions provided in the growth chamber for 20 min between 1130 and 1150 h.

After ^{13}C feeding was completed in the first group, the flag leaf was instantly removed from the plants with a razor blade at the junction between leaf blade and leaf sheath to minimize further import of residual ^{13}C from the leaf to the kernels, and the cut surface was immediately coated with Vaseline to avoid dehydration. The pre-fixed inferior kernels, attached to a similar position as used for the in situ Ψ_p assay, were harvested 24 h after the dry wind treatment for the following ^{13}C isotope analysis. Extraction of the starch fraction prior to ^{13}C isotope analysis was conducted using a modification of the method of Gholipour et al. [15]. Freeze-dried kernels, in a similar position to those used for the in situ Ψ_p assay, were homogenized with water/ethanol (1 4 v/v). After centrifuging for 20 min at 3000×*g*, starch in insoluble residue was washed twice with 50 mM sodium acetate buffer (pH 4.5), and solubilized starch was hydrolyzed to Glc by adding

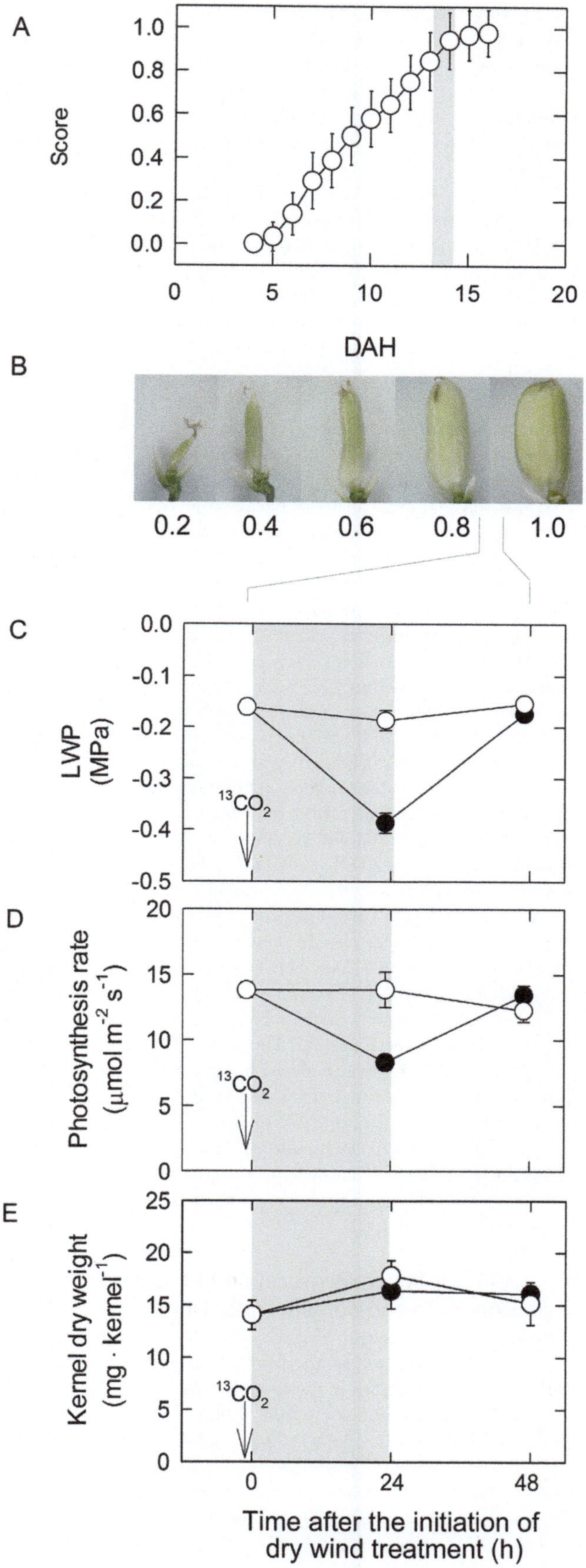

Figure 1. Time course of changes in kernel growth score (A) visually observed through hull; definition of kernel growth scores, 0 through 1.0 (B); and changes in leaf water potential (LWP) (C), photosynthesis rate (D), and kernel weight (E) under 24 h dry wind conditions, starting at 13 days after heading (DAH). Open and closed circles indicate control and dry wind treatments, respectively. Gray area indicates the 24-h dry wind treatment. For A, each point is the mean ± SD of at least 20 kernels from 4 different plants; for C–E, each point is the mean ± SE of 3–6 samples from different plants.

100 L α-amyloglucosidase (2 mg mL^{-1}) dissolved in sodium acetate buffer to 4 mL of starch solution and kept in a water bath at 55°C for 1 h. Sodiated Glc, which was the most abundantly yielded glucose ion including Glc (see Fig. S1 in File S1), was analyzed using a nano-ESI MS (Exactive Orbitrap LC-MS, Thermo Fisher Scientific Inc.), and the isotopic ratio of $m/z = 204$ ($C_5{}^{13}CH_{12}O_6{+}^{23}Na$; Mm = 204.0561693) to $m/z = 203$ ($C_6H_{12}O_6{+}^{23}Na$; Mm = 203.0527693) was determined.

In the second group, plant parts were similarly harvested and separated into six components: flag leaf blade, flag leaf sheath, uppermost internode, whole panicle, inferior kernels, and other organs. Because the preliminary test and reference [20] both indicated little to no ^{13}C distribution in roots at this stage, and because this estimate would not affect the analysis, roots were discarded to facilitate rapid completion of the dissection. The five tissue components were freeze-dried and the other organs were oven-dried with forced air, after which they were ground to a fine powder. For each tissue, approximately 1.0 mg well-mixed powder was used to determine total carbon and the isotopic ratio of ^{12}C:^{13}C using an element analyzer/isotopic ratio MS (ANCA-SL, SerCon Company, Cheshire, UK). Whole-plant ^{13}C abundance was estimated according to Mohapatra et al. [20].

Gene Expression Analysis

Total RNA was extracted from 13–20 mg fresh weight of four kernels per panicle using RNAs-ichi!-S (Rizo Inc., Tsukuba, Japan). The isolated RNA (4 µg) was treated with TURBO DNase (Life Technologies), and then reverse-transcribed using an iScript Advanced cDNA Synthesis kit (Bio-Rad) in a total volume of 20 µL. Quantitative real-time PCR was carried out in a real-time PCR system (CFX 96 Connect, Bio-Rad). Each reaction (5 µL) contained 300 nM of each primer, 0.5 µL of 1:100 diluted cDNA, and 2.5 µL of SsoAdvanced SYBR Green Supermix (Bio-Rad). The thermal cycling conditions were 95°C for 30 s followed by 95°C for 5 s and 60°C for 10 s for 40 cycles. For amplification of -*amylase 1A*, *2A*, *3A*, *3D*, and *invertase 3*, cDNA (1:10 dilution) was used with an annealing/extension time of 30 s. Dissociation curves for each amplicon were then analyzed to verify the specificity of each amplification reaction; the dissociation curve was obtained by heating the amplicon from 65 to 95°C. No evidence of any primer dimer or formation of other non-specific product was detected for any of the primer pairs used. Each PCR was run in duplicate within the same plate, and the Ct values obtained from the technical replicates were averaged. The transcripts of 27 genes known from the RiceXPro gene expression database (http://ricexpro.dna.affrc.go.jp/) to exhibit strong expression in developing rice endosperm [21] were quantified by comparing the cycle threshold (CT) of the target gene with that of eukaryotic elongation factor-1α [22] and ubiquitin-conjugating enzyme E2 [23]. Gene expression was expressed as the mean and standard error of the three biological replicates. Primer sequences used for the amplification are listed in Table S2 in File S1.

Kernel Quality and Weight

Spikelets in each portion in a panicle were further classified according to Matsuba [24]. The appearance of dehulled grains was objectively evaluated with a chalky rice grain predictor (RN-850, Kett Ltd., Tokyo, Japan) that is capable of determining the number of MWR kernels by analysis of scanned images of sliced kernels. Dry weight of kernel samples was determined during the treatment and at harvest [4].

Microscopy

Transverse sections of MWR kernels collected at harvest were observed by scanning electron microscopy (SEM). Sections were placed on double-sided carbon tape and viewed under a field-emission SEM (Hitachi S-3500N; Hitachi Co., Hitachi, Japan) at an accelerating voltage of 8.0 kV. The SEM images taken of the chalky region were converted to 8-bit grayscale and binarized by using the intensity threshold function, and the spatial ratio of air space to total area (*k*) was determined using image software (Image J, NIH). Sections were also photographed at 50x using a digital microscope (KH-3000, Hirox, Tokyo, Japan). Using the images, the volume of kernel (V^{kernel}) and chalky area (V^{chalk}), not including the volume of inner translucent area, was individually calculated as an ellipsoid body according to the equation $V = 4\pi abc/3$, where *a*, *b*, and *c* denote the three semi-axis of an ellipsoid. The volumetric ratio (*l*) of air space observed in the chalky region in the kernel volume was estimated according to;

$$l = V^{chalk} / V^{kernel} xk \qquad (1)$$

Statistical Analysis

Analysis of all physiological data and rice appearance data was performed using Tukey's Honestly Significant Difference (HSD) test, PROC ANOVA or Student's *t*-test, PROC TTEST in SAS version 9.2 (SAS Institute, Cary, NC). A simple linear regression was performed between PWP and kernel water potential at 13 DAH.

Results

Rice Appearance

The rice appearance data showed that the proportion of ring-shaped chalky kernels, known as 'milky-white rice,' was increased substantially by application of dry wind treatment for 24 h, with a corresponding decrease in perfect rice (Table 1). There was no reduction in kernel weight under dry wind conditions up to 48 h, but ring-shaped chalkiness increased significantly (Table 1). A significant decline in final kernel weight was observed when the duration of dry wind extended to 72 h (Table 1). These results suggest that 24-h wind treatment represents a critical time point for initiating metabolic modifications related to the formation of chalkiness. Hence, we focused on the 24-h duration and conducted the subsequent carbon isotope and gene expression assays in the 24-h dry wind-treated kernels.

Effects of Photosynthesis and Kernel Weight under Dry Wind Conditions

When plants were treated with dry wind for 24 h, LWP was significantly reduced but recovered within one day after stopping the wind treatment (Fig. 1C). Rates of photosynthesis declined to 60% that of the control treatment under shoot water deficit conditions, and became similar to that of the control on the day following the dry wind treatment (Fig. 1D). The average thickness of the boundary layer (measured as mean flag leaf length = 20 cm and numerical factor = 4.0) under control and dry wind conditions was 3.9 and 0.7 mm, respectively, according to Eq. (7.8) in Nobel [26], indicating that the boundary layer thickness was substantially reduced under the dry wind condition relative to the control. Kernel dry weight was similar between the control and dry wind treatments (Fig. 1E). There was also no significant difference ($P<0.05$) in fresh weight or water content between the treatments (data not shown). The dry weight of panicles collected at 24 h was 1.70±0.16 g and 1.70±0.15 g (mean ± SD, n = 6) for the control and dry wind treatment, respectively. There was no difference in spikelet numbers per panicle between the control and dry wind treatment (112±2 and 110±2 [mean ± SD, n = 6], respectively), indicating that there was no difference in sink size. These results indicate that rates of metabolite translocation to kernels were not affected by inhibition of photosynthesis in the short term.

Water Relations of Plants and Endosperm

After initiation of the dry wind treatment at 13 DAH, PWP decreased to approximately –0.6 MPa at 6 h, and remained at that value throughout the dry wind treatment (Fig. 2A). After the dry wind was stopped at 24 h, PWP began to increase and recovered to the level of the control on the following day (Fig. 2A). The SWP values were much higher than the PWP values, and SWP exhibited a similar pattern to PWP under dry wind conditions (Fig. 2B). Consequently, the gradient of water potential between stem and panicle under dry wind conditions was approximately 0.3 MPa greater than that of the control, and recovered to the level of the control within 24 h after stopping the treatment (data not shown). In the inner endosperm cells, Ψ_p was maintained overall, even at low PWP under dry wind conditions. The stability of Ψ_p over time was fairly high, with an average value of –0.001 MPa/min ($n = 96$). Maintenance of Ψ_p was observed until 51 h after initiation of the dry wind treatment (Fig. 2C). The tissue-averaged water potential of kernels collected at 13 DAH was highly correlated with PWP ($r = 0.95$, $P<0.0001$), which indicated that the relationship was linear and essentially equivalent across a broad range of PWP, ranging from –0.20 to –0.78 MPa in both treatments (inset in Fig. 2D). In contrast, there was no significant correlation ($r = 0.26$, $p = 0.43$) between the kernel water potential and PWP for the kernels collected on DAH14, presumably due to the lowering of water potential from the onset of kernel dehydration (inset in Fig. 2D). The osmotic potential of endosperm cells (Fig. 2D) was calculated by subtracting endosperm Ψ_p from kernel water potential, according to the correlation between kernel water potential and PWP observed at 13 DAH (inset in Fig. 2D). The osmotic potential of endosperm cells declined significantly as water deficit developed under dry wind conditions, reaching –0.65 MPa in 6 h and remaining at that value until the dry wind was stopped.

No Significant Starch Degradation but Increased Assimilate Import to Kernels under Dry Wind Conditions

Plants were subjected to dry wind conditions after the carbon isotope was pre-fixed and synthesized as ^{13}C-containing Glc on the surface of starch granules in the kernels. The mass spectrum from the starch fraction indicated that sodiated Glc was the most abundant (Fig. S1 in File S1). The isotopic ratio of sodiated $C_5{}^{13}CH_{12}O_6$ ($m/z = 204$) to sodiated $C_6H_{12}O_6$ ($m/z = 203$) partially digested from starch extracted from the unfed kernels was 5.95%, whereas the isotopic ratio in the pre-fixed kernels just before applying the dry wind treatment ($T = 0$ h) was determined to be 6.58%. When the isotopic ratio was determined from the starch fraction collected at 24 h, there was no significant difference ($P = 0.08$) between the treatments (Table 2), with similar ^{13}C distribution percentage in the kernels (data not shown). The

Table 1. Percentage of filled kernels, final kernel weight, and rice appearance of tertiary pedicels attached at the middle primary rachis branches, where an in situ turgor assay was conducted, in a panicle grown under 24, 48, or 72 h dry wind treatment in growth chambers.

Treatments	Filled kernels	Final kernel weight	Rice appearance[a]		
			PR	MWR	OR
	%	$Mg \cdot kernel^{-1}$	%	%	%
Control	90.2	23.0a[§]	95.0a	1.6c	3.5
24 h Wind	84.1	22.5ab	82.9ab	15.3bc	1.8
48 h Wind	84.2	21.8ab	69.0bc	29.4ab	1.6
72 h Wind	86.8	21.2b	56.2c	41.3a	2.5

[a]PR = perfect rice; MWR = milky white rice (ring-shaped chalky kernels); OR = other rice appearance.
Values in each treatment are means ± SE of four biological replications of panicles.
Means within a column followed by different letters are significantly different ($P<0.05$) (Tukey's Honestly Significant Difference [HSD] test).

isotopic ratio of the less-sodiated $C_4{}^{13}C_2H_{12}O_6$ ($m/z = 205$) to sodiated $C_6H_{12}O_6$ ($m/z = 203$) also indicated no difference between treatments (data not shown). Import of ^{13}C to the kernels and panicle appeared to increase by approximately 1.3-fold relative to the control (Figs. 3A and B). Assimilate transport from the flag leaf was enhanced at 24 h after application of dry wind (Fig. 3C), and a temporal increase in assimilate supply from other organs (mostly culms and leaf sheaths) to sink tissues was observed (Fig. 3D). No significant difference between treatments was observed in any organs at maturation (Figs. 3A–D).

Patterns of Gene Expression in Endosperm under Dry Wind

To investigate the underlying molecular basis for chalky formation under dry wind condition, we analyzed 39 genes related to sugar and starch metabolism, hormone metabolism, and stress response. No significant difference was observed between the two treatments for any genes related to amylolytic degradation, including α-amylase and β-amylase-encoding genes (Fig. 4A). For starch biosynthesis, expression of ADP-glucose pyrophosphorylase large subunit 1-encoding gene (*AGPL1*, categorized as endosperm-preferred) [25], was up-regulated at a relatively low level, although expression of the rest of the starch metabolism-related genes was consistently down-regulated under dry wind conditions (Fig. 4B). In particular, the ratio of relative expression of the endosperm-specific genes *AGPS1* and *GBSS1* to the endosperm-preferred gene *ISA1* declined significantly under dry wind conditions compared to the control. Expression of invertase-encoding gene (*CIN2*), three sucrose synthase genes (*SuSy2*, *SuSy3*, and *SuSy4*), ADP-glucose transporter-encoding gene (*BT1-2*), and two sucrose transporter-encoding genes (*SUT1* and *SUT2*) was not significantly altered at 24 h, whereas expression of *cyPPDKB* was significantly reduced under dry wind conditions (Table S1 in File S1). In addition, small heat shock protein-encoding genes (*HSP16.9A*, *HSP17.9A*, *HSP18.0*, and *HSP26.7*), late embryogenesis-abundant protein-encoding genes (*LEA8* and *LEA21*), and ABA 8′-hydroxylase-encoding gene (*ABA8ox2*) were up-regulated, while expression of manganese superoxide dismutase 1 (*MSD1*) was significantly down-regulated (Table S1 in File S1).

Kernel Anatomy and Starch Accumulation

The transverse sections of ring-shaped chalky kernels indicated that inner endosperm cells with loosely packed starch granules typically formed over several cell layers. SEM observation clearly showed that the chalky region was sandwiched between two translucent areas (Fig. 5A). In the chalky region, heterogeneity in the size of amyloplasts caused the presence of air space in the cells (Figs. 5B and C). In contrast, SEM revealed densely packed starch granules in the translucent area, where the cytosol was well filled with the granules (Fig. 5D). The image analysis indicated that the spatial ratio of air space in the chalky cells, k, was 10.6±4.2% (mean ± SD, $n = 11$). Since the volumetric ratio of chalky region to kernel (V^{chalk}/V^{kernel}, Eq. 1) was 13.3±3.5% (mean ± SD, $n = 21$), the average volumetric ratio of air space to kernel in the chalky region, l, was estimated as 1.41% using Eq. 1. In addition, no pits resulting from starch hydrolysis by amylolytic enzymes reported at high temperature (e.g., Reference [9]) were observed on the surface of starch granules in the chalky region when treated under short-term dry wind condition (Fig. 5C). It is noteworthy that an increase in air space in the chalky region did not contribute to weight loss in whole kernels (Table 1). This means that small metabolites accumulated in the symplast except for amyloplasts may not have been microscopically detected by SEM.

Discussion

Our results indicate that hot and dry wind conditions during ripening, for the shortest duration known to have been recorded, induce chalky ring formation, which degrades rice appearance. We showed that a partial reduction in starch biosynthesis occurred concomitantly with osmotic adjustment and prior to chalky ring formation in endosperm cells under dry wind conditions. The combined analysis of dynamic pre-fixed ^{13}C tracer (Table 2) and qPCR gene expression (Fig. 4, Table S2 in File S1) showed an overall reduction in starch biosynthesis and insignificant starch degradation in the short term. Kernel growth was maintained by osmotic adjustment under dry wind conditions, without a detectable reduction in kernel weight (Figs. 3 and 5), consistent with the previous work [4]. Given the similarity between the symptoms of progression of chalky ring formation and starch accumulation patterns in the developing rice endosperm [5,6], it is very likely that the reduction in starch synthesis was temporary and recovered after the treatment was stopped, as the changes in water status occurred temporarily at high VPD during the dry wind treatment, as indicated in Figure 5E. This proposal is consistent with the suggestion about MWR provided earlier [27]. Hence, we propose that a temporary reduction in starch

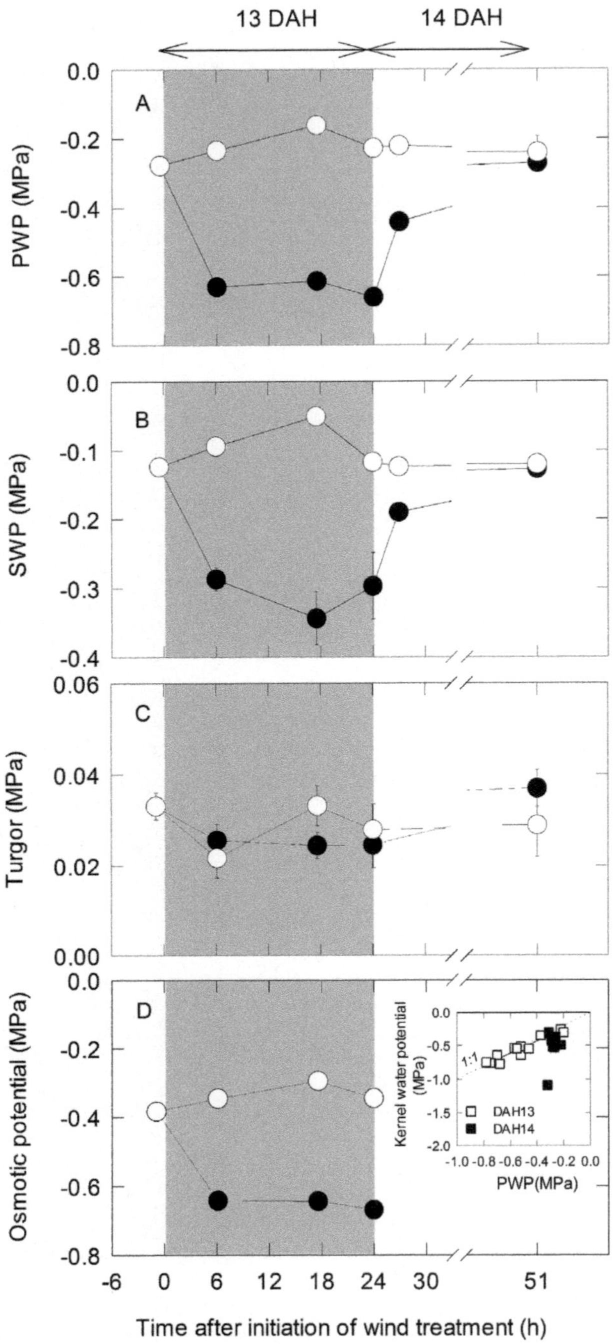

Figure 2. Changes in panicle water potential (PWP) (A), stem water potential (SWP) (B), endosperm cell turgor (C), and the weight-averaged calculated endosperm osmotic potential (D) in developing rice kernels under dry wind conditions. Open and closed circles indicate control and dry wind treatments, respectively. In D, the inset indicates kernel water potential measured with an isopiestic psychrometer as a function of panicle water potential measured with a pressure chamber in both the control and dry wind treatments, collected at 13 days after heading (DAH) (open squares) and 14 DAH (closed squares). The gray zone indicates the duration of dry wind treatment, corresponding to 13 DAH. In the inset in panel D, the slanted dotted line labeled '1:1' indicates an equipotential line. The solid line indicates the regression line at 13 DAH. For A and B, each point is the mean ± SE of 3–19 samples from different plants. For C, data are means ± SE of 4–22 cells measured in 2–8 inferior grains in the same location on a panicle.

biosynthesis during osmotic adjustment is responsible for the ring-shaped chalkiness formed by dry wind-induced moderate water stress developed at high VPD, rather than an effect of high temperature, during the short term.

It has been observed that the occurrence of MWR increases under long-term (generally ≥4 days) high-temperature conditions (e.g., 4 and 15 days in References [28] and [10], respectively), although the duration of dry wind treatment (24–72 h) conducted here was much shorter than that of the high temperature treatment. Therefore, it is noteworthy that such a large occurrence of ring-shaped chalkiness (Table 1) was induced at low water potential under the high VPD environment, rather than by high temperature *per se*. While it has been also recognized that sink-source manipulation has a significant effect on the occurrence of MWR [30], no differences in sink size of plants were observed between treatments in this study. Therefore, other potential indirect effects on chalky rice formation related to the sink-source balance could be ruled out. In addition, there was no significant difference in kernel number and weight between the 24 h dry wind condition and the control, which indicates that kernel growth continued even at low water potential, resulting in no detectable final kernel weight loss (Fig. 1E) even with the increase in ring-shaped chalky kernels (Table 1). This observation is not surprising because the volumetric ratio of air space observed in the chalky region was negligible (≈1.4%) in ring-shaped chalky rice, and was insufficient to contribute to potential weight loss. In addition, small metabolites accumulated during osmotic adjustment must have additional extra weight.

The fact that a high absorption of ^{13}C-labelled assimilates (Fig. 3) and active solute accumulation (Fig. 2D) were both observed at low water potential is strong evidence that the assimilate deficit *per se* does not account for the chalky ring formation under short-term dry wind conditions. Importantly, it was suggested that the onset of ring-shaped chalky formation was closely associated with the water status of the growing sink tissues (Fig. 2). Because the PWP observed here (approximately −0.65 MPa) was considerably higher than that often observed at spikelet desiccation (e.g., −1.3 MPa in reference [30]), the intensity of stress might have been fairly mild in this study. When the dry wind treatment was extended (see 72 h W in Table 1), MWR increased in number and kernel weight loss became significant, presumably as a result of insufficient assimilate delivery [31] and/or direct effects on starch biosynthesis (Fig. 4) in the kernels, as we observed here. The present study suggests another scenario that as water status declines and/or the duration of low water potential is extended, there might be a threshold above which starch metabolism is disturbed but kernel growth is ensured by increasing sugar delivery, at moderately low water potentials under dry wind conditions.

As in the previous work with low light intensity followed by dry wind treatments [4], here we have shown that Ψ_p is maintained by osmotic adjustment in endosperm cells under dry wind conditions regardless of the history of low light intensity. Moreover, this work showed that rates of starch biosynthesis were reduced by dry wind conditions. The observed dynamic changes in water status under dry wind conditions are likely to be closely associated with spatiotemporal metabolic modifications related to starch accumulation that occurred at the midway of the endosperm prior to ring-shaped chalky formation, as shown in Fig. 5E. Our data further showed that PWP was equivalent to the kernel water potential under high evaporative demand, at least before kernel desiccation occurred (inset in Fig. 2D). Since there was no detectable reduction in kernel weight (Table 1) with elevated assimilate delivery to the sink tissues (Fig. 3), even with the increase in

Table 2. Isotopic ratio of ^{13}C-labeled sodiated glucose, (m/z=204 to m/z=203) detected in the ethanol-insoluble fraction in kernels sampled at 24 h after the initiation of dry wind treatment.

Treatments	Isotopic ratio
	%
Control	6.64±0.05
Wind	6.54±0.04
P-value	0.08

The plants were labeled with $^{13}CO_2$ at 12 days after heading (DAH) from the flag leaf to pre-fix ^{13}C in the kernels prior to initiation of the dry wind treatment at 13 DAH. Values indicate the mean ± SE of three biological replications of two pooled kernels.

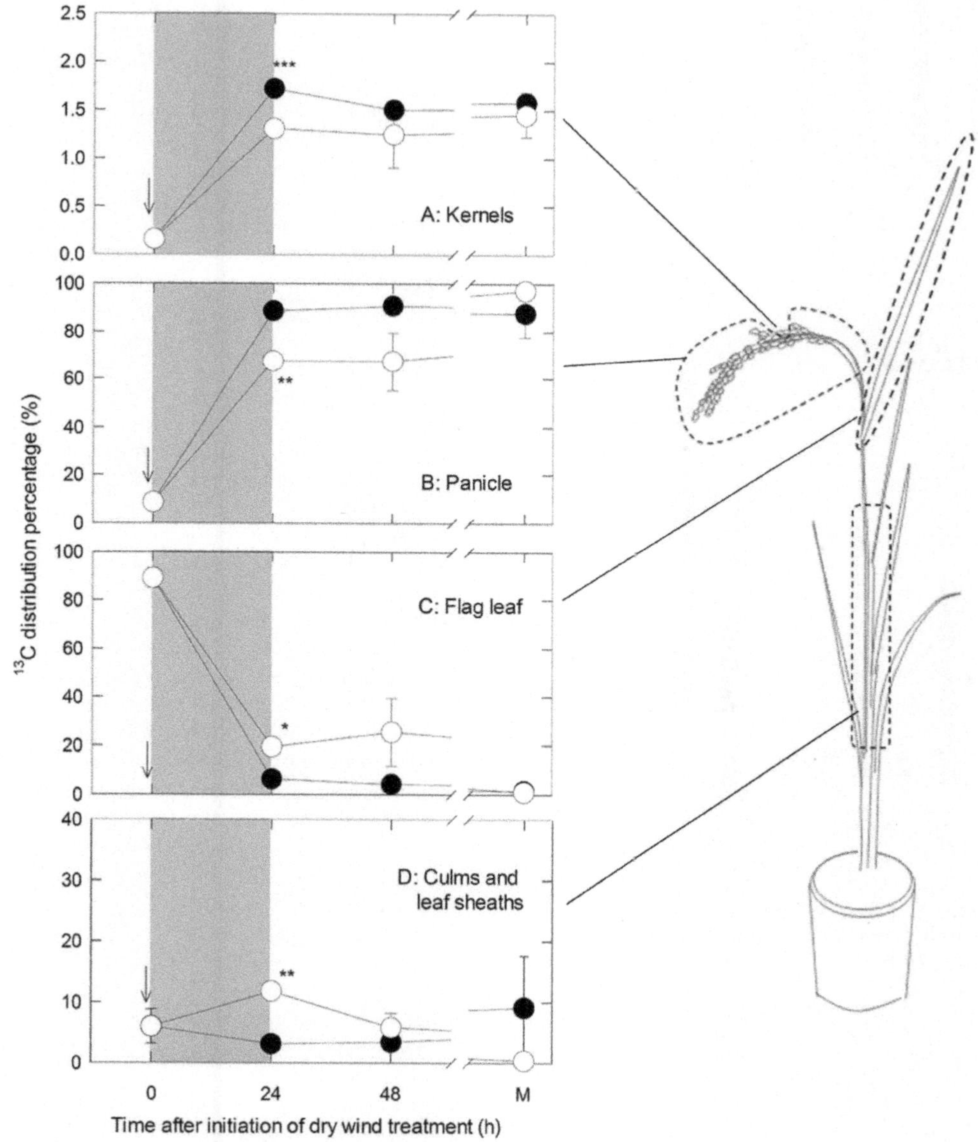

Figure 3. Time course of changes in ^{13}C distribution in kernels located in the same position where an in situ turgor assay was conducted (A); in the panicle excluding the kernels (B); in the flag leaf (C); and in culms and leaf sheaths (determined as pooled organs with flag leaf sheath, uppermost internode, and other tissues) (D), corresponding to the sampled locations shown in the schematic potted plant to the right of the figure panels. 'M' on *x*-axis indicates maturation. The plants were labeled with $^{13}CO_2$ for 20 min from the flag leaf at 13 DAH, just before the dry wind treatment was initiated, as indicated by the arrow. Open and closed circles indicate the control and dry wind treatments, respectively. Gray areas indicate the dry wind treatment. Each point is the mean ± SE of three samples from different plants. Significance at the 0.05, 0.001 and 0.0001 probability levels is indicated by *, **, and ***, respectively.

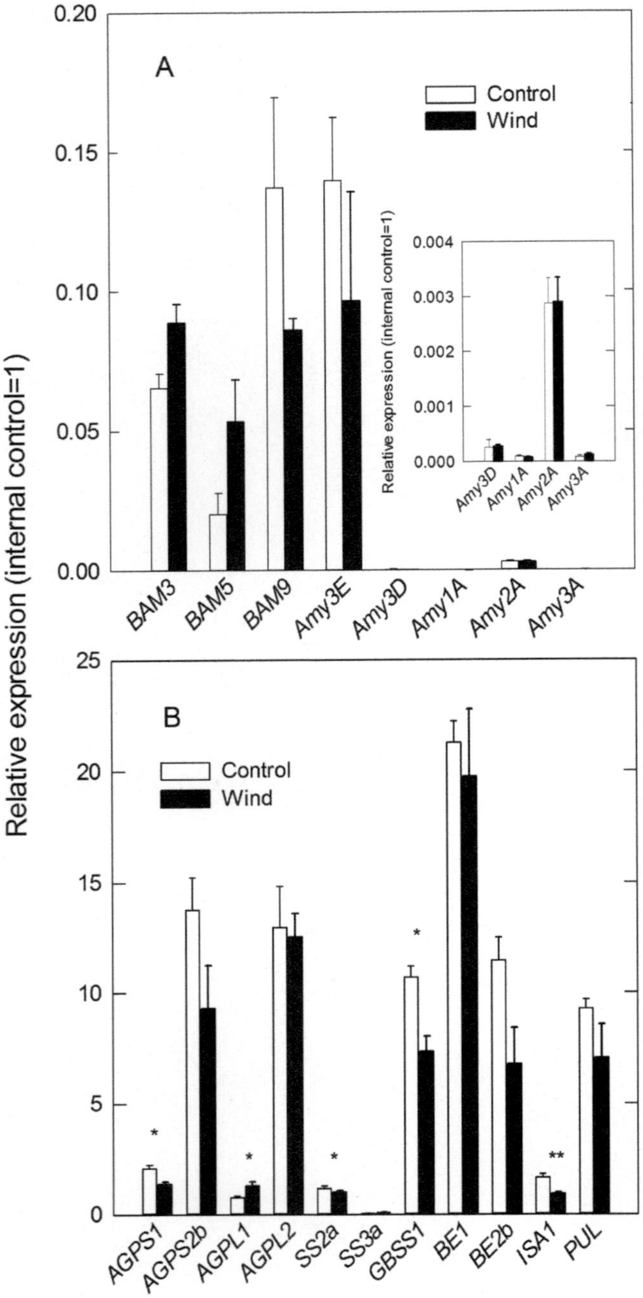

Figure 4. Expression profiling of starch degradation-related genes (A) and starch biosynthesis-related genes (B) in the growing kernels after 24 h of dry wind treatment. Inset in A is an expansion of the figure showing the expression of four α-amylase-encoding genes. Black and white bars indicate dry wind and control treatments, respectively. Data are the mean ± SE ($n = 3$) of four inferior kernels pooled at the position in the panicles where the in situ Ψ_p assay was conducted. Significance at the 0.05 and 0.01 probability levels is indicated by * and **, respectively.

chalkiness in moderately-stressed kernels, we interpret that kernel growth occurred at the control level as a consequence of Ψ_p maintenance (Fig. 2C) by osmotic adjustment while kernel thickness was increasing [4]. When the plants were temporally subjected to high-speed dry wind conditions and the water potential of air reached −98.3 MPa (at 34°C and 50% RH), cuticular transpiration should have increased at the spikelet surface that lacks stomatal control [32]. This would occur as a result of increased conductance of water vapor diffusing across the cuticular layers and increased xylem tension in the panicle according to the steep water potential gradient established between the air and panicle surface. Nonami and Hossain [33] showed that water fluxes for growth and transpiration were superimposed under high-VPD conditions. Based on our findings, it is likely that endosperm growth occurred by establishment of a growth-induced water potential gradient [34,35] under foehn-induced high evaporative demand.

Under adverse climate conditions, such as low light intensity during ripening, a reduction in newly assimilated carbohydrates is compensated by enhanced translocation from non-structural carbohydrates pre-stored in culms and sheaths [36]. Likewise, an approximately 1.3-fold increase in ^{13}C-labelled assimilates in the kernels from mature organs (Fig. 3) could have directly contributed to kernel development under the short-term dry wind conditions in spite of the 40% decline in rates of photosynthesis (Fig. 1D). Similar observations have been made for mature tissues at osmotic adjustment in other crops [37,38]. An analysis of the data presented in the previous work [4] showed that there was a significant interaction between low light intensity and dry wind conditions on chalky ring formation, indicating that a history of low light intensity has an impact on chalky formation (data not shown). Low light intensity would also reduce the availability of assimilates stored prior to lowering of plant water status, and this assimilate supply determines kernel weight and quality, depending on the intensity and duration of the stress [39].

Molecular understanding of chalky formation at high temperatures during grain-filling has recently extended [10,12,40]. Although most work has concentrated on response to long-term high temperature, it is also important to understand how chalkiness induced by the combined stress of heat and water deficit, especially foehn-like short-term hot and dry wind conditions, would impact rice quality under accelerating global warming [3]. Foehn-induced dry wind conditions represent a simultaneous stress consisting of temporary shoot water deficit and high temperature, and therefore plant responses to these two different stressors might overlap. Yamakawa's pioneering work showed that expression of several α-amylase-encoded genes, including *Amy1A* and *Amy3D*, was attributed to chalky formation at high temperature [10,12]. However, this evidence has been called into question by Ishimaru et al. [41], who performed a spatial analysis of α-amylase-encoding genes in endosperm by using a laser micro-dissection method. These researchers seemed to observe different types of chalkiness (MWR with or without chalky ring formation). These inconsistencies might be explained by differences in the sampled sites, potentially depending on differences in the intensity and/or duration of stress.

In this work, we used the Orbitrap-based MS to determine the isotopic ratio of Glc from starch in combination with the partial enzymatic degradation [15] in the prefixed kernels. According to the calculated isotopic ratio of C_6 molecules for carbon based on the ^{13}C content (Reference [42]), the isotopic ratio should correspond to 6.50%. However, what we observed in the unfed kernels was determined to be 5.95%, lower than the theoretical value. The difference between theoretical and observed values might be attributed to the effect of discrimination against the heavier isotope of carbon by Rubisco in leaves during the C_3 photosynthetic process [43]. It has been noted that the isotopic ratio of Glc from the kernels fed under the closed ^{13}C-enriched atmosphere condition increased to be 6.58%, similar to the theoretical value. By imposing the prefixed samples to the dry wind condition, Orbitrap-based MS analysis allowed us to

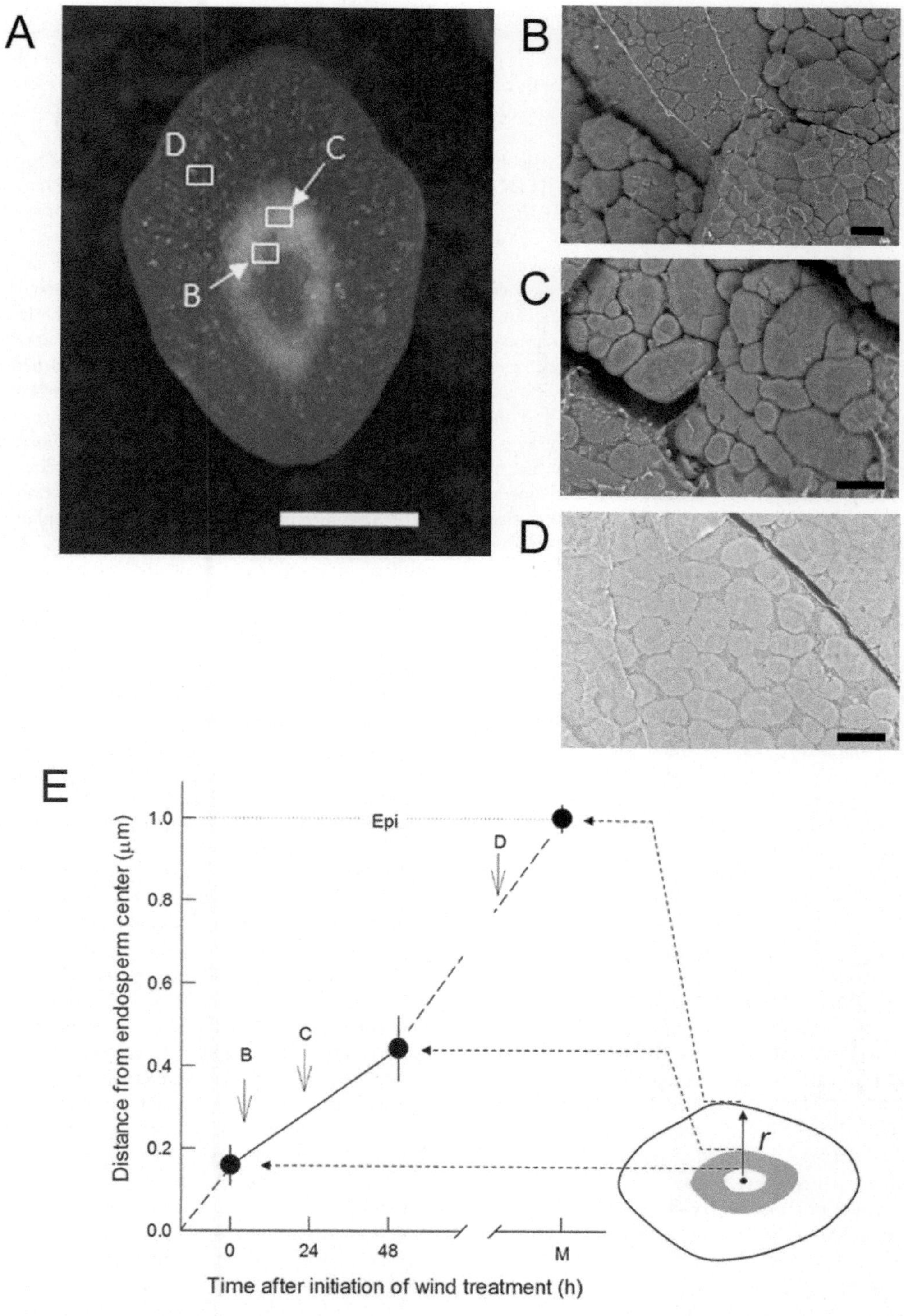

Figure 5. Transverse sections of a typical ring-shaped chalky kernel (A); and expanded SEM images taken at the three zones (B–D) shown in A. B, C, and D indicate the border between the inner translucent and chalky region (chalky cell in the left corner cell), the chalky region with loosely packed starch granules, and the translucent zone, respectively. The distance (*r*) from the endosperm center along the semi-minor axis in ring-shaped chalky kernels as a putative function of the duration of dry wind conditions (E). In E, the inner and outer borders of translucence and chalkiness correspond to the initiation of the dry wind treatment and the recovery point observed at 52 h in Fig. 2A, respectively; B, C, and D are indicated by arrows. 'Epi' indicates epidermis. In B–D, note that fissured fractures in each figure correspond to the cell wall. Bars in A and in B–D indicate 1 mm and 10 μm, respectively.

examine the possibility of starch degradation at osmotic adjustment even with the potential discrimination effect. Consequently, we found, at least during short-term (24 h) dry wind conditions, that there was little effect on the breakdown of starch synthesized in amyloplasts. SEM observations showed no starch granules forming digestive pits at high temperature [9,27]. Therefore, it appears that the 24-h duration examined was too short to explain the formation of chalky rings by starch breakdown alone under dry wind conditions.

Quantitative PCR analysis provided evidence that rates of starch synthesis decreased at moderately low water potential (Fig. 4B and Table S1 in File S1). Considering that sucrose accumulation accompanying decreased starch biosynthesis has been observed under mild water stress in other plants [44,45], there might be an interaction between the reduced starch biosynthesis and sugar accumulation in developing rice endosperm. In starch metabolism, down-regulated expression of *GBSS1*, which is responsible for amylose synthesis [46], has been observed at mild water deficit (Fig. 4B). In addition, down-regulation of *ISA1* expression (for debranching enzyme isoamylase 1, which releases soluble maltoligosaccharides from starch granules) at osmotic adjustment (Fig. 4B), allows us to exclude the effect of starch degradation. It is generally recognized that changes in amylopectin composition are involved in the process of chalky formation [10,46]. Amylopectin composition was not analyzed in this study, although it is possible that amylopectin was structurally modified under the short-term dry wind conditions. The down-regulation of *GBSS1* and *BE* observed here, which was similarly observed at high temperature [10,47], may have some effect on the lack of short-chain amylopectin formation under the short-term stress.

Aside from the reduced starch biosynthesis, the responses of other genes, such as plant hormone metabolism-associated and stress responsive genes, were quite variable. Consistent expression patterns for HSP- and LEA-encoding genes (Table S1 in File S1) showed additive responses to shoot water deficit and high temperature. Among small HSP-encoding genes, upregulation of *HSP16.9A* and *HSP26.7* was detected under dry wind conditions. *HSP16.9A* and *HSP26.7* have been associated with heat and water deficit (and anoxia) in rice kernels, respectively [48], and thus might play different roles in the cells. It has been reported that reactive oxygen species might be involved in chalky grain formation [49]. As the expression of genes that encode manganese superoxide dismutase and pyruvate Pi dikinase, recognized as the proteins targeted by thioredoxin [50], were eventually downregulated at 24 h (see *MSD1* and *cyPPDK* in Table S1 in File S1), dry wind stress might have disturbed the redox homeostasis and modified cellular metabolism at low water potential. In addition, expression of *ABA8ox2*, considered to be one of the key genes in modulating ABA levels [51], increased 2.2-fold in the moderately stressed kernels (Table S1 in File S1). It remains unclear whether the expression of several genes analyzed at the kernel level under dry wind conditions reflect the responses that occurred in endosperm layers that developed chalkiness. Therefore, it is necessary to perform spatial analysis of metabolic changes that occur prior to chalky ring formation in the further study–a subject we are currently investigating.

Early studies of rice endosperm showed that starch accumulation occurs from the center towards outward in the endosperms [5,6], and it is important to keep in mind that our conclusion agrees well with the general consensus on the accumulation pattern. Also, it is reasonable to assume that cell expansion by osmotic adjustment precedes starch accumulation in the rice endosperm cells. In osmotically adjusted cells, imported sugars would be accumulating faster than consumed for growth (including wall biosynthesis and respiration), and the rest might be utilized for the potential substrate(s) of starch and fatty acids to be synthesized in plastids. The fate of sugars at the subcellular level remains unclear, however, the most likely explanation based on our findings is that sugars to be synthesized to starch kept accumulating in vacuoles and cytosol in the cells, so that the cells could be expanded by maintaining water volume in size and cell Ψ_p [37,52], followed by starch accumulation. It is reasonable to assume that vacuole's volume in the cells would be maintained during osmotic adjustment. The early microscopic observations showed that increasing the packing density of amyloplasts in the cells was closely associated with distorting of the volume occupied with other organelles including vacuole which are filled with water containing small molecules [7,53]. The situation might have been altered when osmotic adjustment turned on. It is possible that the vacuolar volume might have preserved in cytosol without replacing their volume to amyloplasts' volume during osmotic adjustment. If this were the case, then in addition to the reduction in starch biosynthesis, the presence of vacuoles stayed in the cytosol as a consequence of osmotic adjustment is anticipated to have an impact on the formation of air space in the cells, which would occur during the kernel dehydration.

In addition to the above-mentioned metabolic changes observed at low water potential, the size heterogeneity of amyloplasts microscopically observed in the dry wind-treated kernels was due to the presence of a number of small amyloplasts formed with the relatively less large amyloplasts (Fig. 5C). Zakaria et al. [9] similarly observed many small amyloplasts containing single starch granules that created large air spaces at high temperature, consistent with our observation. Little is known about the molecular basis for the downsizing of amyloplasts in the chalky area, but several possibilities may be raised. Since plastid volume is responsive to rapid changes in cell osmotic potential [54,55], regulation of plastid (amyloplast) volume in certain endosperm cell layers may be influenced by dynamic changes in cellular water status. Secondly, the size heterogeneity of amyloplasts might be attributed to abnormal development, such as a failure of plastid division [56]. Third, the metabolic speed of fatty acid synthesis in plastids may be retarded under dry wind conditions. Because PEP and pyruvate in cytosol act as precursors for several pathways (including fatty acid biosynthesis and amino acid metabolism) that occur in plastids and require carbon and ATP, the 50% reduction in expression of *cyPPDK* (Table S2 in File S1), which helps to modulate carbon and lipid metabolism [57], suggests that primary metabolism might have been inhibited in the developing plastids. The physiological role of *cyPPDK* has not yet been well clarified; however, the down-regulation of *cyPPDK* detected here and similarly observed at high temperature (e.g., reference [10]) may be involved in similar metabolic deterioration under stress conditions. All of these possibilities will need to be addressed further.

In conclusion, this work has demonstrated that foehn-induced hot and dry wind conditions in ripening rice grains induce chalky ring formation within the shortest time-frame known to have been reported, resulting in deterioration of rice appearance. Combined analyses using a pre-fixed ^{13}C tracer and qPCR assay indicated that partial inhibition of starch biosynthesis without significant starch degradation during osmotic adjustment was strongly associated with ring-shaped chalkiness in the growing endosperm cells. As indicated in the previous work [4], kernel growth was sustained at control levels by osmotic adjustment, with a temporary increase in sugar delivery to the kernels under dry wind conditions. Based on these results, we propose that ring-shaped chalkiness was closely associated with moderately low water status that developed at high VPD, which was increased at high temperature. Independent differences between short-term dry wind and long-term high temperature remain to be further clarified. Considering the different types of chalkiness observed in kernels, it would be useful to conduct a spatial analysis in the developing endosperm to understand the formation of rice chalkiness under environmental stress. The application of Orbitrap-based MS with the stable isotope(s), that we used as a pre-fixed ^{13}C tracer in this work, together with the cellular analysis, such as cell metabolomics (e.g., Reference [58]), will shed additional light and help to provide many valuable insights into environmental plant science.

Supporting Information

File S1 Table S1, *P* values, ratio to control, and relative expression of 39 differentially expressed genes analyzed in the developing caryopsis after 24 h dry wind treatment. Table S2, Primer sequences used in quantitative RT-PCR. Figure S1, (A) Nano-ESI mass spectrum (positive ion mode) of the ethanol-insoluble fraction from ^{13}C-pre-fixed kernels after partial starch-degradation treatment (see Materials and Methods). (B) Expanded mass spectrum of the range $m/z = 203–205$ in (A). Since sodiated glucose ($m/z = 203$) was the most abundant ion among those containing glucose ($[Glc + H]^+$, $[Glc + Na]^+$, and $[Glc + K]^+$) produced by ESI in the Exactive Orbitrap MS, the signal intensity of ions with $m/z = 204$ ($C_5{}^{13}CH_{12}O_6{+}^{23}Na$; Mm = 204.0561693) and $m/z = 203$ ($C_6H_{12}O_6{+}^{23}Na$; Mm = 203.0527693) was used to calculate the isotopic ratio for glucose. (C) Nano-ESI mass spectrum (positive ion mode) of the buffer solution used in the experiment (blank spectrum).

Author Contributions

Conceived and designed the experiments: HW SM. Performed the experiments: HW CM-K FT YG KT. Analyzed the data: HW CM-K YG HN. Contributed reagents/materials/analysis tools: HW CM-K YG RE-B HN KH. Contributed to the writing of the manuscript: HW CM-K SM.

References

1. Ebata M, Ishikawa M (1989) Effects of wind and rain on the fertilization, kernel development and kernel characters of rice plant (In Japanese with English abstract). Japanese Journal of Crop Science 58: 555–561.
2. Ishihara K, Horiguchi T, Mizuno S, Takahashi H (2005) On the occurrence of the milky white like kernels of rice plants under high temperature conditions with reference to water stress and nitrogen fertilizer application in 2003. (In Japanese with English Abstract). Proceedings of the Crop Science Society of Japan 74: 122–123.
3. IPCC (2013) Climate Change 2013: The Physical Science Basis. Working Group I Contribution to the Fifth Assessment Report of the Intergovernmental Panel on Climate Change. Cambridge, United Kingdom and New York.
4. Wada H, Nonami H, Yabuoshi Y, Maruyama A, Tanaka A, et al. (2011) Increased ring-shaped chalkiness and osmotic adjustment when growing rice grains under foehn-induced dry wind condition. Crop Science 51: 1703–1715.
5. Juliano BO (1985) Rice: Chemistry and Technology. St. Paul, Minnesota, USA: American Association of Cereal Chemists, Inc.
6. Hoshikawa K (1989) The Growing Rice Plant: An Anatomical Monograph. Tokyo: Nosan Gyoson Bunka Kyokai.
7. del Rosario A, Briones VP, Vidal AJ, Juliano BO (1968) Composition and endosperm structure of developing and mature rice kernel. Cereal Chemistry 45: 225–235.
8. Tashiro T, Wardlaw IF (1991) The effect of high-temperature on kernel dimensions and the type and occurrence of kernel damage in rice. Australian Journal of Agricultural Research 42: 485–496.
9. Zakaria S, Matsuda T, Tajima S, Nitta Y (2002) Effect of high temperature at ripening stage on the reserve accumulation in seed in some rice cultivars. Plant Production Science 5: 160–168.
10. Yamakawa H, Hirose T, Kuroda M, Yamaguchi T (2007) Comprehensive expression profiling of rice grain filling-related genes under high temperature using DNA microarray. Plant Physiology 144: 258–277.
11. Yamakawa H, Hakata M (2010) Atlas of rice grain filling-related metabolism under high temperature: joint analysis of metabolome and transcriptome demonstrated inhibition of starch accumulation and induction of amino acid accumulation. Plant and Cell Physiology 51: 795–809.
12. Hakata M, Kuroda M, Miyashita T, Yamaguchi T, Kojima M, et al. (2012) Suppression of alpha-amylase genes improves quality of rice grain ripened under high temperature. Plant Biotechnology Journal 10: 1110–1117.
13. Hu QZ, Noll RJ, Li HY, Makarov A, Hardman M, et al. (2005) The Orbitrap: a new mass spectrometer. Journal of Mass Spectrometry 40: 430–443.
14. Zubarev RA, Makarov A (2013) Orbitrap Mass Spectrometry. Analytical Chemistry 85: 5288–5296.
15. Gholipour Y, Nonami H, Erra-Balsells R (2008) In situ analysis of plant tissue underivatized carbohydrates and on-probe enzymatic degraded starch by matrix-assisted laser desorption/ionization time-of-flight mass spectrometry by using carbon nanotubes as matrix. Analytical Biochemistry 383: 159–167.
16. Turner NC (1988) Measurement of plant water status by the pressure chamber technique. Irrigation Science 9: 289–308.
17. Hüsken D, Steudle E, Zimmermann U (1978) Pressure probe technique for measuring water relations of cells in higher-plants. Plant Physiology 61: 158–163.
18. Wada H, Matthews MA, Choat B, Shackel KA (2011) *In situ* turgor stability in grape mesocarp cells and its relation to cell dimensions and microcapillary tip size and geometry. Environmental Control in Biology 49: 61–73.
19. Boyer JS (1995) Measuring the Water Status of Plants and Soils. San Diego, Academic Press.
20. Mohapatra PK, Masamoto Y, Morita S, Takanashi J, Kato T, et al. (2004) Partitioning of ^{13}C-labelled photosynthate varies with growth stage and panicle size in high-yielding rice. Functional Plant Biology 31: 131–139.
21. Sato Y, Antonio BA, Namiki N, Takehisa H, Minami H, et al. (2011) RiceXPro: a platform for monitoring gene expression in japonica rice grown under natural field conditions. Nucleic Acids Research 39: D1141–D1148.
22. Jain M, Nijhawan A, Tyagi AK, Khurana JP (2006) Validation of housekeeping genes as internal control for studying gene expression in rice by quantitative real-time PCR. Biochemical and Biophysical Research Communications 345: 646–651.
23. Li QF, Sun SSM, Yuan DY, Yu HX, Gu MH, et al. (2010) Validation of candidate reference genes for the accurate normalization of real-time quantitative RT-PCR data in rice during seed development. Plant Molecular Biology Reporter 28: 49–57.
24. Matsuba K (1991) The morphogenetic mechanism of formation of the panicle branching system in rice plants (*Oryza sativa* L.). (In Japanese with English abstract). Bulletin of the Chugoku Agricultural Experiment Station 11–58.
25. Ohdan T, Francisco P, Sawada T, Hirose T, Terao T, et al. (2005) Expression profiling of genes involved in starch synthesis in sink and source organs of rice. Journal of Experimental Botany 56: 229–3244.
26. Nobel PS (2009) Physicochemical and Environmental Plant Physiology, Second edition, San Diego, Academic Press.
27. Nagato K, Kobayashi Y (1959) On the development of the tissue of starch-cell in rice kernels. (In Japanese with English abstract). Japanese Journal of Crop Science 27: 204–206.
28. Morita S (2000) Effects of high air temperature on ripening in rice plants: Analysis of ripening performance in growth chamber experiments. Japanese Journal of Crop Science 69: 391–399.
29. Tsukaguchi T, Iida Y (2008) Effects of assimilate supply and high temperature during grain-filling period on the occurrence of various types of chalky kernels in rice plants (*Oryza sativa* L.). Plant Production Science 11: 203–210.
30. O'Toole JC, Hsiao TC, Namuco OS (1984) Panicle water relations during water-stress. Plant Science Letters 33: 137–143.
31. Nagato K, Chaudhry FM. (1970) Influence of panicle clipping, flag leaf cutting and shading on ripening of *Japonica* and *Indica* rice. Proceedings of the Crop Science Society of Japan 39: 204–212.
32. Garrity DP, Vidal ET, O'Toole JC (1986) Manipulating panicle transpiration resistance to increase rice spikelet fertility during flowering stage water-stress. Crop Science 26: 789–795.
33. Nonami H, Hossain MM (2010) Superposition of the transpiration-induced water potential and the growth-induced water potential associated with expanding tomato leaves. Environmental Control in Biology 48: 117–125.
34. Nonami H, Boyer JS (1993) Direct demonstration of a growth-induced water potential gradient. Plant Physiology 102: 13–19.
35. Nonami H, Wu Y, Boyer JS (1997) Decreased growth-induced water potential: Primary cause of growth inhibition at low water potentials. Plant Physiology 114: 501–509.
36. Morita S, Nakano H (2011) Nonstructural carbohydrate content in the stem at full heading contributes to high performance of ripening in heat-tolerant rice cultivar Nikomaru. Crop Science 51: 818–828.
37. Meyer RF, Boyer JS (1981) Osmoregulation, solute distribution, and growth in soybean seedlings having low water potentials. Planta 151: 482–489.
38. Yakushiji H, Morinaga K, Nonami H (1998) Sugar accumulation and partitioning in Satsuma mandarin tree tissues and fruit in response to drought stress. Journal of the American Society for Horticultural Science 123: 719–726.
39. Barnabás B, Jaeger K, Feher A (2008) The effect of drought and heat stress on reproductive processes in cereals. Plant, Cell & Environment 31: 11–38.

40. Liao JL, Zhou HW, Zhang HY, Zhong PA, Huang YJ (2014) Comparative proteomic analysis of differentially expressed proteins in the early milky stage of rice grains during high temperature stress. Journal of Experimental Botany 65: 655–671.
41. Ishimaru T, Horigane AK, Ida M, Iwasawa N, San-oh YA, et al. (2009) Formation of grain chalkiness and changes in water distribution in developing rice caryopses grown under high-temperature stress. Journal of Cereal Science 50: 166–174.
42. Gross JH (2004) Mass spectrometry: a textbook. Berlin; London: Springer. Xviii.
43. Farquhar GD, Ehleringer JR, Hubick KT (1989) Carbon isotope discrimination and photosynthesis. Annual Review of Plant Physiology and Plant Molecular Biology 40: 503–537.
44. Vassey T, Sharkey T (1989) Mild water-stress of *Phaseolus vulgaris* plants leads to reduced starch synthesis and extractable sucrose phosphate synthase activity. Plant Physiology 89: 1066–1070.
45. Chaves MM (1991) Effects of water deficits on carbon assimilation. Journal of Experimental Botany 42: 1–16.
46. Umemoto T, Terashima K (2002) Activity of granule-bound starch synthase is an important determinant of amylose content in rice endosperm. Functional Plant Biology 29: 1121–1124.
47. Zhang G, Cheng Z, Zhang X, Guo X, Su N, et al. (2011) Double repression of soluble starch synthase genes *SSIIa* and *SSIIIa* in rice (*Oryza sativa* L.) uncovers interactive effects on the physicochemical properties of starch. Genome 54: 448–459.
48. Sarkar N, Kim Y, Grover A (2009) Rice sHsp genes: genomic organization and expression profiling under stress and development. BMC Genomics 10: 393.
49. Liu XL, Guo T, Wan XY, Wang HY, Zhu MZ, et al. (2010) Transcriptome analysis of grain-filling caryopses reveals involvement of multiple regulatory pathways in chalky grain formation in rice. BMC Genomics 11: 730.
50. Buchanan B, Balmer Y (2005) Redox regulation: A broadening horizon. Annual Review of Plant Biology 56: 187–220.
51. Saika H, Okamoto M, Miyoshi K, Kushiro T, Shinoda S, et al. (2007) Ethylene promotes submergence-induced expression of *OsABA8ox1*, a gene that encodes ABA 8′-hydroxylase in rice. Plant and Cell Physiology 48: 287–298.
52. Morgan JM (1977) Differences in osmoregulation between wheat genotypes. Nature 270: 234–235.
53. Hoshikawa K (1967) Studies on the development of endosperm in rice: 1. Process of endosperm tissue formation. (In Japanese with English abstract). Proceedings of the Crop Science Society of Japan 36: 151–161.
54. McCain DC (1995) Combined effects of light and water-stress on chloroplast volume regulation. Biophysical Journal 69: 1105–1110.
55. Veley K, Marshburn S, Clure C, Haswell E (2012) Mechanosensitive channels protect plastids from hypoosmotic stress during normal plant growth. Current Biology 22: 408–413.
56. Pyke K (1999) Plastid division and development. Plant Cell 11: 549–556.
57. Kang H, Park S, Matsuoka M, An G (2005) White-core endosperm *floury endosperm-4* in rice is generated by knockout mutations in the C_4 type pyruvate orthophosphate dikinase gene (*OsPPDKB*). Plant Journal 42: 901–911.
58. Gholipour Y, Erra-Balsells R, Hiraoka K, Nonami H (2013) Living cell manipulation, manageable sampling, and shotgun picoliter electrospray mass spectrometry for profiling metabolites. Analytical Biochemistry 433: 70–78.

14

Identification and Molecular Characterization of MYB Transcription Factor Superfamily in C_4 Model Plant Foxtail Millet (*Setaria italica* L.)

Mehanathan Muthamilarasan, Rohit Khandelwal, Chandra Bhan Yadav, Venkata Suresh Bonthala, Yusuf Khan, Manoj Prasad*

National Institute of Plant Genome Research, New Delhi, India

Abstract

MYB proteins represent one of the largest transcription factor families in plants, playing important roles in diverse developmental and stress-responsive processes. Considering its significance, several genome-wide analyses have been conducted in almost all land plants except foxtail millet. Foxtail millet (*Setaria italica* L.) is a model crop for investigating systems biology of millets and bioenergy grasses. Further, the crop is also known for its potential abiotic stress-tolerance. In this context, a comprehensive genome-wide survey was conducted and 209 MYB protein-encoding genes were identified in foxtail millet. All 209 *S. italica* MYB (SiMYB) genes were physically mapped onto nine chromosomes of foxtail millet. Gene duplication study showed that segmental- and tandem-duplication have occurred in genome resulting in expansion of this gene family. The protein domain investigation classified SiMYB proteins into three classes according to number of MYB repeats present. The phylogenetic analysis categorized SiMYBs into ten groups (I - X). *SiMYB*-based comparative mapping revealed a maximum orthology between foxtail millet and sorghum, followed by maize, rice and *Brachypodium*. Heat map analysis showed tissue-specific expression pattern of predominant SiMYB genes. Expression profiling of candidate MYB genes against abiotic stresses and hormone treatments using qRT-PCR revealed specific and/or overlapping expression patterns of *SiMYBs*. Taken together, the present study provides a foundation for evolutionary and functional characterization of MYB TFs in foxtail millet to dissect their functions in response to environmental stimuli.

Editor: Girdhar K. Pandey, University of Delhi South Campus, India

Funding: This work was financially supported by the core grant of National Institute of Plant Genome Research (NIPGR), New Delhi, India. The funder had no role in study design, data collection and analysis, decision to publish, or preparation of the manuscript.

* Email: manoj_prasad@nipgr.ac.in

Introduction

In plants, transcription factors (TFs) play crucial role in regulating the gene expression and thereby guiding various biological processes including growth, development, cell division and response to stresses. Deciphering the molecular and physiological roles of these TFs has become hottest topic of research since understanding the control of gene expression would enable researchers to develop elite varieties of crop plants with desired agronomic traits which will ensure global food security in the scenario of climate change. Hence research on TFs had gained momentum and numerous TFs regulating vital biological processes in plants have been identified. MYB (myeloblastosis) TFs are omnipresent in eukaryotic systems and in plants was first identified in *Zea mays* involving the regulation of anthocyanin biosynthesis [1].

MYB proteins possess a highly conserved MYB DNA-binding domain throughout the eukaryotes at N-terminus and 1–4 imperfect repeats (R0, R1, R2, and R3) [2]. Each of these repeats contains 50–53 amino acids and encodes three α-helices, with second and third helices forming a helix–turn–helix (HTH) structure. Upon binding to DNA, HTH intercalates in the major groove [2,3]. In addition, each MYB repeat comprises three regularly spaced tryptophan residues, which form a cluster in a hydrophobic core of the repeat and stabilize the structure of DNA-binding domain [4]. Contrarily, C-terminus is the activation domain and varies significantly between MYB proteins, which results in wide range of regulatory roles of MYB gene family [3,5,6].

There are numerous reports available demonstrating the role of MYB TFs in various plant processes such as secondary metabolism [7–10], hormone signal transduction [11,12], environmental stresses [13–15], cell structuring and organ development [16–19]. The emergence of next-generation sequencing technologies and high-throughput analysis platforms had expedited genome sequencing projects in crop plants [20]. The availability of genome sequence information facilitated the crop research community to perform genome-wide analysis of TFs which would be more informative in conducting further research on functional characterization of candidate TFs. Of the various TFs, MYBs are most extensively studied in almost all land plants whose genome sequence is available [21–25], though there was no such genome-wide study on MYB TFs conducted in foxtail millet (*Setaria italica* L.).

Foxtail millet is a member of Poaceae family and it is the second most-widely cultivated millets, next to pearl millet (FAOSTAT 2012; http://faostat.fao.org/). Being a C_4 panicoid crop, it is efficient in photosynthesis and better water use efficiency (WUE). Photosynthesis in C_4 crops is more efficient than C_3 plants even under high light intensity and high temperatures since phosphoenolpyruvate carboxylase (PEPC) helps in immediate uptake of carbon di oxide and delivering it to RuBisCO. The immediate quenching of carbon di oxide and delivery by PEPC does not require keeping stomata open for a long period and hence less water is lost by transpiration resulting in better WUE [26,27]. Moreover, foxtail millet has been considered as a model crop for studying systems biology of biofuel crops due to its close phylogenetic relationships with several biofuel crops such as switchgrass (*Panicum virgatum*), napier grass (*Pennisetum purpureum*) and pearl millet (*Pennisetum glaucum*) [26,27]. Hence considering its importance, BGI (Beijing Genomics Institute), China and the Joint Genome Institute (JGI) of Department of Energy, USA has sequenced the genome of two foxtail millet accessions [28,29]. This motivated the present study, to perform (i) genome-wide identification of MYB TFs encoded in foxtail millet genome, (ii) protein and gene structure analysis using proteomic tools as well as phylogenetic analysis, (iii) promoter and miRNA analysis, (iv) *in silico* expression profiling using RNA-Seq data, (v) ortholog identification in related grass genomes, (vi) evolutionary analysis of orthologs and paralogs, and (vii) expression analysis of candidate *MYB* genes using qRT-PCR.

Materials and Methods

Identification, chromosomal location and gene duplication analysis of MYB proteins

The Hidden Markov Model (HMM) profile of MYB DNA-binding domain (PF00249) downloaded from Pfam v27.0 (http://pfam.sanger.ac.uk/family/PF00249) [30] was queried against Phytozome v9.1 (www.phytozome.net) of *Setaria italica*. All hits with expected values less than 1.0 were retrieved and compared with MYB protein dataset available in Foxtail millet Transcription Factor Database (http://59.163.192.91/FmTFDb/) [31]. In addition, *S. italica* MYB (SiMYB) proteins were examined using CDD search (http://www.ncbi.nlm.nih.gov/Structure/cdd/wrpsb.cgi) for presence of one or more MYB DNA-binding domains. The chromosomal location and identification of alternate transcripts of SiMYBs were performed by executing BLASTP search against the foxtail millet protein sequences in Phytozome and redundant sequences with different identification numbers and same chromosome locus were discarded. Based on ascending order of chromosomal position from short arm telomere to long arm telomere, *SiMYBs* were annotated as *SiMYB001* to *SiMYB209* and physical map was constructed using MapChart v2.2 [32]. The map was manually inspected for identifying tandem duplications. Adjacent genes of same sub-family located within 10 predicted genes apart or within 30 kb of each-other are considered as tandem duplicated genes [33–36]. Segmental duplications were predicted using Multiple Collinearity Scan toolkit (MCScanX) [37] as described by Puranik et al. [35] and Mishra et al. [36].

Protein structure, gene organization and phylogenetic analysis

The presence of conserved domains in SiMYB proteins were analysed by employing hmmscan tool (http://hmmer.janelia.org/search/hmmscan). BioEdit tool (http://www.mbio.ncsu.edu/bioedit/bioedit.html) was used to perform multiple sequence alignment (MSA) of SiMYB amino acid sequences. The physicochemical properties of SiMYB proteins were then investigated using ExPASy ProtParam tool (http://web.expasy.org/protparam/). Gene Structure Display Server was used to identify the position of introns and exons in respective SiMYB genes (http://gsds.cbi.pku.edu.cn/) [38]. The SiMYB protein sequences were then imported into MEGA v6.06 [39] and MSA was performed using ClustalW with default parameters. The alignment file was used to construct neighbor-joining phylogenetic tree using a bootstrapping method with 1000 replicates.

Gene Ontology annotation, promoter analysis and prediction of miRNAs targeting SiMYBs

The amino acid sequences of SiMYB are loaded in Blast2GO suite [40] to perform BLASTP, against *Oryza sativa* protein sequences of NCBI, with default parameters. The hits were mapped to retrieve GO terms and annotation of GO terms associated with BLAST hits was executed. For promoter analysis, 2 kb upstream sequences of SiMYB genes were retrieved from Phytozome through in-house perl programming and *cis*-regulatory elements were identified using database of Plant *Cis*-acting Regulatory DNA Elements (PLACE) [41]. Moreover, *Setaria italica* miRNAs (Sit-miRs) targeting *SiMYB*s were predicted by aligning Sit-miRs [42,43] with *SiMYB* transcript sequences using psRNATarget [44].

In silico expression profiling and marker localization

Setaria italica Illumina RNA-HiSeq reads for 4 tissues namely spica, stem, leaf and root were retrieved from European Nucleotide Archive [SRX128226 (spica); SRX128225 (stem); SRX128224 (leaf); SRX128223 (root)] [45]. The RNA-seq data was then filtered by NGS toolkit (http://59.163.192.90:8080/ngsqctoolkit/) [46] to remove low quality reads and mapped onto the gene sequences of *S. italica* using Bowtie2 [47]. The number of reads mapped was normalized by RPKM (reads per kilobase per million) method. The heat map showing tissue specific expression was generated on RPKM values for each gene in all the tissue samples using TIGR MultiExperiment Viewer (MeV4) software [48,49]. The physical positions of DNA markers reported in foxtail millet [50] including simple sequence repeats (SSRs) [51,52], EST-derived SSRs (eSSRs) [53] and intron length polymorphic markers (ILPs) [54,55] were compared with the chromosomal location of SiMYB genes to identify the transcription factor gene-derived functional markers.

Identification of SiMYB orthologs in related grass genomes and evolutionary analysis of paralogs and orthologs

SiMYB protein sequences were BLASTP searched against the peptide sequences of sorghum, maize, rice and *Brachypodium* (http://www.gramene.org/; http://www.phytozome.net/) to predict the MYB orthologs in these grass species. A reciprocal BLAST (BLASTP) was also performed to ensure unique relationship between the orthologous genes and all hits with E-value≤1e-5 and at least 80% homology were considered significant. The orthologous relationships between foxtail millet and these grass species were then visualized using Circos v0.55 (http://circos.ca/). The synonymous (Ks) and non-synonymous (Ka) substitution rates were estimated for paralog and ortholog MYB gene pairs through CODEML program in PAML interface tool of PAL2NAL (http://www.bork.embl.de/pal2nal/) [56]. Time (million years

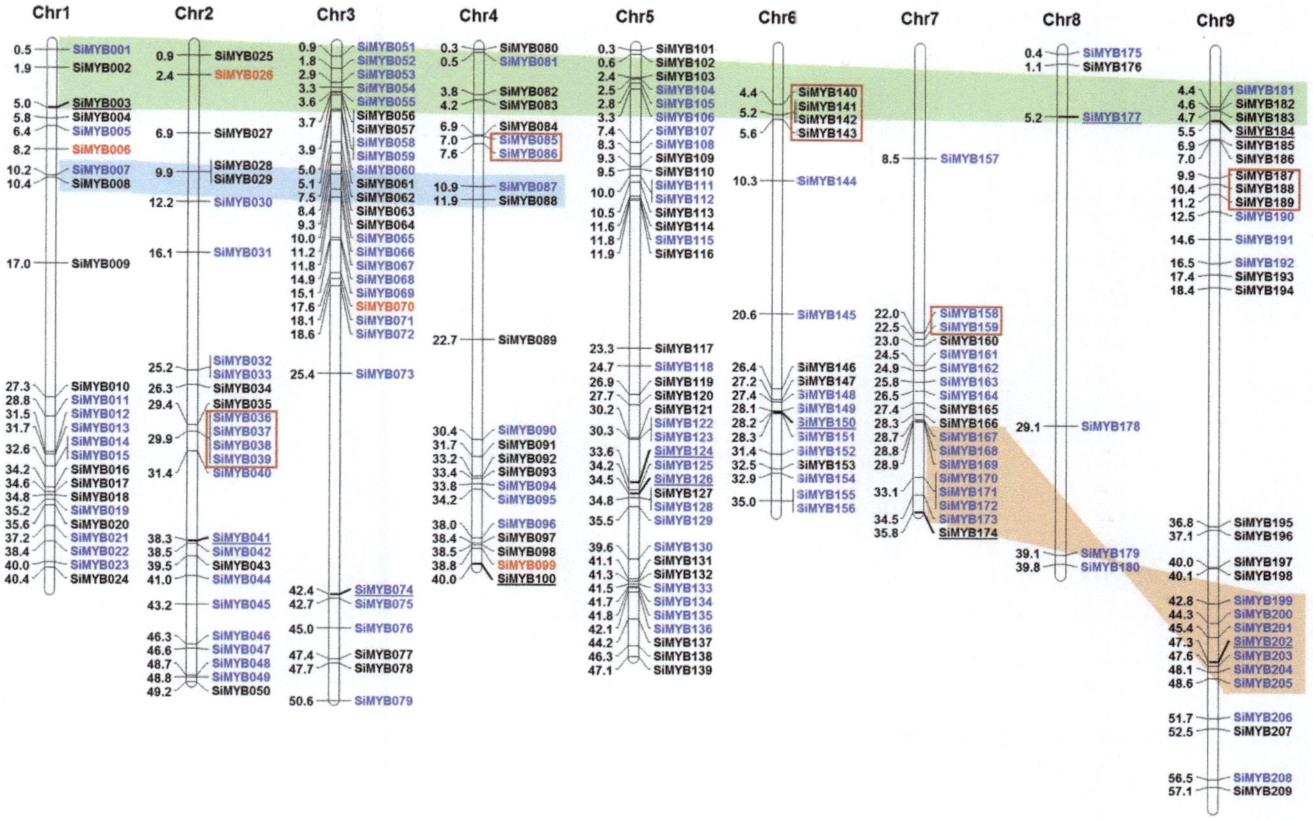

Figure 1. Physical map of 209 SiMYB genes. SiMYB genes were plotted onto nine chromosomes of foxtail millet. The numbers at left indicates the position of SiMYB genes, and the names of the respective genes are given in right. The 'MYB-Related' are given in black, 'MYB-R2R3' are in blue and 'MYB-R1R2R3' are in red. The SiMYB genes used for expression analysis through qRT-PCR are underlined. Segmentally duplicated gene-pairs are highlighted with different colours. Tandemly duplicated gene-pairs are enclosed within red boxes.

ago, Mya) of duplication and divergence of SiMYB gene pairs were estimated using a synonymous mutation rate of λ substitutions per synonymous site per year, as $T = Ks/2\lambda$, where, $\lambda = 6.5 \times 10^{-9}$ [35,57,58,59].

Plant materials and treatments

A salt and dehydration tolerant foxtail millet cultivar 'Prasad' [60,61] was chosen for analysing the expression pattern of candidate *SiMYB*s. The seedlings were grown in a plant growth chamber (PGC-6L; Percival Scientific Inc., USA) for 21 d after germination at 28±1°C day/23±1°C night/70±5% relative humidity with a photoperiod of 14 h and a photosynthetic photon flux density of 500 µmol m^{-2} s^{-1}. The plants were watered daily with one-third strength Hoagland's solution [62]. For stress and hormone treatments, 21-day-old seedlings were individually exposed to 250 mM NaCl (salinity), 20% PEG 6000 (dehydration), 100 µM abscisic acid (ABA), 100 µM methyl jasmonate (MJ) and 100 µM salicylic acid (SA) for 1 h (early) and 24 h (late) following previous reports [35,61,63]. After respective treatment, whole seedlings are immediately frozen in liquid nitrogen and stored at −80°C for RNA isolation. Untreated plants were maintained as control. The above experiments were repeated thrice to ensure precision and reproducibility.

RNA isolation and quantitative real-time PCR analysis

Total RNA was isolated by following the method described by Longeman et al. [64] and treated with RNase-free DNase I (50 U/µl; Fermentas, USA). Its quality and purity was ensured using NanoDrop Spectrophotometer (Thermo Fisher Scientific, USA) [OD_{260}: OD_{280} nm absorption ratio (1.8–2.0)] and integrity of RNA was checked by resolving on 1.2% agarose gel containing 18% formaldehyde. An amount of ~1 µg total RNA was reverse transcribed to first strand cDNA using random primers by Thermo Scientific Verso cDNA kit (Thermo Fisher Scientific, USA) following manufacturer's instructions. The qRT-PCR primers were designed using GenScript Real-time PCR Primer Design tool (https://www.genscript.com/ssl-bin/app/primer) (Table S1). qRT-PCR analysis was performed in three technical replicates for each biological duplicate by StepOne Real-Time PCR Systems (Applied Biosystems, USA). The PCR mixtures and reactions were used as described previously by Kumar et al. [65] Melting curve analysis (60 to 95°C after 40 cycles) and agarose gel electrophoresis were performed to check amplification specificity. A constitutive *Act2* gene-based primer was used as endogenous control [65]. The amount of transcripts accumulated for SiMYB genes normalized to the internal control *Act2* were analysed using $2^{-\Delta\Delta Ct}$ method cDNA synthesis [65]. The PCR efficiency was calculated as: $Efficiency = 10^{(-1/slope)} - 1$ by the default software (Applied Biosystems).

Homology modeling and active site prediction

SiMYB protein sequences were searched against Protein Data Bank (PDB) (http://www.rcsb.org/pdb/) for predicting the best template with homologous amino acid sequence and known structure. The result was then imported in Phyre2 server (Protein Homology/AnalogY Recognition Engine; http://www.sbg.bio.ic.

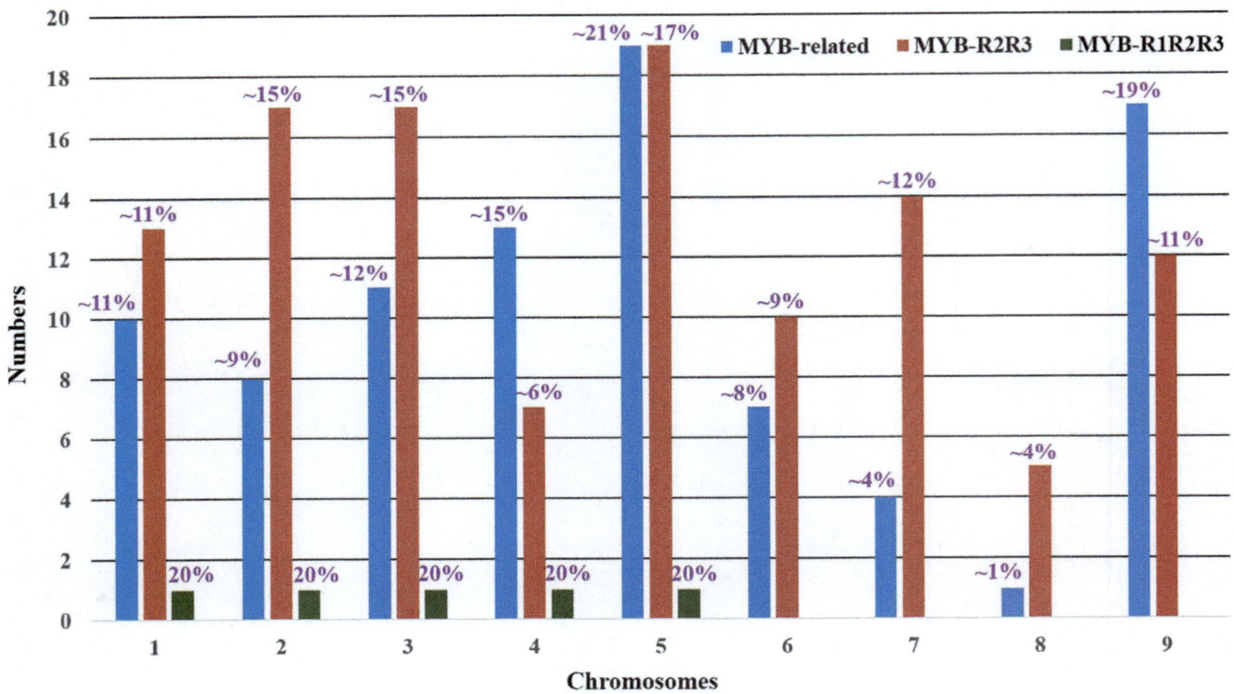

Figure 2. Chromosome-wise distribution of SiMYB genes. The distribution of three different SiMYB genes on individual chromosomes is shown.

ac.uk/phyre2) for predicting three-dimensional structures of proteins by homology modeling under 'normal' mode [66]. Active sites were precited using COACH server (http://zhanglab.ccmb.med.umich.edu/COACH/) and highlighted using UCSF Chimera 1.8.

Results and Discussion

Identification of SiMYB genes

The availability of foxtail millet whole genome sequence [28,29] facilitated genome wide identification of MYB transcription factor (TF) gene family. The HMM search showed the presence of 229 MYB genes in foxtail millet genome (Table S2). The result was then confirmed by comparing the data with Foxtail millet Transcription Factor Database (FmTFDb; http://59.163.192.91/FmTFDb/) [31]. Further, all 229 SiMYB genes were BLAST searched against foxtail millet genome in Phytozome to identify alternate transcripts and chromosomal location. The results showed that 209 *SiMYBs* were primary transcripts and 20 were alternate transcripts. SiMYB017 had a maximum of three alternate transcripts, whereas SiMYB080, SiMYB137, SiMYB143 and SiMYB185 have two splice variants. Thirteen *SiMYBs* were evidenced to have one splice variant each.

The 209 SiMYB genes reported in the present study is higher than the numbers reported in Arabidopsis (197) [67], grape (118) [68] and Populus (192) [69]. On contrary, this number is lesser than soybean (244) [23] and apple (229) [70] (Figure S1). Among sequenced grass genomes, foxtail millet was evidenced to encode for a highest number of MYB TFs. Its closest relative sorghum has 100 MYB genes, whereas maize, rice and *Brachypodium* has 132, 155 and 98 MYB genes, respectively (http://planttfdb.cbi.pku.edu.cn/; Figure S1). Of note, further experimentations are requisite to identify the pseudogenes among the 209 SiMYBs.

SiMYBs location and duplication in foxtail millet genome

All the 209 SiMYB genes were physically mapped onto nine chromosomes of foxtail millet (Figure 1). The map revealed an uneven distribution of SiMYB genes in foxtail millet genome (Table S2). Highest number of SiMYB genes were observed in chromosome 5 (39; ~19%) and lowest in chromosome 8 (6; ~3%), with a density of 0.8 and 0.1 MYB genes/Mb, respectively. The distribution pattern of *SiMYBs* on individual chromosomes also revealed certain physical regions with a relatively higher accumulation of gene clusters. For instance, *SiMYBs* located on chromosomes 1, 3, 4, 5, 6 and 9 appear to be clustered at both upper end and lower end of the arms. Moreover, examination of gene duplication showed that SiMYB genes underwent both tandem and segmental duplications (Figure 1). Twelve gene-pairs were evidenced to be segmentally duplicated, whereas 15 genes were tandemly duplicated.

Structural classification of SiMYB proteins

The CDD search demonstrated the presence of conserved motifs in SiMYB proteins (Figure S2). It also showed the presence of more than one MYB repeats in some of these proteins along with the presence of additional domains. The hmmscan analysis provided comprehensive information on functional domains present in SiMYB proteins (Table S3). Based on the presence of one, two and three MYB repeats, the SiMYBs are classified into 'MYB-related', 'MYB-R2R3' and 'MYB-R1R2R3' (Table S2). The multiple sequence alignment of other three types of SiMYBs confirmed the presence of MYB repeats in the protein sequences (Figure S3–S5). Of the three types, 'MYB-R2R3' genes were found to be in maximum (114; ~55%) followed by 'MYB-related' (90; ~43%). Only 5 'MYB-R1R2R3' were evidenced in foxtail millet distributed one MYB-R1R2R3 encoding gene each in chromosome 1 to 5 (Figs. 1, 2). The predominance of 'MYB-R2R3' type *SiMYBs* suggest that they could have been evolved

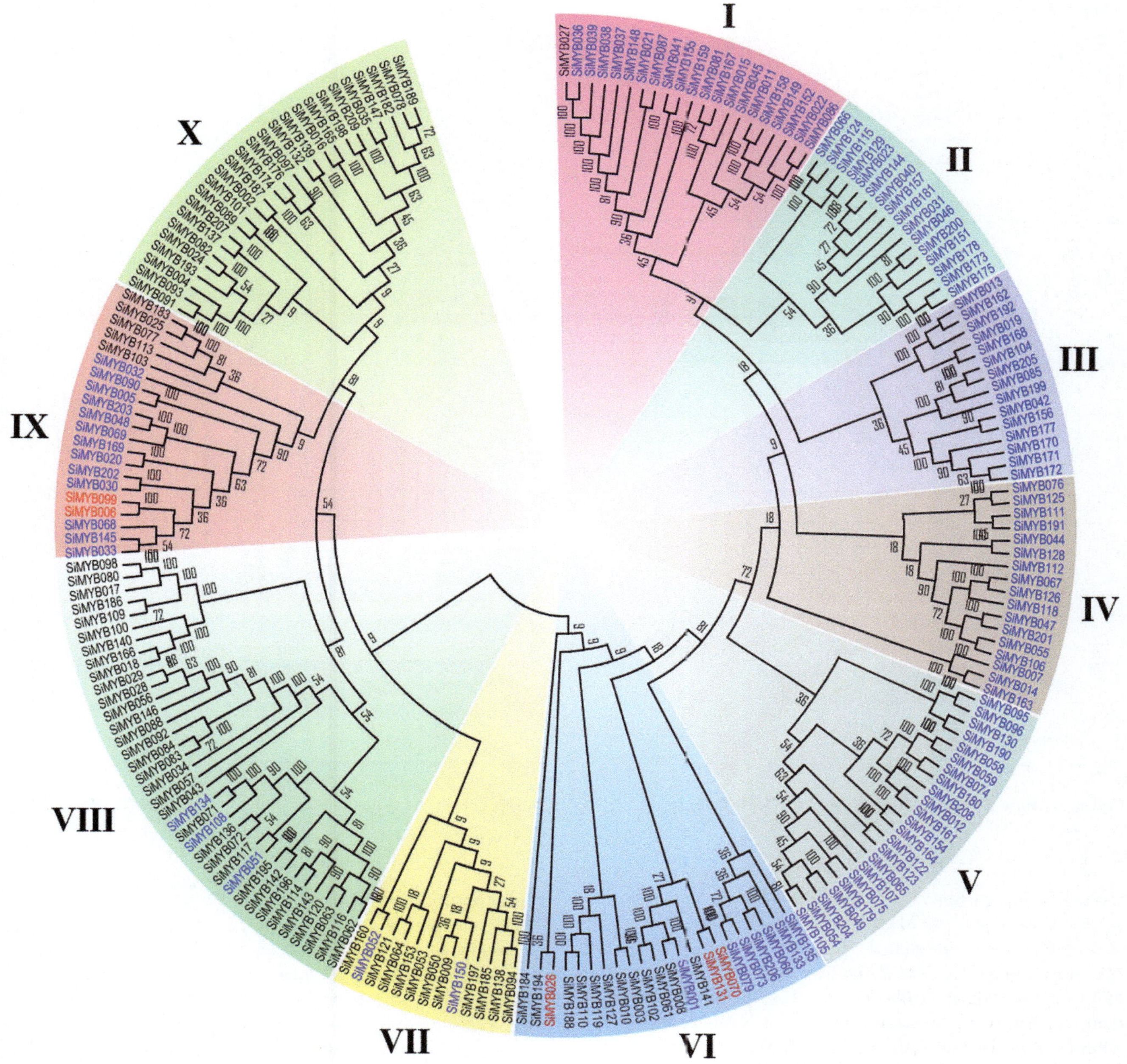

Figure 3. Phylogenetic relationships of foxtail millet MYB transcription factors. The unrooted phylogenetic tree was constructed by neighbor-joining method with 1000 bootstrap replicates. The bootstrap values are shown at the nodes. The tree was divided into ten phylogenetic cluster designated as I to X.

from MYB-R1R2R3 gene progenitor through loss of R1 repeat or from a MYB-R1 gene through duplication of R1 repeat [71,72]. Interestingly, no 'Atypical MYB' was identified in the foxtail millet, though one 'Atypical MYB' was reported in rice, two each in *Arabidopsis* and soybean [23,67]. Among 'MYB-R2R3' and 'MYB-related' genes, highest numbers were found to be encoded in chromosome 5 (19 each). The lowest number of these genes are in chromosome 8 (5; ~4% and 1; ~1%, respectively) (Figs. 1, 2).

Further, hmmscan analysis also revealed the presence of specific, functionally important domains present in C-terminal regions outside MYB domain which are involved in transcriptional activity and protein-protein interactions (Table S3). SiMYB003, SiMYB094, SiMYB160 and SiMYB185 possess SWIRM domain. It is a small alpha-helical domain of about 85 amino acid residues found in chromosomal proteins. It contains a helix-turn-helix motif and binds to DNA [73]. 'Response_reg' is 'response regulator receiver domain', which was first identified to receive signal from the sensor partner in bacterial two-component systems. SiMYB004, SiMYB024, SiMYB082, SiMYB091, SiMYB093, SiMYB137, SiMYB193 and SiMYB207 have this domain. It is usually found at N-terminal to a DNA binding effector domain and its role in plant systems remain elusive [74]. The domain 'P_C' in SiMYB205 denotes 'P protein C-terminus', which represents C-terminus of plant P proteins. The maize P gene is a transcriptional regulator of genes encoding enzymes for flavonoid biosynthesis in the pathway leading to production of a

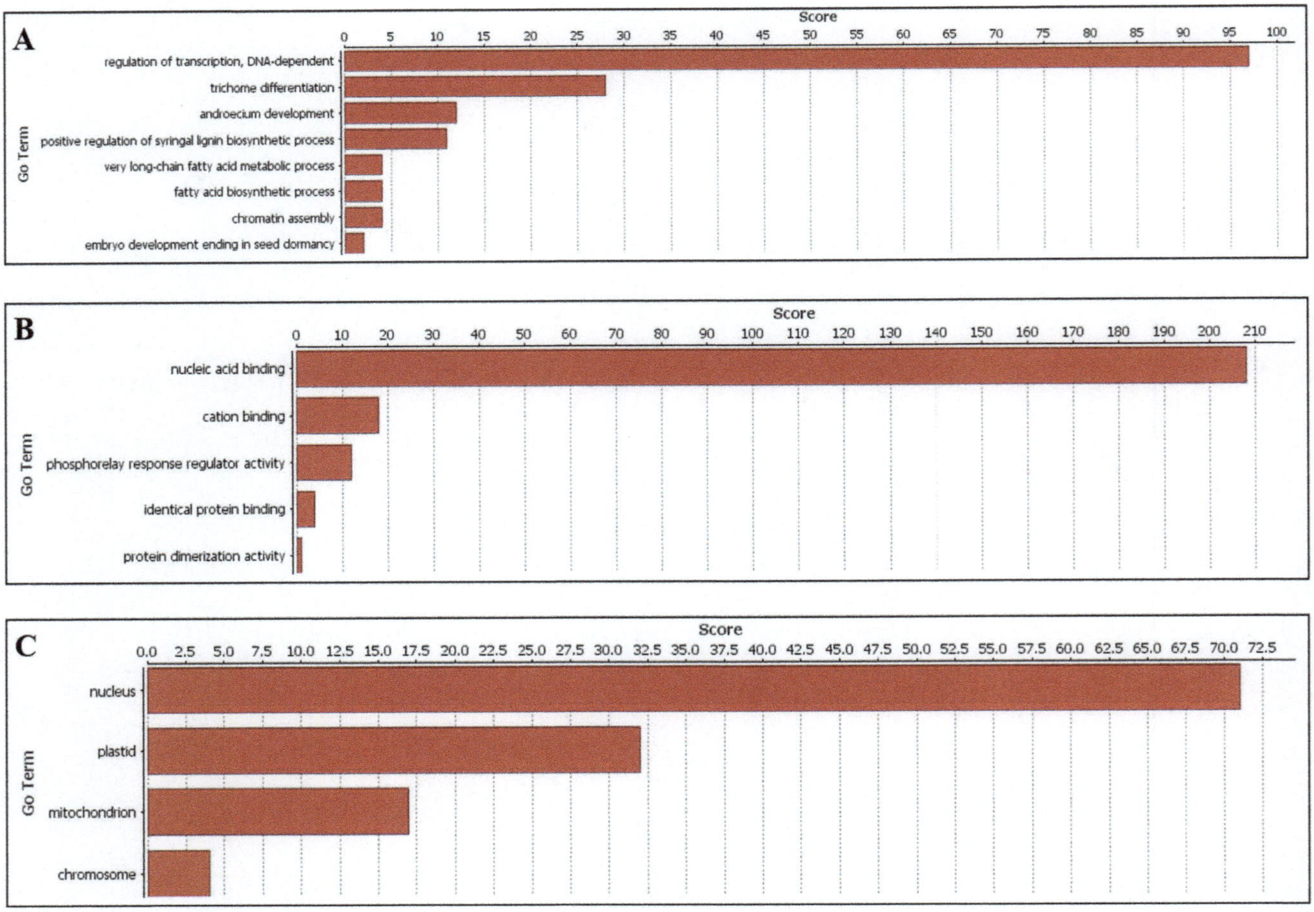

Figure 4. Gene Ontology distribution for SiMYB proteins. Blast2GO program defines the gene ontology under three categories. (**A**) biological processes, (**B**) molecular functions and (**C**) cellular component.

red phlobaphene pigment [75] and P proteins are homologous to DNA-binding domain of MYB-like transcription factors [76]. The 'ZZ' domain is ZZ type-Zinc finger whose binding to DNA is not yet identified [77]. This domain is present in SiMYB094, SiMYB184 and SiMYB185. It has 4–6 cysteine residues in its sequence, which are responsible for coordinating zinc ions and to reinforce the structure [78]. SiMYB153 was evidenced to have 'Bromodomain', which is reported to recognize acetylated lysine residues. This recognition is prerequisite for protein-histone association and chromatin remodelling [79]. The domain 'DUF3651' is present in SiMYB055 and SiMYB106, which denotes 'Domain of Unknown Function'. The 'Linker_histone' present in SiMYB110, SiMYB119, SiMYB127 and SiMYB188, which is an essential component of chromatin structure [80].

Except for the presence of conserved MYB domain, SiMYBs varied notably in size, sequence, intron-exon numbers and distribution and its physicochemical properties (Table S2). Gene structure analysis showed that 10 *SiMYB* were intron-less while *SiMYB026* and *SiMYB184* have maximum of 11 introns (Figure S6). The longest gene was *SiMYB094* with 8194 bases and the shortest was *SiMYB186* (348 bases). The longest protein sequence was SiMYB053 (1741 amino acids) and the least was SiMYB194 (98 amino acids) (Table S2). The enormous diversity with respect to MYB gene and proteins observed in foxtail millet was not reported in any other crop species [21–25], and this shows the presence of putative novel variants.

Phylogenetic classification of SiMYB proteins

The phylogenetic tree constructed using neighbour-joining (NJ) method for all 209 SiMYB proteins conformed to the phylogenetic classification enlisted in FmTFDb (Figure 3). The internal branches were detected to have high bootstrap values and this lead to the derivation of statistically reliable pairs of possible homologous proteins possessing similar functions from a common ancestor. Based on topology and clade support values, the phylogenetic tree was divided into ten clades (I to X) (Figure 3). Although SiMYB proteins tended to cluster together in the tree with respect to their type and they were not equally distributed in the clades which may be due to the occurrence of duplication and divergence of SiMYB genes (Figure 3).

Gene Ontology annotation

Gene Ontology (GO) annotation for all 209 SiMYB proteins was performed using Blast2GO. The SiMYB protein sequences were BLAST searched against *Oryza sativa* protein database and its biological processes, molecular functions and cellular components were identified (Figure 4). The results indicate that SiMYB proteins participate in diverse biological and molecular functions in the cell (Table S4). Noteworthy, predominant of SiMYBs were predicted to function in regulation of transcription in a DNA-dependent manner, followed by playing pivotal role in trichome differentiation and androecium development. SiMYBs are reported to least participate in breaking seed dormancy (Figure 4). The

Table 1. List of *Setaria italica* miRNAs targeting *SiMYB* transcripts.

miRNA ID	Target.	E-value	UPE	Target start	Target end	miRNA aligned fragment	Target aligned fragment	Inhibition
Sit-miR5568.58	SiMYB026	3.0	16.26	1923	1943	UUUCUAGGUUUAUAUCUUUUG	CACAAGAGAUAGAUCUAGAAA	Cleavage
Sit-miR2118d	SiMYB028	2.5	14.91	150	171	UUCCUGAUGCCUCUCAUUCCUA	UUGGAAGGAGAGGCAUCAGGUA	Cleavage
Sit-miR2118d	SiMYB029	2.5	13.89	152	171	UUCCUGAUGCCUCUCAUUCC	GGAAGGAGAGGCAUCAGGUA	Cleavage
Sit-miR159a	SiMYB130	2.5	16.88	960	979	UUUGGAUUGAAGGGAGCUCU	GGAGCUCCCUUCACUCCAAG	Cleavage
Sit-miR159a	SiMYB190	3.0	19.88	884	904	UUUGGAUUGAAGGGAGCUCUG	UGGAGCUCCCUUCAAACCAAU	Cleavage

UPE - unpaired energy.

biological function analysis also revealed that 68 (32.5%) SiMYB proteins function in response to stress (Table S4). The molecular role showed that most of SiMYBs have function in nucleic acid binding followed by cation binding. Cellular component data predicted that SiMYBs are nuclear localized. In addition, SiMYB proteins were also indicated to localize in plastids and mitochondria (Figure 4). In addition, Blast2GO was performed to correlate the domain composition of families/sub-families with functional classes, but no correlation was observed.

Cis-acting elements and *SiMYB* targeting miRNAs

The promoter analysis identified ~300 diverse *cis*-acting elements upstream of *SiMYB* genes (Table S5). Of these, ARR1AT, CAATBOX1, CACTFTPPCA1, DOFCOREZM, EBOXBNNAPA, GATABOX, GT1CONSENSUS, GTGANTG10, MYCCONSENSUSAT, POLLEN1LELAT52, WBOXNTERF3 and WRKY71OS were present in all 209 *SiMYB* genes. ARR1AT (NGATT) is a cytokinin response regulator binding motif [81] and CAATBOX1 (CAAT) are reported to regulate flowering [82]. CACTFTPPCA1 (CACGTG) is responsible for mesophyll-specific gene expression of C4 phosphoenolpyruvate carboxylase gene in C_4 plants [83] and DOFCOREZM (AAAG) is the binding cite of Dof transcription factors [84]. EBOXBNNAPA (CANNTG) is E-box sequence responsible for light responsiveness and is regulated by bHLH and MYB-transcription factor in directing tissue-specific expression [13]. GATABOX (GATA) is the binding site for transcription factors with a zinc finger motif [85], GT1CONSENSUS (GRWAAW) recognizes GT-1 proteins which have tri-helix DNA-binding domains [86] and GTGANTG10 (GTGA) is a pollen-specific *cis*-elements [87]. MYCCONSENSUSAT (CANNTG) is a MYC recognition site which regulates transcription of genes under cold conditions by a MYC-like bHLH transcriptional activator [88], POLLEN1LELAT52 (AGAAA) is a regulatory element responsible for pollen-specific activation of gene expression [89], WBOXNTERF3 (TGACY) is a W-box promoter motif functioning in response to wound signal [90] and WRKY71OS (TGAC) is a binding site of rice WRKY71, a transcriptional repressor of gibberellin signaling pathway [91]. In addition to these, some *cis*-elements were present unique to specific *SiMYBs* such as AAGACGTAGATACL12 in *SiMYB167*, ABADESI1 in *SiMYB025*, ABADESI2 in *SiMYB110*, ABRE3HVA22 in *SiMYB158* and ABREDISTBBNNAPA in *SiMYB122*. Further, *Setaria italica* miRNAs (Sit-miRs) targeting the *SiMYB* transcripts for post-transcriptional gene regulation were also identified. Three Sit-miRs namely Sit-miR5568.58, Sit-miR2118d and Sit-miR159a were evidenced to target *SiMYB026*, *SiMYB028*, *SiMYB029*, *SiMYB130* and *SiMYB190* (Table 1). This data would assist in deciphering post-transcriptional control of gene regulation during physiological and stress-induced cellular responses.

Heat map analysis and identification of transcription factor-derived markers

The publicly available Illumina RNA-Seq data for four tissues of foxtail millet root, leaf, stem and spica were used to analyse the expression pattern of SiMYB genes. The heat map showed differential expression of *SiMYBs* across the tissues (Figure 5). Predominant of *MYB* genes including *SiMYB003*, *SiMYB004*, *SiMYB005*, *SiMYB017*, *SiMYB025*, *SiMYB051*, *SiMYB064*, *SiMYB080*, *SiMYB089*, *SiMYB090*, *SiMYB0100* were predicted to be highly expressed in all the tissues. Interestingly, tissue-specific higher expression was also noted. *SiMYB007* and *SiMYB087* were highly expressed in leaf, while *SiMYB192* showed higher

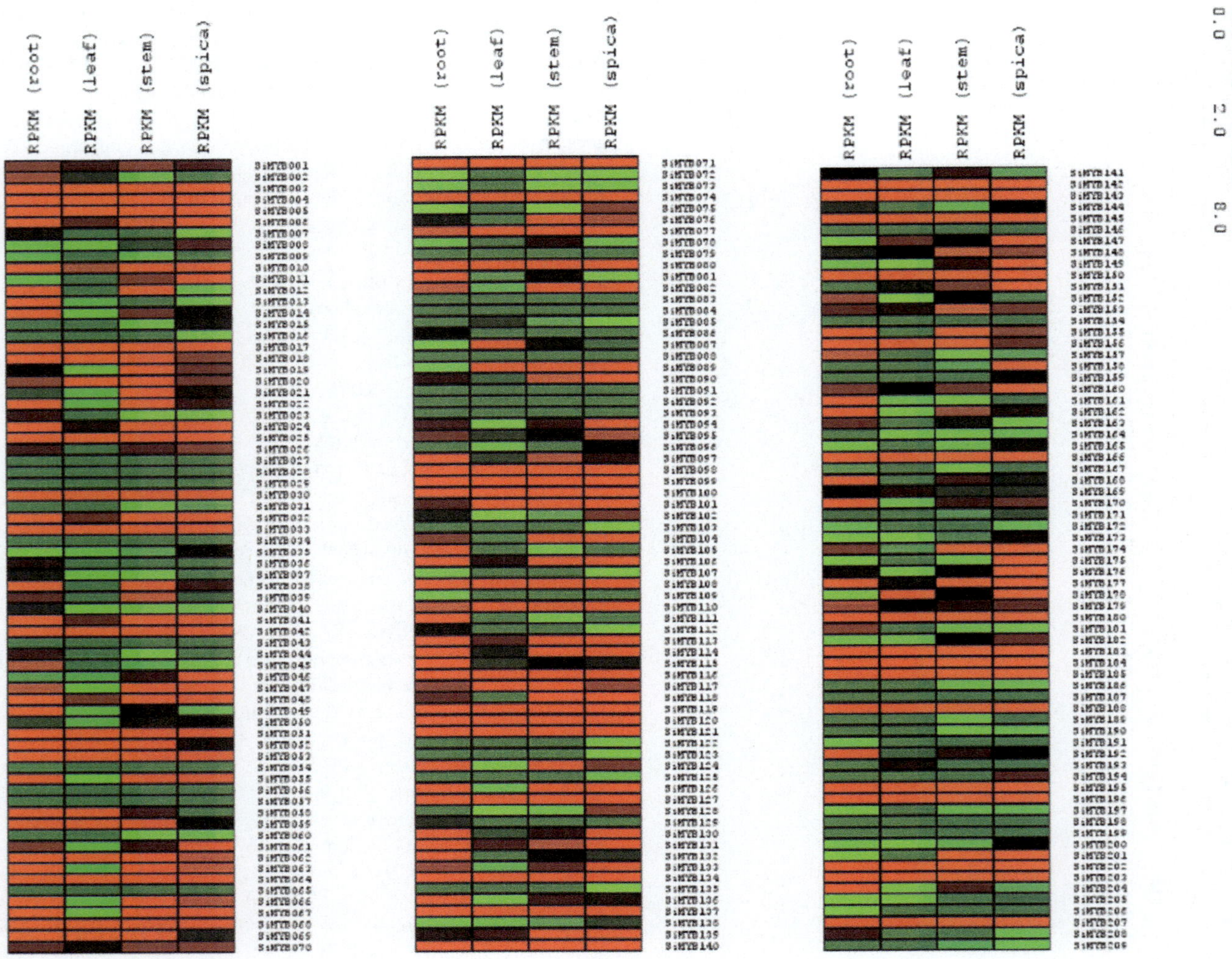

Figure 5. Heat map representation of SiMYB genes in four tissues. The Illumina RNA-seq data were re-analyzed and heat map was generated. Bar at the top represents $\log_2$ transformed values, thereby values 0.0, 2.0 and 8.0 represent low, intermediate and high expression, respectively.

expression in root (Figure 5). This expression profiling would facilitate combinatorial usage of *SiMYBs* in transcriptional regulation of different tissues, whereas ubiquitously expressed *SiMYBs* might regulate the transcription of a broad set of genes. Further, this heat map data also enables selection and designing of studies to impart stress tolerance in both foxtail millet and related crop species.

In addition, molecular markers present in *SiMYBs* were also identified. Seventy three SiMYB genes (~35%) were evidenced to possess DNA markers (Table S6). Of the various types of markers reported in foxtail millet, presence of microsatellites (SSRs and eSSRs) in *SiMYB*s were found predominant (~90%) followed by ILP markers (~10%). Among the SSRs (63), tri-nucleotide repeats were maximum (41) followed by di-nucleotide repeats (20). Only two eSSRs namely SieSSR302 and SieSSR70 were found to be present in SiMYB030 and SiMYB094, respectively, and both the markers were of tri-nucleotide repeat type. SSRs SiGMS12780 and SiGMS8287 were a tetra- and hexa-nucleotide repeats, evidenced to be present in SiMYB179 and SiMYB125, respectively (Table S6). The functional relevance of these informative genic markers needs to be demonstrated by integrating agronomic trait association analysis with genetic mapping, differential expression profiling, protein modelling and haplotype gene evolution study in relation to *MYB* genes.

SiMYB orthologs in sorghum, maize, rice and *Brachypodium*

Of the 209 SiMYB genes, 72 (~34%) showed significant orthologous relationships with sorghum, 55 (~26%) with maize, 23 (~11%) with rice and 13 (~6%) with *Brachypodium* (Figure 6; Tables S7–S10). Foxtail millet chromosome 5 showed highest synteny with sorghum chromosome 3 (14; ~19%), and maize chromosomes 3 and 8 (11; 20%). Chromosome 3 of foxtail millet has highest orthologous pairs with rice chromosomes 4, 5 and 12 (~22%), and *Brachypodium* chromosomes 2, 4 and 5 (~38%). Except the presence of conserved nucleotide and amino acid sequences of MYB domain between foxtail millet-rice and foxtail millet-*Brachypodium*, the regions other than the MYB domain were highly diverse. The degree of decrease in orthologous relationships among foxtail millet-sorghum, -maize, -rice and *Brachypodium* is due to genetic relatedness of these genomes. Foxtail millet along with sorghum and maize belong to Panicoideae, whereas rice and *Brachypodium* belong to Ehrhartoideae and Pooideae, respectively. This pattern of decreasing

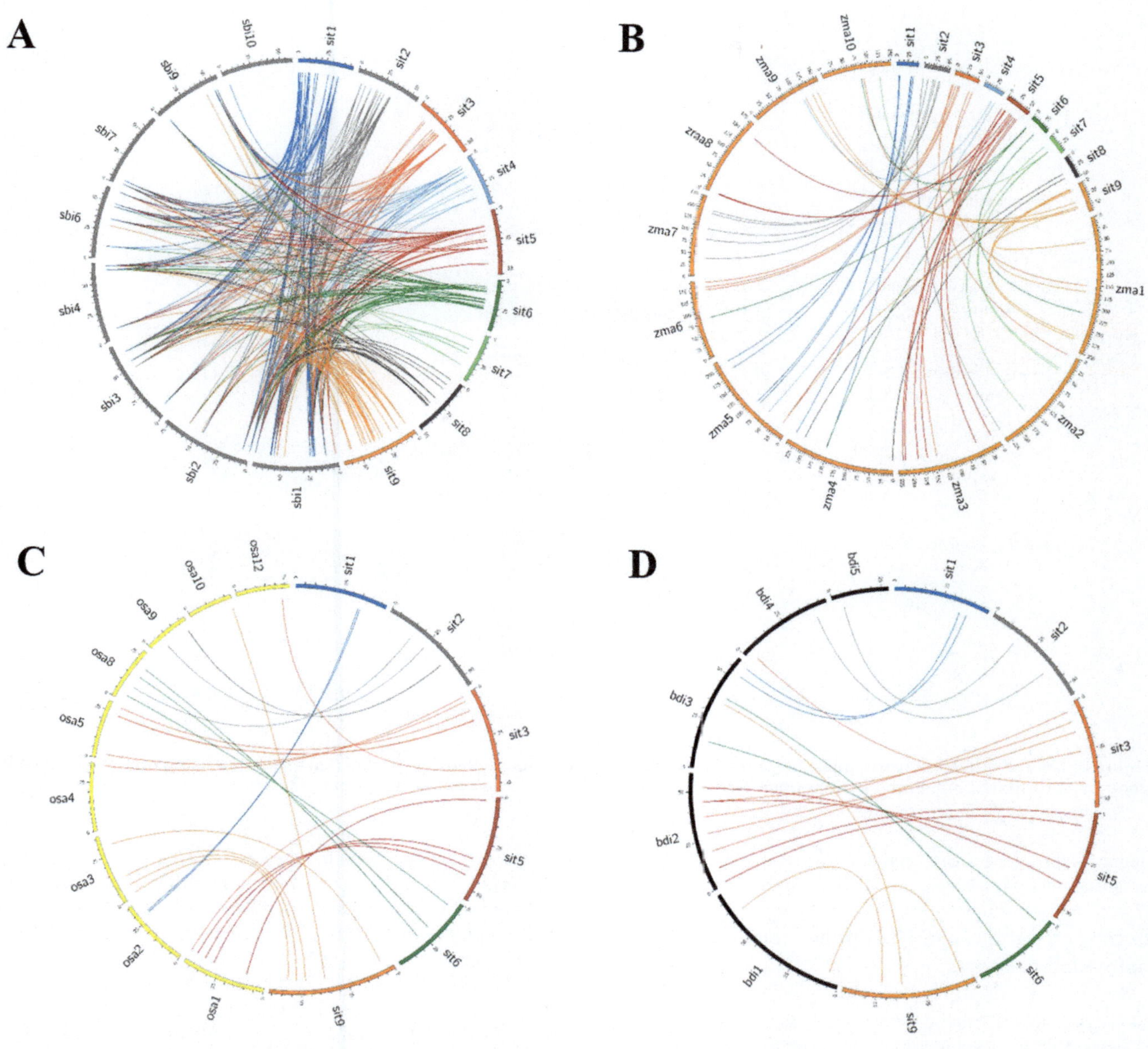

Figure 6. Comparative physical mapping revealed high degree of orthologous relationships of SiMYB genes located on nine chromosomes of foxtail millet. (**A**) *Sit* (foxtail millet) – *Sbi* (sorghum), (**B**) *Sit* – *Zma* (maize), (**C**) *Sit* – *Osa* (rice), and (**D**) *Sit* – *Bdi* (*Brachypodium*).

synteny was also observed in the comparative mapping of NAC TFs [35], WD40 genes [36], Dicer-like, Argonaute, RNA-dependent RNA polymerase genes [92], and C_2H_2-type zinc finger TFs [59]. This comparative map would enable the researchers to select candidate MYB genes from foxtail millet and utilize them in genetic enhancement of related grass family members.

Evolutionary aspects of duplication and divergence

The ratios of non-synonymous (Ka) versus synonymous (Ks) substitution rate (Ka/Ks) were estimated for duplicated and orthologous gene-pairs of *SiMYB*. The ratios of Ka/Ks for segmentally and tandemly duplicated gene-pairs ranged from 0.02 to 0.1 and 0.07 to 0.1, respectively (Tables S11–S12). The estimated Ka/Ks was <1, and this justified that the duplicated SiMYB genes were under strong purifying selection pressure. This is in agreement with previous genome-wide reports on NAC TFs by Puranik et al. [35], WD40 gene family by Mishra et al. [36], and C_2H_2-type zinc finger TFs by Muthamilarasan [59]. In addition, it was also estimated that the duplication event of these segmentally and tandemly duplicated genes occurred ~33 Mya (Figure 7). Although further experimentations are necessary to validate this, the present study suggests that these duplication events must be specific to *Setaria* since divergence of sorghum and maize had occurred ~13 Mya [28,29].

Among orthologous gene-pairs of *SiMYB*, average Ka/Ks value was maximum between foxtail millet and *Brachypodium* followed by foxtail millet and rice (Figure 7). The Ka/Ks was minimum for foxtail millet-sorghum gene-pairs. The relatively higher rate of synonymous substitution between *Brachypodium* and foxtail millet *MYB* genes suggested their earlier divergence around 42 Mya. The evolutionary analysis also predicts the divergence time of foxtail millet-rice as ~39 MYA, foxtail millet-maize as ~29 MYA and foxtail millet-sorghum as ~27 MYA (Figure 7). These Ka/Ks

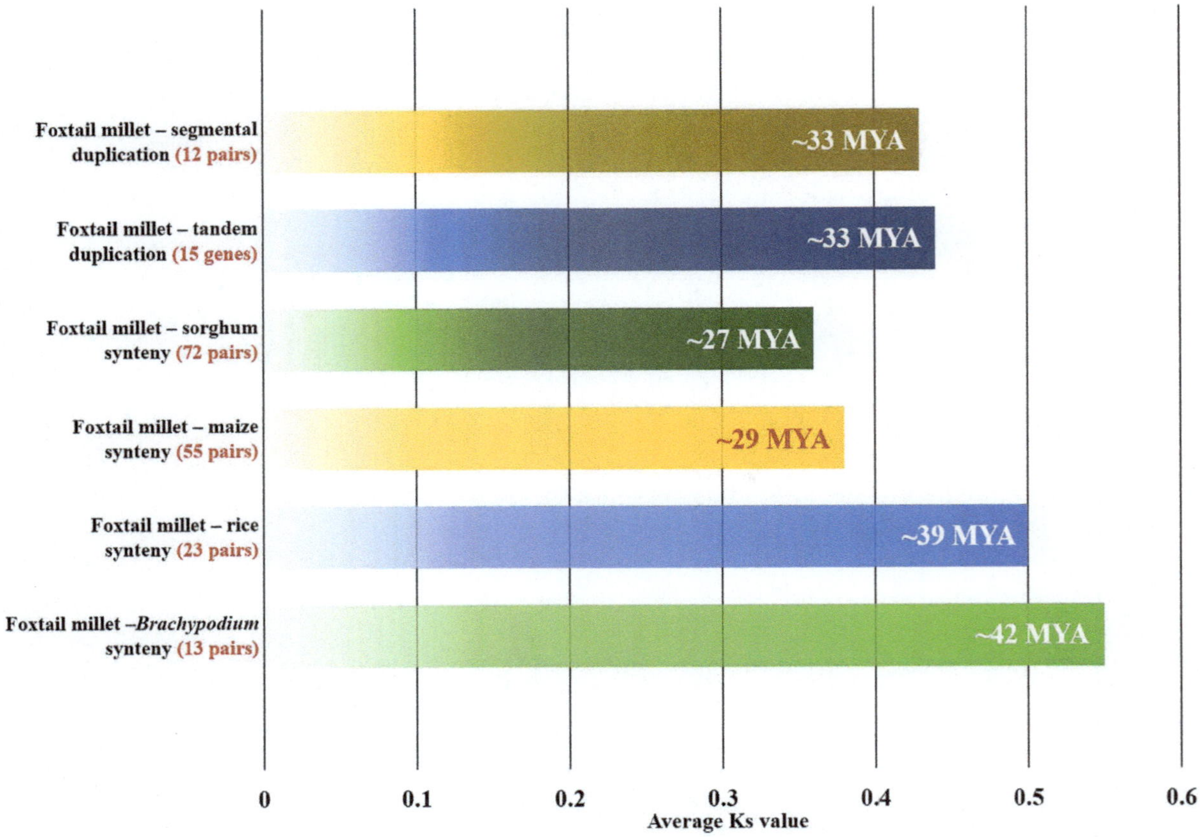

Figure 7. Time of duplication and divergence (MYA) based on synonymous substitution rate (Ks) estimated using duplicated SiMYB gene-pairs of foxtail millet and orthologous *MYB* gene pairs between foxtail millet- sorghum, -maize, -rice and *Brachypodium*.

dating data along with comparative map would assist in understanding the evolution of MYB genes in grasses genomes.

Expression pattern of *SiMYB*s under abiotic stress and hormone treatments

To study the expression patterns of identified *SiMYBs*, 11 candidate SiMYB genes were chosen representing all the nine chromosomes of foxtail millet and all the 10 clades of phylogenetic tree. The relative transcript abundance of these candidate *MYBs* at both early (1 h) and late (24 h) duration of abiotic stresses (salinity and dehydration), and hormone treatments (ABA, MJ, SA) were studied using qRT-PCR (Figure 8). During early phase of salinity stress *SiMYB003*, *SiMYB100*, *SiMYB124* and *SiMYB174* were highly up-regulated whereas *SiMYB041*, *SiMYB074*, *SiMYB126*, *SiMYB150* and *SiMYB202* were down-regulated. Though *SiMYB100* and *SiMYB174* showed significant up-regulation during early hours of salinity stress, a drastic down-regulation was observed at late phase. *SiMYB150* showed ~2 fold higher expression during late phase of salinity stress. In course of dehydration stress, up-regulation during early phase was evidenced in *SiMYB074*, *SiMYB100*, *SiMYB174* and *SiMYB202*. SiMYB124 showed higher expression at both early and late phases. A drastic up-regulation of SiMYB genes at 24 h after dehydrations stress was noticed in *SiMYB126, SiMYB110* and *SiMYB184*.

During hormone treatments, *SiMYB126*, *SiMYB150* and *SiMYB177* were down-regulated in all the treatments and time points, except a considerable higher expression of *SiMYB126* and *SiMYB177* at 24 h post ABA treatment. *SiMYB003* was down-regulated at both the phases of ABA treatment, while significant up-regulation at 24 h post ABA application was observed in *SiMYB184* and *SiMYB202*. The genes *SiMYB003, SiMYB100* and *SiMYB184* were up-regulated during early phase of MJ treatment, whereas up-regulation was observed at late phase in *SiMYB074* and *SiMYB202*. Up-regulation of *SiMYB100* and *SiMYB184* was evidenced in early phase of SA treatment. After 24 h, *SiMYB003* and *SiMYB202* showed higher expression.

Since foxtail millet is well known for its adaptability to abiotic stresses such as dehydration and salinity [60–62], deciphering the role of MYB TFs will provide novel insights into the stress tolerance mechanism of this model crop. This preliminary study on expression profiling of candidate SiMYB genes in response to different stress and hormone treatments would serve as a foundation for downstream characterization of these TFs.

Homology modelling of SiMYB proteins

Three dimensional protein models were constructed for all 209 SiMYBs and of note, the proteins were modelled at 90% confidence (Table S13). Further, its potential active sites were also identified (Figure S7). As mentioned earlier, the SiMYBs were classified into 'MYB-related', 'MYB-R2R3' and 'MYB-R1R2R3' based on the presence of one, two and three MYB repeats in the N-terminus. The homology modelling showed that each MYB repeat comprised of 52 amino acids and encoded three α-helices, with the second and third helices forming a helix-turn-helix (HTH) structure. Further, it included three regularly spaced tryptophan residues, forming a hydrophobic core in the three dimensional HTH structure (Figure S7). The third helix of each repeat formed

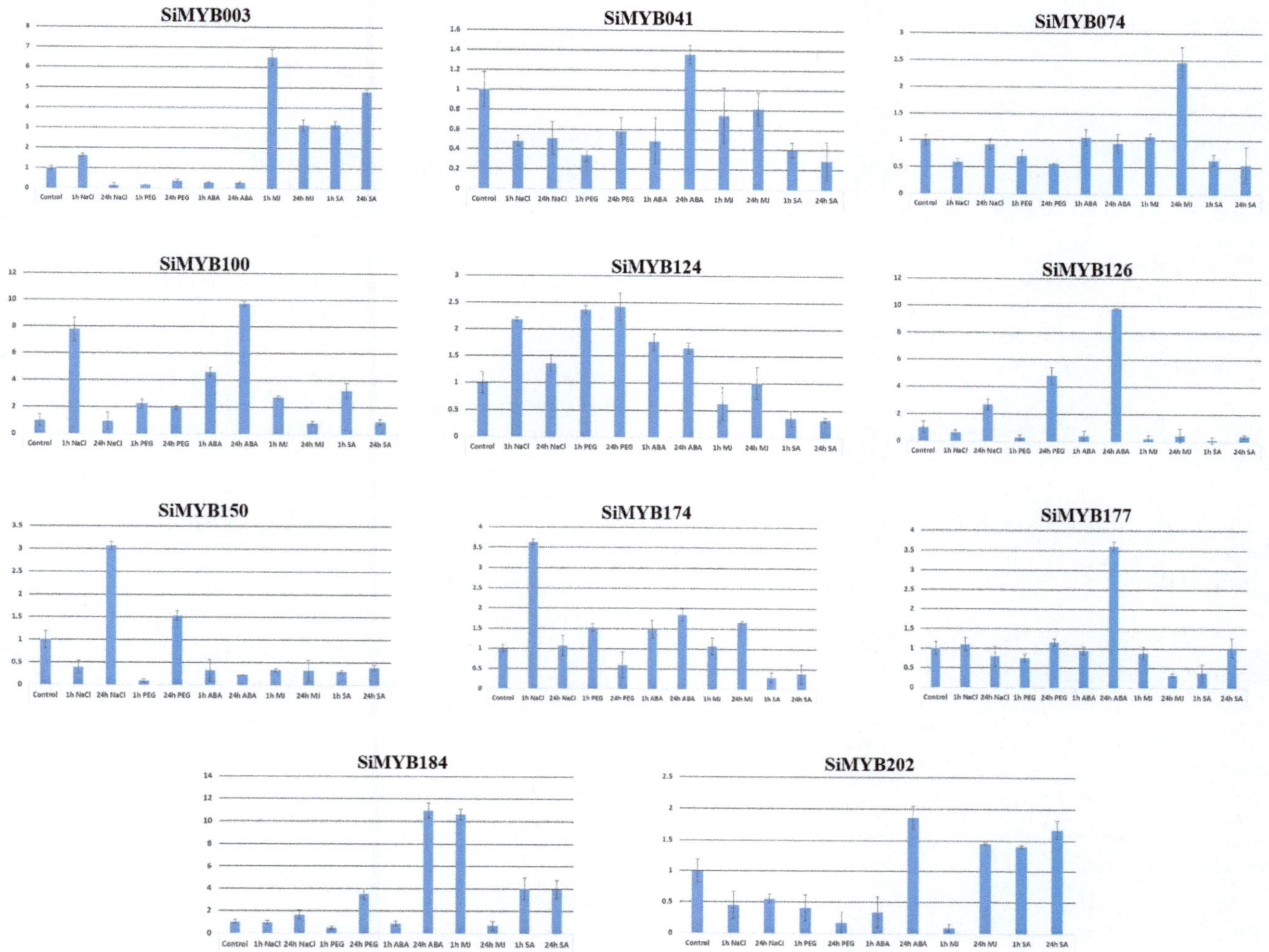

Figure 8. The relative expression ratio of 11 candidate SiMYB genes analysed using qRT-PCR under salinity stress, dehydration stress and hormone treatments for 1 h (early) and 24 h (late). The relative expression ratio of each gene was calculated relative to its expression in control sample (0 h). *Act2* was used as an internal control to normalize the data. Error bars representing standard deviation were calculated based on three technical replicates for each biological duplicate. *ABA - abscisic acid, MJ - methyl jasmonate, SA - salicylic acid.*

'recognition helix', which was reported to interact with DNA and intercalate in the major groove [6]. Contrarily, the C-terminus varied considerably between SiMYB proteins and this could be responsible for the wide range of regulatory roles of SiMYB gene family. This preliminary data would provide a base for further molecular structure analyses and interaction studies of SiMYB proteins.

In summary, the present study identified 209 MYB proteins in model crop foxtail millet and physically mapped them onto the genome. The protein structure and gene organization of individual SiMYBs were analysed and a phylogenetic tree was constructed. The domain analysis showed the presence of diverse protein domains in MYB proteins and the multiple sequence alignment revealed the conserved sequences in these domains. In addition, DNA markers present in *SiMYB* genes were also investigated. Candidate *SiMYB*s were chosen for expression profiling through qRT-PCR and the analysis showed the role of these genes in response to abiotic stresses and hormone treatments. The present study is the first report on genome-wide identification, characterization and expression profiling of MYB transcription factors in foxtail millet and promisingly, the data obtained from this study would contribute to a better understanding of the complexity of MYB TFs in plants and provide a useful reference for further functional analysis of MAPK genes in foxtail millet and related grasses.

Supporting Information

Figure S1 Phylogenetic relationships between all the sequenced plant species. The total number of MYB proteins found in each genome is indicated on the right. The data was retrieved from PlantTFDB (http://planttfdb.cbi.pku.edu.cn/). The data of foxtail millet (*Setaria italica*) excludes alternate transcripts.

Figure S2 Conserved domains in SiMYB proteins, identified using CDD database.

Figure S3 The multiple sequence alignment of 'MYB-related' proteins.

Figure S4 The multiple sequence alignment of 'MYB-R2R3' proteins.

Figure S5 The multiple sequence alignment of 'MYB-R1R2R3' proteins.

Figure S6 Gene structures of SiMYB proteins. Exons and introns are represented by green boxes and black lines, respectively.

Figure S7 Predicated three dimensional structures of all the 209 SiMYB proteins.

Table S1 Details of primers used for quantitative real-time PCR.

Table S2 Characteristic features of 209 MYB Transcription factor gene family members identified in *Setaria italica*.

Table S3 Summary of functional domains present in the 209 SiMYB proteins.

Table S4 Blast2GO annotation details of SiMYB protein sequences.

Table S5 Characteristics of the promoter region of SiMYB genes.

Table S6 Details of DNA markers present in SiMYB genes.

Table S7 The Ka/Ks ratio, estimated divergence time and percentage identity for orthologous SiMYB proteins between foxtail millet and sorghum.

Table S8 The Ka/Ks ratio, estimated divergence time and percentage identity for orthologous SiMYB proteins between foxtail millet and maize.

Table S9 The Ka/Ks ratio, estimated divergence time and percentage identity for orthologous SiMYB proteins between foxtail millet and rice.

Table S10 The Ka/Ks ratio, estimated divergence time and percentage identity for orthologous SiMYB proteins between foxtail millet and *Brachypodium*.

Table S11 The Ka/Ks ratios and estimated divergence time for segmentally duplicated SiMYB genes.

Table S12 The Ka/Ks ratios and estimated divergence time for tandemly duplicated SiMYB genes.

Table S13 List of templates used in homology modelling of 209 SiMYB proteins.

Acknowledgments

MM acknowledges the award of Junior Research Fellowship from University Grants Commission, New Delhi, India. Assistance from Ms Jananee Jaishankar and Mr Subodh Verma, NIPGR is greatly appreciated.

Author Contributions

Conceived and designed the experiments: MP. Performed the experiments: MM RK CBY VSB YK. Analyzed the data: MP MM. Wrote the paper: MM MP.

References

1. Paz-Ares J, Ghosal D, Wienand U, Peterson PA, Saedler H (1987) The regulatory c1 locus of *Zea mays* encodes a protein with homology to myb proto-oncogene products and with structural similarities to transcriptional activators. EMBO J 6: 3553–3558.
2. Lipsick JS (1996) One billion years of Myb. Oncogene 13: 223–235.
3. Stracke R, Werber M, Weisshaar B (2001) The R2R3-MYB gene family in *Arabidopsis thaliana*. Curr Opin Plant Bio 14: 447–456.
4. Ogata K, Morikawa S, Nakamura H, Hojo H, Yoshimura S, et al. (1995) Comparison of the free and DNA-complexed forms of the DNA-binding domain from c-Myb. Nat Struct Biol 2: 309–320.
5. Jin H, Martin C (1999) Multifunctionality and diversity within the plant MYB-gene family. Plant Mol Biol 41: 577–585.
6. Dubos C, Stracke R, Grotewold E, Weisshaar B, Martin C, et al. (2010) MYB transcription factors in Arabidopsis. Trends Plant Sci 15: 573–581.
7. Goicoechea M, Lacombe E, Legay S, Mihaljevic S, Rech P, et al. (2005) EgMYB2, a new transcriptional activator from Eucalyptus xylem, regulates secondary cell wall formation and lignin biosynthesis. Plant J 43: 553–567.
8. Mehrtens F, Kranz H, Bednarek P, Weisshaar B (2005) The Arabidopsis transcription factor MYB12 is a flavonol-specific regulator of phenylpropanoid biosynthesis. Plant Physiol 138: 1083–1096.
9. Bedon F, Grima-Pettenati J, Mackay J (2007) Conifer R2R3-MYB transcription factors: sequence analyses and gene expression in wood-forming tissues of white spruce (*Picea glauca*). BMC Plant Biol 7: 17.
10. Stracke R, Ishihara H, Huep G, Barsch A, Mehrtens F, et al. (2007) Differential regulation of closely related R2R3-MYB transcription factors controls flavonol accumulation in different parts of the *Arabidopsis thaliana* seedling. Plant J 50: 660–677.
11. Gocal GF, Poole AT, Gubler F, Watts RJ, Blundell C, et al. (1999) Long-day up-regulation of a GAMYB gene during *Lolium temulentum* inflorescence formation. Plant Physiol 119: 1271–1278.
12. Abe H, Urao T, Ito T, Seki M, Shinozaki K, et al. (2003) Arabidopsis AtMYC2 (bHLH) and AtMYB2 (MYB) function as transcriptional activators in abscisic acid signaling. Plant Cell 15: 63–78.
13. Hartmann U, Sagasser M, Mehrtens F, Stracke R, Weisshaar B (2005) Differential combinatorial interactions of cis-acting elements recognized by R2R3-MYB, BZIP, and BHLH factors control light-responsive and tissue-specific activation of phenylpropanoid biosynthesis genes. Plant Mol Biol 57: 155–171.
14. Jung C, Seo JS, Han SW, Koo YJ, Kim CH, et al. (2008) Overexpression of AtMYB44 enhances stomatal closure to confer abiotic stress tolerance in transgenic Arabidopsis. Plant Physiol 146: 623–635.
15. Adler G, Blumwald E, Bar-Zvi D (2010) The sugar beet gene encoding the sodium/proton exchanger 1 (BvNHX1) is regulated by a MYB transcription factor. Planta 232: 187–195.
16. Higginson T, Li SF, Parish RW (2003) AtMYB103 regulates tapetum and trichome development in *Arabidopsis thaliana*. Plant J 35: 177–192.
17. Perez-Rodriguez M, Jaffe FW, Butelli E, Glover BJ, Martin C (2005) Development of three different cell types is associated with the activity of a specific MYB transcription factor in the ventral petal of *Antirrhinum majus* flowers. Development 132: 359–370.
18. Baumann K, Perez-Rodriguez M, Bradley D, Venail J, Bailey P, et al. (2007) Control of cell and petal morphogenesis by R2R3 MYB transcription factors. Development 134: 1691–1701.
19. Pattanaik S, Kong Q, Zaitlin D, Werkman JR, Xie CH, et al. (2010) Isolation and functional characterization of a floral tissue-specific R2R3 MYB regulator from tobacco. Planta 231: 1061–1076.
20. Muthamilarasan M, Theriappan P, Prasad M (2013) Recent advances in crop genomics for ensuring food security. Curr Sci 105: 155–158.

21. Dias AP, Braun EL, McMullen MD, Grotewold E (2003) Recently duplicated maize R2R3 Myb genes provide evidence for distinct mechanisms of evolutionary divergence after duplication. Plant Physiol 131: 610–620.
22. Li Q, Zhang C, Li J, Wang L, Ren Z (2012) Genome-wide identification and characterization of R2R3MYB family in *Cucumis sativus*. PLoS ONE 7: e47576.
23. Du H, Yang SS, Feng BR, Liu L, Tang YX (2012) Genome-wide analysis of the MYB transcription factor superfamily in soybean. BMC Plant Biol 12: 106.
24. Du H, Feng BR, Yang SS, Huang YB, Tang YX (2012) The R2R3-MYB transcription factor gene family in maize. PLoS ONE 7: e37463.
25. Du H, Wang YB, Xie Y, Liang Z, Jiang SJ, et al. (2013) Genome-wide identification and evolutionary and expression analyses of MYB-related genes in land plants. DNA Res 20: 437–448.
26. Doust AN, Kellogg EA, Devos KM, Bennetzen JL (2009) Foxtail millet: A sequence-driven grass model system. Plant Physiol 149: 137–141.
27. Lata C, Gupta S, Prasad M (2013) Foxtail millet: a model crop for genetic and genomic studies in bioenergy grasses. Crit Rev Biotechnol 33: 328–343.
28. Zhang G, Liu X, Quan Z, Cheng S, Xu X, et al. (2012) Genome sequence of foxtail millet (*Setaria italica*) provides insights into grass evolution and biofuel potential. Nature Biotech 30: 549–554.
29. Bennetzen JL, Schmutz J, Wang H, Percifield R, Hawkins J, et al. (2012) Reference genome sequence of the model plant *Setaria*. Nature Biotechnol 30: 555–561.
30. Finn RD, Bateman A, Clements J, Coggill P, Eberhardt RY, et al. (2014) The Pfam protein families database. Nucleic Acids Res 42: D222–D230.
31. Bonthala VS, Muthamilarasan M, Roy R, Prasad M (2014) FmTFDb: a foxtail millet transcription factors database for expediting functional genomics in millets. Mol Biol Rep DOI: 10.1007/s11033-014-3574-y.
32. Voorrips RE (2002) MapChart: software for the graphical presentation of linkage maps and QTLs. J Hered 93: 77–78.
33. Du D, Zhang Q, Cheng T, Pan H, Yang W, et al. (2012) Genome-wide identification and analysis of late embryogenesis abundant (LEA) genes in *Prunus mume*. Mol Biol Rep 40: 1937–1946.
34. Shiu S-H, Bleecker AB (2003) Expansion of the Receptor-Like Kinase/Pelle Gene Family and Receptor-Like Proteins in *Arabidopsis*. Plant Physiol 132: 530–543.
35. Puranik S, Sahu PP, Mandal SN, B VS, Parida SK, et al. (2013) Comprehensive genome-wide survey, genomic constitution and expression profiling of the NAC transcription factor family in foxtail millet (*Setaria italica* L.). PLoS ONE 8: e64594.
36. Mishra AK, Muthamilarasan M, Khan Y, Parida SK, Prasad M (2014) Genome-wide investigation and expression analyses of WD40 protein family in the model plant foxtail millet (*Setaria italica* L.). PLoS ONE 9: e86852.
37. Wang Y, Tang H, DeBarry JD, Tan X, Li J, Wang X, et al. (2012) MCScanX: a toolkit for detection and evolutionary analysis of gene synteny and collinearity. Nucleic Acids Res 40: e49.
38. Guo AY, Zhu QH, Chen X, Luo JC (2007) GSDS: a gene structure display server. Yi Chuan 29: 1023–1026.
39. Tamura K, Stecher G, Peterson D, Filipski A, Kumar S (2013) MEGA6: Molecular Evolutionary Genetics Analysis version 6.0. Mol Biol Evol 30: 2725–2729.
40. Götz S, García-Gómez JM, Terol J, Williams TD, Nagaraj SH, et al. (2008) High-throughput functional annotation and data mining with the Blast2GO suite. Nucleic Acids Res 36: 3420–3435.
41. Higo K, Ugawa Y, Iwamoto M, Korenaga T (1999) Plant cis-acting regulatory DNA elements (PLACE) database: 1999. Nucleic Acids Res 27: 297–300.
42. Yi F, Xie S, Liu Y, Qi X, Yu J (2013) Genome-wide characterization of microRNA in foxtail millet (*Setaria italica*). BMC Plant Biol 13: 212.
43. Khan Y, Yadav A, Suresh BV, Muthamilarasan M, Yadav CB, et al. (2014) Comprehensive genome-wide identification and expression profiling of foxtail millet [*Setaria italica* (L.)] miRNAs in response to abiotic stress and development of miRNA database. Plant Cell Tiss Organ Cult 118: 279–292.
44. Dai X, Zhao PX (2013) psRNATarget: a plant small RNA target analysis server. Nucleic Acids Res 39: W155–W159.
45. Cochrane G, Alako B, Amid C, Bower L, Cerdeño-Tárraga A, et al. (2013) Facing growth in the European Nucleotide Archive. Nucleic Acids Res 41: D30–D35.
46. Patel RK, Jain M (2012) NGS QC Toolkit: A toolkit for quality control of next generation sequencing data. PLoS ONE 7: e30619.
47. Langmead B, Salzberg SL (2012) Fast gapped-read alignment with Bowtie 2. Nat Methods 9: 357–359.
48. Saeed AI, Bhagabati NK, Braisted JC, Liang W, Sharov V, et al. (2006) TM4 microarray software suite. Methods Enzymol 411: 134–193.
49. Saeed AI, Sharov V, White J, Li J, Liang W, et al. (2003) TM4: a free, open-source system for microarray data management and analysis. Biotechniques 34: 374–378.
50. Suresh BV, Muthamilarasan M, Misra G, Prasad M (2013) FmMDb: a versatile database of foxtail millet markers for millets and bioenergy grasses research. PLoS ONE 8: e71418.
51. Gupta S, Kumari K, Sahu PP, Vidapu S, Prasad M (2013) Sequence-based novel genomic microsatellite markers for robust genotyping purposes in foxtail millet [*Setaria italica* (L.) P. Beauv]. Plant Cell Rep 31: 323–337.
52. Pandey G, Misra G, Kumari K, Gupta S, Parida SK, et al. (2013) Genome-wide development and use of microsatellite markers for large-scale genotyping applications in foxtail millet [*Setaria italica* (L.)]. DNA Res 20: 197–207.
53. Kumari K, Muthamilarasan M, Misra G, Gupta S, Subramanian A, et al. (2013) Development of eSSR-markers in *Setaria italica* and their applicability in studying genetic diversity, cross-transferability and comparative mapping in millet and non-millet species. PLoS ONE 8: e67742.
54. Gupta S, Kumari K, Das J, Lata C, Puranik S, et al. (2011) Development and utilization of novel intron length polymorphic markers in foxtail millet (*Setaria italica* (L.) P. Beauv.) Genome 54: 586–602.
55. Muthamilarasan M, Venkata Suresh B, Pandey G, Kumari K, Parida SK, et al. (2014b) Development of 5123 intron-length polymorphic markers for large-scale genotyping applications in foxtail millet. DNA Res 21: 41–52.
56. Suyama M, Torrents D, Bork P (2006) PAL2NAL: robust conversion of protein sequence alignments into the corresponding codon alignments. Nucleic Acids Res 34: W609–W612.
57. Lynch M, Conery JS (2000) The evolutionary fate and consequences of duplicate genes. Science 290: 1151–1155.
58. Yang Z, Gu S, Wang X, Li W, Tang Z, et al. (2008) Molecular evolution of the cpp-like gene family in plants: insights from comparative genomics of *Arabidopsis* and rice. J Mol Evol 67: 266–277.
59. Muthamilarasan M, Bonthala VS, Mishra AK, Khandelwal R, Khan Y, et al. (2014) C_2H_2 type of zinc finger transcription factors in foxtail millet define response to abiotic stresses. Funct Integr Genomics 14: 531–543.
60. Lata C, Sahu PP, Prasad M (2010) Comparative transcriptome analysis of differentially expressed genes in foxtail millet (*Setaria italica* L.) during dehydration stress. Biochem Biophys Res Commun 393: 720–727.
61. Puranik S, Jha S, Srivastava PS, Sreenivasulu N, Prasad M (2011) Comparative transcriptome analysis of contrasting foxtail millet cultivars in response to short-term salinity stress. J Plant Physiol 168: 280–287.
62. Lata C, Bhutty S, Bahadur RP, Majee M, Prasad M (2011) Association of an SNP in a novel DREB2-like gene SiDREB2 with stress tolerance in foxtail millet (*Setaria italica* (L.)). J Exp Bot 62: 3387–3401.
63. Le DT, Nishiyama R, Watanabe Y, Mochida K, Yamaguchi-Shinozaki K, et al. (2011) Genome-wide survey and expression analysis of the plant-specific NAC transcription factor family in soybean during development and dehydration stress. DNA Res 18: 263–276.
64. Longeman J, Schell J, Willmitzer L (1987) Improved method for the isolation of RNA from plant tissues. Anal Biochem 163: 16–20.
65. Kumar K, Muthamilarasan M, Prasad M (2013) Reference genes for quantitative Real-time PCR analysis in the model plant foxtail millet (*Setaria italica* L.) subjected to abiotic stress conditions. Plant Cell Tiss Organ Cult 115: 13–22.
66. Kelley LA, Sternberg MJE (2009) Protein structure prediction on the Web: a case study using the Phyre server. Nature Protocols 4: 363–371.
67. Katiyar A, Smita S, Lenka SK, Rajwanshi R, Chinnusamy V, et al. (2012) Genome-wide classification and expression analysis of MYB transcription factor families in rice and Arabidopsis. BMC Genomics 13: 544.
68. Matus JT, Aquea F, Arce-Johnson P (2008) Analysis of the grape MYB R2R3 subfamily reveals expanded wine quality-related clades and conserved gene structure organization across Vitis and Arabidopsis genomes. BMC Plant Biol 8: 83.
69. Wilkins O, Nahal H, Foong J, Provart NJ, Campbell MM (2009) Expansion and diversification of the Populus R2R3–MYB family of transcription factors. Plant Physiol 149: 981–993.
70. Cao ZH, Zhang SZ, Wang RK, Zhang RF, Hao YJ (2013) Genome wide analysis of the apple MYB transcription factor family allows the identification of MdoMYB121 gene confering abiotic stress tolerance in plants. PLoS ONE 8: e69955.
71. Jiang C, Gu J, Chopra S, Gu X, Peterson T (2004) Ordered origin of the typical two– and three-repeat Myb genes. Gene 326: 13–22.
72. Rosinski JA, Atchley WR (1998) Molecular evolution of the Myb family of transcription factors: evidence for polyphyletic origin. J Mol Evol 46: 74–83.
73. Da G, Lenkart J, Zhao K, Shiekhattar R, Cairns BR, et al. (2006) Structure and function of the SWIRM domain, a conserved protein module found in chromatin regulatory complexes. Proc Natl Acad Sci U S A 103: 2057–2062.
74. Pao GM, Saier MH (1995) Response regulators of bacterial signal transduction systems: selective domain shuffling during evolution. J Mol Evol 40: 136–154.
75. Chopra S, Athma P, Peterson T (1996) Alleles of the maize P gene with distinct tissue specificities encode Myb-homologous proteins with C-terminal replacements. Plant Cell 8: 1149–1158.
76. Grotewold E, Drummond BJ, Bowen B, Peterson T (1994) The myb-homologous P gene controls phlobaphene pigmentation in maize floral organs by directly activating a flavonoid biosynthetic gene subset. Cell 76: 543–553.
77. Ponting CP, Blake DJ, Davies KE, Kendrick-Jones J, Winder SJ (1996) ZZ and TAZ: new putative zinc fingers in dystrophin and other proteins. Trends Biochem Sci 21: 11–13.
78. Roberts RG (2001) Dystrophins and dystrobrevins. Genome Biol 2: REVIEWS3006.
79. Dhalluin C, Carlson JE, Zeng L, He C, Aggarwal AK, et al. (1999) Structure and ligand of a histone acetyltransferase bromodomain. Nature 399: 491–496.
80. Ramakrishnan V, Finch JT, Graziano V, Lee PL, Sweet RM (1993) Crystal structure of globular domain of histone H5 and its implications for nucleosome binding. Nature 362: 219–223.

81. Sakai H, Aoyama T, Oka A (2000) Arabidopsis ARR1 and ARR2 response regulators operate as transcriptional activators. Plant J 24: 703–711.
82. Wenkel S, Turck F, Singer K, Gissot L, Le Gourrierec J, et al. (2006) CONSTANS and the CCAAT box binding complex share a functionally important domain and interact to regulate flowering of Arabidopsis. Plant Cell 18: 2971–2984.
83. Gowik U, Burscheidt J, Akyildiz M, Schlue U, Koczor M, et al. (2004) cis-Regulatory elements for mesophyll-specific gene expression in the C4 plant *Flaveria trinervia*, the promoter of the C4 phosphoenolpyruvate carboxylase gene. Plant Cell 16: 1077–1090.
84. Yanagisawa S, Schmidt RJ (1999) Diversity and similarity among recognition sequences of Dof transcription factors. Plant J 17: 209–214.
85. Reyes JC, Muro-Pastor MI, Florencio FJ (2004) The GATA family of transcription factors in Arabidopsis and rice. Plant Physiol 134: 1718–1732.
86. Villain P, Mache R, Zhou DX (1996) The mechanism of GT element-mediated cell type-specific transcriptional control. J Biol Chem 271: 32593–32598.
87. Hobo T, Suwabe K, Aya K, Suzuki G, Yano K, et al. (2008) Various spatiotemporal expression profiles of anther-expressed genes in rice. Plant Cell Physiol 49: 1417–1428.
88. Chinnusamy V, Ohta M, Kanrar S, Lee BH, Hong X, et al. (2003) ICE1: a regulator of cold-induced transcriptome and freezing tolerance in Arabidopsis. Genes Dev 17: 1043–1054.
89. Filichkin SA, Leonard JM, Monteros A, Liu PP, Nonogaki H (2004) A novel endo-beta-mannanase gene in tomato LeMAN5 is associated with anther and pollen development. Plant Physiol 134: 1080–1087.
90. Nishiuchi T, Shinshi H, Suzuki K (2004) Rapid and transient activation of transcription of the ERF3 gene by wounding in tobacco leaves: possible involvement of NtWRKYs and autorepression. J Biol Chem 279: 55355–55361.
91. Zhang ZL, Xie Z, Zou X, Casaretto J, Ho TH, et al. (2004) A rice WRKY gene encodes a transcriptional repressor of the gibberellin signaling pathway in aleurone cells. Plant Physiol 134: 1500–1513.
92. Yadav CB, Muthamilarasan M, Pandey G, Prasad M (2014) Identification, characterization and expression profiling of Dicer-like, Argonaute and RNA-dependent RNA polymerase gene families in foxtail millet. Plant Mol Biol Rep DOI: 10.1007/s11105-014-0736-y.

15

Phytoavailability of Cadmium (Cd) to Pak Choi (*Brassica chinensis* L.) Grown in Chinese Soils: A Model to Evaluate the Impact of Soil Cd Pollution on Potential Dietary Toxicity

Muhammad Tariq Rafiq[1,2]☯, Rukhsanda Aziz[1]☯, Xiaoe Yang[1], Wendan Xiao[1], Peter J. Stoffella[3], Aamir Saghir[4], Muhammad Azam[5], Tingqiang Li[1]*

1 Ministry of Education Key Laboratory of Environmental Remediation and Ecological Health, College of Environmental and Resource Sciences, Zhejiang University, Hangzhou, China, **2** Department of Environmental Science, International Islamic University, Islamabad, Pakistan, **3** Indian River Research and Education Center, Institute of Food and Agricultural Sciences, University of Florida, Fort Pierce, Florida, United States of America, **4** Institute of Statistics, Zhejiang University, Hangzhou, China, **5** College of Agriculture and Biotechnology, Zhejiang University, Hangzhou, China

Abstract

Food chain contamination by soil cadmium (Cd) through vegetable consumption poses a threat to human health. Therefore, an understanding is needed on the relationship between the phytoavailability of Cd in soils and its uptake in edible tissues of vegetables. The purpose of this study was to establish soil Cd thresholds of representative Chinese soils based on dietary toxicity to humans and develop a model to evaluate the phytoavailability of Cd to Pak choi (*Brassica chinensis* L.) based on soil properties. Mehlich-3 extractable Cd thresholds were more suitable for Stagnic Anthrosols, Calcareous, Ustic Cambosols, Typic Haplustalfs, Udic Ferrisols and Periudic Argosols with values of 0.30, 0.25, 0.18, 0.16, 0.15 and 0.03 mg kg^{-1}, respectively, while total Cd is adequate threshold for Mollisols with a value of 0.86 mg kg^{-1}. A stepwise regression model indicated that Cd phytoavailability to Pak choi was significantly influenced by soil pH, organic matter, total Zinc and Cd concentrations in soil. Therefore, since Cd accumulation in Pak choi varied with soil characteristics, they should be considered while assessing the environmental quality of soils to ensure the hygienically safe food production.

Editor: Wenju Liang, Chinese Academy of Sciences, China

Funding: This study was supported by the Ministry of Environmental Protection of China (grant no. 2011467057), the ministry of Science and Technology of China (grant no. 2012AA100605), and by the Fundamental Research Funds for the Central Universities of China. The funders had no role in study design, data collection and analysis, decision to publish, or preparation of the manuscript.

Competing Interests: The authors have declared that no competing interests exist.

* Email: litq@zju.edu.cn

☯ These authors contributed equally to this work.

Introduction

Cadmium (Cd) is an important environmental pollutant toxic to animals and human beings. It is one of the most mobile elements, among all the toxic heavy metals [1]. Cadmium is not required for plants growth or reproduction, however its bioaccumulation and subsequent accrual in the food chain surpasses all other trace elements due to its high mobility in soil [2]. It is the most toxic element in the environment and even at low concentrations is very toxic to living cells and considered as carcinogenic [3]. In humans, Cd exposure can result in multiple adverse effects, such as testicular damage, renal and hepatic dysfunction, etc. [3]. Moreover, Cd is implicated in the development of cancer, phytotoxic at higher levels of concentrations [4] and classified as a type I carcinogen by the International Agency for Cancer Research [5]. Significant quantities of Cd can be transferred from contaminated soil to plants [6]. Therefore, crops produced from Cd contaminated soils may be unsuitable or even detrimental for animal and human consumption [7].

Vegetables are an important component of human diet since they contain proteins, carbohydrates as well as minerals and vitamins [8]. The proportion of vegetables consumed in the total diet has been increased with the improvement of living standards. However, vegetables are also one of the most important pathways through which heavy metals enter the food chain and affect human health. Leafy vegetables can accumulate higher concentrations of Cd than other crops [9,10]. Leafy vegetables are known to accumulate higher concentrations of Cd in the edible parts even when grown in soils containing low concentrations of Cd [11]. Pak choi (*Brassica chinensis* L.), also known as Chinese cabbage, is a popular leafy vegetable, grown and consumed worldwide. Therefore, it is imperative to control Cd concentrations in Pak choi, especially in its edible parts to ensure food safety. To limit the transfer of soil Cd into the edible parts of Pak choi, an understanding of its accumulation characteristics is required. Currently, there is an elevated concern over Cd accumulation in food and its potential risks to human health [12]. Cadmium accumulation and distribution varies among vegetable cultivars and tissues [13]. However, the accumulation and distribution of Cd in vegetables grown in a diversity of soil types were rarely studied [14].

About one fifth of agricultural land is contaminated by Cd, lead (Pb) and arsenic (As) in China [15]. Moreover, it was reported that about 20% of farm lands in China are contaminated with heavy metals and Cd contamination accounts for more than 1.3×10^5 ha of the total affected area [16,17]. Cadmium uptake by rice (*Oryza sativa* L.) and vegetables from soil is the initial source of exposure for human beings [18,19]. Therefore, there are environmental concerns of soils, food safety and human health for the present and future agricultural and environmental sustainability of world vegetable supplies. As, only a small fraction of total trace metals in soil is available for plant absorption, it is widely accepted that the total metal content in soils is neither a viable indicator of phytoavailability nor an adequate tool to assess the potential risk of dietary toxicity [12]. Tracy and Sheila [20] reported that extractable Cd content in soil may be an improved indicator of bioavailability and toxicity than the total contents and toxicity and availability of metals differed among soils types. Metal uptake and translocation studies were conducted for different crops under varying soil conditions, to further understanding uptake and the transport mechanisms [21,22].

To ensure the food safety and environmental quality of soils, guidelines for permissible concentrations of Cd in agricultural soils need to be established. Due to limited number of studies, the soil environmental quality guidelines for heavy metals in farmland soils developed and applied in the world are still based on total metal contents of soil. Minimal attention has been focused on metal accumulation differences among the edible parts of crops, and the relationship between total concentration and phytoavailability of heavy metals in different soil types [4]. Developing the linkage between the bioavailability of Cd in soil and its transfer into the edible plant parts is a key to improving existing soil environmental quality standards. Information is vital on the degree of translocation of heavy metals from soils to plants, which are used as food crops, and absorption of metals in food plants to concentration that does not cause phytotoxicity symptoms [23]. This study was conducted in seven Chinese soil types to establish direct relationship of Cd level in such contaminated soils and Cd uptake in Pak choi. The main objectives were to establish soil Cd thresholds for representative Chinese soils based on human dietary toxicity and to determine the relationships between several soil properties and Cd accumulation in Pak choi. This information will be useful in establishing soil protection guidelines to produce hygienically safe vegetables.

Materials and Methods

Ethics Statement

The soils used in this study were agricultural soils. No specific permissions were required for the described locations. We confirm that the field studies did not involve endangered or protected species.

Soil Collection and Analysis

Seven Chinese soils were selected for this study. Udic Ferrisols, Mollisols, Periudic Argosols, Typic Haplustalfs, Ustic Cambosols, Calcaric Regosols and Stagnic Anthrosols were collected from Chinese cities of Guilin (104°40′–119°45′E, 24°18′– 25°41′N), Harbin (125°42′– 130°10′E, 44°04′–46°40′N), Huzhou (119°14′–120°29′E, 30°22′–31°11′N), Zhanjiang (110°08′–110°77′E, 20°33′–21°62′N), Qufu (116°51′–117°13′E, 35°29′–35°49′N), Ya'an (102°37′–103°12′E, 29°23′–30°37′N) and Jiaxing (120°7′–121°02′E, 30°5′–30°77′N), respectively. Soils samples were taken at a depth of up to 20 cm from the upper horizon. Each sample was air-dried, ground, and screened through two mm sieve before laboratory analysis. Soil pH, cation exchange capacity, organic matter contents, and particle size density were measured by using previously described methods [24–27]. Physicochemical properties of these soils are reported (Table 1).

Cadmium Spiking and Aging

Soil samples of Mollisols, Periudic Argosols, Stagnic Anthrosols and Ustic Cambosols were spiked with Cd as $Cd(NO_3)_2$ in an aqueous solution at loading rates of 1.0, 2.0, 4.0, 6.0 and 8.0 mg Cd kg^{-1} soil along with an untreated control (Ck), the background values of Cd concentration was below 0.50 mg kg^{-1} in these soil. However, the Udic Ferrisols, Typic Haplustalfs and Calcaric Regosols soil samples, with the background values of Cd concentration above 0.50 mg kg^{-1}, were spiked with Cd to establish the contamination levels of 2.0, 4.0, 6.0 and 8.0 mg Cd kg^{-1} soil along with the untreated control (Ck). Soil moisture was maintained up to 70% of its water-holding capacity by using distilled water. All the spiked soils were aged for one year subsequent to greenhouse experimentation. After one year aging period, the concentrations of total Cd, and Mehlich-3 extractable Cd were determined in each of the spiked soils.

Containerized Experiment

A containerized experiment was performed in greenhouse by growing Pak choi (*Brassica chinensis* L.) during March – April, 2012 at Zhejiang University, Hangzhou, China. Seed of Pak choi was obtained from the Zhejiang Seed Co. Hangzhou, China. Seeds were washed with distilled water and air-dried prior to sowing. Seeds were germinated in dark at 25°C and transplanted into quartz sand bed to establish seedlings. Four healthy, uniform and 21-day-old seedlings were transplanted into plastic containers with a diameter of 18 cm and height of 17 cm. Each container had 3 kg of soil. Fertilizers were applied at the rates of 0.4 g of N as $CO(NH_2)_2$ and 0.2 g P as KH_2PO_4 per kg of soil. The experiment was carried out in a completely randomized design (CRD). Treatments were established in triplicate, and the containers were randomly arranged in a greenhouse bench under controlled conditions of 16 h of light at 30°C and 8 h of dark at 22°C. Plants were monitored daily and watered as necessary.

Plant Sample Collection

Pak choi was harvested after 30 days from transplanting. The plants of Pak choi were removed from each container and separated into root and shoots (including stems and leaves). Roots and shoots of Pak choi were first washed with tap water and then with ultrapure distilled water, to remove all visible soil particles. Clean plant samples were first blotted dry, and then dried at 70°C for 72 h in an oven. Dry shoot weight of samples was recorded. Dry plant samples were ground to pass through a 60 mm sieve using an agate mill prior to Cd concentration analysis.

Total Cd of Soil and Plant

For determination of total Cd concentration in soil, 0.20 g of soil samples was digested with HNO_3–HF –$HClO_4$ (5:1:1) [4]. For plant samples, 0.20 g of shoots for each treatment was digested with HNO_3–H_2O_2 (5:1). After cooling the digest was transferred to a volumetric flask, diluted with distilled water to 50 mL [28]. The concentrations of Cd in the filtrate were determined using inductively coupled plasma–mass spectrometry (ICP-MS, Agilent, 7500a, CA, USA). The ICP-MS was operated at the following

Table 1. Basic Chemical and Physical Characteristics of Seven Chinese soils.

Soil Types	Mollisols	Ustic Cambosols	Stagnic Anthrosols	Periudic Argosols	Typic Haplustalfs	Udic Ferrisols	Calcaric Regosols
pH	7.23±0.08	7.80±0.02	6.49±0.03	4.85±0.06	5.16±0.05	4.43±0.07	8.02±0.04
OM (g kg^{-1})	32.2±0.32	7.54±0.20	21.4±0.34	11.6±0.17	6.37±0.56	19.1±0.15	21.8±0.14
CEC (cmol kg^{-1})	34.0±2.51	15.8±1.62	20.2±1.41	12.6±1.52	8.33±2.14	17.3±1.96	25.5±1.46
Total Cd (mg kg^{-1})	0.51±0.02	0.59±0.06	0.79±0.04	0.47±0.02	0.92±0.01	1.06±0.09	0.96±0.06
Total Zn (mg kg^{-1})	31.18±1.47	26.93±0.43	41.36±1.71	15.17±0.88	25.3±1.44	24.6±0.23	28.59±1.38
Sand (%)	20.6±1.54	21.6±1.29	11.4±0.26	24.8±0.65	37.4±0.96	32.75±1.65	31.6±0.57
Silt (%)	60.2±2.21	65.4±2.62	73.0±2.41	58.2±1.04	40.8±1.66	39.8±1.26	44.0±1.26
Clay (%)	19.2±1.24	13.0±1.05	15.6±1.17	17.0±0.34	21.8±0.82	49.6±1.19	24.4±1.32

conditions: the radio frequency power at the torch 1.2 kW, the plasma gas flow 15 L min^{-1}, the auxiliary gas flow 0.89 L min^{-1}, and the carrier gas flow 0.95 L min^{-1} [28]. The same procedure without samples was used as control and three replications were conducted for each sample.

Mehlich-3 Extractable Cd in Soils

Mehlich-3 extractable Cd in soils was determined following the extraction method described by Mehlich [29]. Briefly, 5 g (0.2 mm sieved) of dry soil was shaken with 50 mL of Mehlich-3 solution (0.2 mol L^{-1} CH_3COOH, 0.25 mol L^{-1} NH_4NO_3, 0.015 mol L^{-1} NH_4F, 0.013 mol L^{-1} HNO_3, 0.001 mol L^{-1} EDTA) for 5 min (200 rpm) at 25°C. The suspension was centrifuged at 5000 rpm for 10 min and filtered through 0.45 μm filter paper. The same procedure without samples was used as control and three replications were conducted for each soil sample. The Cd concentration in the filtrate was analyzed by inductively coupled plasma–mass spectrometry (ICP-MS, Agilent 7500a, CA, USA).

Quality Control for Cd Analysis

Quality assurance and quality control (QA/QC) for Cd in soil and Pak choi were conducted by determining Cd contents in the certified reference materials (soil GSBZ 50013-88 and plant GBW-07402) respectively, approved by General Administration of Quality Supervision, Inspection and Quarantine of the People's Republic of China (AQSIQ) and National Center for Reference Materials. The analytical results showed a recovery rate of 97.3% and 102.1% respectively.

Derivation of Soil Cd Thresholds for Potential Dietary Toxicity in Pak choi

For ensuring the environmental and food safety for human beings, an effort has been made to develop guidelines for acceptable concentrations of potentially harmful Cd in seven agricultural soils types of China. In this context, the amounts of Cd in Pak choi above than threshold level of food safety are adversely affecting humans are critical. Since, Cd bioavailability differed among soil types, the focus was on the development of soil Cd thresholds for representative Chinese soils based on food safety, Provisional Tolerable Weekly Intake (PTWI) of Cd recommended by FAO/WHO Joint Expert Committee on Food Additives, is 7 μg kg^{-1} of body weight [30]. Estimated daily intake of metal (EDIM) was determined by the following equation.

$$\mathrm{EDIM} = \frac{C_{cadmium} \times C_{factor} \times D_{daily\ intake}}{B_{average\ weight}}$$

Where, $C_{cadmium}$, C_{factor}, $D_{food\ intake}$ and $B_{average\ weight}$ represent average Cd concentration in Pak choi (mg kg^{-1}), conversion factor, daily consumption of Pak choi (g) and average body weight (kg) of the adult consumers, respectively. Average daily consumption of Pak choi for adults was considered to be 0.345 kg $person^{-1}$ d^{-1} [31] and a conversion factor 0.085 was used to convert fresh Pak choi weight to dry weight [32]. Average body weight of adult was considered to be 60 kg as motioned in previous reports [30]. According to the above equation of EDIM, the provisional tolerable daily tolerable intake of Cd for Pak choi was 2.04 mg kg^{-1} on a dry weight basis. Soil Cd threshold levels for potential dietary toxicity from Pak choi were calculated according to the tolerable daily dietary intake level of Cd (2.04 mg kg^{-1}) and the regression equations.

Statistical Analysis

Stepwise multiple regression analysis, single linear regression and one-way analysis of variance (ANOVA) were performed using the statistical software package SPSS (version 18.0). All values reported in this work are means of three independent replications. Treatment means were separated by least significant difference (LSD) test, at 5% level.

Results

Characteristics of Soils

Soils evaluated were representative of most of Chinese soil types, pH range of soils were strongly acidic to mild alkaline. Chemical and physical characteristics varied among the seven soils. Highest total Cd and Zn concentrations (background value) were observed in Udic Ferrisols and Stagnic Anthrosols respectively. Mollisols contained the highest amount of organic matter and exhibited an elevated cation exchange capacity as well (Table 1).

Mehlich-3 Extractable Cd in Soils after Aging of 1 Year

Mehlich-3 extractable Cd content increased significantly with increasing Cd spiking levels in all the seven soils. Mehlich-3 extractable Cd ranged from 0.16–3.95 mg kg^{-1} in these soils under different Cd levels (Table 2). The Cd contents varied significantly among these soils, decreasing in order: Periudic Argosols> Typic Haplustalfs> Udic Ferrisols> Stagnic Anthrosols> Mollisols> Ustic Cambosols> Calcaric Regosols. Mehlich-3 extractable Cd concentration was greater at higher rates of Cd spiking in each soil. These results indicated that minimum and maximum extractability of Cd was found in Calcaric Regosols and Periudic Argosols, respectively under the highest (8 mg kg^{-1}) level of Cd spiked. Mehlich-3 extractable concentrations were dependent on total Cd in each soil, however the extractability was significantly higher in low pH soils as compared to the medium and high pH soils (Table 2).

Biomass Yield of Pak choi

Generally, Pak choi had tolerance to Cd toxicity in Mollisols, Stagnic Anthrosols and Calcaric Regosols soils, indicating low phytoavailability of Cd in these soils. Shoot biomass of Pak choi under different Cd treatments of these soils did not decrease significantly as compared with their respective controls. However, the shoot biomass of Pak choi grown in Ustic Cambosols, Udic Ferrisols, Periudic Argosols and Typic Haplustalfs decreased significantly as compared with the control indicating higher phytoavailability of Cd in these soils (Table 3). The stimulating effect of Cd on shoot biomass of Pak choi occurred at 1 mg kg^{-1} and 2 mg kg^{-1} in Ustic Cambosols and Mollisols respectively, whereas in Stagnic Anthrosols, it occurred at 4 mg kg^{-1}. The dry weight of Pak choi shoots at 8 mg kg^{-1} Cd generally decreased in order of: Calcaric Regosols> Mollisols, Stagnic Anthrosols> Ustic Cambosols> Udic Ferrisols> Periudic Argosols> Typic Haplustalfs (Table 3).

Accumulation and Distribution of Cadmium in Pak choi

Cadmium concentration in the shoots and roots of Pak choi varied significantly among soils at different Cd levels and soil types. Roots exhibited the higher Cd contents as compared with Pak choi shoots. The content of Cd enhanced with increasing Cd loading rate in the soils. Cd concentration was high in the roots (2.42 to 169. 95 mg kg^{-1} DW), while low in the shoot (0.48 to 89.21 mg kg^{-1} DW) (Table 4). Cd uptake in Pak choi tissues was affected by soil type, primarily due to the variation in Cd

Table 2. Mehlich-3 Extractable Cd Contents (mg kg^{-1}) in Seven Chinese Soils at the onset of Containerized Experiment after Aging of 1 year.

Cd conc. (mg kg^{-1})	Mollisols	Ustic Cambosols	Stagnic Anthrosols	Periudic Argosols	Typic Haplustalfs	Udic Ferrisols	Calcaric Regosols
Ck	0.19±0.05c	0.17±0.07c	0.22±0.08d	0.16±0.06e	0.31±0.12d	0.42±0.09d	0.29±0.07c
1	0.21±0.04c	0.19±0.06c	0.30± 0.04d	0.41±0.13e	-	-	-
2	0.62±0.04bc	0.48±0.10c	0.56±0.27d	0.99±0.29d	0.69±0.17d	0.65±0.11d	0.35±0.06c
4	1.19±0.08b	1.05±0.90bc	1.17±0.24c	1.79±0.54c	1.52±0.26c	1.46±0.60c	0.71±0.10bc
6	2.21±0.91a	2.13±0.90b	2.01±0.47b	2.98±0.46b	2.44±0.64b	2.39±0.90b	1.38±0.29ab
8	2.89±1.04a	3.28±0.84a	3.29±0.73a	3.95±0.41a	3.58±0.53a	3.47 ±0.70a	2.01±0.94a

Mean values followed by different letters within the same column are significantly different at P <0.05.

Table 3. Dry Biomass (g plant^{-1}) of Pak choi Shoots Grown on Seven Chinese Soils with Different Loading Rates of Cd.

Cd conc. (mg kg^{-1})	Mollisols	Ustic Cambosols	Stagnic Anthrosols	Periudic Argosols	Typic Haplustalfs	Udic Ferrisols	Calcaric Regosols
Ck	1.95±0.18a	1.54±0.13a	1.13±0.31a	0.84±0.29a	0.73±0.20a	0.95±0.11a	2.07±0.46a
1	1.91±0.37a	1.66±0.59a	1.10±0.30a	0.63±0.35ab	-	-	-
2	2.22±0.58a	1.02±0.49a	1.14±0.53a	0.59±0.47ab	0.54±0.41ab	0.79±0.23ab	1.98±0.11a
4	1.89±0.49a	1.47±0.08a	1.25±0.28a	0.51±0.03ab	0.43±0.38ab	0.58±0.06abc	1.95±0.06a
6	1.86±0.48a	1.35±0.13ab	1.09±0.09a	0.36±0.02ab	0.38±0.11ab	0.50±0.24bc	1.90±0.15a
8	1.54±0.28a	0.95±0.15b	1.11±0.51a	0.17±0.06b	0.10±0.01b	0.39±0.29c	1.75±0.17a

Mean values followed by different letters within the same column are significantly different at $P < 0.05$.

bioavailability. The lowest and highest Cd concentrations in the Pak choi tissues were at the highest (8 mg kg^{-1}) level of Cd in Calcaric Regosols and Periudic Argosols, respectively. Cadmium concentrations in Pak choi followed an order of: Periudic Argosols> Typic Haplustalfs> Udic Ferrisols> Ustic Cambosols> Mollisols> Stagnic Anthrosols> Calcaric Regosols at 8 mg kg^{-1} soil Cd level (Table 4).

Relationship between Mehlich-3 Extractable Cd in Soils and Pak choi Cd Content

Cadmium concentrations in shoots of Pak choi were significantly correlated to total Cd and Mehlich-3 extractable Cd contents in soils ($R^2 = 0.95$ to 0. 99, and 0.97 to 0.99 respectively). Cadmium concentrations in Pak choi shoots were best related to total Cd content in Mollisols ($R^2 = 0.99$). Whereas, the Cd concentrations of Pak choi shoots were best correlated to Mehlich-3 extractable Cd in Ustic Cambosols, Stagnic Anthrosols Periudic Argosols, Udic Ferrisols, Typic Haplustalfs and Calcaric Regosols with $R^2 = 0.97$, 0.99, 0.99,0.98,0.99 and 0.98, respectively (Table 5).

Soil Cd Thresholds for Potential Dietary Toxicity in Pak choi

Total soil Cd thresholds for potential dietary toxicity from the consumption of Pak choi conformed to an order of: Calcaric Regosols> Stagnic Anthrosols> Ustic Cambosols> Mollisols> Udic Ferrisols> Typic Haplustalfs> Periudic Argosols, and were 1.25, 1.16, 1.02, 0.86, 0.72, 0.70 and 0.12 mg kg^{-1}, respectively. Mehlich-3 extractable Cd thresholds were 0.30, 0.25, 0.23, 0.18, 0.16, 0.15 and 0.03 mg kg^{-1} and decreased in the following order of: Stagnic Anthrosols> Calcareous> Mollisols> Ustic Cambosols> Typic Haplustalfs> Udic Ferrisols>Periudic Argosols, respectively (Table 6).

Discussion

Biomass Yield of Pak choi

Dry weight of Pak choi did not decrease significantly under different Cd levels (Ck to 8.0 mg kg^{-1}) in Mollisols, Stagnic Anthrosols and Calcaric Regosols and even increased at 1, 2 and 4.0 mg kg^{-1}of treatment levels. Similar stimulatory responses of biomass to Cd exposure have also been reported in several plant species [33,34]. The stimulatory effect of Cd on plant biomass may be explained by various mechanisms, for examples, metal ions can serve as enzyme activators in cytokinins metabolism, which stimulates the growth of plants, [35] and a low dose of metal exposure may cause changes in cytokinins and plant hormones that regulate growth and development of plants [36]. Kaminek [36] reported that cytokinins may delay senescence by maintaining chlorophyll production and photosynthetic activity in plant leaves.

Cd exposure may cause changes to various physiological and biochemical processes in plant tissues, such as, reduction in dry biomass may be due to the negative effects of Cd on the roots, and plants could not take up nutrients to continue their normal activities. It has been well reported that Cd can reduce plant growth and development by interfering in various metabolic processes, such as, inhibition of the proton pump, reduction in root elongation, and damage to photosynthetic activity [37,38]. The excess amount of Cd in soil may be responsible for causing disturbances in mineral nutrition and carbohydrate metabolism [39].

Shoot biomass of Pak choi grown in Ustic Cambosols, Udic Ferrisols, Periudic Argosols and Typic Haplustalfs decreased significantly as compared with the control. The inhibitory effect of

Table 4. Cd Concentration (mg kg^{-1} DW) in Pak choi Grown under Different Cd Levels in Seven Chinese Soils.

Cd (mg kg^{-1})	Mollisols		Ustic Cambosols		Stagnic Anthrosols		Periudic Argosols		Typic Haplustalfs		Udic Ferrisols		Calcaric Regosols	
	Root	Shoot	Root	Shoot	Root	Shoot	Root	Shoot	Root	Shoot	Root	Shoot	Root	Shoot
Ck	3.41±0.64e	1.00±0.28e	2.42±0.45f	0.48±0.22e	2.73±0.28e	0.85±0.12e	7.26±1.10f	1.84±0.56f	12.35±1.68e	4.37±1.02e	11.81±0.78e	6.77±0.92e	3.11±0.66e	1.92±0.58b
1	5.09±0.91e	2.53±0.76e	6.69±0.65e	2.06±0.98e	4.12±0.75e	1.92±0.94e	21.3±2.01e	11.00±1.57e	-	-	-	-	-	-
2	8.92±1.09d	4.31±0.54d	9.30±1.29d	4.77±0.22d	8.34±1.11d	3.94±0.83d	39.33±3.86d	28.51±2.61d	25.41±1.79d	12.09±1.34d	19.41±1.27d	10.06±1.10d	4.24±0.99d	2.70±0.99b
4	14.21±0.90c	8.34±1.11c	14.00±1.43c	9.01±1.06c	13.21±1.55c	7.06±1.06c	68.43±3.71c	39.12±3.06c	39.11±2.28c	32.51±2.51c	33.55±1.71c	20.89±1.46c	10.37±1.21c	3.46±1.10b
6	21.24±1.20b	11.75±1.17b	24.36±1.11b	17.23±1.50b	20.21±1.80b	11.63±1.54b	117.76±6.53b	68.21±4.08b	68.00±3.96b	41.11±3.01b	55.53±2.67b	31.45±2.48b	14.41±1.03b	6.57±0.93a
8	38.96±1.39a	18.11±1.12a	43.21±2.21a	22.22±1.51a	28.56±1.42a	17.63±1.37a	169.95±8.11a	89.21±5.70a	118.21±6.24a	69.21±4.11a	85.32±4.28a	53.73±2.88a	18.73±1.54a	8.13±0.87a

Mean values followed by different letters within the same column are significantly different at $P < 0.05$.

Cd on shoot growth is consistent with earlier reports of three Chinese cabbage cultivars exposed to different soil Cd levels of 1, 2.5 and 5 mg kg^{-1}. A significant decrease in the shoot biomass was observed at 2.5 and 5 mg kg^{-1} levels of Cd as compared to their respective controls [40]. Shentu *et al.* [4] found a 46% reduction in root dry weight of radish at 6.31 mg kg^{-1} Cd exposure in red yellow soil, which is in accordance with our results as we also noticed a shoot dry weight reduction of 58.9%, 79.7%, and 86.3% in Udic Ferrisols, Periudic Argosols and Typic Haplustalfs respectively at 8 mg kg^{-1} level of soil Cd as compared to their respective controls.

Accumulation and Distribution of Cadmium in Pak choi

Variations of Cd accumulation in Pak choi grown in different soils with different pH may be due to the difference in bioavailability of Cd in each soil. Liang *et al.* [13] stated that Cd content of spinach plants was highly dependent upon the soil pH being highest at pH 5.3. Lai and Chen [41] reported that Cd concentration in Pak choi shoots was up to 85 mg kg^{-1} DW with an application of soil Cd up to 20 mg kg^{-1}. Moreover, it was observed that accumulation of the Cd in rice shoot ranged from 67.9 to 241.7 μg/pot in different rice genotypes at 5 mg kg^{-1} soil Cd level [42].

Relationship between Mehlich-3 Extractable Cd in Soils and Pak choi Cd Content

Mehlich-3 extraction technique appeared efficient to assess Cd phytoavailability to Pak choi, grown in seven textured soils, as evidenced by high correlation coefficients (R^2>0.97). This is in agreement with our previous studies, [43,28] which reflected a high linear correlation (R^2>0.98) between Mehlich-3 Cr and Cr contents in Pak choi and rice grown under six different textured soils. These results are similar to those reported in which extractable soil metal was an improved indicator for Cd phytoavailability in several vegetable crops [4]. Mehlich-3 extraction method is applicable to a large range of soil types, from acidic to alkaline, and makes it ideal for application at a wide scale [44]. Generally, the extraction techniques are assumed to have a relationship between the extractable fraction of metals and the phytoavailability of the metals to plants, and these metals such as exchangeable, soluble, and loosely adsorbed metals are labile and thus readily available to plants [12,45]. The efficiency of Mehlich-3 extraction method was compared with the EPA 3050 B method (a strong acid digestion method) to assess the predictive capabilities through a lettuce (green specie) bioassay. Mehlich-3 extraction was positively correlated with the more costly EPA test, and could be developed as a less expensive and easily conduct able technique [46].

Soil Cd Thresholds for Potential Dietary Toxicity in Pak choi

Cadmium concentrations in the shoots of Pak choi were significantly correlated to total Cd and Mehlich-3 extractable Cd contents in soils, with R^2 values of 0.95 to 0.99, and 0.97 to 0.99, respectively. From this investigation, Cd contents in Pak choi shoots were correlated to total Cd content in Mollisols (R^2 values of 0.99). Cadmium concentrations of Pak choi shoots were highly correlated to Mehlich-3 extractable Cd in Ustic Cambosols, Stagnic Anthrosols, Periudic Argosols, Udic Ferrisols, Typic Haplustalfs and Calcaric Regosols with R^2 values of 0.97, 0.99, 0.99, 0.98, 0.99 and 0.98, respectively. Total Cd threshold levels for potential dietary toxicity conformed to an order of: Calcaric Regosols> Stagnic Anthrosols> Mollisols> Ustic Cambosols>

Table 5. Regression Correlation between Cd Contents in the Edible Shoots of Pak choi and Different Forms of Cd in Various Soils.

Soil type	Form of Soil Cd	Regression equation	R^2
Mollisols	Total Cd	y = 2.1706x +0.1669	0.99
	Mehlich-3 extractable Cd	y = 5.7207x +0.7037	0.98
Ustic Cambosols	Total Cd	y = 2.9358x −0.9586	0.96
	Mehlich-3 extractable Cd	y = 7.0409x +0.7648	0.97
Stagnic Anthrosols	Total Cd	y = 2.2344x −0.5651	0.98
	Mehlich-3 extractable Cd	y = 5.3848x +0.4458	0.99
Periudic Argosols	Total Cd	y = 11.061x +0.6961	0.98
	Mehlich-3 extractable Cd	y = 22.326x +1.3968	0.99
Typic Haplustalfs	Total Cd	y = 8.7318x −4.0825	0.97
	Mehlich-3 extractable Cd	y = 19.123x −0.8046	0.98
Udic Ferrisols	Total Cd	y = 6.6697x −2.7793	0.97
	Mehlich-3 extractable Cd	y = 14.873x −0.3773	0.99
Calcaric Regosols	Total Cd	y = 0.8936x +0.924	0.95
	Mehlich-3 extractable Cd	y = 3.5937x +1.1512	0.98

Typic Haplustalfs> Udic Ferrisols> Periudic Argosols and were 1.25, 1.16, 1.02, 0.86, 0.72, 0.70 and 0.12 mg kg^{-1}, respectively. Mehlich-3 extractable Cd thresholds decreased in the following order of: Stagnic Anthrosols> Calcareous> Mollisols> Ustic Cambosols> Udic Ferrisols>Typic Haplustalfs> Periudic Argosols and were 0.30, 0.25, 0.23, 0.18, 0.16, 0.15 and 0.03 mg kg^{-1}, respectively (Table 6).

Cadmium concentrations in Pak choi shoots, were highly correlated to total Cd content in Mollisols with the threshold levels of 0.86 mg kg^{-1} with a $R^2 = 0.99$. However, the Cd concentrations of Pak choi shoots were best related to Mehlich-3 extractable Cd in Stagnic Anthrosols, Calcaric Regosols, Mollisols, Ustic Cambosols, Typic Haplustalfs and Periudic Argosols with thresholds values of 0.30, 0.25, 0.23, 0.18, 0.16, 0.15 and 0.03 mg kg^{-1}, (R^2 values of 0.97, 0.99, 0.99, 0.98, 0.98 and 0.98), respectively. Based on the wide range of applicability and the simplicity of extraction method, it is proposed that Mehlich-3 extractable Cd is more suitable to be used as soil Cd thresholds for potential dietary toxicity in Pak choi. Our previous study evaluated the phytoavailability of Cd to rice, and demonstrated the suitability of Mehlich-3 extraction method in different textured soils [47]. Similar to our results, Murakami *et al.* [48] reported that Mehlich-3 extractable Cd was an improved indicator than total soil Cd and HCl-extractable Cd to predict the grain Cd content of japonica rice varieties. Our results are also in agreement with Shentu *et al.* [4] who also concluded that extractable Cd was a better soil test index for Cd phytoavailability of several vegetables and could be used as soil Cd thresholds for food safety. Among the predicted thresholds (total soil Cd) the lowest value (0.12 mg kg^{-1}) was observed for the Periudic Argosols, an acidic soil. Bioavailability and uptake of Cd are very high in this soil. The leafy vegetables like Pak choi can accumulate large quantities of Cd as compared to other crops [9,10]. Therefore the predicted threshold is even lower than background value of Cd in soil; it means that there is a risk for dietary toxicity from Pak choi grown on it, even with the background value of total Cd in soil. This kind of information has been reported in our previous study. The threshold of total soil Cd for rice was 0.21 mg kg^{-1} which was also lower than background value of total Cd in soil [47].

Cd levels (Ck, 1, 2, 4, 6, 8 mg kg^{-1}) used in this investigation represented uncontaminated, lightly contaminated, and moderately Cd polluted soils. Therefore, these levels of Cd contamination are realistic, comparable to those applied in other soil safety risk assessment studies, and thus, the results are applicable in field conditions as well.

Table 6. Soil Cd Threshold Levels for Potential Dietary Toxicity in Edible Part of Pak choi Calculated from the Permissible Limit of Cd (2.04 mg kg^{-1} DW) in Leafy Vegetables and Regression Equations.

Soil Type	Total Cd (mg kg^{-1})	Mehlich-3 extractable Cd (mg kg^{-1})
Mollisols	0.86	0.23
Ustic Cambosols	1.02	0.18
Stagnic Anthrosols	1.16	0.30
Periudic Argosols	0.12	0.03
Udic Ferrisols	0.72	0.16
Typic Haplustalfs	0.70	0.15
Calcaric Regosols	1.25	0.25

Table 7. Stepwise Regression Model for Predicting Cd Concentration (Y) in Edible Part of Pak choi based on Soil Properties.

Stepwise regression model	R^2	F value	T value and R^2 of partial regression coefficient		
				T value	R^2
$Y = -39.256 - 15.516\ pH - 0.944\ Zn_T - 0.379\ OM + 26.752\ Cd_T$	0.977	138.808*	pH	−18.682**	0.964
			Zn_T	−3.788**	0.524
			OM	−3.652*	0.376
			Cd_T	2.67*	0.354

[a] Cd_T and Zn_T refer to the total Cadmium and Zinc concentrations.
[b] Superscripts * and ** indicate significant levels of probability at 0.05 and 0.01, respectively.

Stepwise Regression Model for Predicting Cd Phytoavailability to Pak choi

Many physicochemical properties of soils can influence the heavy metal accumulation in vegetables. For example, the amount of heavy metal uptake from soils was influenced by soil pH, organic matter (OM) content, cation exchange capacity (CEC) and soil texture [49]. The combinations of basic soil properties may explain Cd uptake by plants [50]. By considering this aspect, soil pH, OM, CEC, total soil Cd, total Zn and clay contents were integrated to simulate the combined effects of soil environment on Cd phytoavailability to Pak choi. Stepwise linear regression was conducted and four independent variables pH, total Zn, OM and total Cd significantly influenced the accumulation of Cd in Pak choi plants (Table 7). Both the multiple correlation and partial regression coefficients reached the statistically significant levels at least the 0.05. For multiple linear regression analyses, R^2 values could be used to explain variation of the dependents [12]. It was found that R^2 value was above 0.97, which means that more than 97% of variation in Cd concentration in Pak choi shoots could be attributed to soil pH, total Zinc, OM and total Cd contents in soils (Table 7).

The influence of each factor on Cd concentration of Pak choi (Y) shoots could be further explained by the values of each coefficient [12]. Stepwise regression model revealed that Cd concentration in the Pak choi was enhanced by lower soil pH (negative coefficients showed negative effect and vice versa), total Zinc, OM contents and higher total soil Cd. Lower soil pH, zinc, OM and higher soil total Cd are among the factors which enhance the bioavailability Cd contents in soils. Therefore, these three variables had the contradictory effect on Cd phytoavailability to Pak choi. Wang *et al.* [12] reported that soil characteristics (e.g. pH, CEC and OM) affected the phytoavailability of different heavy metals in soils, and such influences could be considered in the assessment of phytoavailability of heavy metals. There are four parameters involved in this model and then interactions between them were obvious (e.g. Cd concentration in the extractable fraction was correlated with lower soil pH, soil zinc and OM content). Furthermore, the coefficients obtained in the present model can regulate these cross effects and result in an improved model fitting. For example, there was a negative correlation between the soil pH, Zinc and OM, these factors had an inverse effect on Cd phytoavailability and soil Cd was the leading factor influencing Cd phytoavailability to Pak choi (coefficient of soil Cd was greater than those of pH, Zinc and OM). Our results about soil Cd and pH are in accordance with our recent study which developed an empirical model to correlate the Cd phytoavailability to rice with several soil properties. Soil pH and bioavailable soil Cd were major influencing factors which (pH negatively and soil Cd positively) correlate with the Cd phytoavailability, however total Zn and OM were not included in our previously developed model [47]. Eriksson and Soderstrom [51] reported that the Cd concentration of wheat grain grown on non-calcareous soils of Sweden was positively correlated to soil total Cd and negatively to extractable Zn. A study was conducted on Cd contaminated soils in Taiwan, whereas regression equation was developed to predict Cd concentrations in rice roots by available fractions of Cd and Zn in soil [52]. The negative coefficient of Zn indicated that soil Zn suppressed the uptake of Cd by rice roots in all varieties as Zn has an antagonistic effect on Cd uptake by root [53]. Oliver *et al.* [54] also observed a significant decrease of Cd up to 50% in wheat grain when 2.5–5.0 kg Zn ha^{-1} was applied to Cd contaminated Australian soils.

Organic matter content was negatively correlated with the accumulation of Cd in Pak choi shoots (Table 7). Organic matter plays an important role in determining the bioavailability and mobility of heavy metals in soils. Organic matter is involved in supplying organic chemicals to the soil solution, which may act as chelates and increase metal bioavailability to plants [55]. However, OM could reduce the bioavailability of heavy metals in soils by adsorption or forming stable complexes with humic substances [56]. Halim *et al.* [57] reported that addition of humic acid demonstrated a decrease in extractable heavy metal fraction in metal contaminated soils. This could partially explain the negative correlation of organic matter contents and Cd uptake observed in our present study.

Conclusions

The present study concludes that Cd concentration in Pak choi tissues was dependent on soil type. To establish the soil Cd thresholds of potential dietary toxicity from Pak choi, both Cd bioavailability in garden soils and Pak choi tissues should be taken into consideration. The selection of proper soil types for vegetable production can help us to avoid the toxicity of Cd in our daily diet. Stepwise regression model demonstrated that soil pH, organic matter, total Cd and Zinc contents may be the major factors having influence on the phytoavailability of Cd in different textured soils.

Acknowledgments

M.T. Rafiq acknowledges International Islamic University Islamabad Pakistan for providing a PhD scholarship under HEC-IIUI faculty development program.

Author Contributions

Conceived and designed the experiments: MTR XY TL. Performed the experiments: MTR RA WX. Analyzed the data: RA AS MA. Contributed reagents/materials/analysis tools: XY. Wrote the paper: MTR RA PJS.

References

1. Liu JG, Qian M, Cai GL, Yang JC, Zhu QS (2007) Uptake and translocation of Cd in different rice cultivars and the relation with Cd accumulation in rice grain. J Hazard Mater 143: 443–447.
2. Mahler RJ, Bingham FT, Page AL (1978) Cadmium-enriched sewage sludge application to acid and calcareous soils: Effect on yield and cadmium uptake by lettuce and Swiss chard. J Environ Qual 7: 274–281.
3. Stohs SJ, Bagchi D, Hassoun E, Bagchi M (2000) Oxidative mechanisms in the toxicity of chromium and cadmium ions. J Environ Pathol Toxicol Oncol 19: 201–213.
4. Shentu J, He Z, Yang XE, Li TQ (2008) Accumulation properties of cadmium in a selected vegetable -rotation system of south eastern China. J Agric Food Chem 56: 6382–6388.
5. IARC (1993) (International agency for research on cancer), Monographs on the evaluation of the carcinogenic risks to humans beryllium, cadmium, mercury and exposures in the glass manufacturing industry. IARC, Scientific Publications. Lyon, France. 119–238.
6. Li ST, Liu RL, Wang M, Wang XB, Shan H, et al. (2006) Phytoavailability of cadmium to cherry-red radish in soil applied composted chicken or pig manure. Geoderma 136: 260–271.
7. Lebeau T, Bagot D, Jezequel K, Fabr B (2002) Cadmium biosorption by free and immobilized microorganisms cultivated in a liquid soil extract medium: Effects of Cd, pH, and techniques of culture. Sci Total Environ 291: 73–83.
8. Abdola M, Chmtelnicka J (1990) New aspects on the distribution and metabolism of essential trace elements after dietary exposure to toxic metals. Biol Trace Element Res 23: 25–53.
9. Yang JX, Guo HT, Ma YB, Wang LQ, Wei DP, et al. (2010) Genotypic variations in the accumulation of exhibited by different vegetables. J Environ Sci 22: 1246–1252.
10. Yang Y, Zhang FS, Li HF, Jiang RF (2009) Accumulation of cadmium in the edible parts of six vegetable species grown in Cd-contaminated soils. J Environ Manag 90: 1117–1122.
11. Chen HS, Huang QY, Liu LN, Cai P, Liang W, et al. (2010) Poultry manure compost alleviates the phytotoxicity of soil cadmium: Influence on growth of Pak choi (*Brassica chinensis* L.). Pedosphere 20: 63–70.
12. Wang XP, Shan XQ, Zhang SZ, Wen B (2004) A model for evaluation of the phytoavailability of trace elements to vegetables under field conditions. Chemosphere 55: 811–822.
13. Liang Z, Ding Q, Wei D, Li J, Chen S, et al. (2013) Major controlling factors and predictable equations for Cd transfer factor involved in soil-spinach system. Ecotox Environ Saf 93: 180–185.
14. Ge Y, Murray P, Hendershot WH (2000) Trace metal speciation and bioavailability in urban soils. Environ. Pollut. 107, 137–144.
15. Gu JG, Zhou QX, Wang X (2003) Reused path of heavy metal pollution in soils and its research advance. J Basic Sci Eng 11: 143–151 (in Chinese).
16. Gu JG, Zhou QX (2002) Cleaning up through phytoremediation: A review of Cd contaminated soils. Ecol Sci 21: 352–356. (in Chinese with English abstract)
17. Du TP (2005) Food safety and strategy in China. Productivity Res: 6, 139–141. (in Chinese with English abstract)
18. Franz E, Römkens P, Van Raamsdonk L, Van Der Fels-Klerx I (2008) A chain modeling approach to estimate the impact of soil cadmium pollution on human dietary exposure. J Food Protect: 71, 2504–13.
19. Kobayashi E, Suwazono Y, Dochi M, Honda R, Nishijo M, et al. (2008) Estimation of benchmark doses as threshold levels of urinary cadmium, based on excretion of β_2-microglobulin in cadmium polluted and non-polluted regions in Japan. Toxicol Lett 179: 108–12.
20. Tracy S, Sheila M (2006) Cadmium and zinc accumulation in soybean: A threat to food safety. Sci Total Environ 371: 63–73.
21. Ide G, Becker B (1995) Relationship between the arsenic concentration in soil and in the cropped vegetables. International Conference on Heavy Metal in the Environment, Hamburg, vol. 2, CEP Consultants Ltd, Norwich, UK 302–304
22. Szteke B, Jedrzejczak R (1995) The variability of heavy metal contents in plant and soil samples from fields of one farm. International Conference on Heavy Metal in the Environment, Hamburg, vol. 2, CEP Consultants Ltd., Norwich, UK 228–231.
23. Salvatore M, Carratù G, Carafa A (2009) Assessment of heavy metals transfer from a moderately polluted soil into the edible parts of vegetables. J Food Agric Environ 7: 683–688.
24. Chaturvedi R, Sankar K, (2006) Laboratory manual for the physicochemical analysis of soil, water and plant. Wildlife Institute of India, Dehradun, India.
25. Hendershot WH, Duquette M (1986) A simple barium chloride method for determining cation exchange capacity and exchangeable cations. Soil Sci Soc Am J 50: 605–608.
26. Rashid A, Ryan J, Estefan G (2001) Soil and plant analysis laboratory manual. International center for agricultural research in the dry areas (ICARDA), Aleppo, Syria.
27. Day PR (1965) Particle fractionation and particle-size analysis. In: Klute, A. (eds.), Methods of soil analysis. ASA and SSSA, Madison, WI, pp. 545–567.
28. Xiao W, Yang XE, He Z, Rafiq MT, Hou D, et al. (2013) Model for evaluation of the phytoavailability of chromium (Cr) to rice (*Oryza sativa* L.) in representative Chinese soils. J Agric Food Chem 61: 2925–2932.
29. Mehlich A (1984) Mehlich-3 soil test extractant a modification of Mehlich-2 extractant. Commun Soil Sci Plan 15: 1409–1416.
30. FAO/WHO (2003) Report of the sixty first meeting of Joint FAO/WHO expert committee on food additives. Rome.
31. Wang XL, Sato T, Xing BS, Tao S (2005) Health risks of heavy metals to the general public in Tianjin, China via consumption of vegetables and fish. Sci Total Environ 350: 28–37.
32. Rattan R, Datta S, Chhonkar P, Suribabu K, Singh A (2005) A Long-term impact of irrigation with sewage effluents on heavy metal content in soils, crops and groundwater: A case study. Agric Ecosyst Environ 109: 310–322.
33. Peter MC (2002) Ecological risk assessment (ERA) and hormesis. Sci Total Environ 288: 131–140.
34. Liu X, Peng K, Wang A, Lian CL, Shen ZG (2010) Cadmium accumulation and distribution in populations of *Phytolacca Americana* L. and the role of transpiration. Chemosphere 78: 1136–1141.
35. Peter N, Karoly B, Laszlo G (2003) Characterization of the stimulating effect of low-dose stressors in maize and bean seedlings. J Plant Physiol 160: 1175–1183.
36. Kaminek M (1992) Progress in cytokinin research. TIBTECH 10: 159–162.
37. Ali B, Tao QJ, Zhou YF, Gill RA, Ali S, et al. (2013) 5-aminolevolinic acid mitigates the cadmium-induced changes in *Brassica napus* as revealed by the biochemical and ultra-structural evaluation of roots. Ecotox Environ Safe 92: 271–280.
38. Ali B, Wang B, Ali S, Ghani MA, Hayat MT, et al. (2013) 5-Aminolevulinic acid ameliorates the growth, photosynthetic gas exchange capacity and ultrastructural changes under cadmium stress in *Brassica napus* L. J Plant Growth Regul 32: 604–614.
39. Moya JL, Ros R, Picazo I (1993) Influence of cadmium and nickel on growth, net photosynthesis and carbohydrate distribution in rice plants. Photosyn Res 36: 75–80.
40. Weitao L, Zhou Q, Ana J Suna Y, Liu R (2010) Variations in cadmium accumulation among Chinese cabbage cultivars and screening for Cd-safe cultivars. J Hazard Mater 173: 737–743.
41. Lai HY, Chen BC (2013) The dynamic growth exhibition and accumulation of cadmium of Pak choi (*Brassica campestris* L. ssp. chinensis) grown in contaminated soils. Int J Environ Res Public Health 10: 5284–5298.
42. Cui YJ, Zhu YG, Smith SA, Smith SE (2004) Cadmium uptake by different rice genotypes that produce white or dark grains. J Environ Sci 16: 962–967.
43. Xiao W, Yang XE, Zhang Y, Rafiq MT, He Z, et al. (2013) Accumulation of chromium in Pak choi (*Brassica chinensis* L.) grown on representative Chinese soils. J Environ Qual 42: 758–765.
44. De Villiers S, Thiart C, Basson NC (2010) Identification of sources of environmental lead in South Africa from surface soil geochemical maps. Environ Geochem Health 32: 451–459.
45. Kabata-Pendias A, Pendias H (2001) Trace Elements in Soils and Plants. Boca Raton, FloridaCRC Press.
46. Laura W, Wander M, Phillips E (2011) Testing and educating on urban soil lead: A case of Chicago community gardens. J Agric Food Sys Comm Develop, ISSN: 2152–0801 online.
47. Rafiq MT, Aziz R, Yang XE, Wendan X, Rafiq MK, et al. (2014) Cadmium phytoavailability to rice (*Oryza sativa* L.) grown in representative Chinese soils. A model to improve soil environmental quality guidelines for food safety. Ecotox Environ Safe 103:101–107.
48. Murakami M, Nakagawa F, Ae N, Ito M, Arao T (2009) Phytoextraction by rice capable of accumulating Cd at high levels: Reduction of Cd content of rice grain. Environ Sci Technol 43: 5878–5883.
49. Jung MC, Thornton I (1996) Heavy metal contamination of soils and plants in the vicinity of a lead-zinc mine, Korea. Appl Geochem 11: 53–59.
50. McBride M (2002) Cadmium uptake by crops estimated from soil total Cd and pH. Soil Sci 15: 84–92.
51. Eriksson JE, Sderstrom M (1996) Cadmium in soil and winter wheat grain in southern Sweden. Factors influencing Cd levels in soils and grain. Acta Agric Scand Sect B 46: 240–248.
52. Romkens PFAM, Guo HY, Chu CL, Liu TS, Chiang CF, et al. (2009) Prediction of cadmium uptake by brown rice and derivation of soil-plant transfer models to improve soil protection guidelines. Environ Pollut 157: 2435–2444.
53. Giordano PM, Mays DA, Behel AD (1979) Soil temperature effects on the uptake of cadmium and zinc by vegetables grown on sludge amended soil. J Environ Qual 8: 232–236.
54. Oliver DP, Hannam R, Tiller KG, Wilhelm NS, Merry RH, et al. (1994) The effects of zinc fertilization on cadmium concentration in wheat grain. J Environ Qual 23: 705–711.
55. McCauley A, Jones C, Jacobsen J (2009) Soil pH and Organic Matter. Nutrient management modules 8, #4449-8. Montana State University Extension Service, Bozeman, Montana, pp. 1–12.
56. Liu LN, Chen HS, Cai P, Liang W, Huang QY (2009) Immobilization and phytotoxicity of Cd in contaminated soil amended with chicken manure compost. J Hazard Mater 163: 563–567.
57. Halim M, Conte P, Piccolo A (2003) Potential availability of heavy metals to phytoextraction from contaminated soils induced by exogenous humic substances. Chemosphere 52: 265–275.

16

Production of Phytotoxic Cationic α-Helical Antimicrobial Peptides in Plant Cells Using Inducible Promoters

Nuri Company, Anna Nadal, Cristina Ruiz, Maria Pla*

Institute for Food and Agricultural Technology, University of Girona, Girona, Spain

Abstract

Synthetic linear antimicrobial peptides with cationic α-helical structures, such as BP100, have potent and specific activities against economically important plant pathogenic bacteria. They are also recognized as valuable therapeutics and preservatives. However, highly active BP100 derivatives are often phytotoxic when expressed at high levels as recombinant peptides in plants. Here we demonstrate that production of recombinant phytotoxic peptides in transgenic plants is possible by strictly limiting transgene expression to certain tissues and conditions, and specifically that minimization of this expression during transformation and regeneration of transgenic plants is essential to obtain viable plant biofactories. On the basis of whole-genome transcriptomic data available online, we identified the *Os.hsp82* promoter that fulfilled this requirement and was highly induced in response to heat shock. Using this strategy, we generated transgenic rice lines producing moderate yields of severely phytotoxic BP100 derivatives on exposure to high temperature. In addition, a threshold for gene expression in selected tissues and stages was experimentally established, below which the corresponding promoters should be suitable for driving the expression of recombinant phytotoxic proteins in genetically modified plants. In view of the growing transcriptomics data available, this approach is of interest to assist promoter selection for specific purposes.

Editor: Zhengguang Zhang, Nanjing Agricultural University, China

Funding: This work was financially supported by the Spanish Ministerio de Economía y Competitividad (reference AGL2010-17181/AGR). The research group belongs to 2014SGR697, recognized by the Catalonian Government. N.C. and C.R. received fellowships from Generalitat de Catalunya. The funders had no role in study design, data collection and analysis, decision to publish, or preparation of the manuscript.

Competing Interests: The authors have declared that no competing interests exist.

* Email: maria.pla@udg.edu

Introduction

Antimicrobial peptides (AMPs) are key components of innate immunity in plants and animals, and are also produced by microbes in antibiosis processes. A significant proportion are strongly cationic and have linear structures that adopt an amphipathic α-helical conformation that binds to the phospholipid membranes of target microbes before the hydrophobic face is inserted into the membrane bilayer. This unique mode of action explains the lack of resistance in target pathogens, and makes AMPs valuable novel therapeutic agents against bacteria, fungi, viruses, parasites and tumor cells. Improved or synthetic AMPs have been designed with increased potency against selected pathogens [1–5]. As an example, the synthetic undecapeptide BP100 (KKLFKKILKYL-NH_2) is effective against *Xanthomonas vesicatoria* in pepper, *Erwinia amylovora* in apple and *Pseudomonas syringae* in pear [3] with the same efficacy as standard antibiotics, while being biocompatible, as determined by acute oral toxicity tests in mice [6]. Plant expression of recombinant cationic α-helical peptides such as BP100 is preferable for industrial and phytosanitary applications. We have recently showed that active BP100-derived peptides can be expressed as recombinant peptides in plants [7,8], demonstrated by the increased resistance of GM plants to some rice pathogens [7] and by *in vitro* growth inhibition assays [8]. These recombinant BP100-derived peptides include endoplasmic reticulum (ER) retention motifs to minimize toxicity to the host plant. This did not affect the antimicrobial activity of the products *in vitro*, demonstrated using bacterial growth inhibition tests. Following transient expression in *Nicotiana benthamiana* and stable expression in *Arabidopsis thaliana* seedlings, the peptides accumulated in large ER-derived vesicles, along with typical ER luminal proteins [8]. However, the authors found that the recombinant peptides were often toxic to the host during later developmental stages. Similarly, most transgenic *Oryza sativa* plants constitutively expressing recombinant ER-targeted BP100-derived peptides failed to achieve maturity, with only a few peptides coupling potent antimicrobial and low hemolytic activity accumulating in transgenic rice lines, with a yield of up to 0.5% total soluble protein (TSP) [8]. Not many cationic α-helical peptides can be expressed in transgenic plants following this strategy, although they have potent activities against other types of pathogenic cells making them valuable as novel therapeutics and preservatives.

High temperature stress is one of the most common abiotic stresses among many world crops. Plants have evolved various physiological and molecular mechanisms to resist heat stress. Based on the expression data from different plant species, it has been estimated that high temperatures affect approximately 2% of the plant genome (review in [9]). Exhaustive identification of heat stress-responsive genes has been carried out by means of transcriptomics [10–15]. Two groups of genes have been found, signaling components (e.g. protein kinases and transcription factors) and functional genes such as heat shock proteins (Hsps) [16], although many heat shock genes still have unknown functions. Hsps are functionally linked to molecular chaperones that are essential for maintenance and restoration of protein homeostasis. Protein denaturation occurring during stress triggers high transcription of *hsp* genes by the binding of active heat shock factors (Hsfs) to heat shock elements. Alterations in expression of a high number of genes in response to stress occur in complex systems. Cross-talk seems to exist between the regulatory pathways in response to different abiotic stresses such as temperature, and osmotic or oxidative, and biotic stresses [17]. On fusing several heat-shock promoters to reporter genes, they have been found to be regulated during various stress situations, have organ and developmental stage-specific basal expression levels and are induced by stress [18].

Heat-shock induced genes can be highly up-regulated at the transcriptional level by exposure to temperatures above those of normal plant growth. Therefore, induction can be easily carried out, is inexpensive and does not require the use of hormones or chemicals. Here we explored the use of heat-shock inducible promoters with minimal basal activity during normal plant development, to allow expression of recombinant phytotoxic peptides in transgenic rice. Rice has emerged as a powerful platform for large-scale production of recombinant proteins; it has easy cropping conditions and is a self-pollinating crop [19,20]. It is a most suitable crop for genetic manipulation due to its small genome size and the well-established gene transfer technology. Moreover, the availability of the rice complete genome sequence, the exponential growth of profiling studies reporting gene expression data for rice, and the development of databases and bioinformatics tools, provide a possible way to identify genes related to heat stress responses, and with specific expression patterns.

Materials and Methods

In silico identification of candidate heat shock promoters

Miamexpress-EMBL-EBI (http://www.ebi.ac.uk/miamexpress/) and GEO-NCBI (http://nlmcatalog.nlm.nih.gov/geo/) were used to identify published microarray experiments questioning the late response of rice seedlings to heat shock stress. In the experiment with accession number GSE14275, the authors [10] state that rice Affymetrix microarrays were hybridized with RNA extracted from 14-day-old rice seedlings grown in a growth chamber with a daily photoperiodic cycle of 14 h light and 10 h dark, at between 28–30°C, with or without being subjected to 42 °C for 3 h. The data from this experiment was extracted using the RMA software [21], which includes background adjustment, quantile normalization and summarization, and used to identify probes for subsequent analysis.

Rice gene expression data in different organ and developmental stages, and in response to various stress conditions, was obtained from the CREP database (Collection of Rice Expression Profiles; http://crep.ncpgr.cn/crep-cgi/home.pl) and the microarray hybridization results available at the Miamexpress-EMBL-EBI and GEO-NCBI websites.

The NetAffx-analysis center (www.affymetrix.com) allowed identification of the gene associated to every probe, and the sequence was retrieved from GenBank (http://www.ncbi.nlm.nih.gov/genbank). Likely promoter sequences were identified using the PlantProm DB, an annotated and non-redundant database of proximal promoter sequences [22,23].

Primer design and amplification of rice promoter sequences

Primer blast (http://www.ncbi.nlm.nih.gov/tools/primer-blast/) was used to design primer pairs specifically targeting the identified promoter sequences, blocking the reverse primer at the −1 position (immediately upstream of the ATG translation start codon). Two primer pairs (Table S1) were designed to amplify 553 and 1016 bp fragments of the pHsp18 and pHsp82 promoters. They each included an enzyme restriction site (*Kpn*I and *Spe*I at the distal and proximal promoter end, respectively) to facilitate the subsequent cloning steps.

Oryza sativa L. ssp. *japonica*, v. Senia was grown under controlled conditions at 28±1°C and with a 16 h light / 8 h dark photoperiod with fluorescent Sylvania Cool White lamps. Genomic DNA was extracted from 1 g of leaf samples, obtained from young plants, using the commercial NucleoSpin Plant II kit (Macherey-Nagel, Düren, Germany) according to the manufacturer's instructions.

PCR was in a final volume of 50 µl 1x reaction buffer with the appropriate concentrations of $MgCl_2$ and primers (Fisher Scientific SL, Madrid, Spain) (Table S1), 200 µM dNTPs and 1 unit Expand High Fidelity DNA polymerase (Roche Diagnostics GmbH, Mannheim, Germany). The reaction conditions were as follows: 3 min at 94°C; 10 cycles of 15 s at 94°C, 30 s at the appropriate annealing temperature (Table S1) and 1 min at 72°C; 20 cycles of 15 s at 94°C, 30 s at the same annealing temperature and 1 min, plus an additional 5 s for each successive cycle, at 72°C; and a final extension of 10 min at 72°C.

Construction of plant transformation vectors

The pHsp18 and pHsp82 promoters were separately subcloned into the *Kpn*I and *Spe*I sites of a pBluescriptIIKS + derived vector having a polylinker fragment with the *Kpn*I, *Spe*I and *BamH*I restriction sites, followed by the *A. tumefaciens* nopaline synthase *nos* terminator sequence. The sequences encoding BP100.2 [7], BP100-DsRed-tag54 and DsRed-tag54 [8] were subcloned into these plasmids using the *Spe*I and *BamH*I restriction sites.

For the BP100.2 plasmids, the sequence encoding the *N. tabacum* pathogenesis related protein PR1a signal peptide [24] fused to BP100.2 [7] was PCR amplified using pAHC17-bp100.2 [7] as template and the primers PR1a_Spe and BP100KDEL_Bam, with the *Spe*I and *BamH*I restriction sequences respectively. For DsRed-tag54 control plasmids, the sequence encoding the *Petrosilinum hortense* chalcone synthase 5′ untranslated region plus the codon-optimized leader peptide derived from the heavy chain of the murine mAb24 monoclonal antibody [25] fused to DsRed [26], the epitope tag54 [27] and the KDEL ER retention motif, were PCR amplified using pCdsred-tag54 [8] as template and the primers CHS_Spe and TagKDEL_Bam, that included the required restriction sites. As an additional control, a BP100-DsRed-tag54 plasmid was generated similarly to DsRed-tag54, and including the BP100 antimicrobial sequence placed N-terminal to DsRed-tag54. The coding sequence was obtained by amplification of the pCbp100-dsred-tag54 DNA [8] with the same CHS_Spe and TagKDEL_Bam primers. PCR was in a final volume of 50 µl 1x reaction buffer with the appropriate concentrations of $MgCl_2$ and primers (Fisher Scientific SL, Madrid) (Table S1), 200 µM dNTPs and 1 unit Expand High Fidelity DNA polymerase (Roche Diagnostics Corporation). Reaction conditions were as follows: 3 min at 94°C; 10 cycles of 15 s at 94°C, 30 s at 49°C and 1 min at 72°C; 20 cycles of 15 s at 94°C, 30 s at 58°C and 1 min, plus additional 5 s each successive cycle, at 72°C; and a final extension of 10 min at 72°C.

On sequence verification (Macrogen, Seoul, Korea) using the CLCbio software (Aarhus, Denmark), the complete constructs, *bp100.2*, *bp100-dsred-tag54* and *dsred-tag54* plus promoter and terminator elements, were directionally inserted into the *Kpn*I and *Sbf*I sites of pCAMBIA1300. The resulting binary vectors were transferred into *A. tumefaciens* strain EHA105 by cold shock [28].

Production of transgenic rice plants

Commercial *japonica* rice (*Oryza sativa* L.) var. Senia was transformed by *A. tumefaciens* to obtain transgenic rice lines expressing the chimeric proteins mentioned above, using hygromycin resistance as the selection trait. The control plasmid, pCambia1300 (transferring only the *hpt*II selection gene), was transformed in parallel. Embryonic calluses derived from mature embryos were transformed as previously described [29]. Hygromycin resistant T_0 plants were grown to maturity under standard greenhouse conditions to obtain the T_1 generation. Leaf samples of individual T_0 plants at the 5-leaf vegetative stage were used to extract genomic DNA and assess the presence and copy number of the transgene by real-time PCR (qPCR), targeting the coding region of every transgene as previously described [8]. The *actin* endogenous gene was used to normalize the Ct values. The number of fertile genetically modified (GM) plants containing every plasmid was recorded to calculate the transformation efficiency compared to that of the control pCambia 1300 plasmid (number of fertile GM plants obtained per initial callus. This value was then normalized with that obtained for the control plasmid).

Analysis of gene expression: plant material, RNA isolation and RT-PCR

Plant material: heat shock treatment. Wild type rice seeds (var. Senia) were surface sterilized and germinated *in vitro* in sterile MS medium [30], including vitamins (2.2 g/L MS medium, 8 g/L agar and 30 g/L sucrose), under controlled conditions (28±1°C temperature and a 16 h light / 8 h dark photoperiod with fluorescent Sylvania Cool White lamps). After one week (two-leaf vegetative stage, V2), groups of 15 plants were treated at 42°C for 0, 1, 2, 4, 8 and 16 h. For each treatment, three groups of five plants were collected as biological replicates, immediately frozen in liquid nitrogen and stored at −80°C.

For each GM event, 15 T_1 seeds were surface sterilized and *in vitro* cultured, as mentioned above, up to the V2 stage. They were treated at 42°C for 0 and 2 h and their leaves individually frozen in liquid nitrogen. The presence of the transgene was individually assessed in every plant using the Phire Plant Direct PCR Kit (Thermo Scientific, Lithuania) combined with the DsRed_for and Nos.te_rev (constructs with *dsred-tag54* and *bp100-dsred-tag54*) or the P82_for and Nos.te_rev (constructs with *bp100.2*) primer pairs (Table S1), according to the manufacturer's instructions. Samples giving a positive PCR signal (i.e. harboring the transgene) were mixed for subsequent analyses.

Plant material: monitoring of the transformation process. The transformation to obtain transgenic rice was monitored using *A. tumefaciens* strain EHA105, with or without the basic pCambia1300 vector (with *hpt*II). Embryonic calluses derived from rice mature embryos were transformed and, when *hptII* was transformed, selected with hygromycin [29]. Samples of ten independent calluses or events were taken at eight different stages for each transformation.

For embryo extraction, 120 hygromycin-resistant seeds (harboring *hptII*) and 80 non-transformed seeds were surface-sterilized and their embryos were immediately extracted and frozen, or germinated in 500 µL water or 500 µL water with 45 mg/L hygromycin B (40 seeds per treatment and embryo type). Germination was in a culture chamber (28±1°C with a 16 h light/8 h dark photoperiod, under fluorescent Sylvania Cool White lamps) for two and five days prior to embryo extraction and immediate freezing in liquid nitrogen. Untransformed seeds were not treated with hygromycin B.

RNA isolation and reverse transcription (RT) coupled to qPCR. Samples were homogenized in liquid nitrogen and 100 mg was used to extract RNA with the Trizol reagent (Invitrogen, Karlsruhe, Germany) based protocol. The DNase I digestion (Ambion, Grand Island, NY) was carried out according to the manufacturer's protocol, and RNA concentration and quality was assessed by UV absorption at 260 and 280 nm using a NanoDrop ND1000 spectrophotometer (Nanodrop technologies, Wilmington, DE). RT-qPCR was carried out as previously described [7]. For each sample, cDNA was synthesized with random primers in duplicate and qPCR reactions targeting *bp100.2* and *dsred-tag54* transgenes, and *Os.hsp18* and *Os.hsp82* rice genes were carried out in triplicate. The qPCRs were in a final volume of 20 µl containing 1X SYBR Green PCR Master Mix (Applied Biosystems, Foster City, CA, USA), the appropriate concentrations of primers (Fisher Scientific SL, Madrid, Spain) (Table S1) and 1 µl cDNA. The reaction conditions were as follows: 10 min at 95°C for initial denaturation; 50 cycles of 15 s at 95°C and 1 min at 60°C; and a final melting curve program of 60 to 95°C with a heating rate of 0.5°C/s. Melting curve analyses produced single peaks, with no primer-dimer peaks or artifacts, indicating the reactions were specific. All reactions had a linearity coefficient exceeding 0.995 and efficiency values above 0.95. The *ef1α* gene was used for normalization, its suitability having been confirmed using the geNORM v3.4 statistical algorithm ([31]; *M* values below 0.5 in our samples). Triplicate biological samples, each containing at least five plants, were analyzed in each case.

Protein extraction and western blot analyses

For each GM event, 15 T_1 seeds were surface sterilized and *in vitro* cultured as mentioned above, up to the V2 stage. Rice seedlings were treated at 42°C for 2 h, then transferred to 28±1°C with a 16 h light/8 h dark photoperiod under fluorescent Sylvania Cool White lamps for two days, to allow recombinant protein accumulation and DsRed maturation, and then immediately frozen in liquid nitrogen. The Phire Plant Direct PCR Kit (Thermo Scientific, Lithuania) was used in combination with the SYDsRed_for and SYDsRed_rev primers (Table S1), according to the manufacturer's instructions, to individually identify transgenic T_1 plants. Transgenic seedlings were homogenized in liquid nitrogen and TSP extracted as previously described [8]. Insoluble proteins in the samples expressing *bp100-dsred-tag54* were re-extracted by boiling for 10 min in the same buffer supplemented with 8 M urea and 1% SDS. Protein concentration was determined using the Sigma Bradford Reagent and bovine serum albumin as standard. Twenty µg TSP were separated by PAGE, using 18% (w/v) SDS polyacrylamide gels, and electrotransferred to nitrocellulose membranes. Twenty-five to 1 pmol chemically synthesized controltag54 (GQNIRDGIIKAGPAVAVVGQAT-QIAKAGPAKDWEHLKDEL), mixed with 20 µg TSP extracted from untransformed rice seedlings, were included to allow quantification of the recombinant peptides. They were hybridized overnight with the mAb54k monoclonal antibody [27] (1:1,500 dilution) at 4°C, and with horseradish peroxidase-conjugated anti-mouse IgG as the secondary antibody (GE Healthcare Life sciences) (1:10,000 dilution) for 1 h at room temperature. The hybridization signal was detected by ECL chemiluminescence (Luminata Forte HRP Chemiluminescence Detection Reagents, Millipore – Darmstadt, Germany).

Phenotype evaluation

T1 seeds carrying pHsp82::bp100.2, pHsp82::dsred-tag54 (three independent events per construct) or hptII, and conventional Senia seeds, were surface sterilized and *in vitro* cultured as mentioned above, up to the V2 stage (height, 6±1 cm). They were all treated at 42°C for 2 h and allowed to recover for three days under the same conditions. Plant growth was monitored daily by measuring the height of the aerial part. Transgenic plants were identified using the Phire Plant Direct PCR Kit (Thermo Scientific, Lithuania) as mentioned above, and only those harboring the transgene (at least five plants per event) were included in statistical analyses. As an additional control, five untransformed Senia plants were grown in parallel and not subjected to heat shock.

Confocal microscopy

Transgenic rice seedlings, obtained as mentioned above, were treated at 42°C for 2 h and transferred to the standard conditions for two days to allow accumulation of the recombinant protein and DsRed maturation. Radicles were observed under an FV1000 Olympus confocal microscope and red fluorescent images collected with 543 nm excitation using a 550–600 nm emission window. The ImageJ software (http://rsb.info.nih.gov/ij/) was used to calculate the number and size of fluorescent spots and quantify the DsRed fluorescence.

Results

Rationale of the approach and selection of suitable promoters

With the increasingly abundant transcriptomics analytical tools and datasets available, we used an *in silico* approach to initially select a small number of candidate heat shock promoters. The EMBL-EBI Array Express website was searched for datasets on transcription profiling of the *Oryza sativa* in response to heat shock. A single microarray experiment was found (E-GEOD-14275, [10]) that questioned 14-day old seedlings, grown at 28–30°C, and with or without heat stress at 42°C. Expression data was available for a total of 22,905 sequences. Those with the highest expression levels had fluorescence intensities up to 13,800, and 16,300 units in microarrays hybridized with RNA from control and heat-shock treated seedlings, respectively. Only 15 and 17 sequences had values above 10,000 fluorescence units. Up to 19 sequences, corresponding to 16 genes, had fluorescence intensities more than 5,500 units higher in heat-shock treated than in control seedlings (Table S2). This included three sequences (AK071613.1, AK071240.1 and CB621753) with very high expression values under induction conditions (above 10,000 fluorescence units) and sequences with the greatest changes in expression in response to heat stress treatment (up to 615-fold, AK063751.1). These sequences were initially selected to drive the expression of phytotoxic peptides in rice.

The transcription patterns of these 16 genes were subsequently compared with the Collection of Rice Expression Profiles (CREP) database (http://crep.ncpgr.cn/crep-cgi/home.pl), a compilation of quantitative transcriptomes of 39 tissues of various rice genotypes, from the *indica* and *japonica* subspecies, obtained using the commercial Affymetrix Rice GeneChip microarray. We specifically focused on the tissues and developmental stages with special relevance in transformation and regeneration to obtain GM plants. They included callus at different steps of agrotransformation and hygromycin based selection, and imbibed seeds and seedlings at various developmental stages. Multiple gene expression profiles were retrieved using the Multi-genes Chronologer tool (Figure 1). The Os.519.1.S1_at sequence (AK071240.1, encoding a putative 18 kDa Class II Heat Shock Protein) combined low expression levels in all 14 analyzed tissues considered relevant for transformation and very high expression in heat-shock treated seedlings (0.073 and 1.190 normalized fluorescence units). Although it had some expression in control seedlings (0.052 normalized fluorescence units), this promoter was chosen to drive the expression of phytotoxic peptides in plants. A second selected sequence, Os.11039.1.S1_s_at (AK063751.1, encoding a putative Heat Shock Protein 82), had extremely low expression values in the 14 tissues and control seedlings but only moderate expression in response to high temperature treatment (0.033, 0.001 and 0.650 normalized fluorescence units, respectively). There was no induction of Os.519.1.S1_at or Os.11039.1.S1_s_at upon cold, salt or drought stresses, as assessed *in silico* using profiling datasets with GSE6901 as the reference, reporting transcriptome analyses of 7-day old seedlings of Indica rice variety IR64 subjected to 4±1°C, 200 mM NaCl or insufficient moisture in their roots for 3 h (Figure S1).

Expression patterns of the two selected genes in response to heat shock and during the process of *Agrobacterium tumefaciens* mediated stable transformation of rice using the *hpt*II selection gene

Specific real-time PCR (qPCR) assays were developed to target the coding sequences of *Os.hsp18* and *Os.hsp82* and, coupled to reverse transcription, used to experimentally assess the expression patterns of the selected genes in a range of developmental stages and conditions. Treatment of rice plants with high temperature resulted in increased expression of the two genes (Figure 2). *Os.hsp18* and *Os.hsp82* were highly induced by exposure to 42°C for 2 to 4 h, reaching expression levels of 294 and 516-fold that of *ef1α* in leaves and 326 and 796-fold in roots. There was a decrease in gene expression after these time points.

Additionally, their expression levels were determined during *A. tumefaciens*-mediated transformation and selection using the hygromycin resistance phenotype (Figure 3). Senia wild type seeds were induced to generate primary callus and the standard transformation procedure was followed, using *A. tumefaciens* carrying pCAMBIA1300 (with the *hpt*II selection gene) to produce transgenic plants. *A. tumefaciens* not transformed with the binary plasmid was used as control, without hygromycin as a selection agent. Mature and imbibed embryos from untransformed and transgenic S-hptII plants were also analyzed. *Os.hsp18* was clearly expressed at higher levels than *Os.hsp82* in all tested tissues and stages. Expression was up to 28-fold that of the reference gene in transformed callus incubated for two weeks in hygromycin selection medium and that of mature embryos, and up to 15-fold that of *ef1α* upon *A. tumefaciens* infection. Seed imbibition rapidly resulted in mRNA decline. *Os.hsp82* had a similar expression pattern, although the level was less than that of *ef1α* in virtually all tested samples. As a control, we measured the mRNA levels of the *hptII* transgene, regulated by the CaMV 35S strong constitutive promoter. They were substantially higher than those of *Os.hsp18*, and especially *Os.hsp82*, at all stages in which there were uniquely transgenic cells (i.e. after hygromycin selection).

Os.hsp18 and *Os.hsp82* promoter fragments regulate the expression of transgenes in response to heat treatment

The sequences corresponding to the promoters of *Os.hsp18* and *Os.hsp82* were retrieved from the Plant Promoter Database, and 553 and 1,016 bp promoter fragments of *Os.hsp18* and *Os.hsp82*,

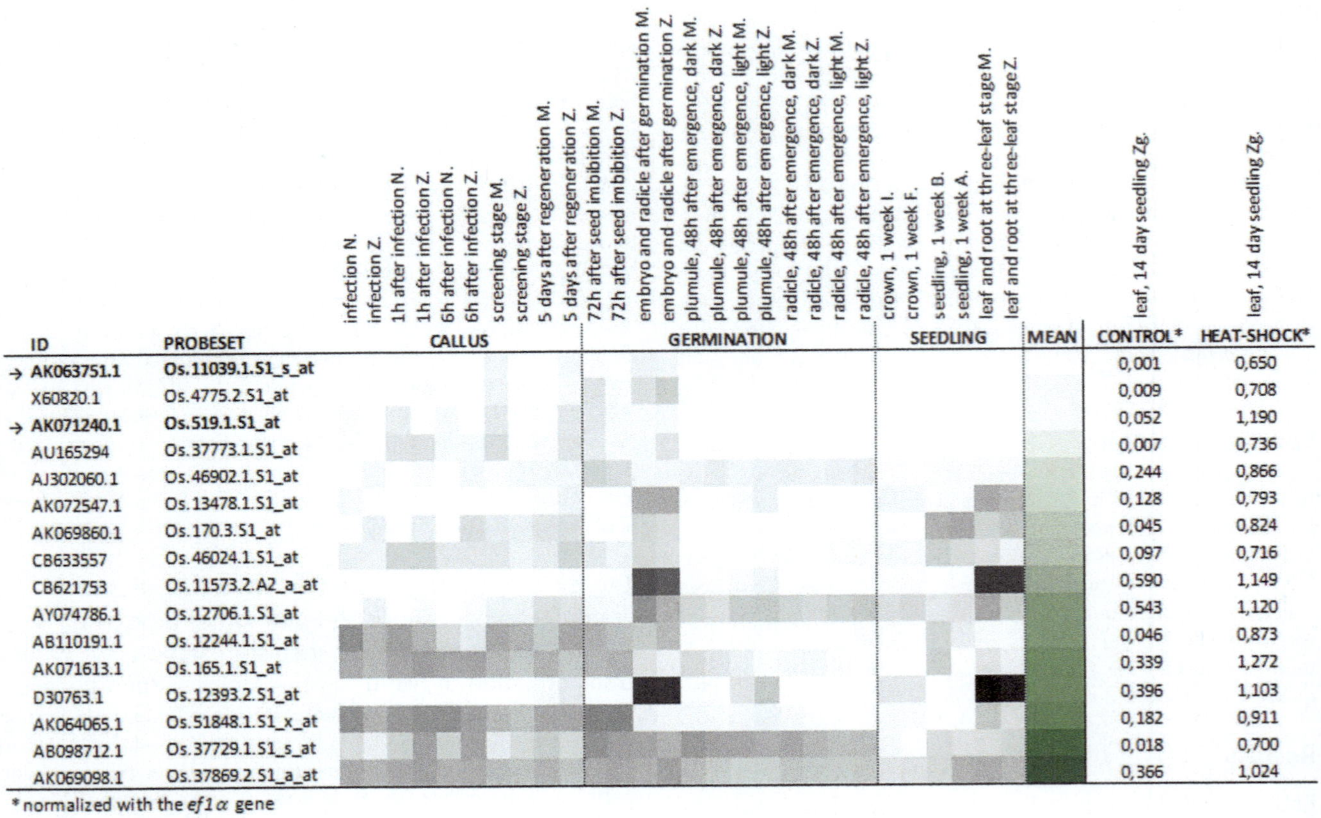

Figure 1. A summary of the expression profiles of a selection of sequences induced by high temperature, in callus and seedlings of eight rice genotypes [Bala (B), FL478 (F), IR29 (I), Minghui 63 (M), Zhenshan 97 (Z), *indica* subspecies, and Azucena (A), Nipponbare (N) and ZhongHua 11 (Zg), *japonica* subspecies], from published *in silico* data. Cluster analysis of 16 genes with the highest expression or induction in response to heat-shock, in a total of 14 tissues considered relevant to the process of obtaining transgenic plants (grey scale). The overall mRNA levels in these tissues and developmental stages were estimated using the mean expression values (green scale). Dark to light color scale represents high to low expression levels, the darkest color corresponding to 90162 (grey scale) and 19788 (green scale) normalized fluorescence units. Arrows indicate the two sequences whose promoters were selected to produce transgenic plants.

respectively, were PCR amplified from rice var. Senia genomic DNA. Their 3′ end was at position −1 relative to the ATG translation start codon. *Os.hsp18* and *Os.hsp82* promoter fragments, pHsp18 and pHsp82, were cloned and sequence verified. The activity of these promoter fragments was experimentally assessed using the *Discosoma* spp. red fluorescent reporter

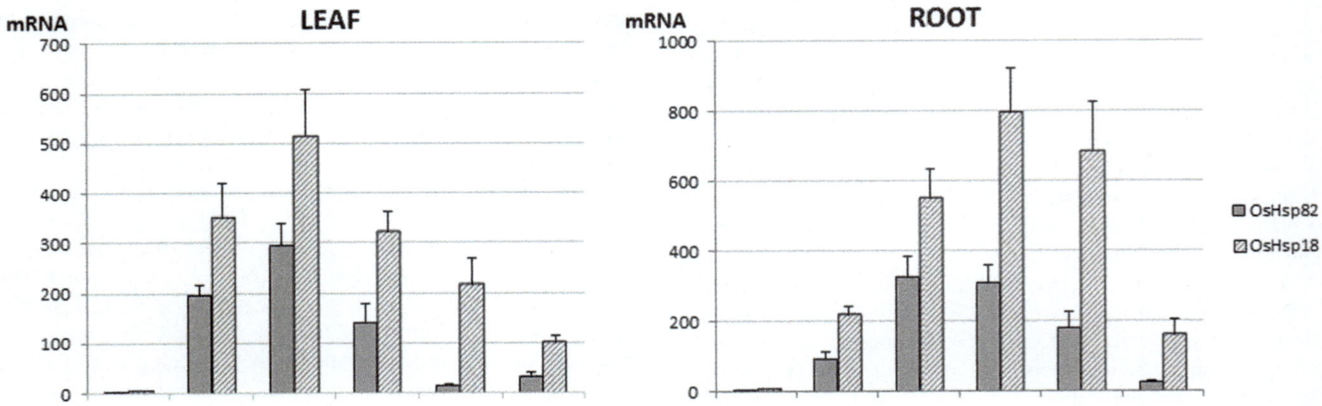

Figure 2. Transcriptional response to high temperature of *Os.hsp18* and *Os.hsp82* in rice var. Senia. Plants at the two leaf vegetative stage were exposed to 42°C for 0, 1, 2, 4, 8 and 16 h, then the mRNA levels of the selected sequences, from leaf and root samples, were analyzed by RT-qPCR. Three biological replicates, each of five plants, with two experimental replicates, were analyzed per treatment. The *ef1α* reference gene (gNORM M value <0.5 in these samples) was used for normalization and the given mRNA values correspond to fold expression vs. the reference gene.

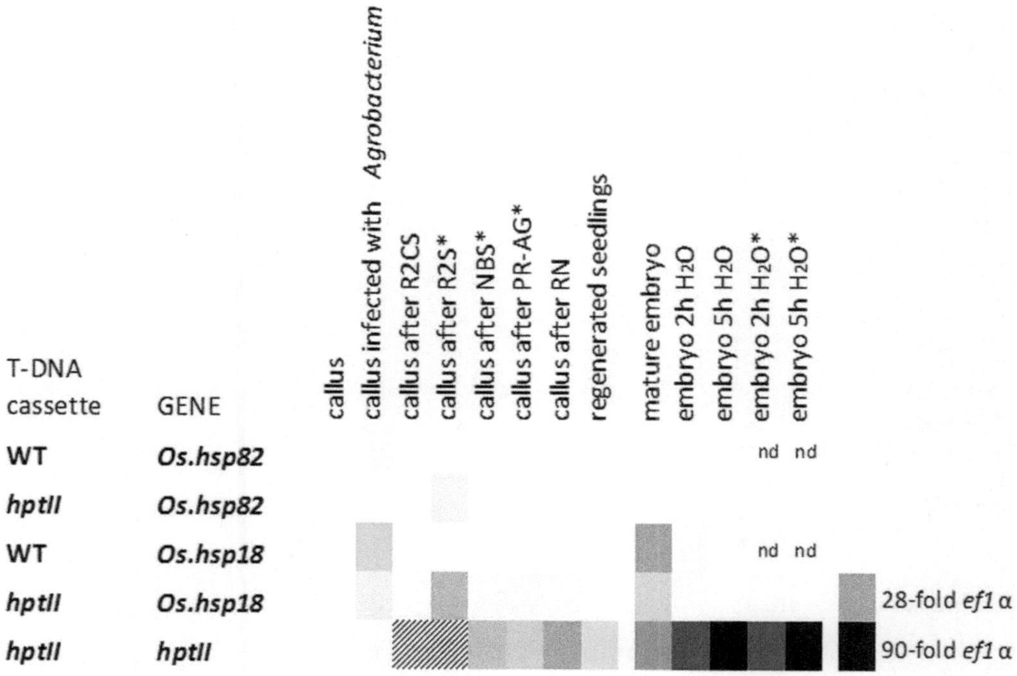

Figure 3. mRNA expression profiles of *Os.hsp18*, *Os.hsp82* and the *hptII* transgene, driven by the CaMV 35S promoter, in rice var. Senia. Different stages during transformation with *A. tumefaciens*, either carrying pCAMBIA1300 (*hptII*) or no T-DNA (WT), and expression in mature and imbibed embryos, are shown. The following stages of callus are shown: 6 weeks after induction of embryogenic callus; immediately after 15 min infection with *A. tumefaciens*; after a 3-day incubation in R2CS medium; after a 2-week incubation in R2S selection medium; 3 weeks after transfer to NBS selection medium; after an 8 to 10-day incubation in PR-AG selection medium; 3–4 weeks after induction of regeneration in RN medium; and two weeks after seedling culture in P medium. Embryos extracted from mature seeds, either before or after imbibition in water for two or five hours, were also analyzed. When the *hptII* DNA cassette was transformed, transgenic cells were selected using hygromycin (asterisk). Three biological replicates, each of five calluses, with two experimental replicates, were analyzed at each stage. The *ef1α* reference gene (M values <0.5 in these samples) was used for normalization. Dark to light color scale represents high to low expression level. All values were in the 90 to 0.01-fold range. nd: not determined. Striped: underestimated values due to callus samples containing a mixture of transgenic and non-transgenic cells.

protein, DsRed [26], and the epitope tag54 sequence to detect recombinant proteins through the specific mAb54 antibody [27]. The sequence encoding the DsRed-tag54-KDEL reporter (hereafter, DsRed-tag54) was placed in-frame with the sequence encoding the signal peptide of the murine monoclonal antibody mAb24 under the control of the *Os.hsp18*, *Os.hsp82* or *Zm.ubi* [32] promoters and *nos* terminator. All constructs were sequence verified, introduced in *A. tumefaciens* and used to transform rice plants, along with the *hpt*II marker for hygromycin selection. All constructs yielded hygromycin-resistant plants and those derived from different calluses were considered independent events.

The transgene copy number and expression profile were assessed in three independent events representing each promoter. The ratio of *dsred-tag54* to the *actin* reference gene sequence was close to 0.5 (mean and SD, 0.63±0.27), as determined by qPCR using leaf genomic DNA from T_0 plants, suggesting single-copy insertions. Transgene expression was quantified by RT-qPCR in leaf samples of transgenic T_1 plants at the two-leaf vegetative stage (V2), with or without exposure to 42°C for 2 h (Figure 4). Significant amounts of *dsred-tag54* mRNA were only observed when plants were kept under control temperature conditions, where the transgene was driven by the constitutive pUbi promoter. The expression values, normalized with the *ef1α* constitutive gene, were 0.47 to 2.5, 0.73 to 18.84 and 603 to 904 for pHsp82, pHsp18 and pUbi, respectively. One-way ANOVA, $P=0.000$, gave two groups in Tukey's b posttest with $\alpha<0.05$. Upon heat treatment, the pHsp18 promoter reached transgene mRNA levels from 214 to 1,117-fold that of *ef1α*, similar to those reached with the pUbi promoter. The levels of expression of the transgene was slightly lower with the pHsp82 promoter, 208 to 636-fold that of *ef1α*, but statistically similar (one-way ANOVA $P=0.279$). As expected, no mRNA encoding DsRed-tag54 was detected in untransformed rice plants. This proved that, after exposure to heat stress, the pHsp82, and especially the pHsp18 promoter fragments, had the capacity to drive the expression of a reporter gene to levels similar to those achieved by a frequently used strong constitutive promoter.

Use of *Os.hsp18* and *Os.hsp82* promoters to drive the expression of phytotoxic BP100 derivatives in rice

The capacity of the pHsp18 and pHsp82 promoter fragments to drive the expression of phytotoxic cationic α-helical antimicrobial peptides in stably transgenic plants was evaluated using BP100.2. The antimicrobial peptide has high phytotoxic and hemolytic activities and its constitutive expression in transgenic rice is incompatible with the survival of GM plantlets [7]. The sequence encoding BP100.2 was placed in-frame with the sequence encoding the signal peptide of the *N. tabacum* pathogenesis related protein PR1a, under the control of the pHsp18 and pHsp82 promoters. After sequence verification, the constructs were introduced in *A. tumefaciens* and used to transform rice plants along with the *hpt*II marker. The empty vector, with only the *hptII* selection gene, was transformed in parallel. Joint analysis of the transformation efficiencies achieved for constructs encoding DsRed-tag54 and BP100.2, under the control of the pHsp18, pHsp82 and pUbi promoters, showed that four out of five

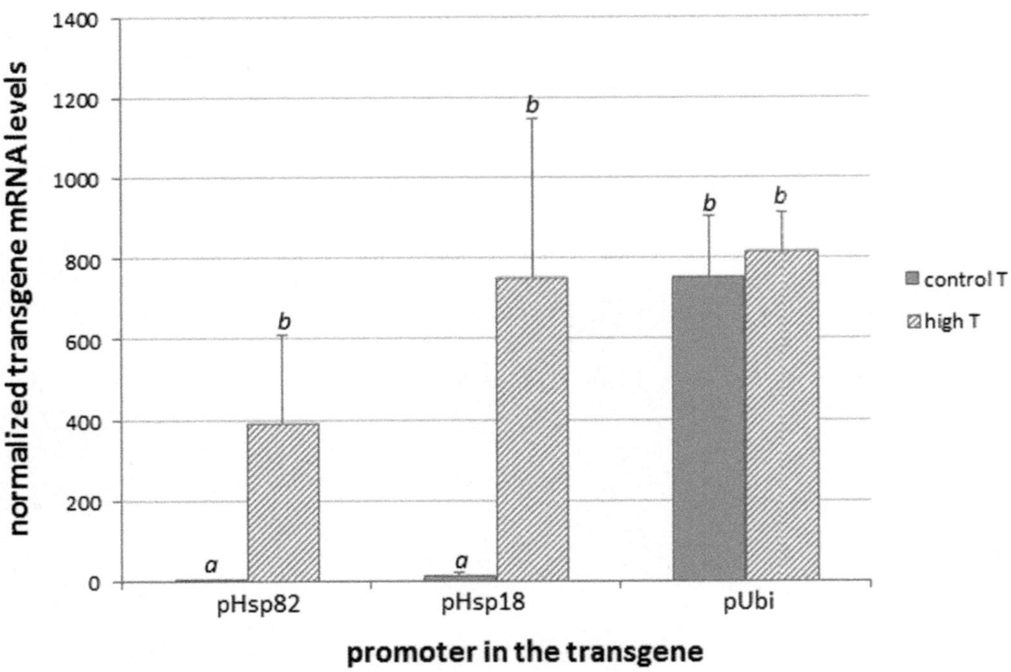

Figure 4. Transgene mRNA expression levels in leaves of GM rice plants in which the reporter *dsred-tag54* was regulated by three different promoters, pHsp82, pHsp18 and pUbi. Three independent events were analyzed carrying every construct. For each event, T_1 seeds were germinated under control conditions and sets of eight plants at the V2 stage were either subjected to 42°C for 2 h, or not, and their leaves were individually sampled. Sets of five plants harboring the transgene (i.e. giving positive transgene signal using the Phire Plant kit) were collectively analyzed by RT-qPCR targeting the sequence encoding DsRed-tag54, and normalized against *ef1α*. Mean and SD values corresponding to each promoter are shown. Letters indicate statistically different transgene mRNA values (One-way ANOVA, Tukey's b posttest $\alpha<0.05$).

constructs yielded hygromycin-resistant plants with efficiencies similar to that of the empty vector (mean efficiency and SD, 97±18%, Table S3). As expected, this included all vectors encoding the non-phytotoxic DsRed-tag54 reporter, irrespective of the specific promoter regulating its expression (constitutive pUbi or heat-shock inducible pHsp18 and pHsp82).

Remarkably, the vector encoding BP100.2 under the control of the high temperature inducible promoter pHsp82 produced fertile transgenic rice plants with the same efficiency. The presence of the *bp100.2* or *dsred-tag54* transgene was confirmed in all T_0 events using qPCR. Conversely, no GM plants were obtained with the construct encoding the highly phytotoxic BP100.2 antimicrobial peptide under the control of the pHsp18 promoter (transformation efficiency <3%). The efficiency of this transformation was at least 30-fold below that of the control plasmid, which can be considered not workable. This is in agreement with the higher activity of the pHsp18 promoter, compared to pHsp82, during transformation.

Transgene expression profiles were assessed in three independent events carrying sequences encoding BP100.2 under the control of pHsp82 (pHsp82::*bp100.2*), using transgenic plants encoding DsRed-tag54 regulated by the same promoter (pHsp82::*dsred-tag54*) as the control. Hygromycin-resistant plants derived from different calluses were considered independent events. The ratio of *bp100.2* to the reference gene sequence was close to 0.5 in these events (mean and SD, 0.53±0.29) as determined by qPCR using leaf genomic DNA from T_0 plants, suggesting single-copy insertions. Transcript levels of the *bp100.2* transgene were, prior to heat shock, 2.18 to 5.70-fold the level of *ef1α* (Figure 5), i.e. similar to those of *dsred-tag54* (One-way ANOVA $P=0.161$). Upon heat shock, pHsp82 drove transcription to similar levels for the phytotoxic *bp100.2* (275 to 357-fold the level of *ef1α*) and the control *dsred-tag54* (One-way ANOVA $P=0.592$). Thus, the pHsp82 promoter is a suitable tool to achieve transcription of transgenes encoding phytotoxic recombinant peptides in stably transformed GM plants.

Recombinant BP100.2 is difficult to detect due to its low extinction coefficient (low content of aromatic amino acids) and lack of immunogenicity [8]. We used its phytotoxic character to indirectly confirm the synthesis of BP100.2 in GM plants in response to heat treatment. We predicted that, in the presence of the phytotoxic recombinant BP100.2, transgenic plants would show altered phenotype after heat shock as compared to control plants. As shown in Figure 6, untransformed plants not subjected to heat treatment were taller than any temperature-stressed groups. Twenty-four hours after heat shock, plants expressing *bp100.2* were shorter than those either untransformed or expressing *dsred-tag54* or *hptII* (One-way ANOVA $P=0.000$, three groups after Tukey b posttest with $\alpha<0.05$) and had more serious symptoms of leaf wilting. This strongly suggested that phytotoxic recombinant BP100.2 was produced in these plants.

Production of recombinant reporter proteins in GM rice plants using high temperature inducible promoters

Phytotoxic recombinant BP100 derivatives in transgenic plants were quantified by fusion of BP100 to a reporter moiety encompassing the fluorescent DsRed and the tag54 epitope. A construct was obtained in which the pHsp82 promoter drove the expression of BP100-DsRed-tag54-KDEL (hereafter, BP100-DsRed-tag54), with the same signal peptide and regulatory elements as the pHsp82::*bp100.2* construct. It was sequence verified and used to transform rice plants using the same *Agrobacterium*-based strategy, achieving similar transformation efficiency as the control (109%). Synthesis of *bp100-dsred-tag54* mRNA was assessed by RT-qPCR in three independent events

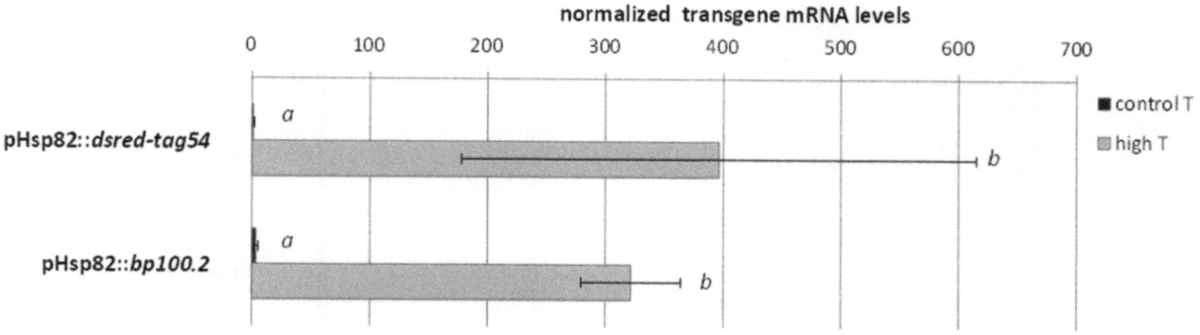

Figure 5. Transgene mRNA expression levels in leaves of GM rice plants harboring pHsp82::*bp100.2* or pHsp82::*dsred-tag54* constructs. Three independent events of each construct were analyzed. For each event, T_1 seeds were germinated under control conditions and sets of eight plants at the V2 stage were either subjected to 42°C for 2 h (black bars) or not (grey bars), and their leaves individually sampled. Sets of five plants harboring the transgene (i.e. giving positive transgene signal using the Phire Plant kit) were collectively analyzed by RT-qPCR targeting the sequence encoding DsRed-tag54 or BP100.2, and normalized against *ef1α*. Mean and SD values corresponding to every transgene are shown. Different letters indicate statistically different transgene mRNA values (One-way ANOVA, Tukey's b posttest $\alpha<0.05$).

with single copy insertions (as determined by qPCR of T_0 leaves using *ef1α* as reference, 0.52±0.15). Residual mRNA levels in leaves of V2 plants grown under control temperature were in the range of 0.07 to 0.3-fold actin mRNA levels, but increased to 142 to 275-fold after a 2 h exposure to heat shock. These values were not statistically distinguishable from those of pHsp82-regulated transgenes expressing *bp100.2* and *dsred-tag54* in the same conditions (One-way ANOVA, $P=0.524$ and $P=0.205$ for control and heat treated plants, respectively).

Transgenic plants from three independent events carrying pHsp82::*bp100-dsred-tag54* were grown to the V2 stage under controlled conditions, heat-induced for 2 h at 42°C and allowed to

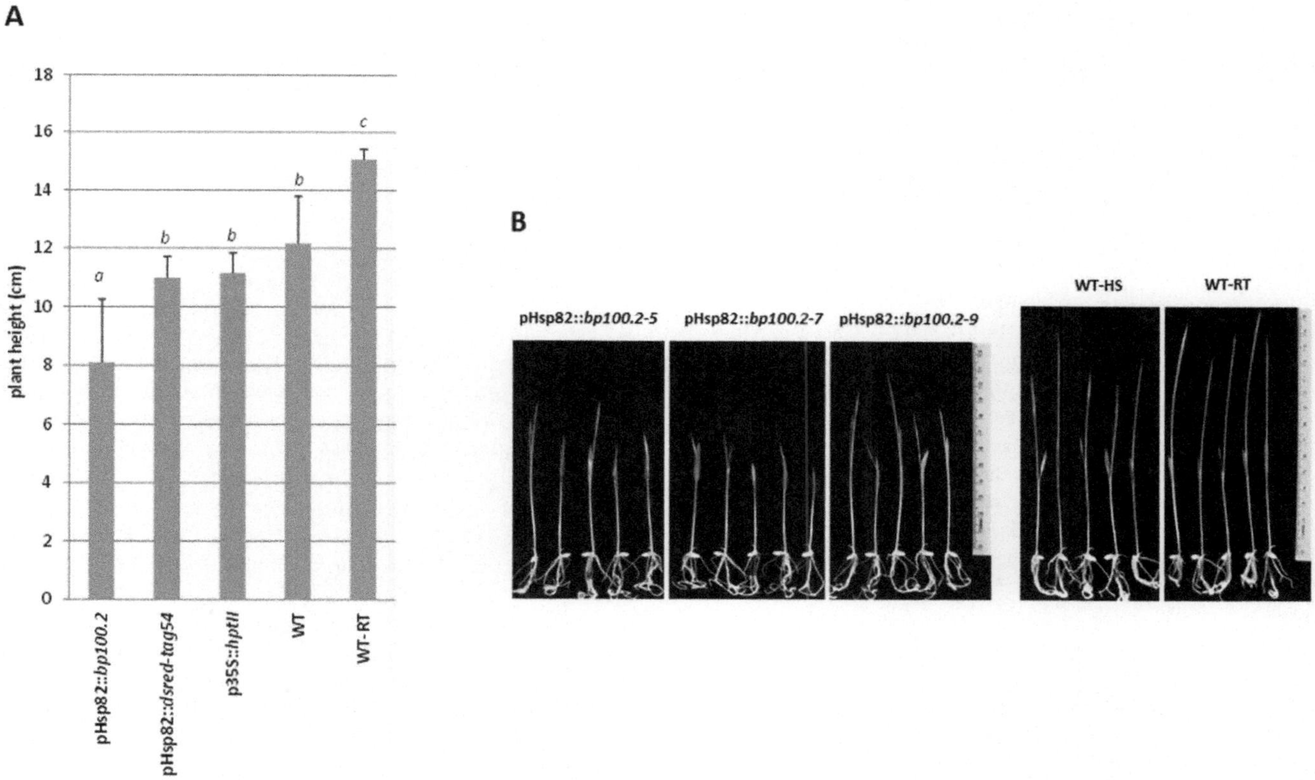

Figure 6. Growth of transgenic rice plants carrying pHsp82::*bp100.2* after heat shock. T_1 seeds of rice plants carrying pHsp82::*bp100.2*, pHsp82::*dsred-tag54* (three independent events per construct) and p35S::*hptII* were germinated for 1 day in the presence of hygromycin, and five plants per event were then *in vitro* cultured, without the selection agent, up to the V2 stage (height, 6±1 cm). Five untransformed Senia plants were grown in parallel (WT). They were all treated at 42°C for 2 h and the length of their aerial parts was measured after 24 h incubation under standard conditions. Five untransformed plants were not subjected to heat stress, as an additional control (WT-RT). (**A**) Means and SD of the height of the different groups are shown. Letters indicate statistically significant differences (One-way ANOVA, Tukey b posttest with α value <0.05). (**B**) The phenotypic effects of heat treatment on transgenic plants carrying pHsp82::*bp100.2* (three independent events) and untransformed plants (WT-HS) are shown. WT-RT is shown as control.

accumulate recombinant BP100-DsRed-tag54 for an additional two days. Three events of transgenic plants expressing *dsred-tag54* under the regulation of the pHsp82 and pHsp18 promoters were used as control. Plants with the constitutive pUbi promoter driving the expression of the same recombinant protein were included in the assay. As shown in Figure 7A, TSP extracts from all transgenic plants expressing *dsred-tag54* produced a band of 29.9 kDa in western blot assays, which is similar to the DsRed-tag54 anticipated size, and a band of about 36.8 kDa. There was an additional secondary band of 22.1 kDa that probably corresponds to a cleavage product originated by boiling TSP before SDS-PAGE [33]. TSP extracted from all three samples expressing *bp100-dsred-tag54* under the control of pHsp82 produced a band of an apparent MW of 37.9 kDa (Figure 7B), which is in agreement with the 36.8 kDa band detectable in extracts of plants producing recombinant DsRed-tag54. When higher amounts of protein extracts were analyzed, a faint band of the expected size (31.5 kDa) was also visible (Figure S2). The BP100 portion of the recombinant molecule seemed to strongly enhance the proportion of the slower band, which was probably the result of incomplete denaturation or interaction of different molecules. Subsequent analysis of the insoluble protein fraction showed that a significant proportion of immune-reactive proteins, with apparent MW of 37.9, 48.5 and 70.6 kDa, could only be extracted in an SDS and urea-based, strongly denaturing buffer (Figure 7B). As expected, no immune-reactive bands were observed upon analysis of TSP from untransformed plants, those uniquely expressing the hygromycin resistance selection marker, or those transformed with pHsp82::*dsred-tag54* and not subjected to high temperature (Figure 7A). Thus, phytotoxic recombinant BP100 derivatives were produced and accumulated after heat induction in transgenic rice using the pHsp82 heat inducible promoter. Quantification of the immune-reactive bands (Figure 7C) showed that the pUbi promoter led to approximately 0.9% TSP of DsRed-tag54, while promoters derived from *Os.hsp18* and *Os.hsp82* led to recombinant DsRed-tag54 levels up to 0.3 and 0.15% TSP, respectively, after heat shock. Finally, use of the same pHsp82 promoter to drive the expression of the phytotoxic BP100 fusion protein led to similar accumulation levels, up to 0.12% TSP in these conditions.

One week after germination, pHsp82::*bp100-dsred-tag54* and control pHsp82::*dsred-tag54* transgenic plantlets from three different events per construct were induced by heat shock, and the recombinant chimeric protein allowed to accumulate and mature for three days prior to confocal microscopy of the plantlet radicles. All analyzed transgenic plants displayed red fluorescence, further confirming the accumulation of the recombinant proteins. Fluorescence intensities were measured in four fields per radicle. The p82::*bp100-dsred-tag54* and control p82::*dsred-tag54* plants expressed similar levels of recombinant proteins, with mean and SD values of 1258 ± 420 and 981 ± 286 fluorescence units per field, respectively (One-way ANOVA $P = 0.135$). This demonstrated that recombinant phytotoxic BP100-derived peptides could be accumulated to levels similar to those of the DsRed-tag54 reporter in stable GM plants constructed with the pHsp82 temperature inducible promoter. The DsRed-tag54 fluorescence showed a typical ER reticular pattern accompanied by a few spherical structures, attributable to the known effects of high temperatures on the internal organization of the cell, such as fragmentation of the Golgi system and the ER [34]. In contrast, fluorescence in pHsp82::*bp100-dsred-tag54* radicles was essentially in numerous and widely distributed vesicles, obscuring the ER network (Figure S3), in agreement with BP100 inducing ER-derived protein bodies [8].

Discussion

Biotechnological production of cationic α-helical antimicrobial peptides that have potent activities against pathogenic microorganisms is of great interest [35], but their constitutive expression is not compatible with the viability and fertility of host plants [7]. This problem has been previously approached by expression of peptides (i) with suboptimal antimicrobial activity linked to lower phytotoxicity [36], and (ii) with rational sequence modification to achieve a targeted reduction in phytotoxicity without affecting antimicrobial potency [8]. Modifications often alter the properties of the peptides not only in terms of antibacterial activity but also the range of target bacterial species [3,5,37]. This makes the design of a peptide with optimal properties difficult [38]. A tool to produce highly active peptides of interest in plants, irrespective of their phytotoxic character, is desirable. Inducible promoters could potentially reduce the negative impact of toxic recombinant peptides on the transgenic plant by forcefully restraining the presence of the peptide during plant development, while allowing high expression upon induction. The probable harmful effects of accumulation of recombinant phytotoxic compounds after induction, to the plant biofactory, are irrelevant since the main objective is recovery of the recombinant product. We predicted that it would only be possible to successfully generate transgenic plants carrying transgenes that encoded phytotoxic peptides by using promoters extremely inactive throughout transformation and regeneration, due to the vulnerability of the plant material during this process. Due to the practicality and low cost of using high temperature as an inducing agent in plants, and the extensive literature on the plant response to heat shock [39–44], we focused on heat shock promoters to assess the feasibility of obtaining recombinant phytotoxic antimicrobial peptides in rice, using as model, highly active and phytotoxic peptides derived from BP100.

Profiling technologies such as microarray hybridization have been widely used to systematically investigate transcriptomic changes during plant development, in different tissues and in response to diverse environmental conditions, including heat shock [45–47]. We selected candidate promoters *in silico* on the unique basis of the massive transcriptome information on publically available web sites (MIAMExpress, CREP), specifically focusing on studies on prolonged heat shock transcriptional response [10]. A similar strategy has recently been used to identify abiotic stress-inducible promoters intended to drive the overexpression of genes encoding nontoxic, stress resistance proteins in transgenic plants, with the aim of reducing the stunted growth and decrease in yield that may be caused by constitutive overexpression of these transgenes [48]. In contrast, here we selected candidate promoters for the expression of phytotoxic recombinant peptides, in plants, to drive high gene expression after heat treatment with minimal gene expression under any other condition, especially focusing on callus during agrotransformation and different tissues in the germination and seedling developmental stages.

The four sequences which best fulfilled these requirements [AK063751.1 (*Os.hsp82*), X60820.1, AU165294 and AK071240 (*Os.hsp18*), Figure 1] encoded heat shock proteins belonging to the Hsp90 (*Os.hsp82*) and the small Hsp (sHsp) families (X60820.1, AU165294 and *Os.hsp18*, UniProt database). Heat shock proteins are found in every organism [49], and play important roles in cell protection against deleterious effects of stress e.g. acting as molecular chaperones [49–55] and in cellular functions related to growth and development [49,56]. Small Hsps are the most abundant, heterogeneous and stress-responsive group of Hsp in higher plants [57–60]. The genes for more than 20 sHsps have been identified in rice, with specific spatial and temporal

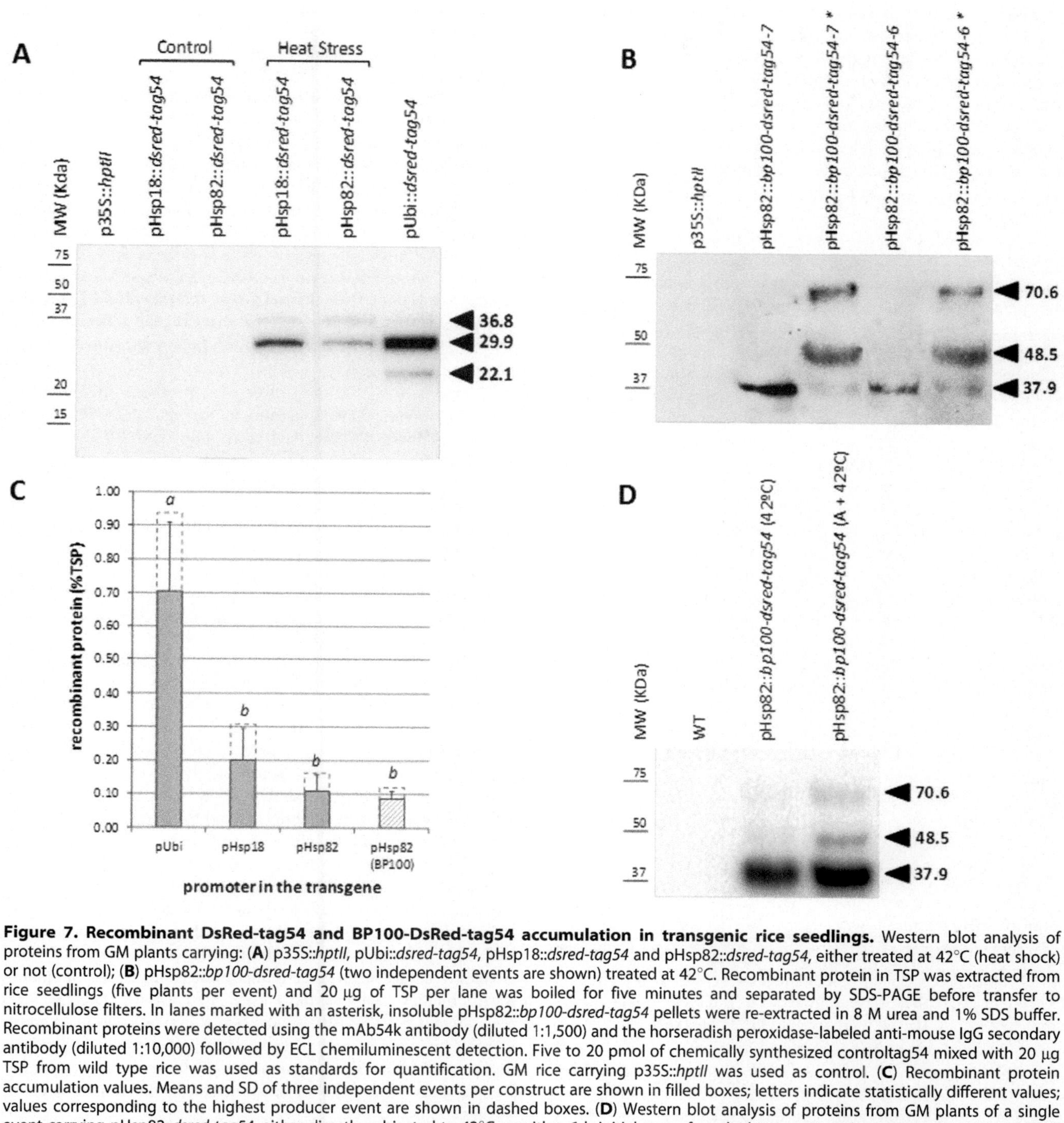

Figure 7. Recombinant DsRed-tag54 and BP100-DsRed-tag54 accumulation in transgenic rice seedlings. Western blot analysis of proteins from GM plants carrying: (**A**) p35S::*hptII*, pUbi::*dsred-tag54*, pHsp18::*dsred-tag54* and pHsp82::*dsred-tag54*, either treated at 42°C (heat shock) or not (control); (**B**) pHsp82::*bp100-dsred-tag54* (two independent events are shown) treated at 42°C. Recombinant protein in TSP was extracted from rice seedlings (five plants per event) and 20 µg of TSP per lane was boiled for five minutes and separated by SDS-PAGE before transfer to nitrocellulose filters. In lanes marked with an asterisk, insoluble pHsp82::*bp100-dsred-tag54* pellets were re-extracted in 8 M urea and 1% SDS buffer. Recombinant proteins were detected using the mAb54k antibody (diluted 1:1,500) and the horseradish peroxidase-labeled anti-mouse IgG secondary antibody (diluted 1:10,000) followed by ECL chemiluminescent detection. Five to 20 pmol of chemically synthesized controltag54 mixed with 20 µg TSP from wild type rice was used as standards for quantification. GM rice carrying p35S::*hptII* was used as control. (**C**) Recombinant protein accumulation values. Means and SD of three independent events per construct are shown in filled boxes; letters indicate statistically different values; values corresponding to the highest producer event are shown in dashed boxes. (**D**) Western blot analysis of proteins from GM plants of a single event carrying pHsp82::*dsred-tag54*, either directly subjected to 42°C or with a 6 h initial step of gradual temperature increase up to 42°C (A+42°C) (five plants per condition). Recombinant proteins were extracted in 8 M urea and 1% SDS buffer. Senia (WT) was used as control.

regulation under stress and in developmental stages [42,61]. Specifically, the *Os.hsp18* gene, encoding the cytosolic class II Hsp18.0, had very high expression upon induction, although our microarray analysis showed that it retained some activity in developmental stages and conditions that are fundamental to the production of GM plants. Hsp90 are, in general, constitutively expressed and among the most abundant proteins in cells, although their expression increases in response to stress [39]. *Os.hsp82* had extremely low expression in control tissues and reached high levels of expression in response to heat, but *Os.hsp18* and especially *Os.hsp82* were not significantly induced by other stress conditions, such as cold, NaCl and drought, in agreement with previous publications [10,62,63]. Their moderate mRNA levels in seeds, as determined using the CREP platform, would most probably allow fertility of the plants. Experimental assessment of the expression patterns of these two genes in the rice genotype of interest, Senia, allowed ratification of the *Os.hsp18* and *Os.hsp82* candidate sequences [10,62,64]. In a thorough analysis of *A. tumefaciens*-mediated transformation of Senia rice and selection using the hygromycin resistance trait, *Os.hsp18* and

Os.hsp82 mRNA levels were consistently 20 and 60-fold lower, respectively, than in leaves under induction conditions (the highest values were a transient response to *Agrobacterium* infection and hygromycin treatment) and about 10 and 300-fold lower, respectively, than those of the *hptII* transgene regulated by the p35S constitutive promoter. We therefore considered that *Os.hsp82*, and perhaps *Os.hsp18*, promoters would be able to drive the expression of transgenes encoding phytotoxic peptides in rice, possibly the latter with higher yields, and allow survival of GM plants.

Five hundred to 1,000 base pair promoter fragments of *Os.hsp18* and *Os.hsp82* regulated the expression of a reporter sequence in leaves of transgenic rice plants, in a similar manner as the endogenous genes. The low transgene mRNA levels measured prior to heat shock in GM plants carrying pHsp18::*dsred-tag54* and pHsp82::*dsred-tag54* (about 10 and 1-fold those of the reference gene, respectively) increased to approximately 700 and 400-fold that of the reference gene 2 hours after heat treatment. Even though different GM events had different transgene mRNA levels, globally the expression of endogenous *Os.hsp18* and *Os.hsp82* genes was similar to, respectively, that of the *dsred-tag54* transgene driven by the pHsp18 and pHsp82 promoters (One-way ANOVA $P = 0.465$ and $P = 0.463$, respectively). In addition, western blot analyses and confocal microscopy showed accumulation of recombinant DsRed-tag54 specifically upon heat shock in leaves and roots carrying pHsp18::*dsred-tag54* (data not shown) and pHsp82::*dsred-tag54*. The selected pHsp18 and pHsp82 promoter fragments retained the essential elements to drive heat shock induction in these tissues, and they performed as expected in GM plants. This is in agreement with previous publications where regulatory elements including the TATA box were *in silico* predicted in the two 5′ proximal regions [65,66].

Use of the pHsp18 promoter to drive the expression of the phytotoxic peptide BP100.2 did not allow survival of fertile transgenic plants. Remarkably, transgenic plants producing BP100.2 and BP100-DsRed-tag54, also phytotoxic [8], were obtained with pHsp82, with the same efficiency as those expressing a non-toxic recombinant protein such as the reporter DsRed-tag54. After heat treatment, leaves of plants carrying pHsp82::*bp100.2* and pHsp82::*bp100-dsred-tag54* produced levels of transgene mRNA similar to those harboring the *dsred-tag54* reporter under the control of the same pHsp82 promoter, indicating that these transgenes were fully functional. As BP100 is difficult to detect due to lack of antigenicity and a low extinction coefficient [8], a phenotypic approach was indirectly used to prove the synthesis of BP100.2 in these cells. High temperatures are known to cause wilting and a significant decline in relative growth rate [67]. Growth of pHsp82::*bp100.2* plants after a high temperature shock was slower than that of untransformed or transgenic control plants and there was considerably more wilting of their leaves, which supported the production of the phytotoxic recombinant BP100.2 peptide in these cells.

In view of the somewhat higher expression of *Os.hsp18* than *Os.hsp82* during transformation and regeneration (about 5–10 fold), the finding that pHsp82, but not pHsp18, was able to drive the expression of phytotoxic recombinant peptides in plants supports our initial hypothesis and experimental approach. It is essential that promoter sequences to regulate the expression of phytotoxic transgenes in plants are chosen on the basis of their extremely low basal activity in the tissues, the developmental stages and under the conditions for production of GM plants and germination. No viable transgenic plants were obtained that expressed BP100.2 under the control of 1000-bp promoter fragments of the *Os.37773.1.s1_at* (AU165294) and *Os.165.1.s1_at* genes, with mRNA levels above those of *Os.hsp18* during the transformation process (Figure 1). These results place the threshold of basal expression of a phytotoxic transgene during production of the GM plant at very low levels, similar to those of the *Os.hsp82* gene. We then re-evaluated the initially selected heat stress-responsive genes on the basis of this threshold. Five additional sequences (AB110191.1, AU165294, AK069860.1, X60820.1 and AB098712.1, Table S2) had higher expression than *Os.hsp82* after heat shock, and mRNA levels below *Os.hsp18* in control conditions. Four of them were expressed at levels above the *Os.hsp18* threshold during transformation (Figure 1), so their promoters would not be suitable for producing GM plants by driving the expression of phytotoxic components. As X60820.1 had expression patterns similar to *Os.hsp82*, its promoter most probably would be suitable to drive this expression. However, it is unlikely that yields would be substantially above those of pHsp82, as deduced from its moderate transcriptional response to heat shock.

Recombinant BP100 derivatives were produced in transgenic plants in response to heat treatment through regulation of the pHsp82 promoter. Three days after induction, BP100-DsRed-tag54 represented 0.12% of seedling proteins solubilized in urea and SDS containing buffer, determined by western blot using the tag54 epitope. Using the same promoter to drive the expression of DsRed-tag54 yielded up to 0.15% TSP, harsher denaturing conditions not significantly increasing the yield (data not shown). Although an anti-tag54 reactive band of the anticipated size was visible in TSP from heat-treated pHsp82::*bp100-dsred-tag54* leaves, the major bands, especially those only extracted under strongly denaturing conditions, had slower mobility, which is compatible with remaining BP100-driven molecular interactions. This could be explained by the unusual physicochemical properties of BP100 (highly cationic, $pI = 11.5$), known to result in interactions with many components of the cell and reduce the efficiency of isolation from complex matrices [8]. In agreement, after the same temperature treatment, BP100-DsRed-tag54 was accumulated at statistically similar levels to the reporter DsRed-tag54, as determined by quantification of the fluorescence signal in confocal micrographs (i.e. when no extraction is needed). We have previously described [8] how the BP100 sequence induces the formation of ER-derived vesicles or protein bodies (PB), functioning similarly to the tandemly repeated VPGXG elastin-like motif [68], hydrophobins [69] and the γ-zein derived Zera polypeptide [70]. Recombinant BP100-DsRed-tag54 accumulates in these vesicles, together with other luminal ER proteins, in transiently transformed *N. benthamiana* leaves [8]. Here we observed that recombinant BP100-DsRed-tag54 had a similar vesicular pattern in radicles of transgenic rice after high temperature induction, further establishing that recombinant BP100 derivatives accumulate in newly formed ER-derived vesicles in different host plant species.

It is therefore clear that phytotoxic compounds such as BP100 can be expressed in stably transformed plants using a strategy based on inducible promoters. Constitutive expression of this type of transgene has only previously been achieved in transient systems, the harmful effects of the recombinant proteins strongly interfering with normal growth and development of the transgenic plants. With the pHsp82 promoter, we estimate a yield of about 0.1% TSP in our initial induction conditions. Larkindale and Vierling [71] have showed that different acclimation schemes resulted in different yields of Hsp in *Arabidopsis*. Notably, a gradual temperature increase led to higher transcript levels than heat shock without acclimation. Here, use of an acclimation treatment (6 h gradual increase prior to 2 h at 42°C) increased the recombinant protein yield 3-fold (Figure 7D). Thus, this approach allowed obtaining up to about 0.3% total extracted phytotoxic recombinant proteins. This is about half the yield of the BP100.gtag peptide, specifically designed to display extremely low toxicity, achieved using the constitutive maize pUbi promoter in the same rice system [8]. Recombinant Cry1B has been produced in rice at similar yields (0.2% TSP) under the control of the maize proteinase inhibitor (Mpi) wound-inducible promoter [72]. Higher yields of recombinant *Acidothermus cellulolyticus* endoglucanase E1, 1.3% TSP, have been achieved in transgenic tobacco using the tomato Rubisco small subunit (RbcS-3C) light-inducible promoter [73]. Note that this type of highly inducible promoter does not allow production of phytotoxic compounds.

In a prospective exercise, our experimental approach was used to identify alternative promoter sequences with increased performance to drive the expression of phytotoxic peptides in plants. Transcriptome-wide datasets are available for rice seedlings under control and a range of abiotic stress conditions, such as salinity, desiccation, suboptimal temperature (4°C), and various minerals, elicitor and hormone treatments, and were analyzed according to the above mentioned criteria. A total of 23 sequences were identified with high expression under induction conditions (above that of *Os.hsp82*) and minimal expression levels in control conditions (below that of *Os.hsp18*) (Figure S4). On assessment of their expression patterns in callus and during the transformation process and production of transgenic plants (taken as the mean of the microarray normalized data in these tissues), there were only four sequences with mRNA levels below *Os.hsp82*, strongly suggesting that their promoters would serve to drive the expression of phytotoxic proteins in transgenic plants. They were induced by drought, NaCl, phosphorus and the heavy metals chromium (VI) and arsenic. A single sequence, Os.49245.1.S1_at, reached very high levels after drought and NaCl induction (similar to *Os.hsp18* after heat shock), making its promoter the best candidate. Five additional sequences had mRNA levels below *Os.hsp18* but above *Os.hsp82* in the analyzed tissues and developmental stages. Their promoters might also be suitable to drive the expression of toxic compounds in GM plants, but their expression levels after induction were below that of *Os.hsp18*. The best use of our promoter-selection strategy would most likely be achieved by using longer-term expression data to select sequences with the longest span of transcriptional response to the stimulus, which we could speculate would result in higher accumulation of the recombinant proteins.

In conclusion, we produced phytotoxic α-helical antimicrobial peptides derived from BP100 in plants on the basis of strict regulation of transgene transcription. A requirement was that transcription was below a rigorous threshold in specific plant tissues and developmental stages, particularly during transformation and regeneration of GM plants. The heat shock induced promoter, pHsp82, fulfilled this condition and was capable of driving production of phytotoxic recombinant peptides in plants, although the yield is rather low for most commercial applications. In addition, we demonstrate that, thanks to the increasing information available, *in silico* analysis of transcriptome profiles is a suitable and inexpensive approach to select promoters with specific activity patterns.

Supporting Information

Figure S1 A summary of the expression profiles of 16 genes selected as having the highest expression or induction in response to heat-shock in 7-day IR64 (*indica*) seedlings subjected to cold, salt and drought stress for 3 h. Data on 14-day ZhongHua 11 (*japonica*) seedlings subjected to heat shock for 3 h are also shown. Dark to light color scale represents high to low expression levels, black corresponding to 11,309 normalized fluorescence units. Arrows indicate the two sequences whose promoters were selected to produce transgenic plants.

Figure S2 Recombinant BP100-DsRed-tag54 accumulation in transgenic rice seedlings. Western blot analysis of proteins from GM plants carrying pHsp82::*dsred-tag54* and pHsp82::*bp100-dsred-tag54* (two independent events taken as example) treated at 42°C. Recombinant protein in TSP was extracted from rice seedlings (five plants per event) and 30 μg of TSP per lane was boiled for five minutes and separated by SDS-PAGE before transfer to nitrocellulose filters. Recombinant proteins were detected using the mAb54k antibody (diluted 1:1,500) and the horseradish peroxidase-labeled anti-mouse IgG secondary antibody (diluted 1:10,000) followed by ECL chemiluminescent detection.

Figure S3 Confocal micrographs of rice radicles from transgenic plantlets carrying pHsp82::*bp100-dsred-tag54* (A and C) and control pHsp82::*dsred-tag54* (B and D), at the V2 developmental stage, subjected to 42°C for 2 hours and further incubated under control growth conditions for three days in a culture chamber. (A and B), DsRed fluorescence; (C and D), bright field. Scale bars: 0.5 μm.

Figure S4 A summary of the expression profiles of a selection of sequences induced by temperature, drought and NaCl stress, and hormone treatment, in callus and seedlings of rice, based on the following *in silico* published data [74–79]. Genes with mRNA levels below that of *Os.hsp18* in control seedlings (control) and above that of *Os.hsp82* after induction (induced) are shown. Normalized expression levels in 28 tissues, relevant to the process of obtaining transgenic plants, of three rice genotypes: Minghui 63 (M, *indica*), Zhenshan 97 (Z, *indica*) and Nipponbare (N, *japonica*) (grey scale). The mRNA levels in these tissues and developmental stages were estimated using the mean expression values (green scale). Dark to light color scale represents high to low expression levels, the darkest corresponding to 37,405 (grey scale) and 12,480 (green scale) normalized fluorescence units.

Table S1 Primer sequences and PCR conditions used. Additional restriction sites are underlined.

Table S2 Sequences with the greatest expression changes in response to heat stress in rice, extracted from publically available microarray hybridization data. Details on the sequence (representative public ID, Affymetrix code and description) and the mRNA expression in response to treatment at 42°C for 3 h [normalized fluorescence units in rice seedlings under control (control) and heat-shock (heat-shock) conditions; fold change (fold) and difference (HS-C) of normalized fluorescence intensities in the two conditions are also indicated]. Note that the sequences with the same Affymetrix number correspond to the same gene. Dark to light scale of shading represents high to low expression level. The promoters of the two sequences indicated in bold were selected to produce transgenic plants.

Table S3 Transformation efficiencies of the different constructs encoding the DsRed-tag54 reporter or the phytotoxic BP100.2 AMP, under the control of a rice heat-shock (pHsp18 and pHsp82) or a maize constitutive (pUbi) promoter.

Acknowledgments

We thank J. Messeguer, E. Melé (IRTA, Cabrils) and E. Montesinos (UdG) for valuable suggestions; M. Amenós (CRAG, Barcelona) for technical assistance in confocal microscopy; and the scientific writer S. Burgess for revision of the manuscript.

Author Contributions

Conceived and designed the experiments: MP AN. Performed the experiments: NC CR. Analyzed the data: AN MP NC CR. Contributed reagents/materials/analysis tools: AN NC MP. Wrote the paper: MP AN NC.

References

1. Monroc S, Badosa E, Besalú E, Planas M, Bardají E, et al. (2006) Improvement of cyclic decapeptides against plant pathogenic bacteria using a combinatorial chemistry approach. Peptides 27: 2575–2584.
2. Cavallarín L, Andreu D, San Segundo B (1998) Cecropin A-derived peptides are potent inhibitors of fungal plant pathogens. Mol Plant Microbe Interact 11: 218–227.
3. Badosa E, Ferré R, Planas M, Feliu L, Besalú E, et al. (2007) A library of linear undecapeptides with bactericidal activity against phytopathogenic bacteria. Peptides 28: 2276–2285.
4. Marcos JF, Muñoz A, Pérez-Paya E, Misra S, López-García B (2008) Identification and rational design of novel antimicrobial peptides for plant protection. Annu Rev Phytopathol 46: 273–301.
5. López-García B, Pérez-Paya E, Marcos JF (2002) Identification of novel hexapeptides bioactive against phytopathogenic fungi through screening of a synthetic peptide combinatorial library. Appl Environ Microbiol 68: 2453–2460.
6. Montesinos E, Bardaji E (2008) Synthetic antimicrobial peptides as agricultural pesticides for plant-disease control. Chem Biodivers 5: 1225–1237.
7. Nadal A, Montero M, Company N, Badosa E, Messeguer J, et al. (2012) Constitutive expression of transgenes encoding derivatives of the synthetic antimicrobial peptide BP100: impact on rice host plant fitness. BMC Plant Biol 12: 159.
8. Company N, Nadal A, La Paz J-L, Martínez S, Rasche S, et al. (2014) The production of recombinant cationic α-helical antimicrobial peptides in plant cells induces the formation of protein bodies derived from the endoplasmic reticulum. Plant Biotechnol J 12: 81–92.
9. Qu A-L, Ding Y-F, Jiang Q, Zhu C (2013) Molecular mechanisms of the plant heat stress response. Biochem Biophys Res Commun 432: 203–207.
10. Hu W, Hu G, Han B (2009) Genome-wide survey and expression profiling of heat shock proteins and heat shock factors revealed overlapped and stress specific response under abiotic stresses in rice. Plant Sci 176: 583–590. doi:10.1016/j.plantsci.2009.01.016.
11. Qin D, Wu H, Peng H, Yao Y, Ni Z, et al. (2008) Heat stress-responsive transcriptome analysis in heat susceptible and tolerant wheat (Triticum aestivum L.) by using Wheat Genome Array. BMC Genomics 9: 432. doi:10.1186/1471-2164-9-432.
12. Ginzberg I, Barel G, Ophir R, Tzin E, Tanami Z, et al. (2009) Transcriptomic profiling of heat-stress response in potato periderm. J Exp Bot 60: 4411–4421. doi:10.1093/jxb/erp281.
13. Mittal D, Madhyastha DA, Grover A (2012) Genome-wide transcriptional profiles during temperature and oxidative stress reveal coordinated expression patterns and overlapping regulons in rice. PLoS One 7. doi:10.1371/journal.pone.0040899.
14. Li Y-F, Wang Y, Tang Y, Kakani VG, Mahalingam R (2013) Transcriptome analysis of heat stress response in switchgrass (Panicum virgatum L.). BMC Plant Biol 13: 153.
15. Zhang X, Rerksiri W, Liu A, Zhou X, Xiong H, et al. (2013) Transcriptome profile reveals heat response mechanism at molecular and metabolic levels in rice flag leaf. Gene 530: 185–192. doi:10.1016/j.gene.2013.08.048.
16. Todaka D, Nakashima K, Shinozaki K, Yamaguchi-Shinozaki K (2012) Toward understanding transcriptional regulatory networks in abiotic stress responses and tolerance in rice. Rice 5: 6. doi:10.1186/1939-8433-5-6.
17. Hadiarto T, Tran L-SP (2011) Progress studies of drought-responsive genes in rice. Plant Cell Rep 30: 297–310. doi:10.1007/s00299-010-0956-z.
18. Khurana N, Chauhan H, Khurana P (2013) Wheat Chloroplast Targeted sHSP26 Promoter Confers Heat and Abiotic Stress Inducible Expression in Transgenic Arabidopsis Plants. PLoS One 8: e54418. doi:10.1371/journal.pone.0054418.
19. Ramessar K, Capell T, Christou P (2008) Molecular pharming in cereal crops. Phytochem Rev 7: 579–592. doi:10.1007/s11101-008-9087-3.
20. Wakasa Y, Takaiwa F (2013) The use of rice seeds to produce human pharmaceuticals for oral therapy. Biotechnol J 8: 1133–1143. doi: 10.1002/biot.201300065.
21. Irizarry RA, Hobbs B, Collin F, Beazer-Barclay YD, Antonellis KJ, et al. (2003) Exploration, normalization, and summaries of high density oligonucleotide array probe level data. Biostatistics 4: 249–264.
22. Yamamoto YY, Obokata J (2008) ppdb: a plant promoter database. Nucleic Acids Res 36: D977–D981. doi:10.1093/nar/gkm785.
23. Hieno A, Naznin HA, Hyakumachi M, Sakurai T, Tokizawa M, et al. (2013) Ppdb: Plant Promoter Database Version 3.0. Nucleic Acids Res: 1–5.
24. Cornelissen BJ, Horowitz J, van Kan JA, Goldberg RB, Bol JF (1987) Structure of tobacco genes encoding pathogenesis-related proteins from the PR-1 group. Nucleic Acids Res 15: 6799–6811.
25. Vaquero C, Sack M, Chandler J, Drossard J, Schuster F, et al. (1999) Transient expression of a tumor-specific single-chain fragment and a chimeric antibody in tobacco leaves. Proc Natl Acad Sci USA 96: 11128–11133.
26. Matz MV, Fradkov AF, Labas YA, Savitsky AP, Zaraisky AG, et al. (1999) Fluorescent proteins from nonbioluminescent Anthozoa species. Nat Biotechnol 17: 969–973.
27. Rasche S, Martin A, Holzem A, Fischer R, Schinkel H, et al. (2011) One-step protein purification: Use of a novel epitope tag for highly efficient detection and purification of recombinant proteins. Open Biotechnol J 5: 1–6.
28. Sambrook J, Russell D (2001) Molecular Cloning: a laboratory manual. Cold Spring Harbor, N.Y.: Cold Spring Harbor Laboratory Press.
29. Sallaud C, Meynard D, van Boxtel J, Gay C, Bes M, et al. (2003) Highly efficient production and characterization of T-DNA plants for rice (Oryza sativa L.) functional genomics. Theor Appl Genet 106: 1396–1408. doi:10.1007/s00122-002-1184-x.
30. Murashige T, Skoog F (1962) A revised medium for rapid growth and bioassays with tobacco tissue cultures. Physiol Plant 15: 473–497.
31. Vandesompele J, De Preter K, Pattyn F, Poppe B, Van Roy N, et al. (2002) Accurate normalization of real-time quantitative RT-PCR data by geometric averaging of multiple internal control genes. Genome Biol 3: RESEARCH0034. doi:10.1186/gb-2002-3-7-research0034.
32. Christensen AH, Quail PH (1996) Ubiquitin promoter-based vectors for high-level expression of selectable and/or screenable marker genes in monocotyledonous plants. Transgenic Res 5: 213–218.
33. Gross LA, Baird GS, Hoffman RC, Baldridge KK, Tsien RY (2000) The structure of the chromophore within DsRed, a red fluorescent protein from coral. Proc Natl Acad Sci USA 97: 11990–11995. doi:10.1073/pnas.97.22.11990.
34. Richter K, Haslbeck M, Buchner J (2010) The Heat Shock Response: Life on the Verge of Death. Mol Cell 40: 253–266. doi:10.1016/j.molcel.2010.10.006.
35. Rajasekaran K, Cary J, Jaynes J, Montesinos E (2012) Small wonders: peptides for disease control. Washington, DC: Oxford University Press: American Chemical Society: Oxford University Press.
36. Osusky M, Zhou G, Osuska L, Hancock RE, Kay WW, et al. (2000) Transgenic plants expressing cationic peptide chimeras exhibit broad-spectrum resistance to phytopathogens. Nat Biotechnol 18: 1162–1166.
37. Monroc S, Badosa E, Feliu L, Planas M, Montesinos E, et al. (2006) De novo designed cyclic cationic peptides as inhibitors of plant pathogenic bacteria. Peptides 27: 2567–2574.
38. Badosa E, Moiset G, Montesinos L, Talleda M, Bardají E, et al. (2013) Derivatives of the Antimicrobial Peptide BP100 for Expression in Plant Systems. PLoS One 8: e85515.
39. Efeoğlu B (2009) Heat Shock Proteins and Heat Shock Response in Plants. J Sci 22: 67–75.
40. Kotak S, Larkindale J, Lee U, von Koskull-Döring P, Vierling E, et al. (2007) Complexity of the heat stress response in plants. Curr Opin Plant Biol 10: 310–316. doi:10.1016/j.pbi.2007.04.011.
41. Sarkar NK, Thapar U, Kundnani P, Panwar P, Grover A (2013) Functional relevance of J-protein family of rice (Oryza sativa). Cell Stress Chaperones 18: 321–331.
42. Sarkar NK, Kim Y-K, Grover A (2009) Rice sHsp genes: genomic organization and expression profiling under stress and development. BMC Genomics 10: 393. doi:10.1186/1471-2164-10-393.
43. Sarkar NK, Kundnani P, Grover A (2013) Functional analysis of Hsp70 superfamily proteins of rice (Oryza sativa). Cell Stress Chaperones 18: 427–437.
44. Singh A, Singh U, Mittal D, Grover A (2010) Genome-wide analysis of rice ClpB/HSP100, ClpC and ClpD genes. BMC Genomics 11: 95. doi:10.1186/1471-2164-11-95.
45. Sarkar NK, Kim Y-K, Grover A (2014) Coexpression network analysis associated with call of rice seedlings for encountering heat stress. Plant Mol Biol 84: 125–143.
46. Jung Y-J, Lee S-Y, Moon Y-S, Kang K-K (2012) Enhanced resistance to bacterial and fungal pathogens by overexpression of a human cathelicidin

antimicrobial peptide (hCAP18/LL-37) in Chinese cabbage. Plant Biotechnol Rep 6: 39–46. doi:10.1007/s11816-011-0193-0.

47. Yamakawa H, Hakata M (2010) Atlas of rice grain filling-related metabolism under high temperature: joint analysis of metabolome and transcriptome demonstrated inhibition of starch accumulation and induction of amino acid accumulation. Plant Cell Physiol 51: 795–809. doi:10.1093/pcp/pcq034.
48. Rerksiri W, Zhang X, Xiong H, Chen X (2013) Expression and promoter analysis of six heat stress-inducible genes in rice. ScientificWorldJournal 2013: 397401.
49. Lindquist S, Craig E (1988) The heat-shock proteins. Annu Rev Genet 22: 631–677.
50. Lee GJ, Roseman AM, Saibil HR, Vierling E (1997) A small heat shock protein stably binds heat-denatured model substrates and can maintain a substrate in a folding-competent state. EMBO J 16: 659–671. doi:10.1093/emboj/16.3.659.
51. Blumenthal C, Bekes F, Wrigley C, Barlow E (1990) The Acquisition and Maintenance of Thermotolerance in Australian Wheats. Aust J Plant Physiol 17: 37. doi:10.1071/PP9900037.
52. Feder ME, Hofmann GE (1999) Heat-shock proteins, molecular chaperones, and the stress response: evolutionary and ecological physiology. Annu Rev Physiol 61: 243–282. doi:10.1146/annurev.physiol.61.1.243.
53. Iba K (2002) Acclimative response to temperature stress in higher plants: approaches of gene engineering for temperature tolerance. Annu Rev Plant Biol 53: 225–245. doi:10.1146/annurev.arplant.53.100201.160729.
54. Sorensen JG, Kristensen TN, Loeschcke V (2003) The evolutionary and ecological role of heat shock proteins. Ecol Lett 6: 1025–1037.
55. Young R, Elliott T (1989) Stress proteins, infection, and immune surveillance. Cell 59: 5–8.
56. Vinocur B, Altman A (2005) Recent advances in engineering plant tolerance to abiotic stress: Achievements and limitations. Curr Opin Biotechnol 16: 123–132. doi:10.1016/j.copbio.2005.02.001.
57. Vierling E (1991) The Roles of Heat Shock Proteins in Plants. Annu Rev Plant Physiol Plant Mol Biol 42: 579–620.
58. Siddique M, Gernhard S, von Koskull-Döring P, Vierling E, Scharf K-D (2008) The plant sHSP superfamily: five new members in Arabidopsis thaliana with unexpected properties. Cell Stress Chaperones 13: 183–197. doi:10.1007/s12192-008-0032-6.
59. Waters ER, Lee GJ, Vierling E (1996) Evolution, structure and function of the small heat shock proteins in plants. J Exp Bot 47: 325–338.
60. Sun W, Van Montagu M, Verbruggen N (2002) Small heat shock proteins and stress tolerance in plants. Biochim Biophys Acta 1577: 1–9.
61. Ouyang Y, Chen J, Xie W, Wang L, Zhang Q (2009) Comprehensive sequence and expression profile analysis of Hsp20 gene family in rice. Plant Mol Biol 70: 341–357. doi:10.1007/s11103-009-9477-y.
62. Sarkar NK, Kim YK, Grover A (2009) Rice sHsp genes: genomic organization and expression profiling under stress and development. BMC Genomics 10: 393.
63. Ham D-J, Moon J-C, Hwang S-G, Jang CS (2013) Molecular characterization of two small heat shock protein genes in rice: their expression patterns, localizations, networks, and heterogeneous overexpressions. Mol Biol Rep: 1–12.
64. Chang PFL, Jinn TL, Huang WK, Chen Y, Chang HM, et al. (2007) Induction of a cDNA clone from rice encoding a class II small heat shock protein by heat stress, mechanical injury, and salicylic acid. Plant Sci 172: 64–75. doi:10.1016/j.plantsci.2006.07.017.
65. Siván P, Svec M (2009) Genome-wide analysis of rice (Oryza sativa L. subsp. japonica) TATA box and Y Patch promoter elements. Genome 52: 294–297. doi:10.1139/G09-001.
66. Yamamoto YY, Ichida H, Matsui M, Obokata J, Sakurai T, et al. (2007) Identification of plant promoter constituents by analysis of local distribution of short sequences. BMC Genomics: 8:8:67. doi:10.1186/1471-2164-8-67.
67. Wahid A, Gelani S, Ashraf M, Foolad MR (2007) Heat tolerance in plants: An overview. Environ Exp Bot 61: 199–223. doi:10.1016/j.envexpbot.2007.05.011.
68. Conley AJ, Joensuu JJ, Menassa R, Brandle JE (2009) Induction of protein body formation in plant leaves by elastin-like polypeptide fusions. BMC Biol 7: 48.
69. Joensuu JJ, Conley AJ, Lienemann M, Brandle JE, Linder MB, et al. (2010) Hydrophobin fusions for high-level transient protein expression and purification in Nicotiana benthamiana. Plant Physiol 152: 622–633. doi:10.1104/pp.109.149021.
70. Torrent M, Llompart B, Lasserre-Ramassamy S, Llop-Tous I, Bastida M, et al. (2009) Eukaryotic protein production in designed storage organelles. BMC Biol 7: 5.
71. Larkindale J, Vierling E (2008) Core genome responses involved in acclimation to high temperature. Plant Physiol 146: 748–761. doi:10.1104/pp.107.112060.
72. Breitler JC, Vassal JM, del Mar Catala M, Meynard D, Marfà V, et al. (2004) Bt rice harbouring cry genes controlled by a constitutive or wound-inducible promoter: protection and transgene expression under Mediterranean field conditions. Plant Biotechnol J 2: 417–430. doi:10.1111/j.1467-7652.2004.00086.x.
73. Dai Z, Hooker BS, Anderson DB, Thomas SR (2000) Expression of Acidothermus cellulolyticus endoglucanase E1 in transgenic tobacco: biochemical characteristics and physiological effects. Transgenic Res 9: 43–54. doi:10.1023/A:1008922404834.
74. Jain M, Khurana JP (2009) Transcript profiling reveals diverse roles of auxin-responsive genes during reproductive development and abiotic stress in rice. FEBS J 276: 3148–3162. doi:10.1111/j.1742-4658.2009.07033.x.
75. Jain M, Nijhawan A, Arora R, Agarwal P, Ray S, et al. (2007) F-box proteins in rice. Genome-wide analysis, classification, temporal and spatial gene expression during panicle and seed development, and regulation by light and abiotic stress. Plant Physiol 143: 1467–1483. doi:10.1104/pp.106.091900.
76. Sharma R, Mohan Singh RK, Malik G, Deveshwar P, Tyagi AK, et al. (2009) Rice cytosine DNA methyltransferases - gene expression profiling during reproductive development and abiotic stress. FEBS J 276: 6301–6311. doi:10.1111/j.1742-4658.2009.07338.x.
77. Zheng L, Huang F, Narsai R, Wu J, Giraud E, et al. (2009) Physiological and transcriptome analysis of iron and phosphorus interaction in rice seedlings. Plant Physiol 151: 262–274. doi:10.1104/pp.109.141051.
78. Dubey S, Misra P, Dwivedi S, Chatterjee S, Bag SK, et al. (2010) Transcriptomic and metabolomic shifts in rice roots in response to Cr (VI) stress. BMC Genomics 11: 648. doi:10.1186/1471-2164-11-648.
79. Kang K, Park S, Natsagdorj U, Kim YS, Back K (2011) Methanol is an endogenous elicitor molecule for the synthesis of tryptophan and tryptophan-derived secondary metabolites upon senescence of detached rice leaves. Plant J 66: 247–257. doi:10.1111/j.1365-313X.2011.04486.x.

17

Estimated Dietary Intake of Radionuclides and Health Risks for the Citizens of Fukushima City, Tokyo, and Osaka after the 2011 Nuclear Accident

Michio Murakami[1,2]*, Taikan Oki[1,2]

1 Institute of Industrial Science, The University of Tokyo, Meguro, Tokyo, Japan, 2 Japan Science and Technology Agency, Core Research for Evolutionary Science and Technology (CREST), Chiyoda, Tokyo, Japan

Abstract

The radionuclides released from the Fukushima Daiichi nuclear power plant in 2011 pose a health risk. In this study, we estimated the 1st-year average doses resulting from the intake of iodine 131 (^{131}I) and cesium 134 and 137 (^{134}Cs and ^{137}Cs) in drinking water and food ingested by citizens of Fukushima City (~50 km from the nuclear power plant; outside the evacuation zone), Tokyo (~230 km), and Osaka (~580 km) after the accident. For citizens in Fukushima City, we considered two scenarios: Case 1, citizens consumed vegetables bought from markets; Case 2, citizens consumed vegetables grown locally (conservative scenario). The estimated effective doses of ^{134}Cs and ^{137}Cs agreed well with those estimated through market basket and food-duplicate surveys. The average thyroid equivalent doses due to ingestion of ^{131}I for adults were 840 μSv (Case 1) and 2700 μSv (Case 2) in Fukushima City, 370 μSv in Tokyo, and 16 μSv in Osaka. The average effective doses due to ^{134}Cs and ^{137}Cs were 19, 120, 6.1, and 1.9 μSv, respectively. The doses estimated in this study were much lower than values reported by the World Health Organization and the United Nations Scientific Committee on the Effects of Atomic Radiation, whose assessments lacked validation and full consideration of regional trade in foods, highlighting the importance of including regional trade. The 95th percentile effective doses were 2–3 times the average values. Lifetime attributable risks (LARs) of thyroid cancers due to ingestion were $2.3–39\times10^{-6}$ (Case 1) and $10–98\times10^{-6}$ (Case 2) in Fukushima City, $0.95–14\times10^{-6}$ in Tokyo, and $0.11–1.3\times10^{-6}$ in Osaka. The contributions of LARs of thyroid cancers due to ingestion were 7.5%–12% of all exposure (Case 1) and 12%–30% (Case 2) in Fukushima City.

Editor: Bart O. Williams, Van Andel Institute, United States of America

Funding: This study was supported by a "Core Research for Evolutionary Science and Technology (CREST)" grant from the Japan Science and Technology Agency, http://www.jst.go.jp/kisoken/crest/en/index.html. MM and TO receive the funding. The funder had no role in study design, data collection and analysis, discussion to publish, or preparation of the manuscript.

Competing Interests: The authors have declared that no competing interests exist.

* Email: michio@iis.u-tokyo.ac.jp

Introduction

Radionuclides were released from the Tokyo Electric Power Company's Fukushima Daiichi nuclear power plant, mainly on 15 March 2011, after the Great East Japan Earthquake and tsunami on 11 March. They were diffused into the atmosphere, deposited mainly through precipitation, and incorporated into surface waters, drinking waters, agricultural crops, and aquatic organisms. Radionuclides have been detected in drinking water and foods in Fukushima and other prefectures in Japan, including Tokyo [1,2,3]. Radioactive iodine 131 (^{131}I), which has a short half-life (8.04 d [4]), can contaminate drinking water and foods immediately after its release, whereas radioactive cesium 134 and 137 (^{134}Cs and ^{137}Cs), which have longer half-lives (2.06 and 30 y, respectively [4]), cause contamination over longer periods of time. Although other radionuclides such as strontium 90 (^{90}Sr), ruthenium 106 (^{106}Ru), and plutonium have half-lives of >1 y, in the year after the accident ^{134}Cs and ^{137}Cs contributed an estimated 84% to 88% of the total radiation dose from radionuclides with half-lives of >1 y in the diet [5].

In response to the accident, the Japanese government announced provisional "indices relating to limits on food and drink ingestion" on 17 March 2011 [1]. The government began releasing monitoring data on radionuclide concentrations in foods on 19 March and restricted the distribution of foods collected in some municipalities from 21 March, as the foods exceeded the limits. Some municipalities in Fukushima prefecture voluntarily imposed more stringent limits. In addition, some local governments, including the Tokyo Metropolitan Government, distributed bottled water to protect the health of citizens, particularly infants, against the high concentrations of ^{131}I detected in drinking water.

Among the exposure pathways, internal exposure from dietary intake has not been fully elucidated. External exposure can be determined from personal dosimeters [6], calculated from ambient dose monitoring and shielding factors [7], or estimated by using numerical dispersion models [8]. Internal exposure from inhalation can be estimated by air monitoring [9] or by numerical dispersion models [8,10]. Internal exposure to Cs can be measured by whole-body-counter surveys; however, the Cs burdens in most inhabitants of Fukushima have been below the detection limit

[11,12]. Internal exposure from the intake of water and food can be determined from market basket surveys (i.e., foods are purchased from the market according to consumption patterns) or food-duplicate surveys (i.e., cooked foods are collected from home) [1,13], but exposure immediately after the accident could not be evaluated by these methods because of the lack of samples. On the other hand, individual foods were monitored just after the accident for restriction of the distribution of foods, as described above. On the basis of the monitoring data, working groups of the Ministry of Health, Labour and Welfare [14] and the World Health Organization (WHO) [15] approximated the nationwide average internal exposure to radionuclides from foods. The United Nations Scientific Committee on the Effects of Atomic Radiation (UNSCEAR) also recently reported that an increased risk of thyroid cancer for infants and children in particular is of concern, that most of the absorbed dose to the thyroid was received during the first year after the accident, and that the major route of exposure was ingestion; e.g., absorbed dose to thyroid in the first year for 1-year-olds in Fukushima Prefecture (districts not evacuated): external + inhalation, 0.2–19 mGy; ingestion, 33 mGy [16]. However, the regional trade in foods was not fully considered in these assessments despite the low food self-sufficiency rates in Japan (54% for major grains, 39% on basis of total energy [17]), and no validations based on the results of market basket, food-duplicate, or whole-body-counter surveys were done. More detailed estimates and validation are required on account of regional variations in concentrations in drinking water and foods, the regional trade in foods, and variations in individual exposure. Furthermore, recent screening of all children in Fukushima Prefecture identified thyroid cancer in 90 of more than 280 000 children (an incidence of 313 per 1 million) by March 2014 [18]. Although this incidence might be attributable to increased screening by advanced ultrasound techniques [18], an increase incidence of cancer in future, especially thyroid cancer in children, is a matter of concern. The the contribution of cancer risk from ingestion needs to be understood.

To provide more detail, we estimated the 1st-year average doses due to intake of drinking water and foods after the accident from radionuclide concentrations measured in water and foods and the effects of countermeasures (i.e., restriction of food distribution, voluntary withholding of rice, and distribution of bottled water for infants). We previously reported the average thyroid doses resulting from the intake of ^{131}I by citizens of Tokyo and the effects of countermeasures [19]. Here, we estimated the intake of ^{131}I, ^{134}Cs, and ^{137}Cs by the citizens of Fukushima City, Tokyo, and Osaka, which differ in the regional trade in foods. The results were validated against observations in market basket, food-duplicate, and whole-body-counter surveys. By comparing the results with reports by WHO and UNSCEAR, whose assessments lacked validation and full consideration of regional trade in foods, we evaluated the role of regional trade in foods. We also estimated variations in intakes by using a Monte Carlo simulation. Finally, we assessed the cancer risks due to intake. We considered all sources of exposure for citizens in Fukushima City, and evaluated the contributions of cancer risks due to ingestion to the overall risk.This information will help to address current concerns of citizens and be useful for establishing the regulation of foods and drinking water.

Methods

Intake pathway

To assess intake via the ingestion of drinking water and foods, we estimated the thyroid equivalent dose and effective dose of ^{131}I and the effective dose of ^{134}Cs and ^{137}Cs among citizens of Fukushima City (~50 km from the nuclear power plant; capital of prefecture), Tokyo (~230 km), and Osaka (~580 km). All three cities, including Fukushima City, were outside the evacuation zone. Citizens were classified into ten groups based on age, sex, and pregnancy: <1-y-old infants, males and females aged 1–6, 7–12, 13–18, and ≥19 y old, and pregnant females.

Monitoring data on drinking water and foods were obtained from ref [1,2,3,20]. Drinking water and foods were classified into 18 categories [21]: drinking water, rice, other grains, potato, leafy vegetables, root crops, beans, fruit vegetables, milk, dairy products, formula milk, beef, pork, chicken, chicken eggs, fresh fisheries products, marine products, and others. "Fruit vegetables" included vegetable-like fruits such as tomatoes and fruits such as persimmons and apples. They were then further classified into sub-categories based on governmental restrictions [1]: leafy vegetables were divided into "spinach", "garland chrysanthemum and ging-geng-cai", "mustard spinach and non-heading lettuce", "heading leafy vegetables", "broccoli and cauliflower", and "naganegi onion, chive, and asparagus"; root crops into "turnip", "bamboo shoots", and "other root crops"; fruit vegetables into "kiwifruit", "chestnut", and "other fruit vegetables"; fresh fisheries products into "wild *ayu*, wild Japanese dace, and wild landlocked *masu* salmon" and "other fresh fisheries products"; and other into "tea", "shiitake mushroom (virgin wood)", and "other mushrooms".

Drinking water

The average doses (thyroid equivalent dose of ^{131}I and the effective dose of ^{131}I, ^{134}Cs, and ^{137}Cs) from drinking water were estimated as:

$$Dose_k(\mu Sv) = A_k \times B \times \sum_t C_{kt} \quad (1)$$

where k is the radionuclide, A_k is the dose coefficient for radionuclide k (μSv/Bq) (Table S1), B is the daily consumption of drinking water per person (g/d) (Table S2), t is the number of days after the accident (consumption date), and C_{kt} is the concentration of radionuclide k in drinking water at t days after the accident (Bq/g).

So as not to underestimate the dose, we set A_k for ^{131}I on the assumption that the fractional thyroid uptake from blood is 0.3, as used in the determination of control index levels by the Nuclear Safety Commission of Japan [22]. All the dose coefficients are reference values taken from the publications of the International Commission on Radiological Protection (ICRP) (Table S1).

B was set at 710 g/d for <1 y, 1000 g/d for 1–6 y, and 1650 g/d for >6 y [22]. The average daily consumption of water in soup and rice was then added [19]: The consumption of water in soup was calculated from the consumption of soup by adults [23] and the ratio of miso (soybean paste) consumption by each age group to that by adults [24]. The consumption of water in rice was calculated from the consumption of rice [21] and the ratio of cooking water to rice [25].

The radionuclide concentrations in drinking water were the values monitored in tap water in Fukushima City [3]; Shinjuku, Tokyo [2]; and Osaka (not detected) [20]. As details of ^{134}Cs and ^{137}Cs concentrations in tap water in Fukushima City were not available, we estimated them from the ^{131}I concentrations and the ratios of ^{134}Cs and ^{137}Cs to ^{131}I in tap water in Tokyo between 19 March and 8 April. Intake was estimated from the date when radionuclides were first detected (16 March in Fukushima City, 18 March in Tokyo). As they were not detected in tap water after 4

May (except on 2 July in Tokyo), intake after then was considered to be negligible.

Foods

The average doses from foods were estimated by the following equation [19], which is modified to include ^{134}Cs and ^{137}Cs, product origins, and food categories:

$$Dose_k(\mu Sv) = A_k \times \sum_i \sum_j \sum_t (B_i \times C_{kijt} \times D_{ij}/100) \quad (2)$$

where k is the radionuclide, A_k is the dose coefficient for radionuclide k (μSv/Bq) (Table S1), i is the individual food category, j is the individual area (source prefecture or prefectural sub-group), t is the number of days after the accident (consumption date), B_i is the daily consumption of food i per person (g/d) (Table S2), C_{kijt} is the arithmetic mean concentration of radionuclide k in food i in area j at t days after the accident (Bq/g), and D_{ij} is the arrival share in an area (the fraction of food i in the market that comes from prefecture or area j) (%).

Daily consumption. For the daily consumption of each food recently reported detailed values were adopted [21]. The daily consumption of the sub-categorized foods was estimated from the daily consumption of the foods in the main categories and the ratio of the amount of each sub-category arriving at the Tokyo Metropolitan Central Wholesale Market to the total in 2010 [26]. The national fishery yield was used to apportion the daily consumption of fresh fisheries products [27]. The daily consumption of mushrooms followed reported values [24]. The daily consumption of tea by ≥19-y-old females was 439 g/d [28], and that by other groups was calculated from that and the ratio of the daily consumption of drinking water excluding water in soup and rice in each group to that in ≥19-y-old females. We assumed that 10 g of leaf is used to make 300 g of tea and that 60% of radionuclides in the leaf enter the tea [14]. The ratio of virgin wood to other substrates used to grow shiitake was used to estimate the daily consumption of "shiitake mushroom (virgin wood)" [29].

Radionuclide concentrations. The radionuclide concentrations in food were drawn from >130 000 food monitoring data [1] and rice monitoring data [3]. Details of the number of samples analyzed each month and in each prefecture are given in the Supporting Information and Tables S3 and S4. Foods were washed with water and the edible parts were then analyzed for radionuclides by Ge, NaI or CsI detectors. We considered the intake of radionuclides via the consumption of foods collected in 16 prefectures: Hokkaido, Aomori, Iwate, Miyagi, Akita, Yamagata, Fukushima, Ibaraki, Tochigi, Gunma, Saitama, Chiba, Kanagawa, Tokyo (Metropolis), Yamanashi, and Shizuoka (Figure S1). Analysis of the variations in radionuclide concentrations among foods in Fukushima Prefecture showed that spinach, broccoli and cauliflower, and marine products were heavily contaminated (Figures S2 and S3). Analysis of the regional differences in radionuclide concentrations in spinach showed that foods collected in Fukushima Prefecture and in prefectures in the Kanto region (Ibaraki, Tochigi, Gunma, Saitama, Chiba, Kanagawa, and Tokyo) had much higher levels than those in the other 8 prefectures (Figure S4). The amounts of radionuclides originating from other prefectures or from overseas were regarded as negligible. When radionuclides were not detected in a food, the contribution of radionuclides from that food to the dose was considered to be negligible. The concentrations in each food were classified by source prefecture and date. The prefectures where the distribution of the food was restricted (for example, rice in Fukushima Prefecture) were further classified into prefectural sub-groups. The arithmetic mean concentration in each area on each day was calculated. When only the sum of ^{134}Cs and ^{137}Cs concentrations was available, individual concentrations were estimated from the average ratio of ^{134}Cs to the sum of ^{134}Cs and ^{137}Cs, namely 0.49 (March–June 2011), 0.46 (July–September 2011), 0.45 (October–December 2011), or 0.42 (January–March 2012), as calculated from foods in which both the ^{134}Cs and the ^{137}Cs concentration exceeded 25 Bq/kg [1]. The radionuclide concentrations in crude tea and manufactured tea were converted to those in tea leaf by multiplying by the weight ratio of crude tea to tea leaf (0.22) [30]. The concentrations of ^{131}I in foods on days before monitoring began or for which there were no data were approximated from the concentrations measured on the closest day and from the half-life of ^{131}I (8.04 d [4]). The concentrations of ^{134}Cs and ^{137}Cs on days before monitoring began or for which there were no data were regarded as the same as the concentrations measured on the closest day. The consumption dates and concentrations in foods were assumed to be as reported [1,3].

Arrival shares. Since citizens generally purchase food from markets, which are supplied from the whole country, the arrival share is an important parameter for estimating the doses due to ingestion. A few citizens in Fukushima City consumed vegetables produced in local fields or home gardens (4.5% of workers in Fukushima City were agricultural workers [31]). We therefore considered two scenarios: Case 1, citizens consumed foods from markets; Case 2, citizens consumed vegetables produced in Fukushima Prefecture and other foods from markets. Case 2 means the arrival shares of vegetables from Fukushima Prefecture are 100%, so this scenario can be regarded as conservative. For citizens in Tokyo and Osaka, only Case 1 was considered. The proportions of targeted foods produced in each area for the markets in each city were calculated as follows. The arrival shares of vegetables and fisheries products were based on the amounts arriving from each prefecture at each central wholesale market in 2010 [26,32,33]. We used the 2010 data so as to evaluate the effects of countermeasures. Since data on the importation of fisheries products from overseas were not available, we used the Japanese food self-sufficiency rates based on quantities of production, imports, and exports [17]. The production of marine products in Fukushima Prefecture was regarded to be nil owing to the tsunami damage and the suspension of fishing. The arrival shares of beef, pork, and chicken eggs were based on the quantities transported from each prefecture to Fukushima, Tokyo, and Osaka prefectures [34,35]. The arrival shares of beef and pork were corrected to the food self-sufficiency rates of Japan. The arrival shares of milk and dairy products were based on the production and transport of raw milk in each prefecture, the amount of milk transported from each prefecture to Fukushima, Tokyo, and Osaka prefectures [36], and the food self-sufficiency rates. For rice, other grains, tea, soybeans, and chicken, we used the nationwide production rates [30,35,37,38] and the food self-sufficiency rates of Japan. The use of these sources is reasonable, especially for dried foods. To estimate the ratios of restricted food production among the prefectural sub-groups, we used the number of cows in each municipality for milk and dairy products [39], the cultivated areas of vegetables and tea [40], the areas of rice paddies [40] and the number of rice farm households [3], and the production [41] or the number of management bodies [42] for shiitake mushrooms.

Calculation periods. It is likely that rainfall on the night of 15–16 March 2011 [43] caused heavy contamination of leafy vegetables and of milk and dairy products by radionuclides, so we estimated intake via these products before the first release of data

(19 March 2011): from 17 March 2011 in Fukushima City, on the assumption that delivery to citizens takes 1 day; and from 18 March 2011 in Tokyo and Osaka, on the assumption that it takes 2 days. The intake of other foods was estimated from 21 March 2011, when the Japanese government restricted the distribution of foods. Since radionuclide contamination in other foods was not attributable to direct deposition [22], the intake before 21 March was not included. This was supported by the lower contamination of other foods by ^{131}I and ^{134}Cs and ^{137}Cs than of leafy vegetables (Figures S2, S3). The intake was estimated from these days to 20 March 2012.

Effects of countermeasures

We evaluated the effects of countermeasures (restrictions on the distribution of foods, voluntary withholding of rice, and distribution of bottled water for infants) on reducing intake. We assumed the consumption of foods harvested in the areas where distribution was restricted or rice was withheld to be nil: citizens ate alternative foods containing negligible radionuclides. In this regard, however, for Case 2, the conservative scenario, we considered the effects of restrictions on the distribution of other foods except vegetables, because we assumed that some citizens consumed vegetables grown locally. Since the Tokyo Metropolitan Government distributed bottled water for infants on the morning of 24 March, we evaluated the effect of this countermeasure on 24 and 25 March. (Fukushima City distributed bottled water on the night of 11 March following the earthquake, but this was too early to consider).

Variations in individual doses

The variations in ingestion doses for ≥19-y-old males (with countermeasures in place) were estimated by Monte Carlo simulation using the commercially available software Crystal Ball (Oracle, California, USA). The drinking water and food variation data input into the Monte Carlo simulation were obtained as below.

We took into account variations in daily consumption of drinking water [44] (Table S2). The drinking water in Tokyo is derived from several sources: 38.0% of tap water comes from the Edogawa River, 41.3% from the Arakawa River, 14.4% from the Tamagawa River, and 6.3% from other sources such as ground water [45,46]. Therefore, we also took into account the radionuclide concentrations in each source. We assumed that citizens drank water from a single source. The radionuclide concentration factors were estimated from the ratios of ^{131}I concentrations in representative purification plants (Kanamachi purification plant for the Edogawa River, Asaka purification plant for the Arakawa River, Ozaku purification plant for the Tamagawa River) to those in tap water in Shinjuku [2,46]. We used the arithmetic mean concentrations on 22–28 March in the purification plants and on 23–29 March in Shinjuku, on the assumption that delivery to the tap takes 1 day. Concentrations were corrected for half-life. Concentration factors were 3.51 for the Edogawa River, 1.31 for the Arakawa River, and 0.46 for the Tamagawa River (i.e., the ^{131}I concentration in the Ozaku purification plant on the Tamagawa River was lower than that in tap water in Shinjuku). The ^{131}I concentration in other water systems was regarded as negligible. Fukushima City has only one drinking water treatment plant, and radionuclides were not detected in the drinking water in Osaka. We therefore did not need to take into account variations in the water resources in these two cities.

For input into the Monte Carlo simulation, variations in the dose from drinking water in Fukushima City were estimated from the variation in daily consumption under the assumption of a log-normal distribution. Variations in Tokyo were estimated from variations in daily consumption, the dependence on water source, and the concentration factor.

Variations in the dose from foods for input into the model were estimated under the assumption that citizens bought each food once a week, and the production area was randomly selected according to the distribution of the arrival shares. The weekly average concentrations of radionuclides in each food were calculated. The variations in the daily consumption of each food were as reported [28,47]. Since these values are based on 1-day surveys, not long-term surveys, their use could have caused us to overestimate the results. The variations in doses obtained by ingesting each food from the same area in the same week at the same daily rate of consumption were considered to indicate variations in the radionuclide concentrations. The relative standard deviations of the variations were estimated from the effective doses calculated from data measured on the same day in Tochigi Prefecture for "ging-geng-cai," in Ibaraki Prefecture for "marine products," in Shizuoka Prefecture for "tea," and in Fukushima Prefecture for other foods ($n \geq 4$ for each food; Table S2). Daily consumption of foods and the radionuclide concentrations in foods were assumed to follow a log-normal distribution. The variations in the dose from each food were estimated from the input data for the Monte Carlo simulation, namely, the variation in daily consumption, the distribution of arrival shares, and the variation in doses received by ingesting the foods in the same areas. The simulation was performed 10 000 times for drinking water and for each food.

Estimation of cancer risk

The lifetime attributable risks (LARs) of cancer incidences up to the age of 89 y were estimated in accordance with the method described in a WHO report [48] and Harada et al. [49]. We estimated those due only to the ingestion of ^{131}I, ^{134}Cs, and ^{137}Cs for citizens in each city and those due to three pathways–ingestion, inhalation in the radioactive cloud, and external exposure to material deposited on the ground and in the cloud–for citizens in Fukushima City. Estimation of doses due to inhalation and external exposure followed previous references [15,50] as described in the Supporting Information. The risk models were based on data from survivors of the atomic bombing in Japan; the validity of the thyroid cancer incidence was also confirmed by a cohort study performed after the Chernobyl accident [51] in the WHO report [48]. A linear-quadratic dose-response model was used for leukemia [52], and linear non-threshold (LNT) models were used for all solid cancers, breast cancer, and thyroid cancer [53]. LAR was calculated from cancer-free survival rates ($S(a, g)$), an excess absolute risk (EAR) model, and an excess relative risk (ERR) model, as follows:

$$LAR(D,e,g)= \int_{e+L}^{89} [w \times EAR(D,e,a,g) + (1-w) \times ERR(D,e,a,g) \times m(a,g)] \frac{S(a,g)}{S(e,g)} da \qquad (3)$$

where D is the annual dose (Sv), e is the age at exposure, g is sex, L is the minimum latency period, w is the weight of the EAR and the ERR models combined, a is the age attained, and $m(a, g)$ is the baseline cancer incidence rate in the unexposed population.

The organ (colon, bone marrow, breast, and thyroid) doses were calculated from the effective doses and the organ dose-to-effective dose ratios for three age groups (<1 y, 10 y, and 20 y in the first year) and for the two sexes in the three cities [48]. Doses in the second and subsequent years were calculated from the doses of ^{134}Cs and ^{137}Cs in March 2012 and from the physical decay of ^{134}Cs and ^{137}Cs (half-lives of 2.06 and 30 y, respectively [4]).

The cancer-free survival rates of males and females were derived from the age- and sex-stratified all-cause mortality in Japan in 2010 [54] plus the difference between the all-cancer incidence in Japan in 2008 [55,56] and the all-cancer mortality in Japan in 2010 [54]. Cancer incidences in Japan in 2008 [55,56] were used as baseline incidence rates for all solid cancers, leukemia, breast cancer, and thyroid cancer.

The minimum latency period was set at 2 y for leukemia, 3 y for thyroid cancer, and 5 y for breast and all solid cancers. The weights of the EAR model and the ERR model combined were set at 0.5 for all solid cancers, leukemia, and thyroid cancer, and 1 for breast cancer.

The details of the EAR and ERR models and their parameters for leukemia, all solid cancers, breast cancer, and thyroid cancer are described in the Supporting Information.

The relationship between cancer risk and dose is still uncertain [57]. However, for the same dose, the risks at low dose rates are known to be lower than those at high dose rates [58]. In particular, coordination of DNA repair processes plays a critical role in allowing proper development and survival of organisms [59].To correct the risk at low dose levels, the International Commission on Radiological Protection (ICRP) and the BEIR VII committee adopted "dose and dose-rate effectiveness factors" of 2 [60] and 1.5 [61]. We followed the ICRP lead and set a factor of 2 for LNT models from the perspective of radiological protection.

Sources of uncertainty

This study included sources of uncertainty: ^{134}Cs and ^{137}Cs concentrations in tap water in Fukushima City, limited data on foods in the early stages, data that were less than detection limits, individual behaviors (such as purchasing of bottled water or not purchasing products from Fukushima Prefecture), and assessment of doses in the second and subsequent years for cancer estimation.

^{134}Cs and ^{137}Cs concentrations in tap water in Fukushima City were estimated from the ^{131}I concentrations and the ratios of ^{134}Cs and ^{137}Cs to ^{131}I in tap water in Tokyo by assuming that the ratios of ^{134}Cs and ^{137}Cs deposition to ^{131}I deposition and the removal efficiencies in drinking water treatment plants were similar between Fukushima City and Tokyo. The deposition ratios are known to be similar in the two cities [62], and drinking water treatment plants in both cities use sedimentation and rapid sand filtration, which are effective for removing ^{134}Cs and ^{137}Cs [63]. This uncertainty did not have a large effect, because the contributions of ^{134}Cs and ^{137}Cs from drinking water were minor (see "Comparison of doses among ages, effects of countermeasures, and change of intake over time).

In the early stages of monitoring, data were not available for some foods in some prefectures in the Kanto region, but these contributions were judged to be small (see Supporting Information). The number of samples was limited for some foods in the early stages even in Fukushima Prefecture. We therefore used a Monte Carlo simulation and show the variations in dose (e.g., 95th percentile value), including those resulting from radionuclide concentrations, in foods in the early stages.

We regarded radionuclide concentrations that were less than detection limits as nil. There were differences in detection limits among periods and institutions surveyed. We therefore confirmed the results through validation against observations in market basket, food-duplicate, and whole-body-counter surveys.

Differences in individual behaviors, such as not purchasing products from Fukushima Prefecture, could have influenced the variation in estimates. However, as the volume of major crops shipped from Fukushima Prefecture did not decrease after the accident [64,65], we ignored differences in behaviors. The nationwide average of bottled soft drink consumption was 410 g/d in 2011 [66], and some citizens also purchased bottled water so as to avoid tap water. We ignored the consumption of bottled water and soft drinks: the assumption that people drank only tap water would conservatively overestimate the dose from drinking water.

The doses due to ingestion in the second and subsequent years for cancer estimation were calculated from the physical decay, although the actual doses might be lower because of tighter regulations. The doses due to external exposure from September 2014 were also calculated from the physical decay. These assumptions can be regarded as conservative.

Results and Discussion

Validation of results against market basket and food-duplicate surveys

The effective doses due to ingestion of ^{134}Cs and ^{137}Cs in the diet in Fukushima City (Case 1), Tokyo, and Osaka agreed within a factor of 2, in general, with those calculated in the market basket [1] and food-duplicate surveys [1,13,67] in five periods from July 2011 to March 2012 (Table 1). The market basket survey includes drinking water as well as foods. This good agreement supports the accuracy and reliability of our results. The doses in the diet in Fukushima City (Case 2) were higher than those in the market basket and food-duplicate surveys. This result is reasonable because Case 2 is conservative. Validation for the period from March to June, including for the thyroid equivalent doses due to ^{131}I, was not done, because no data from market basket and food-duplicate surveys were available.

Comparison of doses among cities

The average thyroid equivalent doses (and effective doses) due to ingestion of ^{131}I for a ≥19-y-old male were 840 μSv (43 μSv) in Fukushima City (Case 1), 2700 μSv (140 μSv) in Fukushima City (Case 2), 370 μSv (19 μSv) in Tokyo, and 16 μSv (0.82 μSv) in Osaka (Table 2). The average effective doses due to ^{134}Cs and ^{137}Cs were 19, 120, 6.1, and 1.9 μSv, respectively. The average effective doses due to total radionuclides were 62, 260, 25, and 2.7 μSv, respectively. There were no large differences between the average and median (within a factor of <1.5, with the exception of the thyroid equivalent dose due to ^{131}I in Osaka). The slightly lower thyroid equivalent dose due to ^{131}I for citizens of Tokyo in this study (370 μSv) than in our previous study (410 μSv) [19] is attributable to the modification of daily consumption rates.

The dose due to ingestion of ^{131}I in Case 2 was 3 times that in Case 1 and the dose due to ^{134}Cs and ^{137}Cs was 6 times that in Case 1. The doses in Fukushima City (Case 1) were lower than the values reported in the WHO preliminary assessment (adults in Fukushima City, thyroid equivalent dose, 800–8000 μSv; effective dose due to total radionuclides, 500–5000 μSv) [15] and the UNSCEAR report (absorbed dose to thyroid, 7800 μGy; effective dose due to total radionuclides, 900 μSv) [16]. Regional trade in foods was considered in this study, whereas the WHO preliminary assessment and UNSCEAR report assumed that consumers consumed mainly food produced in Fukushima and neighboring prefectures. Inclusion of regional trade in foods is a key to accurate dose assessment.

In both doses, Fukushima City had the highest values and Osaka the lowest. This was consistent with the distances from the nuclear power plant (i.e. Fukushima City, ~50 km; Tokyo,

Table 1. Comparison of effective doses of ^{134}Cs and ^{137}Cs in the diet between this study and the market basket and food-duplicate surveys (μSv/month).

	Jul.2011	Sep.-Nov. 2011	Dec. 2011	Feb.-Mar. 2012	Mar. 2012
Fukushima City (Case 1; this study)[a]	2.36	1.18	0.96	0.51	0.44
Fukushima City (Case 2; this study)[a]	4.15	2.85	3.08	6.84	2.60
Fukushima Prefecture (ref[67])[b]	0.53 ± 1.04[f]	-	-	-	-
Fukushima Prefecture (ref[1])[c]	-	1.6	-	0.33-0.55	-
Fukushima Prefecture (ref[1])[b]	-	-	-	-	0.18[g]
Fukushima Prefecture (ref[13])[b]	-	-	2.17± 1.67[f]	-	-
Tokyo (this study) [a]	-	0.44	0.24	0.18	0.19
Tokyo (ref[1])[c]	-	0.22	-	-	-
Kanto (ref[1])[c, d]	-	-	-	0.28-0.33	-
Kanto (ref[1])[b, d]	-	-	-	-	0.15[g]
Kanto (ref[13])[b, e]	-	-	0.92± 1.42[f]	-	-
Osaka (this study)[a]	-	0.15	0.09	0.07	0.07
Osaka (ref[1])[c]	-	-	-	0.13	-
Osaka (ref[1])[b]	-	-	-	-	0.1[g]

Case 1, citizens consumed vegetables bought from markets. Case 2, citizens consumed vegetables grown locally.
[a]with countermeasure (≥19 y old male).
[b]food-duplicate survey.
[c]market basket survey.
[d]including Saitama, and Kanagawa prefectures.
[e]including Tochigi, Gunma, Ibaraki, Saitama, Chiba, Tokyo, Kanagawa, and Nagano prefectures.
[f]arithmetic mean ± standard deviation.
[g]Mar.-May, 2012.

~230 km; Osaka, ~580 km). The contribution of effective dose from ^{131}I to total radionuclides was higher than that from ^{134}Cs and ^{137}Cs in Fukushima City and Tokyo, and lower in Osaka. Except in Case 2, drinking water contributed the highest thyroid equivalent dose due to ingestion of ^{131}I in Fukushima City and Tokyo, followed by vegetables (Figures 1, S5, and S6). ^{131}I was not

Table 2. Average doses and variations due to countermeasures resulting from the intake of ^{131}I, ^{134}Cs and ^{137}Cs from drinking water and foods by a ≥19-y-old male from 16 March 2011 to 20 March 2012.

	Average	5th percentile	25th percentile	Median	75th percentile	95th percentile
^{131}I (thyroid equivalent dose: μSv)						
Fukushima City (Case 1)	840	360	520	690	940	1700
Fukushima City (Case 2)	2700	940	1500	2200	3300	6200
Tokyo	370	140	290	470	780	1400
Osaka	16	<1	1	4	13	74
^{134}Cs and ^{137}Cs (effective dose: μSv)						
Fukushima City (Case 1)	19	8.1	11	15	22	43
Fukushima City (Case 2)	120	60	86	110	150	220
Tokyo	6.1	3.0	4.4	5.8	7.7	12
Osaka	1.9	0.49	0.88	1.4	2.1	4.5
Total (effective dose: μSv)						
Fukushima City (Case 1)	62	29	39	50	66	110
Fukushima City (Case 2)	260	120	170	230	310	530
Tokyo	25	11	21	30	47	80
Osaka	2.7	0.59	1.2	1.8	3.0	7.1

Case 1, citizens consumed vegetables bought from markets. Case 2, citizens consumed vegetables grown locally.

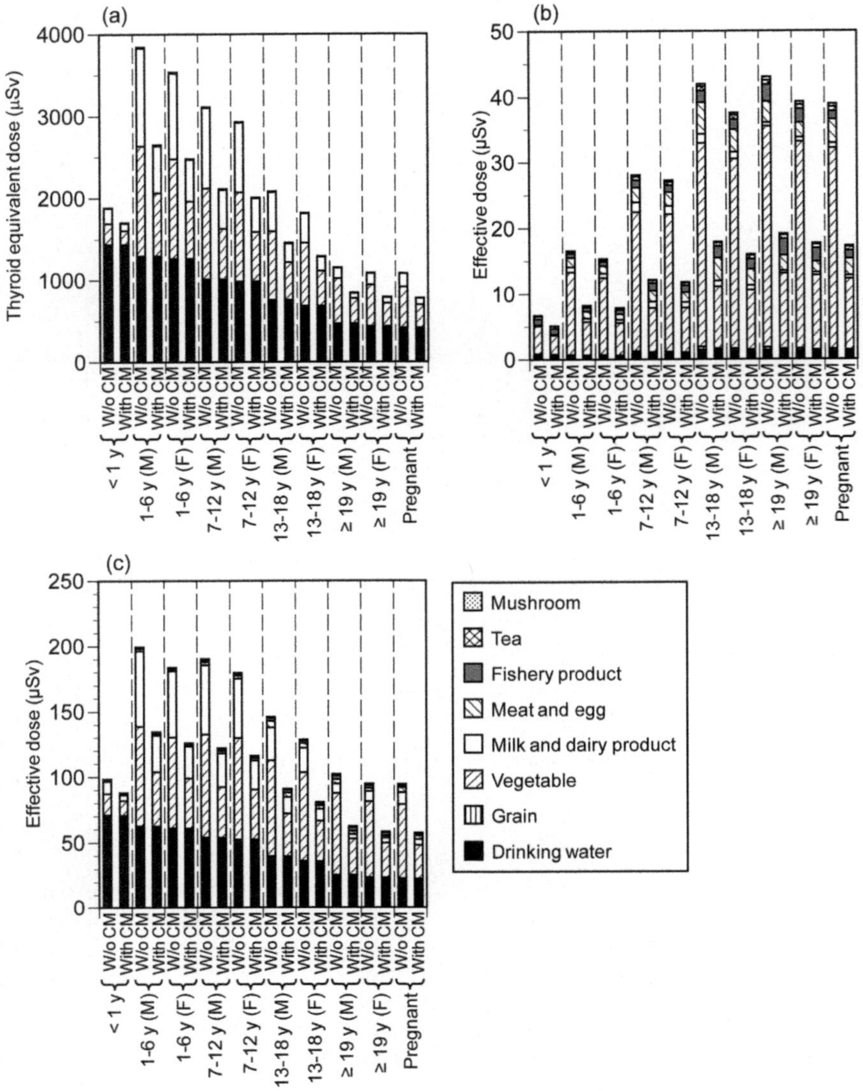

Figure 1. Average doses with and without countermeasures in Fukushima City (Case 1): (a) ^{131}I, (b) ^{134}Cs and ^{137}Cs, (c) total. CM, countermeasures; M, male; F, female. Case 1, citizens consumed vegetables bought from markets.

detected in drinking water in Osaka, and vegetables contributed the highest effective dose due to ^{134}Cs and ^{137}Cs (Figure S7). These results indicate that local contamination of drinking water caused higher intake of ^{131}I by citizens in Fukushima City and Tokyo than in Osaka, and that regional trade in foods played an important role in the intake of ^{134}Cs and ^{137}Cs by citizens in Osaka.

The 95th percentile effective dose due to ingestion of ^{134}Cs and ^{137}Cs in Fukushima City (Case 1) was 43 μSv; this was 4 times the 97th percentile internal exposure to ^{134}Cs and ^{137}Cs (i.e., 10 μSv, the value below which the exposures of 328 out of 337 individuals fell) in inhabitants of the village of Kawauchi in Fukushima Prefecture, as monitored by whole-body-counter surveys in November 2011 [12]. The effects of differences in monitoring time were small, because our average effective dose due to ^{134}Cs and ^{137}Cs in November 2011 (1.3 μSv/month) was similar to the annual average (19 μSv/y = 1.6 μSv/month). Note that the whole-body-counter surveys had sampling bias and may have underestimated the doses received by the general population [11]. The 95th percentile effective doses due to total radionuclides were 110 μSv in Fukushima City (Case 1), 530 μSv in Fukushima City (Case 2), 80 μSv in Tokyo, and 7.1 μSv in Osaka, 2 to 3 times the average values (Table 2). These values were much lower than the annual effective dose due to other natural radionuclides in the diet; e.g., potassium 40 (^{40}K), 130–217 μSv [68]; polonium 210 (^{210}Po), 730 μSv [69]. The provisional limits of radionuclides in drinking water and foods, which were announced by the Japanese government just after the accident, were determined from the intervention levels of 50 000 μSv of an annual thyroid equivalent dose due to ^{131}I and 5000 μSv of an annual effective dose due to ^{134}Cs, ^{137}Cs, ^{89}Sr, and ^{90}Sr [22]. Standards released in April 2012 were determined from the intervention level of 1000 μSv of annual effective dose [21]. The 95th percentile doses due to ingestion of individual and total radionuclides in Fukushima City (Cases 1 and 2) and Tokyo were greater than or equal to 1 order of magnitude lower than the provisional limits and also lower than the new limits. Those in Osaka were greater than 2 orders of magnitude lower.

Comparison of doses among ages, effects of countermeasures, and change of intake over time

The thyroid equivalent dose due to ingestion of ^{131}I in Fukushima City (Case 1) decreased with increasing age from 1 y (Figure 1, Tables S5, S7), although the daily consumption of drinking water and most foods increases with age (Table S2). This discrepancy between dose and daily consumption is attributable to the thyroid ingestion dose coefficient, which depends largely on age. In contrast, the effective doses due to ^{134}Cs and ^{137}Cs increased with age. Overall, the effective doses due to total radionuclides were higher for younger people except for <1-y-old infants (Table S6, S8). The difference in dose between sexes was small. Because the dose coefficients are the same for males and females (Table S1), the differences in doses can be attributed to differences in consumption patterns. Similar results were found in Case 2, Tokyo and Osaka, with the exception that there were no large differences in effective dose due to total radionuclides among ages in Osaka (Figures S5–S7, Tables S9–S20).

The countermeasures reduced the intake of ^{131}I by >1-y-olds by 27%–32% in Fukushima City (Case 1), by 11%–15% in Tokyo, and by 33%–37% in Osaka, and reduced the intake of ^{134}Cs and ^{137}Cs by 49%–57% in Fukushima City (Case 1), by 18%–25% in Tokyo, and by 16%–24% in Osaka (Table S21). The reductions of intake were generally lower for <1-y-old infants because the contributions of drinking water were higher for them.

The cumulative intakes of ^{131}I for ≥19-y-old males in Fukushima City (Case 1) in the presence of countermeasures were 72% of the total in week 1 and 87% in week 2 (Figure 2) and stopped increasing within 1 month. This was consistent with our previous result [19] and is attributable to the short half-life of ^{131}I and the absence of new serious releases from the nuclear power plant to the atmosphere. The intake of ^{131}I was dominant within the first 2 weeks: rapid countermeasures are therefore important in reducing intake. On the other hand, the cumulative intakes of ^{134}Cs and ^{137}Cs were 25% of the total in week 1 and 27% in week 2. The longer-term intake of ^{134}Cs and ^{137}Cs is due to the longer half-lives. In particular, the intake via consumption of vegetables and fisheries products was continuous. Similar results were found in Case 2, Tokyo and Osaka (Figures S8–S10).

LARs of cancer incidence

The LARs of cancer from ingestion are summarized in Table 3. These values do not include risks from external exposure and inhalation. The LARs of all the solid cancer risks were 8.4–12×10^{-6} in Fukushima City (Case 1), 45–71×10^{-6} in Fukushima City (Case 2), 3.0–5.0×10^{-6} in Tokyo, and 1.1–2.0×10^{-6} in Osaka, whereas the LARs of leukemia were >1 order of magnitude lower than those for all the solid cancers (i.e., 0.34–0.59×10^{-6} in Fukushima City (Case 1), 2.0–3.5×10^{-6} in Fukushima City (Case 2), 0.14–0.25×10^{-6} in Tokyo, and 0.05–0.09×10^{-6} in Osaka). The risk of all solid cancers combined together with leukemia is intended to provide an overall indication of the lifetime risk of cancer; however, in circumstances where the tissue doses are highly heterogeneous, such as with doses of ^{131}I to the thyroid, the risk of all solid cancers underestimates the cancer risk in specific tissues [48]. This is because the LARs of thyroid cancer in some young groups (<1 y and 10 y in the first year) in Fukushima City and Tokyo exceeded those of all solid cancers. The LARs of all solid cancers estimated from the colon dose are not suitable in the case of ingestion, although they may be still useful as indicators of the risk of deadly cancers. The LARs of thyroid cancers were 2.3–39×10^{-6} in Fukushima City (Case 1), 10–98×10^{-6} in Fukushima City (Case 2), 0.95–14×10^{-6} in Tokyo, and 0.11–1.3×10^{-6} in Osaka, and the maximum LAR in Fukushima City (Case 1) was found in females <1 y old. These values were calculated from the average doses. The 95th percentile doses were 2× the average values for Fukushima City (Cases 1 and 2) and 3× those for Tokyo and Osaka. Because of the variation of doses, a factor of at least 2–3 could be applied to the 95th percentiles of LARs. In addition, these values cannot be directly compared with the ongoing results of diagnosis of thyroid cancers in Fukushima Prefecture because of the screening effects of diagnosis and the use of advanced ultrasound techniques [18]. The contributions of ^{131}I to the lifetime-exposure thyroid doses were 92%–96% in Fukushima City (Case 1), 84%–92% in Fukushima

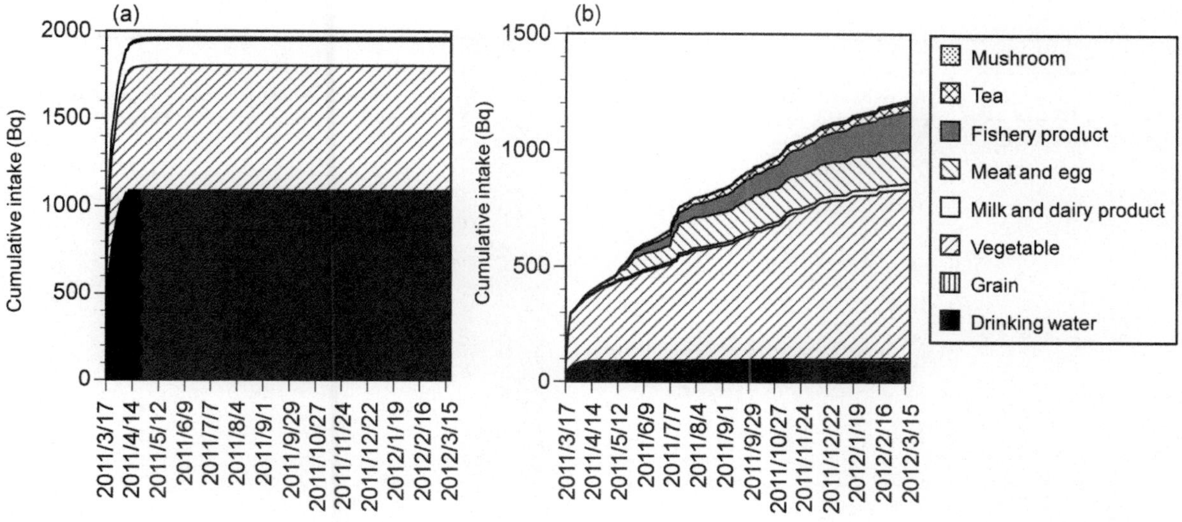

Figure 2. Cumulative intakes for ≥19-y-old male in Fukushima City (Case 1) with countermeasures: (a) ^{131}I, (b) ^{134}Cs and ^{137}Cs. Case 1, citizens consumed vegetables bought from markets.

Table 3. LARs for all solid cancers, leukemia, breast cancer, and thyroid cancer ($\times 10^{-6}$) due to ingestion.

	All solid cancers	Leukemia	Breast cancer	Thyroid cancer
Fukushima City (Case 1)				
<1 y (M)	8.6	0.57	-	9.5
<1 y (F)	12	0.36	1.5	39
10 y (M)	8.4	0.59	-	7.6
10 y (F)	11	0.36	1.5	30
20 y (M)	7.4	0.56	-	2.3
20 y (F)	9.8	0.34	1.1	9.0
Fukushima City (Case 2)				
<1 y (M)	47	3.1	-	15
<1 y (F)	66	2.0	8.4	62
10 y (M)	51	3.5	-	25
10 y (F)	71	2.2	9.3	98
20 y (M)	45	3.4	-	10
20 y (F)	61	2.1	7.1	38
Tokyo				
<1 y (M)	3.6	0.24	-	3.4
<1 y (F)	5.0	0.15	0.63	14
10 y (M)	3.6	0.25	-	3.1
10 y (F)	4.7	0.15	0.61	12
20 y (M)	3.0	0.23	-	0.95
20 y (F)	3.9	0.14	0.44	3.6
Osaka				
<1 y (M)	1.4	0.09	-	0.15
<1 y (F)	2.0	0.06	0.25	0.62
10 y (M)	1.3	0.09	-	0.20
10 y (F)	1.8	0.06	0.23	1.3
20 y (M)	1.1	0.08	-	0.11
20 y (F)	1.4	0.05	0.16	0.42

Ages represent ones in the first year. Case 1, citizens consumed vegetables bought from markets. Case 2, citizens consumed vegetables grown locally.

Table 4. LARs for all solid cancers, leukemia, breast cancer, and thyroid cancer ($\times 10^{-6}$) due to external exposure, inhalation, and ingestion.

	All solid cancers	Leukemia	Breast cancer	Thyroid cancer
Fukushima City (Case 1)				
<1 y (M)	2700	300	-	120
<1 y (F)	4100	210	1200	490
10 y (M)	2000	170	-	66
10 y (F)	3000	110	690	270
20 y (M)	1400	120	-	29
20 y (F)	2100	79	390	120
Fukushima City (Case 2)				
<1 y (M)	2700	310	-	130
<1 y (F)	4200	210	1200	520
10 y (M)	2100	170	-	83
10 y (F)	3100	120	700	340
20 y (M)	1400	120	-	37
20 y (F)	2100	81	390	150

Ages are those in the first year. Case 1, citizens consumed vegetables bought from markets. Case 2, citizens consumed vegetables grown locally.

City (Case 2), 91%–96% in Tokyo, and 62%–78% in Osaka. As described above, most of the ^{131}I was taken up within the first 2 weeks. The LARs of thyroid cancers were attributed mainly to ^{131}I in the first 2 weeks, highlighting again the fact that rapid implementation of countermeasures after a nuclear accident is important. On the other hand, Japanese dietary exposure gives lifetime cancer risks of 310×10^{-6} from inorganic arsenic (for the sum of skin, liver, and lung cancers) [70] and 1400×10^{-6} from acrylamide [71]. Cancer risks from ingestion–especially in the case of the thyroid cancer risk in young females in Fukushima City–might not be negligible, but they are 1 to 2 orders of magnitude lower than the cancer risks from ubiquitous carcinogens in the daily diet.

The total effective doses due to ingestion, inhalation, and external exposure in the first year and over a lifetime (up to 89 y) for Fukushima City are summarized in Table S22. The LARs of all solid cancers due to the three pathways were $1400–4100\times10^{-6}$ in Fukushima City (Case 1) and $1400–4200\times10^{-6}$ in Fukushima City (Case 2); those of thyroid cancer were $29–490\times10^{-6}$ in Fukushima City (Case 1) and $37–520\times10^{-6}$ in Fukushima City (Case 2) (Table 4). The contributions of LARs of all solid cancer due to ingestion were 0.3%–0.5% (Case 1) and 1.6%–3.1% (Case 2); those of thyroid cancer were 7.5%–12% (Case 1) and 12%–30% (Case 2). Even when the 95th percentile values were 2× the average, the contributions of LARs of all solid cancers due to ingestion were minor. However, for the 95th percentile values in Case 2, the contribution of LARs of thyroid cancer due to ingestion was approximately half of the total. Except in such extreme situations, the contributions of LARs of thyroid cancer due to ingestion cannot be regarded as either negligible or predominant but as minor.

As described above, the contributions of ^{131}I to the lifetime-exposure thyroid doses were >84% for Fukushima City, and the contributions of LARs from foods in the second and subsequent years were negligible. However, in Japan and other countries, some people are still concerned about artificial radionuclides in the diet. Although a greater understanding of risks alone may not lead to reasonable countermeasures, the doses and cancer risks calculated in this study are essential information.

Supporting Information

Figure S1 Locations of each prefecture in Japan.

Figure S2 ^{131}I concentrations in leafy vegetables, fruit vegetables, milk and dairy products, meat and eggs, and marine products in Fukushima Prefecture in March 2011. a: Gunma; b: April.

Figure S3 ^{134}Cs and ^{137}Cs concentrations in foods in Fukushima Prefecture in 2011. a: August; b: July; c: June; d: March; e: Gunma in March; f: March–April; g: May; h: April; i: October; j: September; k: Tochigi; December; l: April–May; m: Ibaraki in May.

Figure S4 Radionuclide concentrations in each prefecture in March 2011. (a) ^{131}I; (b) ^{134}Cs and ^{137}Cs.

Figure S5 Average doses with and without countermeasures in Fukushima City (Case 2): (a) ^{131}I, (b) ^{134}Cs and ^{137}Cs, (c) total. CM, countermeasures; M, male; F, female. Case 2, citizens consumed vegetables grown locally.

Figure S6 Average doses with and without countermeasures in Tokyo: (a) ^{131}I, (b) ^{134}Cs and ^{137}Cs, (c) total. CM, countermeasures; M, male; F, female.

Figure S7 Average doses with and without countermeasures in Osaka: (a) ^{131}I, (b) ^{134}Cs and ^{137}Cs, (c) total. CM, countermeasures; M, male; F, female.

Figure S8 Cumulative intakes for ≥ 19 y old male in Fukushima City (Case 2) with countermeasures: (a) ^{131}I, (b) ^{134}Cs and ^{137}Cs. Case 2, citizens consumed vegetables grown locally.

Figure S9 Cumulative intakes for ≥ 19 y old male in Tokyo with countermeasures: (a) ^{131}I, (b) ^{134}Cs and ^{137}Cs.

Figure S10 Cumulative intakes for ≥ 19 y old male in Osaka with countermeasures: (a) ^{131}I, (b) ^{134}Cs and ^{137}Cs.

Methods S1 Numbers of samples analyzed; Dose estimation for inhalation and external exposure; Risk models and their parameters.

Table S1 Thyroid equivalent dose coefficients for ingestion of ^{131}I and the effective dose coefficients for ingestion of ^{131}I, ^{134}Cs and ^{137}Cs (μSv/Bq).

Table S2 Daily consumption rates of drinking water and foods, and relative standard deviations (RSDs) of doses received by ingesting the foods in the same areas.

Table S3 Numbers of samples analyzed each month.

Table S4 Numbers of samples (leafy vegetables, fruit vegetables, milk and dairy products, meat and eggs, and marine products) analyzed in each prefecture in March and April 2011.

Table S5 Average thyroid equivalent doses of ^{131}I without countermeasures in Fukushima City (Case 1) in the first year after the accident (μSv). M, male; F, female. Case 1, citizens consumed vegetables bought from markets.

Table S6 Average effective doses of ^{134}Cs and ^{137}Cs without countermeasures in Fukushima City (Case 1) in the first year after the accident (μSv). M, male; F, female. Case 1, citizens consumed vegetables bought from markets.

Table S7 Average thyroid equivalent doses of ^{131}I with countermeasures in Fukushima City (Case 1) in the first year after the accident (μSv). M, male; F, female. Case 1, citizens consumed vegetables bought from markets.

Table S8 Average effective doses of ^{134}Cs and ^{137}Cs with countermeasures in Fukushima City (Case 1) in the first

year after the accident (μSv). M, male; F, female. Case 1, citizens consumed vegetables bought from markets.

Table S9 Average thyroid equivalent doses of ^{131}I without countermeasures in Fukushima City (Case 2) in the first year after the accident (μSv). M, male; F, female. Case 2, citizens consumed vegetables grown locally.

Table S10 Average effective doses of ^{134}Cs and ^{137}Cs without countermeasures in Fukushima City (Case 2) in the first year after the accident (μSv). M, male; F, female. Case 2, citizens consumed vegetables grown locally.

Table S11 Average thyroid equivalent doses of ^{131}I with countermeasures in Fukushima City (Case 2) in the first year after the accident (μSv). M, male; F, female. Case 2, citizens consumed vegetables grown locally.

Table S12 Average effective doses of ^{134}Cs and ^{137}Cs with countermeasures in Fukushima City (Case 2) in the first year after the accident (μSv). M, male; F, female. Case 2, citizens consumed vegetables grown locally.

Table S13 Average thyroid equivalent doses of ^{131}I without countermeasures in Tokyo in the first year after the accident (μSv). M, male; F, female.

Table S14 Average effective doses of ^{134}Cs and ^{137}Cs without countermeasures in Tokyo in the first year after the accident (μSv). M, male; F, female.

Table S15 Average thyroid equivalent doses of ^{131}I with countermeasures in Tokyo in the first year after the accident (μSv). M, male; F, female.

Table S16 Average effective doses of ^{134}Cs and ^{137}Cs with countermeasures in Tokyo in the first year after the accident (μSv). M, male; F, female.

Table S17 Average thyroid equivalent doses of ^{131}I without countermeasures in Osaka in the first year after the accident (μSv). M, male; F, female.

Table S18 Average effective doses of ^{134}Cs and ^{137}Cs without countermeasures in Osaka in the first year after the accident (μSv). M, male; F, female.

Table S19 Average thyroid equivalent doses of ^{131}I with countermeasures in Osaka in the first year after the accident (μSv). M, male; F, female.

Table S20 Average effective doses of ^{134}Cs and ^{137}Cs with countermeasures in Osaka in the first year after the accident (μSv). M, male; F, female.

Table S21 Reductions of doses by countermeasures (%). M, male; F, female. Case 1, citizens consumed vegetables bought from markets.

Table S22 Effective doses in the first year and over the total lifetime (up to 89 y) due to three pathways (μSv). Ages are those in the first year. M, male; F, female. Case 1, citizens consumed vegetables bought from markets. Case 2, citizens consumed vegetables grown locally.

Author Contributions

Conceived and designed the experiments: MM TO. Performed the experiments: MM. Analyzed the data: MM. Contributed reagents/materials/analysis tools: MM. Wrote the paper: MM TO.

References

1. Ministry of Health Labour and Welfare (2012) Available: http://www.mhlw.go.jp/shinsai_jouhou/shokuhin.html. Accessed: 7 Oct. 2013. [in Japanese]
2. Tokyo Metropolitan Institute of Public Health (2011) Available: http://monitoring.tokyo-eiken.go.jp/monitoring/w-past_data.html. Accessed: 7 Oct. 2013.
3. Fukushima Prefecture (2013) Available: http://www.new-fukushima.jp/monitoring/. Accessed: 13 Sep. 2013. [in Japanese]
4. ICRP (1983) Radionuclide transformations - Energy and intensity of emissions. ICRP Publication 38, Ann. ICRP 11–13.
5. Working Group on Radionuclides Pharmaceutical Affairs and Food Sanitation Council (2011) Estimation of contribution of radiocesium to internal exposure from dietary intake. [in Japanese]
6. Yoshida-Ohuchi H, Hirasawa N, Kobayashi I, Yoshizawa T (2013) Evaluation of personal dose equivalent using optically stimulated luminescent dosemeters in Marumori after the Fukushima nuclear accident. Radiat Prot Dosim 154: 385–390.
7. Yasutaka T, Iwasaki Y, Hashimoto S, Naito W, Ono K, et al. (2013) A GIS-based evaluation of the effect of decontamination on effective doses due to long-term external exposures in Fukushima. Chemosphere 93: 1222–1229.
8. Evangeliou N, Balkanski Y, Cozic A, Moller AP (2013) Global transport and deposition of Cs-137 following the Fukushima nuclear power plant accident in Japan: Emphasis on Europe and Asia using high-resolution model versions and radiological impact assessment of the human population and the environment using interactive tools. Environ Sci Technol 47: 5803–5812.
9. Amano H, Akiyama M, Bi C, Kawamura T, Kishimoto T, et al. (2012) Radiation measurements in the Chiba Metropolitan Area and radiological aspects of fallout from the Fukushima Dai-ichi nuclear power plants accident. J Environ Radioactiv 111: 42–52.
10. Christoudias T, Lelieveld J (2013) Modelling the global atmospheric transport and deposition of radionuclides from the Fukushima Dai-ichi nuclear accident. Atmos Chem Phys 13: 1425–1438.
11. Hayano RS, Tsubokura M, Miyazaki M, Satou H, Sato K, et al. (2013) Internal radiocesium contamination of adults and children in Fukushima 7 to 20 months after the Fukushima NPP accident as measured by extensive whole-body-counter surveys. P Jpn Acad B-Phys 89: 157–163.
12. Tsubokura M, Kato S, Nihei M, Sakuma Y, Furutani T, et al. (2013) Limited internal radiation exposure associated with resettlements to a radiation-contaminated homeland after the Fukushima Daiichi Nuclear Disaster. PLOS ONE 8: e81909.
13. Harada KH, Fujii Y, Adachi A, Tsukidate A, Asai F, et al. (2013) Dietary intake of radiocesium in adult residents in Fukushima prefecture and neighboring regions after the Fukushima nuclear power plant accident: 24-h food-duplicate survey in December 2011. Environ Sci Technol 47: 2520–2526.
14. Ministry of Health Labour and Welfare (2011) Available: http://www.mhlw.go.jp/stf/shingi/2r98520000001ip01-att/2r98520000001ipae.pdf. Accessed: 7 Oct. 2013. [in Japanese]
15. World Health Organization (2012) Preliminary dose estimation from the nuclear accident after the 2011 Great East Japan Earthquake and Tsunami. World Health Organization, Geneva.
16. United Nations Scientific Committee on the Effects of Atomic Radiation (2014) Sources, effects and risks of ionizing radiation. UNSCEAR 2013 Report to the General Assembly with Scientific Annexes. United Nations, New York.
17. Ministry of Agriculture Forestry and Fisheries (2013) Food demand and supply 2010: http://www.maff.go.jp/j/tokei/kouhyou/zyukyu/index.html. Accessed: 13 Sep. 2013. [in Japanese]
18. Nagataki S, Takamura N (2014) A review of the Fukushima nuclear reactor accident: radiation effects on the thyroid and strategies for prevention. Curr Opin Endocrinol Diabetes Obes 21: 384–393.
19. Murakami M, Oki T (2012) Estimation of thyroid doses and health risks resulting from the intake of radioactive iodine in foods and drinking water by the citizens of Tokyo after the Fukushima nuclear accident. Chemosphere 87: 1355–1360.

20. Osaka Water Supply Authority (2013) Available: http://www.wsa-osaka.jp/gaiyou/jigyoukanri/keikakuka/housyanou.html. Accessed: 13 Sep. 2013. [in Japanese]
21. Working Group on Radionuclides Pharmaceutical Affairs and Food Sanitation Council (2011) Revision of new standard for radionuclides in foods. [in Japanese]
22. Nuclear Safety Commission of Japan (1998) Control index levels on the ingestion of food and water. [in Japanese]
23. Kuroda M, Ohta M, Okufuji T, Takigami C, Eguchi M, et al. (2011) Frequency of soup intake and amount of dietary fiber intake are inversely associated with plasma leptin concentrations in Japanese adults. Appetite 54: 538–543.
24. Izumo Y (2001) Ministry of Health Labour and Welfare Grants 200000078A. [in Japanese]
25. Nakamura T, Yurimoto M, Matsumoto K, Kawasaki N, Tanada S (1996) Sensory taste evaluation of milled rice cooked with water different in hardness. Jpn J Food Chem 3: 141–144.
26. Tokyo Metropolitan Central Wholesale Market (2011) Available: http://www.shijou-tokei.metro.tokyo.jp/index.html. Accessed: 7 Oct. 2013. [in Japanese]
27. Ministry of Agriculture Forestry and Fisheries (2013) Fishery and aquaculture industrial production statistics 2011: http://www.e-stat.go.jp/SG1/estat/List.do?lid=000001104479. Accessed: 13 Sep. 2013. [in Japanese]
28. Suzuki K, Okuda T, HIgasine Y (2006) The Daily intakes of drinks and folate from the drinks by young women. Mem Osaka Kyoiku Univ Ser 2 Soc Sci Home Econ 54: 27–34. [in Japanese]
29. Ministry of Agriculture Forestry and Fisheries (2011) Basic data on specific forest products industry: http://www.e-stat.go.jp/SG1/estat/List.do?lid=000001085178. Accessed: 13 Sep. 2013. [in Japanese]
30. Ministry of Agriculture Forestry and Fisheries (2011) Crop production statistics 2009: http://www.e-stat.go.jp/SG1/estat/List.do?lid=000001069768. Accessed: 7 Oct. 2013. [in Japanese]
31. Statistics Bureau of Japan (2011) National Population Census 2010.
32. Fukushima City Central Wholesale Market (2013) Available: http://www.city.fukushima.fukushima.jp/soshiki/24/1030.html. Accessed: 13 Sep. 2013. [in Japanese]
33. Osaka Central Wholesale Market (2013) Available: http://www.pref.osaka.jp/fuichiba/toukeijouhou/nenpou22.html. Accessed: 13 Sep. 2013. [in Japanese]
34. Ministry of Agriculture Forestry and Fisheries (2011) Statistics of animal product distribution 2009: http://www.e-stat.go.jp/SG1/estat/List.do?lid=000001072695. Accessed: 13 Sep. 2013. [in Japanese]
35. Ministry of Agriculture Forestry and Fisheries (2011) Statistics of animal product distribution 2010: http://www.e-stat.go.jp/SG1/estat/List.do?lid=000001076341. Accessed: 7 Oct. 2013. [in Japanese]
36. Ministry of Agriculture Forestry and Fisheries (2010) Statistical survey on milk and dairy products 2008: http://www.e-stat.go.jp/SG1/estat/List.do?lid=000001063504. Accessed: 13 Sep. 2013. [in Japanese]
37. Ministry of Agriculture Forestry and Fisheries (2013) Crop statistics 2011: http://www.e-stat.go.jp/SG1/estat/List.do?lid=000001087011. Accessed: 13 Sep. 2013. [in Japanese]
38. Ministry of Agriculture Forestry and Fisheries (2012) Crop production statistics: http://www.e-stat.go.jp/SG1/estat/List.do?lid=000001081820. Accessed: 13 Sep. 2013. [in Japanese]
39. Ministry of Agriculture Forestry and Fisheries (2007) Statistical survey on livestock 2006: http://www.e-stat.go.jp/SG1/estat/List.do?lid=000001060567. Accessed: 7 Oct. 2013. [in Japanese]
40. Ministry of Agriculture Forestry and Fisheries (2011) Statistical survey on cultivated acreage (FY 2009): http://www.e-stat.go.jp/SG1/estat/List.do?lid=000001061777. Accessed: 7 Oct. 2013. [in Japanese]
41. Fukushima Prefecture (2010) Statistics of forest industry in Fukushima 2009. [in Japanese]
42. Ministry of Agriculture Forestry and Fisheries (2008) Census of agriculture and forestry 2005: http://www.maff.go.jp/j/tokei/census/afc/2010/report05_archives.html. Accessed: 13 Sep. 2013. [in Japanese]
43. Japan Meteorological Agency (2011) Available: http://www.data.jma.go.jp/obd/stats/etrn/index.php. Accessed: 7 Oct. 2013. [in Japanese]
44. Murakami M, Takeda H, Okaneya M, Kobayashi Y, Oki T (2012) Classification and motives of drink intakes based on the situations using behavioral record. SEISAM-KENKYU 64: 359–366. [in Japanese]
45. Japan Water Works Association (2011) Statistics of tap water (2009 FY). [in Japanese]
46. Bureau of Waterworks Tokyo Metropolitan Government (2013) Available: https://www.waterworks.metro.tokyo.jp/press/shinsai22/press01.html. Accessed: 13 Sep. 2013.
47. Ministry of Health and Welfare (2012) The National Health and Nutrition Survey in Japan, 2010. [in Japanese]
48. World Health Organization (2013) Health risk assessment from the nuclear accident after the 2011 Great East Japan Earthquake and Tsunami based on a preliminary dose estimation.
49. Harada KH, Niisoe T, Imanaka M, Takahashi T, Amako K, et al. (2014) Radiation dose rates now and in the future for residents neighboring restricted areas of the Fukushima Daiichi nuclear power plant. P Natl Acad Sci USA 111: E914–E923.
50. Akahane K, Yonai S, Fukuda S, Miyahara N, Yasuda H, et al. (2013) NIRS external dose estimation system for Fukushima residents after the Fukushima Dai-ichi NPP accident. Sci Rep 3: 1670.
51. Brenner AV, Tronko MD, Hatch M, Bogdanova TI, Oliynik VA, et al. (2011) I-131 dose response for incident thyroid cancers in Ukraine related to the Chornobyl accident. Environ Health Persp 119: 933–939.
52. United Nations Scientific Committee on the Effects of Atomic Radiation (2008) Annex A: Epidemiological studies of radiation and cancer. UNSCEAR 2006 Report to the General Assembly: Effects of Ionizing Radiation (United Nations Scientific Committee on the Effects of Atomic Radiation, Vienna), Vol. 1.
53. Preston DL, Ron E, Tokuoka S, Funamoto S, Nishi N, et al. (2007) Solid cancer incidence in atomic bomb survivors: 1958–1998. Radiat Res 168: 1–64.
54. Statistics Bureau of Japan (2011) General mortality in 2010: http://www.e-stat.go.jp/SG1/estat/ListE.do?lid=000001101825. Accessed: 5 Mar. 2014.
55. Center for Cancer Control and Information Services National Cancer Center Japan (2013) Cancer incidence (1975–2008): http://ganjoho.jp/professional/statistics/statistics.html. Accessed: 5 Mar. 2014. [in Japanese]
56. Matsuda A, Matsuda T, Shibata A, Katanoda K, Sobue T, et al. (2013) Cancer incidence and incidence rates in Japan in 2007: A study of 21 population-based cancer registries for the monitoring of cancer incidence in Japan (MCIJ) project. Jpn J Clin Oncol 43: 328–336.
57. Mullenders L, Atkinson M, Paretzke H, Sabatier L, Bouffler S (2009) Assessing cancer risks of low-dose radiation. Nat Rev Cancer 9: 596–604.
58. Russell WL, Kelly EM (1982) Mutation frequencies in male-mice and the estimation of genetic hazards of radiation in men. P Natl Acad Sci–Biol 79: 542–544.
59. Ciccia A, Elledge SJ (2010) The DNA damage response: Making it safe to play with knives. Mol Cell 40: 179–204.
60. ICRP (2007) The 2007 recommendations of the international commission on radiological protection. ICRP Publication 103, Ann. ICRP. 37 (2–4).
61. National Research Council of the National Academies (2006) Health risks from exposure to low levels of ionizing radiation: BEIR VII Phase 2. Board on Radiation Effects Research.
62. Kinoshita N, Sueki K, Sasa K, Kitagawa J-i, Ikarashi S, et al. (2011) Assessment of individual radionuclide distributions from the Fukushima nuclear accident covering central-east Japan. P Natl Acad Sci USA 108: 19526–19529.
63. Kosaka K, Asami M, Kobashigawa N, Ohkubo K, Teracia H, et al. (2011) Removal of radioactive iodine and cesium in water purification processes after an explosion at a nuclear power plant due to the Great East Japan Earthquake. Water Res 46: 4397–4404.
64. Diamond Online (2013) 8th July. http://diamond.jp/articles/-/38458. [in Japanese]
65. Hangui S (2012) Enormous influences on agriculture in Fukushima prefecture caused by the Great East Japan Earthquake and nuclear disaster: Focused on distribution and consumption stage. Jpn J Farm Manag 49: 93–96. [in Japanese]
66. Japan Soft Drink Association (2010) Available: http://j-sda.or.jp/statistically-information/stati01.php. Accessed: 5 Sep. 2013.
67. Koizumi A, Harada KH, Niisoe T, Adachi A, Fujii Y, et al. (2012) Preliminary assessment of ecological exposure of adult residents in Fukushima Prefecture to radioactive cesium through ingestion and inhalation. Environ Health Prev Med 17: 292–298.
68. Sugiyama H, Terada H, Takahashi M, Iijima I, Isomura K (2007) Contents and daily intakes of gamma-ray emitting nuclides, ^{90}Sr, and ^{238}U using market-basket studies in Japan. J Health Sci 53: 107–118.
69. Ota T, Sanada T, Kashiwara Y, Morimoto T, Sato K (2009) Evaluation for committed effective dose due to dietary foods by the intake for Japanese adults. Jpn J Health Phys 44: 80–88.
70. Oguri T, Yoshinaga J, Tao H, Nakazato T (2012) Daily intake of inorganic arsenic and some organic arsenic species of Japanese subjects. Food Chem Toxicol 50: 2663–2667.
71. Murakami M, Nagai T (2013) Cancer risks from micro-pollutants and their acceptable levels. J Jpn Soc Water Environ 36(A): 322–326. [in Japanese]

18

Whole Genome Association Mapping of Plant Height in Winter Wheat (*Triticum aestivum* L.)

Christine D. Zanke[1], Jie Ling[1], Jörg Plieske[2], Sonja Kollers[3], Erhard Ebmeyer[3], Viktor Korzun[3], Odile Argillier[4], Gunther Stiewe[5], Maike Hinze[5], Kerstin Neumann[1], Martin W. Ganal[2], Marion S. Röder[1]*

1 Leibniz Institute of Plant Genetics and Crop Plant Research (IPK), Gatersleben, Germany, **2** TraitGenetics GmbH, Gatersleben, Germany, **3** KWS LOCHOW GMBH, Bergen, Germany, **4** Syngenta France S.A.S., Orgerus, France, **5** Syngenta Seeds GmbH, Bad Salzuflen, Germany

Abstract

The genetic architecture of plant height was investigated in a set of 358 recent European winter wheat varieties plus 14 spring wheat varieties based on field data in eight environments. Genotyping of diagnostic markers revealed the *Rht-D1b* mutant allele in 58% of the investigated varieties, while the *Rht-B1b* mutant was only present in 7% of the varieties. *Rht-D1* was significantly associated with plant height by using a mixed linear model and employing a kinship matrix to correct for population stratification. Further genotyping data included 732 microsatellite markers, resulting in 770 loci, of which 635 markers were placed on the ITMI map plus a set of 7769 mapped SNP markers genotyped with the 90 k iSELECT chip. When Bonferroni correction was applied, a total of 153 significant marker-trait associations (MTAs) were observed for plant height and the SSR markers ($-\log_{10}$ (P-value) $\geq$4.82) and 280 ($-\log_{10}$ (P-value) $\geq$5.89) for the SNPs. Linear regression between the most effective markers and the BLUEs for plant height indicated additive effects for the MTAs of different chromosomal regions. Analysis of syntenic regions in the rice genome revealed closely linked rice genes related to gibberellin acid (GA) metabolism and perception, i.e. GA20 and GA2 oxidases orthologous to wheat chromosomes 1A, 2A, 3A, 3B, 5B, 5D and 7B, *ent*-kaurenoic acid oxidase orthologous to wheat chromosome 7A, *ent*-kaurene synthase on wheat chromosome 2B, as well as GA-receptors like DELLA genes orthologous to wheat chromosomes 4B, 4D and 7A and genes of the GID family orthologous to chromosomes 2B and 5B. The data indicated that besides the widely used GA-insensitive dwarfing genes *Rht-B1* and *Rht-D1* there is a wide spectrum of loci available that could be used for modulating plant height in variety development.

Editor: Pilar Hernandez, Institute for Sustainable Agriculture (IAS-CSIC), Spain

Funding: This research was funded by the Plant Biotechnology program of the German Federal Ministry of Education and Research (BMBF) within the frame of the projects GABI-Wheat and VALID (project numbers 0315067 and 0315947). The funders had no role in study design, data collection and analysis, decision to publish, or preparation of the manuscript.

Competing Interests: SK,VK and EE are employed by the company KWS-LOCHOW GMBH, OA is employed at Syngenta France SAS, GS and MH are employed by Syngenta Seeds GmbH and MG, JP are employed by the company TraitGenetics GmbH. The companies have commercial interest in the results for application in variety development and for the provision of genotyping services. The authors declare that the research was conducted in the absence of any commercial or financial relationships that could be construed as a potential conflict of interest.

* Email: roder@ipk-gatersleben.de

Introduction

Plant height (PH) is an intrinsic component of plant architecture with effects on lodging stability, harvest index and yield. The introduction of dwarfing genes was a major factor during the 'Green Revolution' in the 1970 s, which led to a dramatic increase of production in the staple crops rice, wheat and maize at reduced costs [1,2].

In wheat, the major dwarfing genes are *Rht-B1* and *Rht-D1* (old nomenclature *Rht1* and *Rht2*), which reduce plant height by reducing the response to gibberellin (gibberellin insensitive dwarfing genes) with pleiotropic effects on grain number and yield. The molecular identification of the *Rht-B1* and *Rht-D1* genes in wheat [3] confirmed them as orthologs to the gibberellin response regulator *GAI* in Arabidopsis [4] and the dwarfing genes *dwarf-8* (*d8*) and *dwarf-9* (*d9*) in maize [5]. Mutations in the DELLA-domain of these proteins often cause a dominant gibberellin-insensitive phenotype by increasing the stability of this negative regulator of gibberellin signal transduction [6]. Other orthologous DELLA genes are the *Slender* rice 1 [7] and the barley Slender1 [8,9] mutants. The gibberellin-insensitive dwarf mutant *Dwarf 1* in rice encodes the α-subunit of GTP-binding protein which may be associated with gibberellin signal transduction [10]. The *GID*1 family in rice encodes a soluble receptor for gibberellin [11], while the *gid2* mutant is caused by the loss of function of a putative F-box protein [12].

Besides the gibberellin insensitive dwarfing genes also gibberellin sensitive dwarfing genes exist. A prominent example is the semi-dwarf (sd-1) "green revolution" rice, which contains a defective gibberellin 20-oxidase gene [13,14]. Mutations in *GA20ox1* also caused semi-dwarf phenotypes in Arabidopsis [15] and an orthologous gene is likely to cause the *sdw1/denso* mutant in barley [16]. Like the gibberellin 20-oxidase mutants, the maize

Dwarf3 gene, encoding a protein of the cytochrome P450 superfamily, is involved in gibberellin biosynthesis [17].

A different mechanism for height reduction was described in the maize *brachytic2* (*br2*) mutants, where the loss of a P-glycoprotein modulated polar auxin transport in the stalk [18]. An orthologous mutant *dwarf3* was discovered in sorghum and the ABCB1 P-glycoprotein is a likely candidate for the *d2* dwarfing gene in pearl millet [19].

In wheat, besides the dwarfing genes with known function, e.g. *Rht-B1* and *Rht-D1* [20], a number of additional dwarfing genes were mapped in bi-parental populations [21–23]. In the current study, we were interested to gain a comprehensive picture about the loci involved in the regulation of plant height in European winter wheat, by applying a whole genome association mapping approach. Association mapping can be carried out with candidate genes [24,25] or with anonymous markers covering the genome [26,27]. In fact, the associations of the maize *Dwarf8* polymorphisms with variation in flowering time and plant height were regarded as first association mapping approach reported for crop plants [28], though the validity of this assumption recently has been questioned [29]. In comparison to bi-parental mapping populations, association panels have the advantage that the genetic diversity of larger germplasm panels can be monitored and a refined resolution can be obtained by monitoring the "historical" recombinations, which occurred during evolution or breeding of a line [30]. Association mapping and genomic selection approaches for plant height were conducted in barley [31], sorghum [32], maize [33] and wheat [34].

The goals of the current study were (i) to establish marker-trait associations (MTAs) for plant height in a panel of European winter wheat varieties, (ii) to compare the MTAs obtained with SSR markers to those obtained with SNP markers, and (iii) to connect the obtained MTAs to known genes of the gibberellic acid (GA) metabolism by exploiting the synteny to the annotated genome sequence of rice [35]. Our wheat association panel previously was analyzed for resistance to fungal pathogens [36–38] and heading date [39].

Materials and Methods

Plant material and phenotyping

No specific permits were required for the described field trials. The field trials were performed by companies. The field trials did not involve endangered or protected species. The plant material, consisting of 358 European winter wheat varieties plus 14 spring wheat varieties as a outgroup, is described in more detail in [36]. Field trials were conducted in the season 2008/2009 in Andelu/France (09.AND), Seligenstadt/Germany (09.SEL) and Wohlde/Germany (09.WOH) and in the season 2009/2010 in Andelu/France (10.AND), Janville/France (10.JAN), Saultain/France (10.SAU), Seligenstadt/Germany (10.SEL) and Wohlde/Germany (10.WOH) by applying an alpha design with two replications per site. Plot sizes were 5 to 6.75 m^2 and climatic factors are described in Table S1. Both winter and spring varieties were sown in autumn and common agricultural practices were applied, which included the applications of growth regulators (Cycocel CL at a rate of 2.5L/ha 09.AND. 10.AND, 10.JAN and 10.SAU; CCC720 at a rate of 1.5L/ha at 09.WOH and 10.WOH; 1.0 L/ha CCC720 and 0.3 L/ha Moddus at 09.SEL and 1.3 L/ha CCC720 at 10.SEL). The active substance for CCC720 is Chlormequatchlorid (720 g/L), for Moddus it is Trinexapacethyl (250 g/L) and for Cycocel CL it is Chlormequat (460 g/L) and Imazaquine (10 g/L). Cycocel is based on the action of chlormequat chloride which is an inhibitor of *ent*-kaurene synthase involved in the early steps of GA-biosynthesis [40]. As untreated control, all varieties were grown in a nursery in rows with 6 plants in Gatersleben 2012 (12.GAT). Average plant height of each plot was measured before harvest with exclusion of awns.

Molecular data analysis and genetic mapping

For marker-trait analysis a set of 732 microsatellite markers, resulting in 770 different loci spred across all chromosomes of wheat was used. Of these 770 loci, 635 loci were mapped and 135 loci were unmapped. Since the microsatellites are multi-allelic, they amounted to 3176 alleles. More details about this data set and the description of linkage disequilibrium (LD) and population structure are provided in [36]. For SNP-analysis, all 372 varieties were genotyped on a novel 90 k Infinium chip (90k iSELECT) [41,42]. This resulted in a total of 21742 scorable and polymorphic markers on our association panel by considering all polymorphic markers with a minor allele frequency (MAF) >0.03. Of these markers, only the 7769 mapped markers were included in the association analysis, while the unmapped markers were not used for association analysis. The SSR-markers were mapped on the ITMI-population (International Triticeae Initiative) based on recombinant inbred lines between the parents W7984 (synthetic wheat) × Opata M85 [43,44], while the SNP markers were mapped on 138 lines of a newly created doubled-haploid population of the same parents [45,46]. Map construction was performed using the software package Joinmap 4.1. Both maps have different recombination values, and currently only few common markers are available, which makes comparisons difficult. For display a reduced version of the SNP-map was used containing most relevant markers with MTAs for PH. As candidate genes the photoperiodism gene *Ppd-D1* [47] and the dwarfing genes *Rht-B1* and *Rht-D1* [48] were genotyped on all varieties.

Statistical analysis and association mapping

Each year-location combination was considered as an environment in our study. For each environment, adjusted entry mean of the genotypes were estimated using GenStat 13th edition and following model:

$$y = \mu + replication + genotype + block + e$$

with replication and genotype as fixed factors and block as random effect nested within replication; μ represents an overall mean and e is a residual term; y represents the single plot values within each environment.

Adjusted entry means were used to calculate Best Linear Unbiased Estimators (BLUEs) across all eight environments using the software package GenStat 14th edition [VSN International, Hemel Hempstead, Hertfordshire, UK] as described in [36]:

$$y = \mu + genotype + environment + e$$

with genotype and environment as fixed effects; μ represents an overall mean and e is a residual term.

For calculating genotype-phenotype associations a minor allele frequency (MAF) threshold of 3% (equalling 11 varieties) was used for all markers. A mixed model for association mapping was applied based on the BLUEs across environments using the software package GenStat 14th edition as described in [36] by applying a kinship matrix as relationship model:

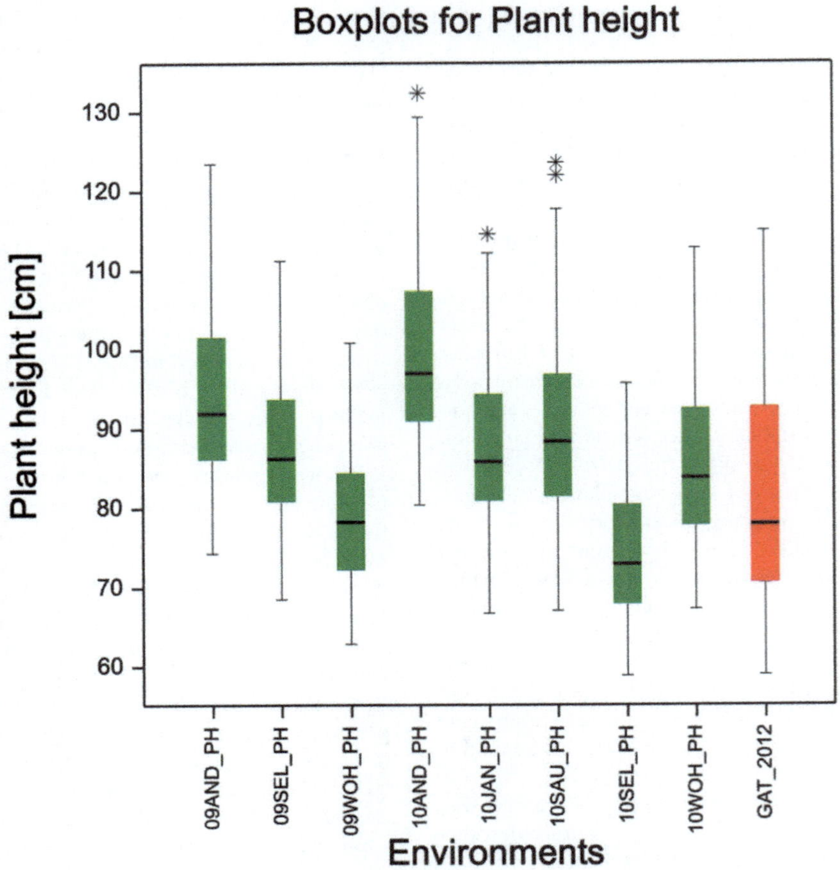

	09AND_PH	09SEL_PH	09WOH_PH	10AND_PH	10JAN_PH	10SAU_PH	10SEL_PH	10WOH_PH	GAT_2012
Min.	74.4	68.6	62.9	80.4	66.8	67.1	75.8	67.3	59.0
Max.	123.5	111.2	100.9	132.4	114.6	123.6	74.0	112.8	115.0
Mean	94.2	87.5	78.7	99.7	88.0	89.6	74.5	85.5	81.5

Figure 1. Boxplots for plant height. Boxplots for eight field environments are depicted in green, the untreated control in a nursery is shown in red. Asterisks mark outlier varieties.

$$y = \mu + marker + genotype + e$$

with $genotype \sim N(0, 2K\sigma^2_{genotype}), error \sim N(0, \sigma^2_e)$.

Marker refers to a fixed effect of every marker, μ represents an overall mean, e is a residual term, and K denotes the kinship matrix among all genotypes. The Loiselle kinship matrix was calculated for 155 SSR markers, equally distributed on the genome, by using the software package SPAGeDi [49]. This kinship matrix was applied to correct for false positives when calculating MTAs with SSR as well as with SNP markers as described by [50]. The threshold of Bonferroni correction for multiple testing was calculated by dividing $P<0.01$ with the number of SSR or SNP markers used for the analysis. Marker effects (r^2) were estimated using the software package TASSEL 3.0. The additive effects were calculated by using GenStat 14th edition, where for the bi-allelic SNP-markers the most frequent alleles was set to zero and the difference of the phenotypic effect towards the less frequent allele was calculated. Therefore negative additive effects mean a decrease in plant height and positive additive effects mean an increase in plant height.

Spearman rank order correlations and ANOVA using the adjusted means of the eight environments were calculated with the software package SigmaPlot 11.0. The broad sense heritability was calculated from the variance components according to the formula:

$$H^2 = Var(genotype)/(Var(genotype) + Var(error)/number\_of\_locations)$$

with variance components calculated with the software package SPSS v. 19. This software was also used to conduct a trait Posthoc test according to Tukey B.

Results

Description of phenotypic data

The phenotypic data were based on eight field environments with various distributions of the trait plant height (Figure 1). The

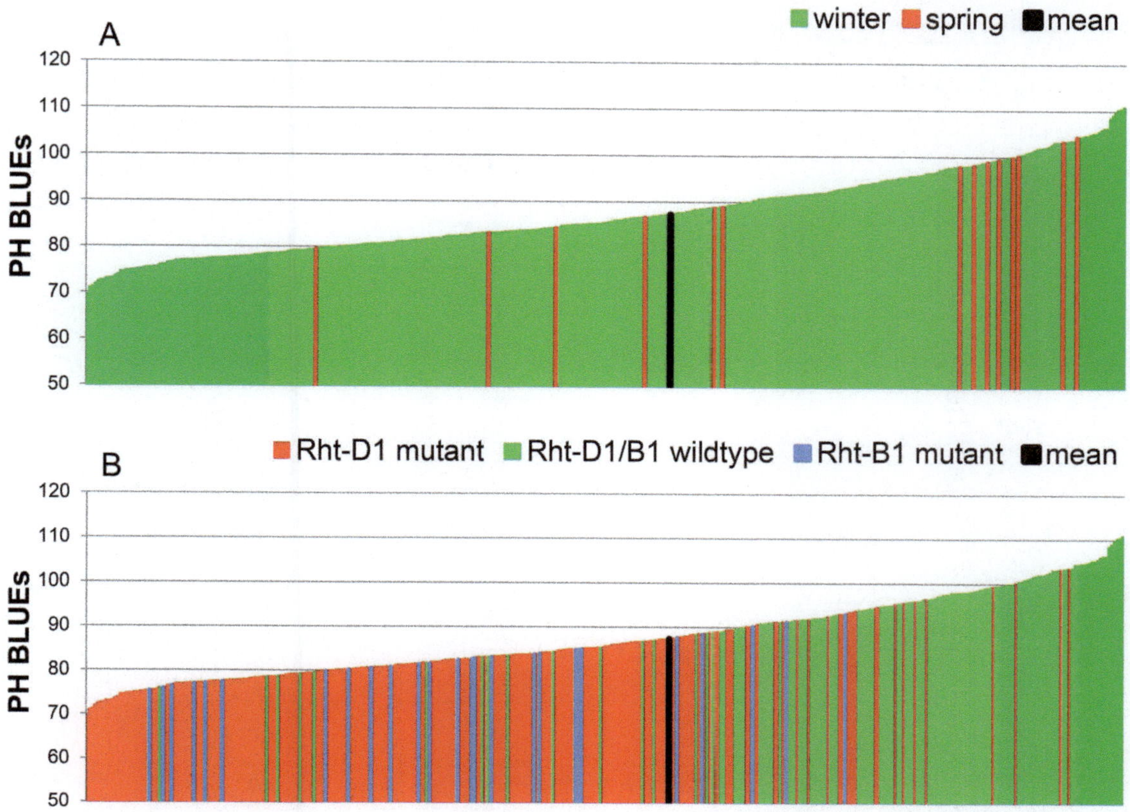

Figure 2. Phenotypic distribution of plant height BLUEs in 372 wheat varieties. The BLUEs for plant height were calculated across eight environments. Plant height BLUEs were arranged according to growth habit (A) or the distribution of the mutant alleles of the dwarfing genes *Rht-B1* and *Rht-D1* (B).

BLUEs for PH across the eight environments ranged from 69.5 cm to 110.7 cm with a mean of 87.2 (Table S2; Figure 2). The untreated control 12.GAT had very similar values ranging from 59.0 cm to 115.0 cm (Figure 1). The broad sense heritability among the eight field environments was 0.89. An ANOVA test revealed a significant effect of environment and a Tukey B test discriminated seven phenotypic classes (Table S3). The Spearman rank order correlations among the eight field environments ranged from 0.862 to 0.943 indicating a good reproducibility of the ranking of the varieties in the different environments (Table S4). The correlation between the BLUEs of the eight field environments and the untreated control 12.GAT was 0.882. The analysis of variance was significant for genotype as well as environment (Table S5).

Analysis of marker trait associations

For analysis of MTAs, the eight field environments and the BLUEs across these eight environments were taken into account. By considering all MTAs found for all eight environments with $-\log_{10}$ (P-value) $\geq$4.0, in total 1873 MTAs were found for the SSRs and 6245 for the SNPs, of those 239 MTAs were detected for calculating with BLUEs across all environments in the SSRs, and 788 MTAs detected in the SNPs (Table 1; Tables S6, S7). After applying Bonferroni-correction, $-\log_{10}$ (P-value) $\geq$4.82 was significant for the SSRs and $-\log_{10}$ (P-value) $\geq$5.89 was significant for SNPs. Then the numbers shifted from 1104 significant MTAs for SSRs associated with BLUEs to 153 MTAs, and from 1904 significant MTAs for the SNPs associated with BLUEs to 280 MTAs (Figure 3). However, many markers detected MTAs in several environments and also especially the significant SNPs co-segregated or were closely linked in many cases, so that the number of real loci was significantly lower (Figure S1). By considering significant MTAs for BLUEs only and by combining marker loci with distance in the genetic map of $\leq$6 cM, the number of combined marker loci was estimated to be 109 for the SSRs and 87 for the SNPs (Tables S6B, S7B, S8).

Analysis of candidate genes

The *Rht-D1b* mutant allele on chromosome 4D was detected in 216 varieties, while only 26 *Rht-B1b* mutant alleles on chromosome 4B were present (Table S2). No double-dwarfs were found. In the spectrum of varieties the *Rht-D1b* and *Rht-B1b* mutant alleles were mainly represented in the first half of the distribution containing the shorter varieties (Figure 2). The MTA with *Rht-D1* candidate gene was highly significant in all environments (Table S9). No significant MTAs with $-\log_{10}$ (P-value) $\geq$3.0 were detected for *Rht-B1*, which may be due to the low representation of the mutant allele in the germplasm.

The varieties were also genotyped for the *Ppd-D1* gene on chromosome 2DS regulating the photoperiodic sensitivity of the lines [46]. The mutant allele *Ppd-D1a* rendering a photoperiod insensitive phenotype was detected in 53 varieties, which mostly originated from South France. The *Ppd-D1* candidate gene was significantly associated with plant height in all environments. A possible reason is that *Ppd-D1* is closely linked to the dwarfing gene *Rht8* [51]. Allelic variants of microsatellite marker GWM261 were reported as diagnostic for *Rht8* [22,52]. The 192-bp allele of GWM261, which is considered to be diagnostic for reduced height

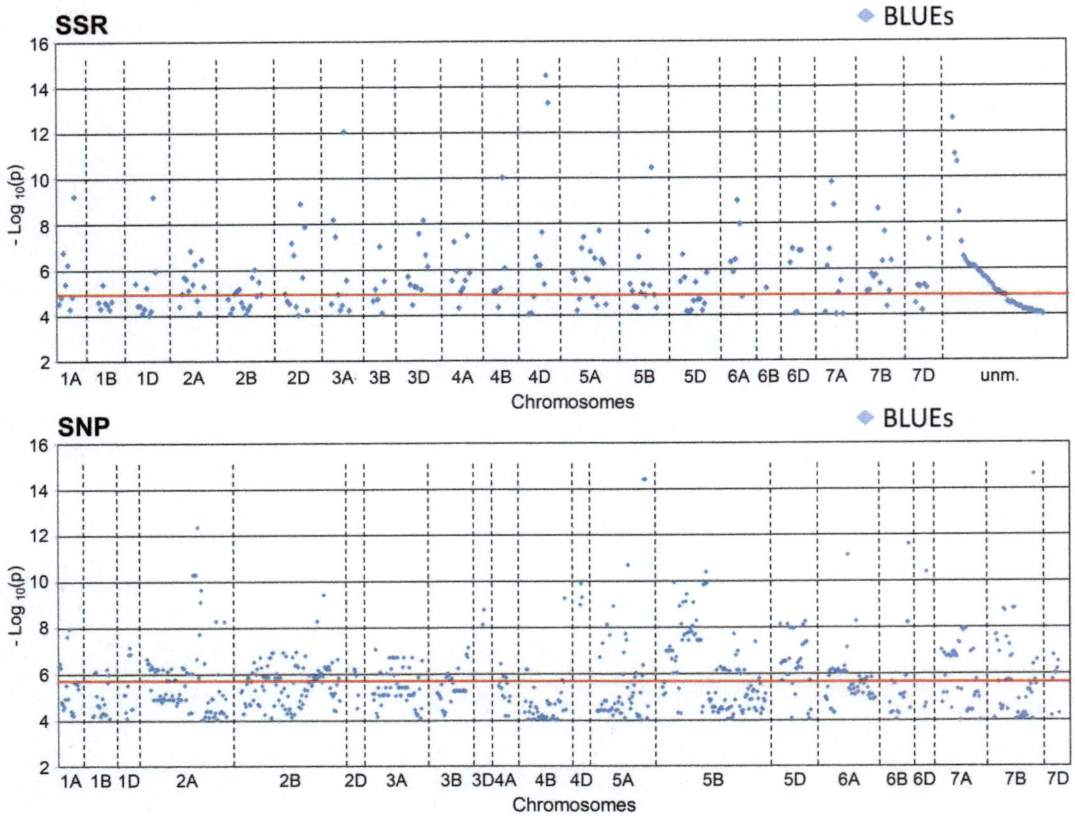

Figure 3. Manhattan plots of SSR or SNP marker alleles associated with plant height. The plot presents significant MTAs at a threshold of $-\log_{10}$ (P-value) ≥4.0 for BLUEs sorted according to their chromosomal location. The red line indicates the threshold for Bonferroni correction, $-\log_{10}$ (P-value) ≥4.82 (SSR) and ≥5.89 (SNP).

at the *Rht8* locus, was present in 35 varieties (Table S2). Moderately significant MTAs for PH were found with $-\log_{10}$ (P-value) ≥3.0 in four environments and with $-\log_{10}$ (P-value) ≥ 4.0 in two environments (Table S9) for this marker. However, no linkage disequilibrium based on r^2 was detected between the *Ppd-D1* candidate gene and marker GWM261, indicating that the significant MTA detected for *Ppd-D1* is independent of the effect based on *Rht8* (Figure S2). The Spearman rank order correlation between BLUEs for PH and BLUEs for heading date in the same environments was moderately significant with r = 0.158 (P = 0.00231).

Additive effects of SSR and SNP markers

The varieties carried between 19 and 170 PH reducing SSR alleles and between 4 to 101 PH promoting alleles (Figure S3). The Spearman Rank order correlation between the number of PH

Table 1. Number of MTAs per environment for the SSR markers and the SNPs of the 90K iSelect chip.

	SSR		**90K iSelect**	
Environments	**$-\log_{10}$ (P-value) ≥4.0**	**$-\log_{10}$ (P-value) ≥4.82**	**$-\log_{10}$ (P-value) ≥4.0**	**$-\log_{10}$ (P-value) ≥5.89**
Andelu (2009)	209	127	660	176
Seligenstadt (2009)	225	132	761	229
Wohlde (2009)	220	118	686	195
Andelu (2010)	221	130	705	273
Janvielle (2010)	169	87	566	134
Saultain (2010)	163	105	620	154
Seligenstadt (2010)	208	120	707	229
Wohlde (2010)	219	132	752	234
BLUEs	239	153	788	280
Sum	**1873**	**1104**	**6245**	**1904**

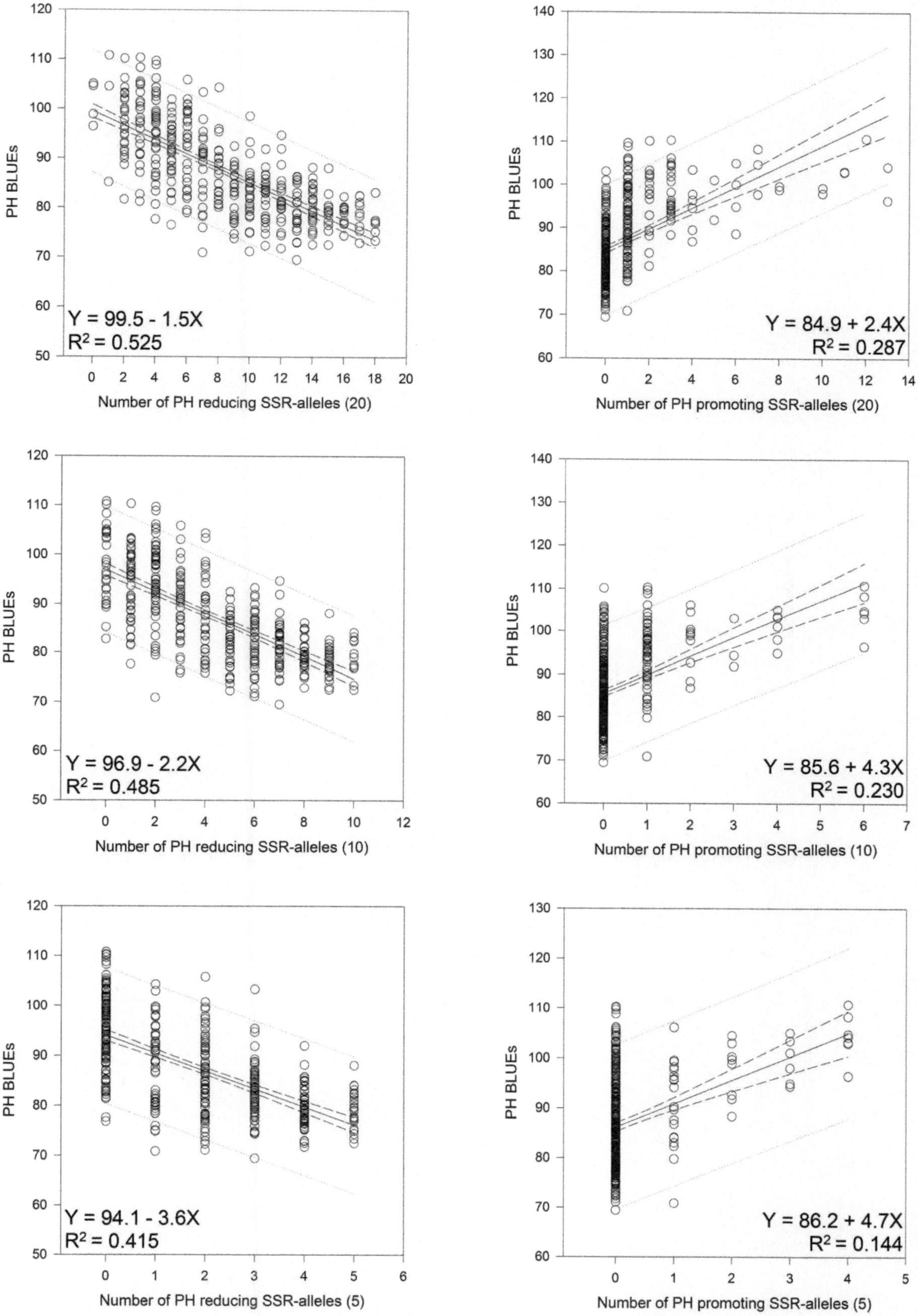

Figure 4. Linear regressions of the most PH reducing ("best") and the most PH promoting ("worst") SSR-alleles with PH-BLUEs. Linear regression resulted in a relationship between PH-BLUEs and the 20, 10 or 5 "best" or "worst" SSR-alleles in 372 varieties.

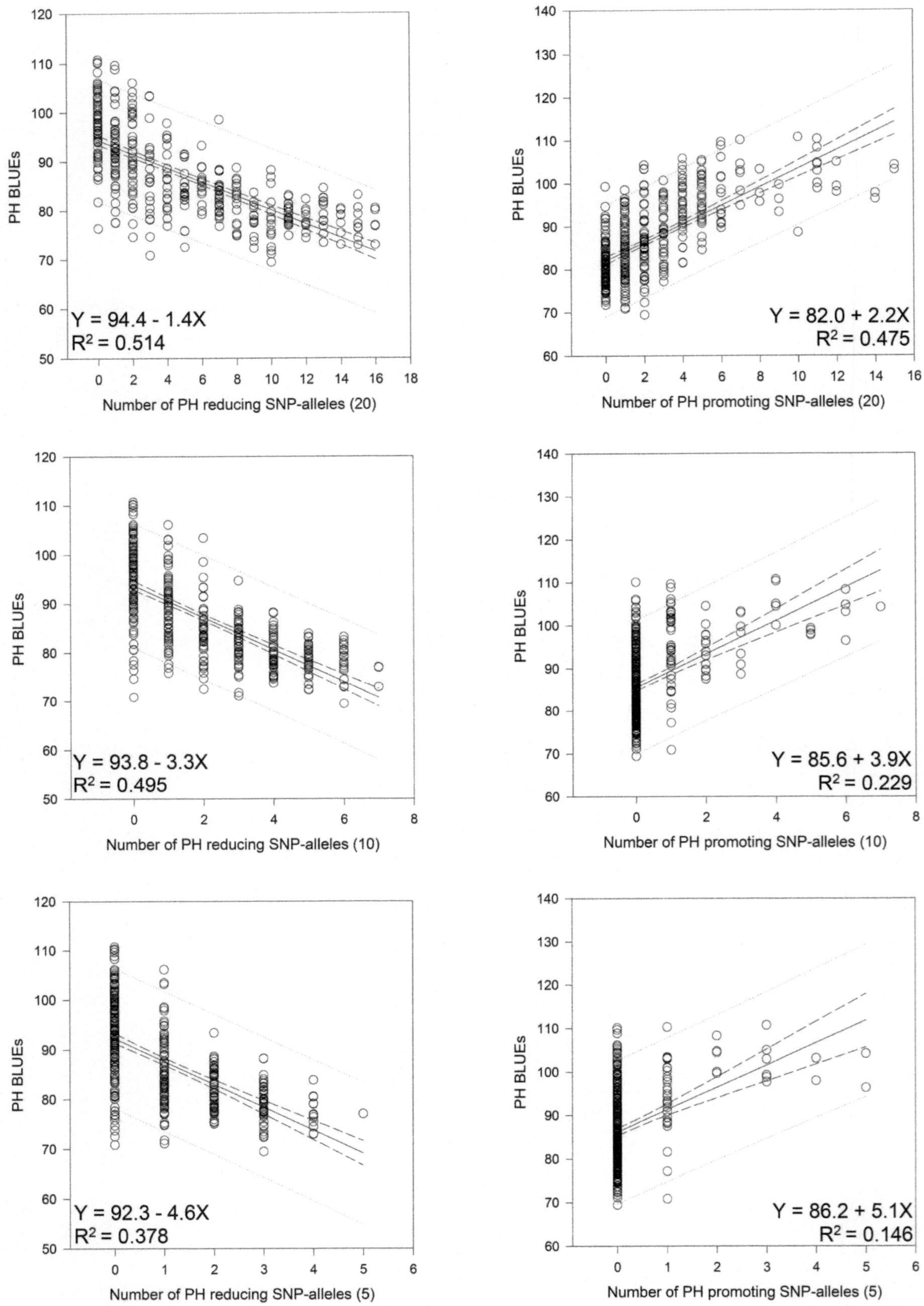

Figure 5. Linear regressions of the most PH reducing ("best") and the most PH promoting ("worst") SNP-alleles with PH-BLUEs. Linear regression resulted in a relationship between PH-BLUEs and the 20, 10 or 5 "best" or "worst" SNP-alleles from different loci in 372 varieties.

Table 2. List of the most PH reducing ("best") and most PH enhancing ("worst") SSR-alleles.

Marker alleles	Chromosome (linked gene)	Position (cM)	alleles belong to the					
			20 best	10 best	5 best	20 worst	10 worst	5 worst
GWM0357_128	1A	92.3	x	x				
GDM0126_196	1B	124.9				x		
GWM4780b_140	1D	89.1	x					
GWM4795_195	1D	130.8	x					
GWM1053_134	2A	0.2				x		
WMC0522_204	2A	88.7				x		
WMC0154_179*	2B	0.4	x					
GWM1419_197	2D	118.4				x		
GWM0539_151	2D	136.0				x	x	
GWM1204_279	2D	140.4	x					
GWM0349_232	2D	192.8				x	x	x
GWM1235_122	2D	215.0	x	x				
GWM1507_135*	3A	61.0	x	x	x			
CFA2193_239	3A	155.5	x	x	x			
GWM0938_144	3B	176.5				x	x	x
GWM1564_235	3B	238.5				x		
GWM0383a_183	3D	131.1				x	x	x
WMC0262_188	4A (*KAO-B1*)	165.1	x					
BARC0163_161	4B	81.2				x	x	x
GWM0819_183*	4D (*Rht-D1*)	34.8	x					
GWM4346_219*	4D (*Rht-D1*)	41.5	x	x	x			
GWM4001a_211	4D	100.6	x	x	x			
BARC0100b_162*	5A	88.1				x		
CFA2163a_187	5A (*Rht9*)	163.3				x	x	
WMC0415b_176*	5B	97.6				x	x	x
GWM1475a_122*	5B	108.2	x	x				
BARC0107_187*	6A	93.5	x					
GWM1241_126*	6D	138.5				x	x	
GWM1207_206	7A	230.3				x		
GWM0400_137	7B	3.4				x		
BARC0267a_145	7B	36.0	x	x				
GWM1650a_191	7D	29.1				x	x	
GWM0885_172	7D	52.1				x		
BARC0149_148	unm.	13.1	x					

Table 2. Cont.

Marker alleles	Chromosome (linked gene)	Position (cM)	alleles belong to the					
			20 best	10 best	5 best	20 worst	10 worst	5 worst
BARC1144_268	unm.	31.0				x	x	
CFD0045_177	unm.	36.0	x	x				
WMC0032_null	unm.	50.0	x					
WMC0054_162	unm.	51.4	x					
WMC0410_115	unm.	74.1	x	x	x			
WMJ3103b_164	unm.	113.1				x		

*Coincides with meta-QTL according to Griffiths et al. (2012) [54].

reducing alleles per variety and the PH BLUEs was −0.750 (P = 0.0000002), and for the number of PH promoting alleles it was 0.808 (P = 0.0000002).

We sorted the significant SSR alleles with $-\log_{10}$ (P-value) ≥4.0 according to their additive effects and tested the 20, 10 and 5 most significant alleles for their effects on the PH BLUEs (Table 2). Linear regression showed a dependence of the PH BLUEs from the number of PH reducing alleles with Y = 99.5–1.5X with $R^2 = 0.525$; for the PH promoting alleles the regression formula was Y = 84.9+2.4X with $R^2 = 0.287$ for the 20 most significant SSR-markers based on the additive effects (Figure 4). Even with only 5 alleles, an $R^2 = 0.415$ was found for the PH reducing alleles. Similar observations were found for the SNP-markers, where the 20 "best" and "worst" markers were selected which represented different combined loci, i.e. cluster of markers in LD were represented by only one marker (Table 3, Figure 5).

Analysis of syntenic regions in the rice genome

We developed a strategy to search for potential candidate genes for plant height regulation in the complete set of 21742 scorable wheat SNPs based on synteny to the annotated rice genome. For this purpose, a keyword search was conducted with keywords 'gibberellin', 'auxin', 'cytochrome P450', 'kaurene', 'kaurenoic' and 'DELLA' in the MSU rice database (http://rice.plantbiology.msu.edu/). The key words were based on gene ontologies of known dwarfing genes from the literature. We then used the BlastX-annotations in rice of the wheat SNP markers, as provided by [42], to relate the rice loci detected in the keyword search to the complete set of wheat SNPs. A total of 694 hits for all keywords were found in the MSU rice database (Table 4). Of those 190 hits were present, scorable and polymorphic in the data of the 90 k iSELECT chip and 44 hits were significant for PH-BLUEs with $-\log_{10}$ (P-value) ≥3.0 (Table 4, Table S10). The significant markers included 20 hits for 'cytochrome P450', 14 hits for 'auxin' and 10 hits for 'gibberellin'. Most significant with $-\log_{10}$ (P-value) ≥11.0 was wheat marker Kukri_c82296_367 on chromosome 5B with synteny to rice locus LOC_Os09g37500, annotated as OsSAUR55, an auxin responsive SAUR gene family. The SAUR (= small auxin-up RNAs) gene family in rice represents early auxin-responsive genes with largely unknown function [53]. Additionally, nine wheat markers with homology to auxin responsive factors on chromosomes 2B, 3A, 3B, 5A, 6A, 7A and 7D were significant. The significant hits for 'gibberellin' included a gibberellin response modulator protein on chromosome 5B and a chitin-inducible gibberellin-responsive protein on chromosome 3A (Table S10).

Out of 84 hits for the keyword 'gibberellin' in the rice database only 20 hits were present and polymorphic in our 90 k iSELECT chip. Therefore, we wanted to test if any rice-genes from the keyword search 'gibberellin' are linked to highly significant genomic regions in the wheat-genome. For this purpose, the syntenic rice loci were extracted from [42] for the significant wheat SNP-markers of Table S7B and this list was compared to the rice loci of the keyword search 'gibberellin'. The results of coinciding genomic regions are depicted in Table 5 and Table S11. The list of genes included several GA20 and GA2 oxidases orthologous to wheat chromosomes 1A, 2A, 3A, 3B, 5B, 5D and 7B, and other genes involved in GA synthesis, like ent-kaurenoic acid oxidase orthologous to wheat chromosome 7A and *ent*-kaurene synthase orthologous to wheat chromosome 2B. But there are also potential GA-receptors, like DELLA-genes orthologous to wheat chromosomes 4B, 4D and 7A and genes of the GID family orthologous to wheat chromosomes 2B and 5B (Table 5, Figure S1).

Table 3. List of the most PH reducing ("best") and most PH enhancing ("worst") SNP-alleles.

SNP-Marker	Chromosome (linked orthologous locus in rice)	Position (cM)	alleles belong to the					
			20 best	10 best	5 best	20 worst	10 worst	5 worst
BS00086680_51	1A	62.6	x	x	x			
Kukri_c54467_100	1A	65.9	x					
tplb0025b13_2687	1D	3.4				x	x	x
wsnp_Ex_c1358_2602235	1D	12.1				x	x	
wsnp_CAP8_c2677_1394934	2A	66.6				x		
BS00022896_51	2A	69.8	x					
RAC875_c38018_278	2A	69.8	x	x				
wsnp_CAP11_c1711_934478	2A (*OsGA 20 oxidase 2*)	95.8	x					
RAC875_c4609_1756	2A	105.3	x					
BS00023075_51	2A	117.2	x					
Ra_c23048_474	2B	66.0	x					
BobWhite_rep_c64068_241	2B	131.7				x		
Tdurum_contig91865_242	3A	38.3	x	x	x			
Kukri_c20889_526	3A (*OsGA2ox3**)	61.4				x	x	
BobWhite_c19725_1329	3B	57.4				x	x	
RAC875_c403_2247	3B (*OsGA2ox2**)	66.8				x	x	x
Jagger_c3839_60	3D	147.1	x					
IAAV7221	4B	2.2	x	x	x			
RAC875_c1357_860	4B	66.4				x		
RFL_Contig5341_816	4B	79.1				x	x	x
wsnp_BE444644A_Ta_2_1	5A	11.0	x					
wsnp_BF293620A_Ta_2_3	5A	57.7	x	x				
wsnp_Ex_c23795_33033959	5A	101.2				x		
RAC875_c25756_279	5B (*Os GA receptor GID1L2*)	50.5	x					
IAAV7207	5B	57.4				x	x	x
Excalibur_c5329_1335	5B	104.1				x		
wsnp_Ex_c65985_64188864	5D	27.9				x		
wsnp_Ex_c1278_2449191	5D	81.7	x	x				
BS00089597_51	5D (*OsGA20ox1**)	166.2	x	x				
BS00036878_51	6A	57.0				x		
BS00084314_51	6B	42.8	x	x	x			
RAC875_rep_c71463_98	6B	95.7	x	x				
Kukri_c73802_205	6D	2.3				x	x	
Excalibur_c15844_1470	6D	19.3				x		

Table 3. Cont.

SNP-Marker	Chromosome (linked orthologous locus in rice)	Position (cM)	alleles belong to the					
			20 best	10 best	5 best	20 worst	10 worst	5 worst
Ex_c101666_634	7B	42.9				x	x	
wsnp_Ex_c24376_33618864	7B	50.6				x		
BS00024215_51	7B	54.5	x					
Tdurum_contig42584_1190	7B	140.4	x	x	x			
Excalibur_c22419_460	7D	56.8				x	x	x
Ku_c884_791	7D	83.9				x		

*Nomenclature according to Sakamoto et al. 2004 [58].

Discussion

The pattern of inheritance for plant height was characterized by many closely linked markers with high LD, especially for the SNPs. When these closely linked marker loci were combined, we estimated 109 loci for the SSRs and 87 for the SNPs. The higher numbers for SSRs may be caused by a better coverage of the D-genome for this marker type. Eventually, further loci could be combined. A meta-QTL analysis of four doubled-haploid populations in European elite winter wheat germplasm estimated sixteen meta-QTL loci [54]. Several of these meta-QTL coincided with our detected MTAs (Table 6). In the population Avalon × Cadenza, a QTL distal to marker GWM261 on chromosome 2DS was described [54], which may coincide with the MTAs detected by markers GWM4815 and GWM1418 in our study. This QTL could be due to the effect of various *Rht8* alleles [22,52] or be a pleiotropic effect of *Ppd-D1* with a reduced time to flowering resulting in reduced plant height [55]. *Rht-D1* was included in a block of highly significant markers in LD including BARC98, WMC473b, GWM819, GDM129, GWM4346 for the SSRs and RAC875_rep_c105718_304 for the SNPs (Figure 6) and coincided with the respective meta-QTL on chromosome 4D [54]. Several of the markers coinciding with meta-QTL regions were found in our list of the 20 most PH reducing or PH promoting alleles based on the additive effects. These included besides the markers linked to *Rht-D1*, GWM4346 and GWM819, WMC154 on chromosome 2B, GWM1507 on 3A, BARC100 on 5A, WMC415b and GWM1475a on 5B, BARC107 on 6A and GWM1241 on chromosome 6D (Table 2).

Of the known mapped *Rht*-genes in wheat, besides *Rht-D1* and *Rht8*, possibly an allelic effect of *Rht4* was discovered in our study. *Rht4* was mapped on chromosome arm 2BL with marker WMC317 explaining most of the phenotypic variance in the cross Vigour18 × Burt *ert* (*Rht4*) [23]. WMC317 was not significant in our association study, however significant MTAs were found for the flanking markers GWM1399, GWM619 and GWM3038 (Figure S1). The *Rht5*-gene was mapped on chromosome 3BS in the Vigour18 × Marfed M (*Rht5*) population and linked to marker BARC102. BARC102 is located in a distance of 1.2 cM (http://wheat.pw.usda.gov/GG2/index.shtml) from marker WMC808, which was significant in our study. Therefore the MTAs discovered by WMC808 and GWM4145 are possibly caused by an allele of *Rht5*. Gene *Rht9* was mapped to chromosome arm 5AL in the Chuan Mai 18 (*Rht8*) × Mara (*Rht8Rht9*) population and linked to marker BARC151 [23]. While BARC151 was not significant in our association panel, the neighboring markers GWM1342 and CFA2163 discovered highly significant MTAs (Figure S1), which may be an allelic effect of *Rht9*. The *Rht13* gene was reported to be linked to marker GWM577 on the distal long arm of chromosome 7B [23]. GWM577 was not significant in our association panel, but marker BARC182 in a distance of 12 cM was highly significant, an effect which may be caused by an allele of *Rht13*.

For a number of gibberellin metabolic pathway genes the chromosomal location is known in wheat [56]. Three *ent*-kaurenic acid oxidase (KAO) genes were mapped to the distal ends of chromosome arms 7AS, 4AL and 7DS corresponding to the 7BS/4AL translocation region [57]. The significant markers GWM681 and GWM735 correspond to the location of *KAO-A1* south of marker GWM471 on chromosome arm 7AS, while the significant markers WMC262 and GWM1251 correspond to *KAO-B1* on chromosome 4AL north of marker GWM160. WMC262 is listed among the 20 most effective SSR-markers for plant height reduction based on the additive effects (Table 2). The significant

Table 4. Keyword search in rice for PH candidate genes.

	Keywords in rice				
	gibberrellin	auxin	cytochrome P450	kaurene, kaurenoic	DELLA
Number of hits in MSU rice database	84	177	421	9	3
Number of syntenic wheat SNPs present and polymorphic in the 90 k iSELECT chip	20	61	103	6	0
Number of wheat SNPs significant for PH-BLUEs with LOD >3.0	10	14	20	0	0

MTAs at both sites, on chromosomes 7AS and 4AL, may be caused by the respective *KAO*-genes, since a *KAO*-gene has been described as *Dwarf3* gene in maize [17]. On chromosome 7AS a large significant cluster of SNP markers was detected which had synteny to the rice gene *OsKAO* [58] on rice chromosome 6 (Table 5, Figure S1), the syntenic locus to *KAO-A1* in wheat.

Three GA 20-oxidase genes, *TaGA20ox1*, were located on the distal long arms of chromosomes 5B and 5D [59,56] and chromosome arm 4AL in the region of the known 5AL/4AL translocation [60]. These loci may correspond to the significant MTAs detected in our study on chromosome arm 5DL by markers BARC177, WMC765, BARC144, GWM1454, GWM272, on chromosome arm 5BL by marker WMC 783 and on chromosome 4A by marker BARC 343. This assumption is supported by the fact that a cluster of SNP-markers on chromosome arm 5DL (wsnp_RFL_Contig2402_1924907, Excalibur_c24145_1643, wsnp_Ex_c11055_179228283, wsnp_Ex_c24145_33394561, wsnp_Ex_c24145_33394644) and an SNP marker on chromosome 5BL (Kukri_c1214_2565) spanned a region on rice chromosome 3, where *OsGA20ox1*, the syntenic locus to *TaGA20ox1*, is located (Table 5, Figure S1, Table S11).

Unfortunately, the 21742 scorable SNPs in our study did only include 20 hits for the keyword 'gibberellin' based on synteny to the rice genome (Table 4) and no GA20-oxidases were present. Therefore, we could only rely on indirect evidence based on mapping in similar genomic regions to assess the impact of gibberellin pathway genes (Table 5). Syntenic relationships to rice genomic regions containing GA oxidases were found for chromosomes 1A, 3A, 3B and 5B (Table 5). The regions on chromosomes 3A and 3B corresponded to *OsGA2ox3* and *OsGA20ox2* [58], both on rice chromosome 1. For both chromosomal regions large clusters of SNP markers were significant (Figure S1) and representative markers (Kukri_c20889_526 on chromosome 3A and RAC875_c403_2247 on chromosome 3B) were in the list of the makers with highest additive effects (Table 3). The semidwarfing gene *sdw1/denso* on chromosome 3HL in barley was described as most likely orthologous gene of the *sd1* gene in rice, which carries the mutation in the gene *OsGA20ox2* [16,61] encoding an oxidase enzyme involved in the biosynthesis of gibberellin [13,14]. Therefore, the highly significant SNP cluster on chromosome 3BL corresponding to *OsGA20ox2* in our study may be the syntenic locus to gene *sdw1/denso* in barley. Since co-location of loci is not a direct proof for the involvement of the mentioned genes in regulating PH, further sequence analysis and SNP-discovery in the wheat genes themselves is necessary to estimate their contribution to the expression of PH.

Syntenic relationships were also discovered for rice genomic regions carrying genes for GA perception, like DELLA proteins on wheat chromosomes 4B, 4D and 7A and GID receptors on wheat chromosomes 2B and 5B (Table 5). The cluster of SNP-markers

Table 5. List of orthologous loci in rice with genes of the GA metabolism or related to dwarfing phenotypes.

Wheat SNPs, chromosomal position	Orthologous locus in rice	Gene annotation in rice
1A; 69.8 cM	LOC_Os05g34854	OsGA2ox4*; GA 20 oxidase 2
2A; 95.8–98.1 cM	LOC_Os04g55070	GA 20 oxidase 2
2B; 64.3 cM	LOC_Os07g44850	GA receptor GID1L2
2B; 75.7–85.1 cM	LOC_Os04g52230	Ent kaurene synthase
3A; 59–61.4 cM	LOC_Os01g55240	OsGA2ox3*; GA 2-beta dioxygenase
3B; 59.7–66.8 cM	LOC_Os01g66100	OsGA20ox2*; GA 20 oxidase 2
4B; 50.8 cM	LOC_Os03g49990	DELLA, SLR1
4D; 47 cM	LOC_Os03g49990	DELLA, SLR1
5B; 45.8–50.5 cM	LOC_Os09g28230	GA receptor GID1L2
5B; 95.4–96.2 cM	LOC_Os03g42130	GA 20 oxidase 2
5B; 165 cM	LOC_Os03g63970	OsGA20ox1*; GA 20 oxidase 1
5D; 163.9–167.8 cM	LOC_Os03g63970	OsGA20ox1*; GA 20 oxidase 1
7A; 1.6–5.7 cM	LOC_Os06g02019	OsKAO*; ent-kaurenoic acid oxidase (cytochrom P450)
7A; 80.3 cM	LOC_Os06g02019	DELLA, RGL3
7B; 68.6 cM	LOC_Os08g44590	GA 20 oxidase 2

*Nomenclature according to Sakamoto et al. 2004 [58]

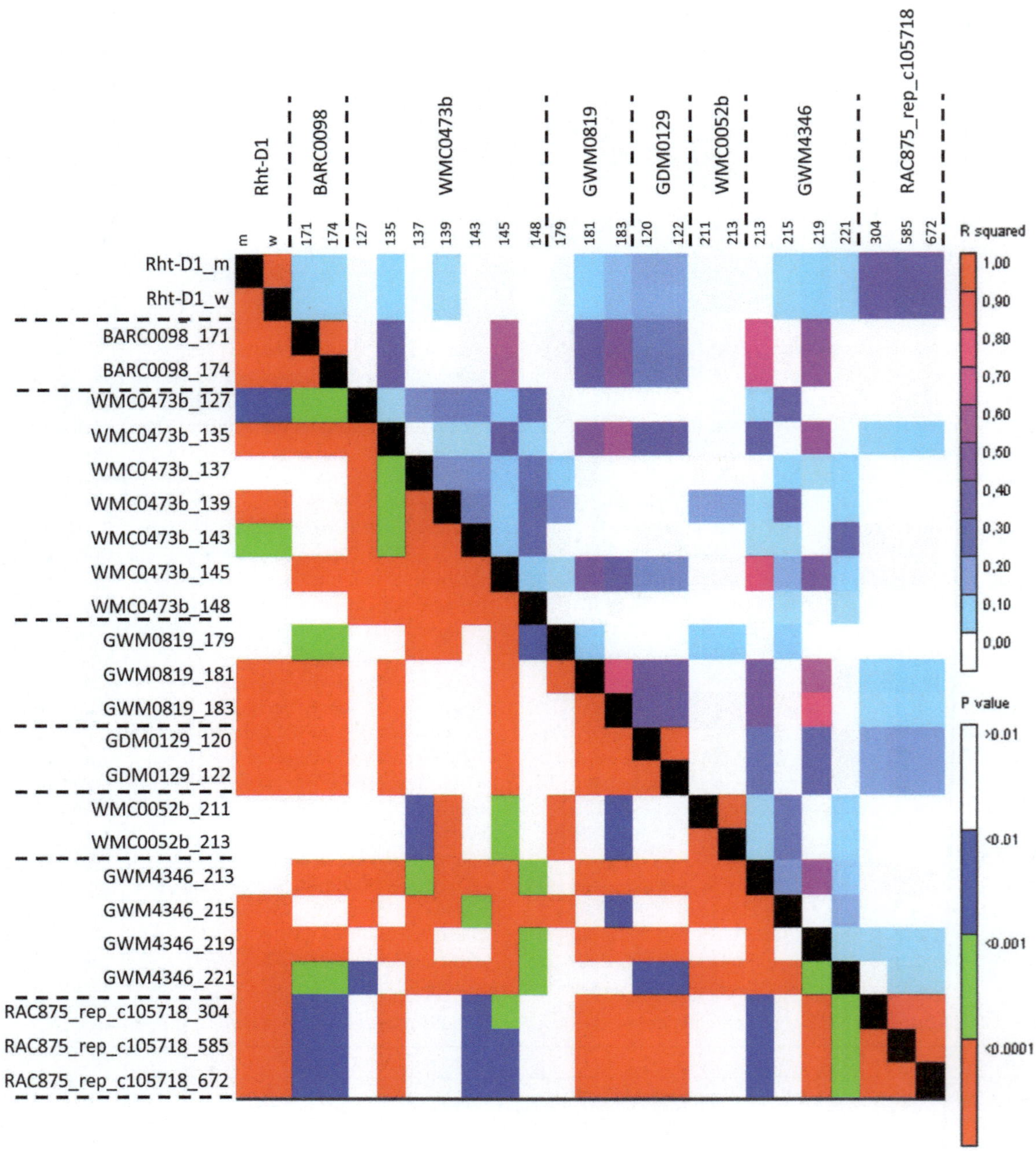

Figure 6. Linkage disequilibrium in the region of *Rht-D1*. The LD around *Rht-D1* on chromosome 4D extended to linked SSR and SNP markers with significant MTAs for plant height. In the upper triangle r^2 and in the lower triangle the P-values are depicted.

on wheat chromosome 4D with synteny to *OsDELLA* on rice chromosome 3 corresponds to the *Rht-D1* gene, since the respective SNP-Markers (RAC875_rep_c105718_304, RAC875_rep_c105718_672, RAC875_rep_c105718_585) are in LD with *Rht-D1* (Figure 6).

The significant candidate genes of the keyword search (Table 4) included 20 hits for 'cytochrome P450'. Cytochrom P450 is part of *ent*-kaurenoic acid oidase, an enzyme of the GA-metabolism, but it also is present in enzymes of other functions including the involvement in brassinosteroid biosynthesis [62]. Highly significant hits were found for auxin-related genes as well. These included markers with homology to auxin responsive factors and the SAUR gene family, the function of which is largely unknown in both cases [53,63].

Our genotyping results indicated, that the *Rht-D1b* mutant allele on chromosome 4D was present in 58% of the varieties analyzed, while the *Rht-B1b* mutant allele on chromosome 4B was found in only 7% of our varieties. The detection method employed [48] did not distinguish from further alleles, such as *Rht-B1b* from *Rht-B1d*, and *Rht-D1b* from *Rht-D1c* and *Rht-D1d* [64,65]. The high percentage of *Rht-D1*mutant allele in our germplasm set contradicted the results of [66], where 12% *Rht-B1b* and 16% *Rht-D1b* mutant alleles were reported in 103 French varieties, while in the total set of 368 worldwide accessions for both loci, *Rht-B1* and *Rht-D1*, a frequency of 12% mutant alleles was found. In contrast to our study, where only registered varieties were investigated, the study of [66] included landraces. Another study reported frequencies of 40% for *Rht-B1b* and 22% for *Rht-D1b* in

Table 6. Coincidences of significant MTAs in our study with QTL or meta-QTL regions described by Griffiths et al. 2012 [54].

Significant markers in our study	Chromosome	Coincidences with QTL regions described in [54]
GWM164	1A	WMC336 - PSP3027
GWM337	1D	GWM337 - WMC36
GWM359	2A	wPt6207 - WMC827
WMC154	2B	Stm17tcac - WMC154
GMW4815, GWM1418	2D	CFD36 - GWM261
GWM1507, GWM4018	3A	BARC45 - WMC269
GWM4660, GWM251, BARC163	4B	GWM149 - GWM6
BARC98, WMC473b, GWM819, GDM129, GWM4346	4D	WMC285 - *Rht-D1*
BARC4, BARC100	5A	wPt3794 - GWM617
GWM1475, WMC415b, BARC109a, GWM1165, GWM1108, GWM335	5B	GWM213 - GWM408
BARC107, GWM4608, WMC672	6A	BARC23 - GWM570
GWM1241	6D	CFD42 - GWM325
GWM1397, GWM3062, GWM4335	7D	wPt6263 - GWM44

a 172 genotypes originating from 20 countries [67]. The data indicated, that *Rht-D1b* may be prevalent in the Middle European wheat varieties, while *Rht-B1b* seems to occur mainly in the world wide germplasm including landraces.

The *Rht1* mutant phenotypes were found to increase harvest index and grain yield [68] and were therefore introduced in many commercial wheat varieties around the world. But also some negative effects, like decreased seedling vigour and shorter coleoptiles were reported for these GA-insensitive dwarfing genes [69,70]. Beneficial effects for the *Rht1* wild type alleles were reported in drought environments [71]. Therefore, the use of alternative height reducing genes was suggested [70,72]. Our data indicate a wealth of alternative loci that could be employed for modulating plant height reduction and which are already present in the germplasm. With only five SNP- or SSR-markers obvious additive effects for PH BLUEs were observed (Figures 4 and 5). Similar results were reported for pyramiding of elite alleles for plant height in two winter wheat populations [73].

Supporting Information

Figure S1 Chromosomal location of marker-trait associations for plant height. The linkage map for the SNP-markers of the 90 k iSELECT chip based on the ITMI-DH population is depicted on the left side, while the linkage map for the SSR-markers based on the ITMI-RIL population is shown on the right side for each linkage group. Each MTA for a single environment or the BLUEs are depicted by an icon as explained on the first page of the file. The locations of known *Rht* genes and known mapping positions for gibberellin metabolism genes for wheat are indicated in the SSR maps. Orthologous sites to rice (*Oryza sativa* = *Os*) loci with genes related to GA metabolism or perception are indicated in the SNP maps.

Figure S2 No Linkage disequilibrium was detected between SSR-marker GWM0261 and the candidate gene *Ppd-D1* on chromosome 2D.

Figure S3 Frequency of (A) PH reducing and (B) PH promoting SSR-alleles per variety.

Table S1 Climatic factors at trial sites.

Table S2 List of varieties, genotyping data of candidate genes and phenotypic data.

Table S3 Estimation of variance components and broad sense heritability and estimation of differences between groups (= environments) using ANOVA and a Tukey B test.

Table S4 Spearman rank order correlation of PH scores in 372 varieties among eight environments and the BLUEs. GAT_2012 represents an untreated control environment in a plant nursery.

Table S5 Analysis of variance (ANOVA) of plant height in 372 varieties in eight environments.

Table S6 List of significant ($-\log_{10}(P)$ >4.0) marker trait associations of the SSR markers. Sheet 6A: List of significant ($-\log_{10}(P)$ >4.0) marker trait associations of the SSR markers. Columns F to U describe the additive effect and the R^2 for each environment. Colum V indicates the R2 of the BLUEs values. Positive additive effects promote plant height and negative additive effects reduce plant height. **Sheet 6B: List of significant ($-\log_{10}(P)$ >4.0) marker trait associations of the SSR markers for BLUEs only.** Markers with distances <6.0 cM were combined by colour coding. Columns F to U describe the additive effect and the R^2 for each environment. Colum V indicates the R^2 of the BLUEs values. Positive additive effects promote plant height and negative additive effects reduce plant height.

Table S7 List of significant ($-\log_{10}(P)$ >4.0) marker trait associations of the SNP markers. Sheet 7A: List of significant ($-\log_{10}(P)$ >4.0) marker trait associations of the SNP markers. Columns G to V describe the additive effect and the R^2 for each environment. Colum W indicates the R^2 of the BLUEs values. Positive additive effects promote plant height and negative additive effects reduce plant height. **Sheet 7B: List of significant (–log10(P) >4.0) marker trait associations of the SNP markers for the BLUEs only.** Markers with distances <6.0 cM were combined by colour coding. Columns G to V describe the additive effect and the R^2 for each environment. Colum W indicates the R^2 of the BLUEs values. Positive additive effects promote plant height and negative additive effects reduce plant height.

Table S8 Number of MTAs per chromosomes for the SSR and SNP markers.

Table S9 List of significant (–log10(P) >2.0) marker trait associations of the candidate genes. Columns F to U describe the additive effect and the R^2 for each environment. Colum V indicates the R^2 of the BLUEs values. Positive additive effects promote plant height and negative additive effects reduce plant height.

Table S10 List of significant (–log10(P) >3.0) marker trait associations of the candidates from the keyword search. Columns Y to AF describe the additive effect for each environment. Column AG indicates the additive effect of the BLUEs values. Positive additive effects promote plant height and negative effects reduce plant height.

Table S11 List of wheat SNP with GA related orthologous loci in rice.

Author Contributions

Conceived and designed the experiments: MWG EE VK OA GS MH MSR. Performed the experiments: EE OA VK JP SK. Analyzed the data: CDZ JL KN. Contributed to the writing of the manuscript: MSR CDZ.

References

1. Khush GS (2001) Green revolution: the way forward. Nat Rev Genetics 2: 815–822.
2. Hedden P (2003) The genes of the Green Revolution. Trends Genet 19: 5–9.
3. Peng J, Richards DE, Hartley NM, Murphy GP, Devos KM, et al. (1999) 'Green revolution' genes encode mutant gibberellin response modulators. Nature 400: 256–261.
4. Peng J, Carol P, Richards DE, King KE, Cowling RJ, et al. (1997) The Arabidopsis GAI gene defines a signaling pathway that negatively regulates gibberellin responses. Genes Dev 11: 3194–3205.
5. Winkler RG, Freeling M (1994) Physiological genetics of the dominant gibberellin-nonresponsive maize dwarfs, *Dwarf8* and *Dwarf9*. Planta 193: 341–348.
6. Lawit SJ, Wych HM, Xu D, Kundu S, Tomes DT (2010) Maize DELLA proteins dwarf plant8 and dwarf plant9 as modulators of plant development. Plant Cell Physiol 51: 1854–1868.
7. Ikeda A, Ueguchi-Tanaka M, Sonoda Y, Kitano H, Koshioka M, et al. (2001) *Slender* rice, a constitutive gibberellin response mutant, is caused by a null mutation of the SLR1 gene, an ortholog of the height-regulating gene *GAI/RGARHT/D8*. Plant Cell 13: 999–1010.
8. Chandler PM, Marion-Poll A, Ellis M, Gubler F (2002) Mutants at the *Slender*1 locus of barley cv. Himalaya. Molecular and Physiological characterization. Plant Phys 129: 181–190.
9. Chandler PM, Harding CA (2013) 'Overgrowth' mutants in barley and wheat: new alleles and phenotypes of the 'Green revolution' DELLA gene. Journal Experimental Botany 64: 1603–1613.
10. Ashikari M, Wu J, Yano M, Sasaki T, Yoshimura A (1999) Rice gibberellin insensitive dwarf mutant gene Dwarf 1 encodes the α-subunit of GTP-binding protein. Proc Natl Acad Sci USA 96: 10284–10289.
11. Ueguchi-Tanaka M, Ashikari M, Nakajima M, Itoh H, Katoh E, et al. (2005) *GIBBERELLIN INSENSITIVE DWARF1* encodes a soluble receptor for gibberellin. Nature 437: 693–698.
12. Sasaki A, Itoh H, Gomi K, Ueguchi-Tanaka M, Ishiyama K, et al. (2003) Accumulation of phosphorylated repressor for gibberellin signaling in an F-box mutant. Science 299: 1896–1898.
13. Spielmeyer W, Ellis MH, Chandler PM (2002) Semidwarf (*sd-1*), "green revolution" rice, contains a defective gibberellin 20-oxidase gene. Proc Natl Acad Sci USA 99: 9043–9048.
14. Sasaki A, Ashikari M, Ueguchi-Tanaka M, Itoh H, Nishimura A, et al. (2002) A mutant gibberellin-synthesis gene in rice. Nature 416: 701–702.
15. Barboza L, Effgen S, Alonso-Blanco C, Kooke R, Keurentjes JJB, et al. (2013) *Arabidopsis* semidwarfs evolved from independent mutations in *GA20ox1*, ortholog to green revolution dwarf alleles in rice and barley. Proc Natl Acad Sci USA 110: 15818–15823.
16. Jia Q, Zhang J, Westcott S, Zhang X-Q, Bellgard M, et al. (2009) GA-20 oxidase as a candidate for the semidwarf gene *sdw1/denso* in barley. Funct Integr Genomics 9: 255–262.
17. Winkler RG, Helentjaris T (1995) The maize Dwarf3 gene encodes a cytochrome P450-mediated early step in gibberellin biosynthesis. Plant Cell 7: 1307–1317.
18. Multani DS, Briggs SP, Chamberlin MA, Blakeslee JJ, Murphy AS, et al. (2003) Loss of an MDR transporter in compact stalks of maize *br2* and sorghum *dw3* mutants. Science 302: 81–84.
19. Parvathaneni RK, Jakkula V, Padi FK, Faure S, Nagarajappa N, et al. (2013) Fine-mapping and identification of a candidate gene underlying the *d2* dwarfing phenotype in pearl millet, *Cenchrus americanus* (l.) Morrone. G3-Genes/Genomes/Genetics 3: 563–572.
20. Börner A, Röder M, Korzun V (1997) Comparative molecular mapping of GA insensitive *Rht* loci on chromosomes 4B and 4D of wheat (*Triticum aestivum* L.). Theor Appl Genet 95: 1133–1137.
21. Korzun V, Röder M, Worland AJ, Börner A (1997) Intrachromosomal mapping of the dwarfing (*Rht12*) and vernalisation response (*Vrn1*) genes in wheat by using RFLP and microsatellite markers. Plant Breed 116: 227–232.
22. Korzun V, Röder MS, Ganal MW, Worland AJ, Law CN (1998) Genetic analysis of the dwarfing gene (*Rht8*) in wheat. Part I. Molecular mapping of the *Rht8* on the short arm of chromosome 2D of bread wheat (*Triticum aestivum* L.). Theor Appl Genet 96: 1104–1109.
23. Ellis MH, Rebetzke GJ, Azanza F, Richards RA, Spielmeyer (2005) Molecular mapping of gibberellin-responsive dwarfing genes in bread wheat. Theor Appl Genet 111: 423–430.
24. Harjes CE, Rocheford TR, Bai L, Brutnell TP, Kandianis CB, et al. (2008) Natural genetic variation in *Lycopene epsilon cyclase* tapped for maize biofortification. Science 319: 330–333.
25. Matthies IE, Weise S, Röder MS (2009) Association of haplotype diversity in the α-amylase gene *amy*1 with malting quality parameters in barley. Mol Breed 23: 139–152.
26. Waugh R, Jannink JL, Muehlbauer GL, Ramsay L (2009) The emergence of whole genome association scans in barley. Curr Opin Plant Biol 12: 1–5.
27. Rafalski JA (2010) Association genetics in crop improvement. Curr Opin Plant Biol 13: 1–7.
28. Thornsberry JM, Goodman MM, Doebley J, Kresovich S, Nielsen D, et al. (2001) *Dwarf*8 polymorphisms associate with variation in flowering time. Nat Genet 28: 286–289.
29. Larsson SJ, Lipka AE, Buckler ES (2013) Lessons from Dwarf8 on strengths and weaknesses of structured association mapping. PLOS Genetics 9: e1003246.
30. Hamblin MT, Buckler ES, Jannink J-L (2011) Population genetics of genomics-based crop improvement methods. Trends in Genet 27: 98–106.
31. Pasam RK, Sharma R, Malosetti M, van Eeuwijk FA, Haseneyer G, et al. (2012) Genome-wide association studies for agronomical traits in a world wide spring barley collection. BMC Plant Biol 12: 16.
32. Brown PJ, Rooney WL, Franks C, Kresovich S (2008) Efficient mapping of plant height quantitative trait loci in a Sorghum association population with introgressed dwarfing genes. Genetics 180: 629–637.
33. Weng J, Xie C, Hao Z, Wang J, Liu C (2011) Genome-wide association study identifies candidate genes that affect plant height in Chinese elite maize (*Zea mays* L.) inbred lines. PLoS ONE 6: e29229.
34. Zhao Y, Mette MF, Gowda M, Longin CFH, Reif JC (2014) Bridging the gap between marker-assisted and genomic selection of heading time and plant height in hybrid wheat. Heredity 112: 638–645.
35. International Rice Genome Project (2005) The map-based sequence of the rice genome. Nature 436: 793–800.
36. Kollers S, Rodemann B, Ling J, Korzun V, Ebmeyer E, et al. (2013) Whole genome association mapping of *Fusarium* head blight resistance in European winter wheat (*Triticum aestivum* L.). Plos One 8: e57500.

37. Kollers S, Rodemann B, Ling J, Korzun V, Ebmeyer E, et al. (2013) Genetic architecture of resistance to *Septoria tritici* blotch (*Mycosphaerella graminicola*) in European winter wheat. Mol Breed 32: 411–423.
38. Kollers S, Rodemann B, Ling J, Korzun V, Ebmeyer E, et al. (2014) Genome wide association mapping of tan spot resistance (*Pyrenophora tritici-repentis*) in European winter wheat. Mol Breed 34: 363–371.
39. Zanke C, Ling J, Plieske J, Kollers S, Ebmeyer E, et al. (2014) Genetic architecture of main effect QTL for heading date in European winter wheat. Frontiers in Plant Science 5: 217.
40. Rademacher W (2000) Growth Retardants: Effects on gibberellin biosynthesis and other metabolic pathways. Ann Rev Plant Phys Plant Mol Biol 51: 501–531.
41. Cavanagh CR, Chao S, Wang S, Huang BE, Stephen S, et al. (2013) Genome-wide comparative diversity uncovers multiple targets of selection for improvement in hexaploid wheat landraces and cultivars. Proc Natl Acad Sci USA 110: 8057–8062.
42. Wang S, Wong D, Forrest K, Allen A, Chao S, et al. (2014) Characterization of polyploid wheat genomic diversity using a high-density 90,000 SNP array. Plant Biotechnology Journal 12: 787–796.
43. Röder MS, Korzun V, Wendehake K, Plaschke J, Tixier M-H, et al. (1998) A microsatellite map of wheat. Genetics 149: 2007–2023.
44. Ganal MW, Röder MS (2007) "Microsatellite and SNP markers in wheat breeding", in Genomics Assisted Crop Improvement: Vol. 2: Genomics Applications in Crops, ed RK Varshney and R Tuberosa, pp. 1–24.
45. Sorrells ME, Gustafson JP, Somers D, Chao S, Benscher D, et al. (2011) Reconstruction of the synthetic W7984 × Opata M85 wheat reference population. Genome 54: 875–882.
46. Poland JA, Brown PJ, Sorrells ME, Jannink J-L (2012) Development of high-density genetic maps for barley and wheat using a novel two-enzyme genotyping-by-sequencing approach (2012) PLoS One 7: 1–8.
47. Beales J, Turner A, Griffiths S, Snape J, Laurie DA (2007) A pseudo-response regulator is misexpressed in the photoperiod insensitive *Ppd-D1a* mutant of wheat (*Triticum aestivum* L.). Theor Appl Genet 115: 721–733.
48. Ellis MH, Spielmeyer W, Gale KR, Rebetzke GJ, Richards RA (2002) "Perfect" markers for the *Rht-B1b* and *Rht-D1b* dwarfing genes in wheat. Theor Appl Genet 105: 1038–1042.
49. Hardy OJ, Vekemans X (2002) SPAGEDi: a versatile computer program to analyse spatial genetic structure at the individual or population levels. Mol Ecol Notes 2: 618–620.
50. Matthies IE, van Hintum T, Weise S, Röder MS (2012) Population structure revealed by different marker types (SSR or DArT) has an impact on the results of genome-wide association mapping in European barley cultivars. Mol Breed 30: 951–966.
51. Pestsova E, Röder MS (2002) Microsatellite analysis of wheat chromosome 2D allows the reconstruction of chromosomal inheritance in pedigrees of breeding programmes. Theor Appl Genet 106: 84–91.
52. Worland AJ, Korzun V, Röder MS, Ganal MW, Law CN (1998) Genetic analysis of the dwarfing gene *Rht8* in wheat. Part II. The distribution and adaptive significance of allelic variants at the *Rht8* locus of wheat as revealed by microsatellite screening. Theor Appl Genet 96: 1110–1120.
53. Jain M, Tyagi AK, Khurana JP (2006) Genome-wide analysis, evolutionary expansion, and expression of early auxin-responsive *SAUR* gene family in rice (*Oryza sativa*). Genomics 88: 360–371.
54. Griffiths S, Simmonds J, Leverington M, Wang Y, Fish L, et al. (2012) Meta-QTL analysis of the genetic control of crop height in elite European winter wheat germplasm. Mol Breed 29: 159–171.
55. Börner A, Worland AJ, Plaschke J, Schumann E, Law CN (1993) Pleiotropic effects of genes for reduced height (Rht) and day-length insensitivity (Ppd) on yield and its components for wheat grown in middle Europe. Plant Breed 111: 204–216.
56. Spielmeyer W, Ellis M, Robertson M, Ali S, Lenton JR, et al. (2004) Isolation of gibberellin metabolic pathway genes from barley and comparative mapping in barley, wheat and rice. Theor Appl Genet 109: 847–855.
57. Khlestkina EK, Kumar U, Röder MS (2010) Ent-kaurenoic acid oxidase genes in wheat. Mol Breed 25: 251–258.
58. Sakamoto T, Miura K, Itoh H, Tatsumi T, Ueguchi-Tanaka M, et al. (2004) An overview of gibberellin metabolism enzyme genes and their related mutants in rice. Plant Phys 134: 1642–1653.
59. Appleford NEJ, Evans DJ, Lenton JR, Gaskin P, Croker SJ, et al. (2006) Function and transcript analysis of gibberellin-biosynthetic enzymes in wheat. Planta 223: 568–582.
60. Linkiewicz AM, Qi LL, Gill BS, Ratnasiri A, Echalier B, et al. (2004) A 2500-locus bin map of wheat homoeologous group 5 provides insights on gene distribution and colinearity with rice. Genetics 168: 665–676.
61. Jia Q, Zhang X-Q, Westcott S, Broughton S, Cakir M, et al. (2011) Expression level of a gibberellin 20-oxidase gene is associated with multiple agronomic and quality traits in barley. Theor Appl Genet 122: 1451–1460.
62. Chapple C (1998) Molecular-genetic analysis of plant cytochrome P450-dependent monooxygenases. Ann Rev Plant Physiol Plant Mol Biol 49: 311–343.
63. Okushima Y, Overvoorde PJ, Arima K, Alonso JM, Chan A, et al. (2005) Functional genomic analysis of the *AUXIN RESPONSE FACTOR* gene family members in *Arabidopsis thaliana*: Unique and overlapping functions of *ARF7* and *ARF19*. Plant Cell 17: 444–463.
64. Pestsova EG, Korzun V, Börner A (2008) Validation and utilization of Rht dwarfing gene specific markers. Cereal Res Commun 36: 235–246.
65. Wilhelm EP, Mackay IJ, Saville RJ, Korolev AV, Balfourier F, et al. (2013) Haplotype dictionary for the *Rht-1* loci in wheat. Theor Appl Genet 126: 1733–1747.
66. Wilhelm EP, Boulton MI, Al-Kaff N, Balfourier F, Bordes J, Greenland AJ, Powell W, Mackay IJ (2013) *Rht-1* and *Ppd-D1* associations with height, GA sensitivity, and days to heading in a worldwide bread wheat collection. Theor Appl Genet 126: 2233–2243.
67. Tosovic-Maric B, Kobiliskij B, Obreht D, Vapa L (2008) Evaluation of wheat *Rht* genes using molecular markers. Genetika 40: 31–38.
68. Flintham JE, Börner A, Worland AJ, Gale MD (1997) Optimizing wheat grain yield: effects of Rht (gibberellin-insensitive) dwarfing genes. Journal of Agricultural Science 128: 11–15.
69. Rebetzke GJ, Appels R, Morrison AD, Richards RA, McDonald G, et al. (2001) Quantitative trait loci on chromosome 4B for coleoptile length and early vigour in wheat (*Triticum aestivum* L.). Australian Journal of Agricultural Research 52: 1221–1234.
70. Ellis MH, Rebetzke GJ, Chandler P, Bonnett D, Spielmeyer W, et al. (2004) The effect of different height reducing genes on the early growth of wheat. Functional Plant Biology 31: 583–589.
71. Zhang J, Dell B, Biddulph B, Drake-Brockman F, Walker E, et al. (2013) Wild-type alleles of *Rht-B1* and *Rht-D1* as independent determinants of thousand-grain weight and kernel number per spike in wheat. Mol Breed 32: 771–783.
72. Rebetzke GJ, Richards RA (2000) Gibberellic acid-sensitive dwarfing genes reduce plant height to increase kernel number and grain yield of wheat. Australian Journal of Agricultural Research 51: 235–246.
73. Zhang B, Shi W, Li W, Chang X, Jing R (2013) Efficacy of pyramiding elite alleles for dynamic development of plant height in common wheat. Mol Breed 32: 327–338.

OsARF16 Is Involved in Cytokinin-Mediated Inhibition of Phosphate Transport and Phosphate Signaling in Rice (*Oryza sativa L.*)

Chenjia Shen[1]*⁹, Runqing Yue[2]⁹, Yanjun Yang[1], Lei Zhang[3], Tao Sun[1], Shuanggui Tie[2]*, Huizhong Wang[1]*

1 College of Life and Environmental Sciences, Hangzhou Normal University, Hangzhou 310036, China, **2** Henan Academy of Agricultural Sciences, Zhengzhou 450002, China, **3** Department of Plant Pathology, Washington State University, Pullman, WA 99164-6430, United States of America

Abstract

Background: Plant responses to phytohormone stimuli are the most important biological features for plants to survive in a complex environment. Cytokinin regulates growth and nutrient homeostasis, such as the phosphate (Pi) starvation response and Pi uptake in plants. However, the mechanisms underlying how cytokinin participates in Pi uptake and Pi signaling are largely unknown. In this study, we found that OsARF16 is required for the cytokinin response and is involved in the negative regulation of Pi uptake and Pi signaling by cytokinin.

Principal Findings: The mutant *osarf16* showed an obvious resistance to exogenous cytokinin treatment and the expression level of the *OsARF16* gene was considerably up-regulated by cytokinin. Cytokinin (6-BA) application suppressed Pi uptake and the Pi starvation response in wild-type *Nipponbare* (NIP) and all these responses were compromised in the *osarf16* mutant. Our data showed that cytokinin inhibits the transport of Pi from the roots to the shoots and that OsARF16 is involved in this process. The Pi content in the *osarf16* mutant was much higher than in NIP under 6-BA treatment. The expressions of *PHOSPHATE TRANSPORTER1* (*PHT1*) genes, *phosphate (Pi) starvation-induced* (*PSI*) genes and *purple PAPase* genes were higher in the *osarf16* mutant than in NIP under cytokinin treatment.

Conclusion: Our results revealed a new biological function for OsARF16 in the cytokinin-mediated inhibition of Pi uptake and Pi signaling in rice.

Editor: Keqiang Wu, National Taiwan University, Taiwan

Funding: This work was funded by the National Natural Science Foundation of China (31401935) and Zhejiang Provincial Natural Science Foundation of China under Grant No. LQ14C060001. The funders had no role in study design, data collection and analysis, decision to publish, or preparation of the manuscript.

Competing Interests: The authors have declared that no competing interests exist.

* Email: shencj@hznu.edu.cn (CS); tieshuanggui@126.com (ST); whz62@163.com (HW)

⁹ These authors contributed equally to this work.

Introduction

The phytohormone, auxin, governs many aspects of growth, development and nutrient homeostasis in plants. Auxin perception and signal transduction regulate the expressions of many downstream genes via the AUXIN (AUX)/INDOLE-3ACETIC ACID (IAA) and AUXIN RESPONSE FACTOR (ARF)-mediated auxin signaling pathways [1]. ARFs, which regulate the expression of auxin-responsive genes, are involved in all the developmental processes from embryogenesis to senescence [2]. Many plant physiological functions depend on ARFs, including apical hook formation in seedlings [3], abaxial identity in the gynoecium [4], embryo development and vascular formation [5], hypocotyl tropisms [6], fruit development from fertilization [7], lateral root formation [8] and root system growth [9].

Auxin interacts with various phytohormones to regulate many physiological functions, either synergistically or antagonistically [10–13]. Among them, cytokinin is known to act antagonistically to auxin and plays an essential role in controlling plant morphogenesis [14]. Two primary cytokinin response transcription factors, *ARR1* and *ARR12*, activate the *SHORT HYPOCOTYL2* (*SHY2*) gene, which encodes a repressor that regulates the transcription of *PIN-FORMED* (*PIN*) auxin transporter genes [15]. Cytokinin inhibits the expression of auxin transport genes and prevents the establishment of an auxin gradient during lateral root initiation [16–18]. *WOX11*, a WUSCHEL-related homeobox gene, is induced by both auxin and cytokinin and represses a cytokinin-signaling negatively regulated gene, *OsRR2* [19]. The mRNA accumulation of *AtARF19*, an auxin response factor involved in cell differentiation, has been shown to rely on the cytokinin-dependent transcription factor, AtARR12 [20]. Embryogenesis and post-embryonic root organ development all depend on the antagonistic interaction between cytokinin and auxin [21].

It has been shown that cytokinin is involved in the phosphate starvation response and phosphate uptake. The expressions of the phosphate transport gene, *AtPT1*, and the *AtPHO1* family genes are suppressed by both auxin and cytokinin [22,23]. Cytokinin also regulates many Pi deprivation-induced genes. For example, *At4*, which is expressed in the vascular tissue and is responsible for Pi accumulation in the shoots, is regulated by cytokinin [24]. *AtIPS1* and other Pi starvation responsive genes are also repressed by exogenous cytokinin [25]. A global analysis of gene expression events in rice showed that the expression changes caused by Pi starvation were reversed by exogenous cytokinin treatment. A large increase in intracellular phosphate levels may reduce the phosphate starvation signaling triggered by exogenous cytokinin [26]. CRE1, the receptor for cytokinin, has been found to participate in suppressing the regulation of many genes following Pi deficiency [22,27]. This implies that a close crosstalk exists between Pi and the cytokinin signaling transduction pathway [23].

Recently, it has been reported that OsARF12 was involved in Pi homeostasis and that OsARF16 facilitated the efficient utilization of Pi in rice [28,29]. However, the underlying mechanism linking auxin/cytokinin and Pi signaling is largely unknown. In this study, we analyzed the differences in Pi transport and Pi signaling between the *osarf16* mutant and NIP when subjected to cytokinin treatment. We found that OsARF16 was associated with cytokinin regulation of phosphate uptake and phosphate signaling in rice.

Materials and Methods

Plant materials and growth conditions

Rice plants (*Oryza sativa L.*) wild-type (WT) NIP and *osarf16* mutant (a knock-out mutant) [29] were grown in culture solution combined with phytohormone treatments in a glasshouse with a light: dark cycle of 12: 12 h at 30: 24°C (day: night) and 60–70% humidity, and the pH was adjusted to 5.8 with HCl. For hydroponics, seedlings were transferred to a plastic net floating on the nutrient solution containing 0.323 mM NaH_2PO_4 or without NaH_2PO_4 for Pi deficiency treatment (−Pi or +Pi). Phytohormone treatments were performed with 0.01 μM to 1 μM of 6-BA for 7days, respectively. The hydroponic experiments were performed in a standard rice culture solution containing 1.425 mM NH_4NO_3, 0.323 mM NaH_2PO_4, 0.513 mM K_2SO_4, 0.998 mM $CaCl_2$, 1.643 mM $MgSO_4$, 0.009 mM $MnCl_2$, 0.075 mM $(NH_4)_6Mo_7O_{24}$, 0.019 mM H_3BO_3, 0.155 mM $CuSO_4$, 0.036 mM $FeCl_3$, 0.070 mM citric acid and 0.152 mM $ZnSO_4$.

β-glucuronidase (GUS) staining and analysis of GUS activity

The construction of *OsARF16* promoter–GUS was performed according to a published method [29]. Exactly 2641 bp of the OsARF16 promoter sequence upstream of its ATG were cloned in front of the GUS gene and was introduced into the *Agrobacterium tumefaciens* strain EHA105. To examine the progression of cell division, the transgenic plant carrying *pCYCB1;1:Dbox-GUS* was crossed with *osarf16* mutant and homozygous plants in the *osarf16* background were assayed for GUS staining [30]. GUS staining of seedlings was performed using 100 mM sodium phosphate buffer (pH 7.0) containing 0.1% v/vTriton X-100 and 2 mM X-Gluc (Sangon, Shanghai, China), and samples were incubated at 37°C overnight. Stained tissues were observed using a Carl Zeiss laser scanning system LSM510 (http://www.zeiss.com/) and a Leica MZ95 stereomicroscope (Leica Instrument, Nusslosh, Germany). The measurement of GUS activity was performed as described by Jefferson [31].

QRT-PCR analysis

Total RNA was isolated from leaves or roots of 7-day-old seedlings. The methods for RNA extraction, reverse transcription and qRT-PCR were conducted as described in a previous report [29]. The sequences of the corresponding primers for qRT-PCR are listed in Table S1, S2, S3 and S4. *OsACTIN* was used as an internal standard to calculate the relative fold differences based on the comparative *Ct* method. $2^{-\Delta\Delta Ct}$ refers to the fold difference in expression of cytokinin-related genes under cytokinin treatment (3 hr) compared with the untreated seedlings. Heat map representation was performed using the normalized *Ct* value with Treeview 1.6 to visualize the analysis data. The different colors correspond to the values of the gene change-fold ratio shown in the bar. The data were analyzed by three independent repeats.

Measurement of Pi contents

NIP and *osarf16* mutants were analyzed to determine their total P contents. P measurements were performed using Flow Analyser SAN++ (Skalar Analytical B.V., Breda, Netherlands). For Pi transport experiment, the rice plants were grown for 7days under Pi deficiency solution, and then were transferred to Pi supply solution (1 mM Pi) with or without cytokinin (6-BA). Long time Pi starvation depressed total Pi level in rice seedlings and resulted in activity of Pi transport during the first 24 hour periods after Pi resupply. For each sample in the experiment, 0.1 g tissue was dried at 80°C for 48 hr and digested with HNO_3/H_2O_2 at 110°C for 0.5 hr using a microware 3000 digestor (Anton Paar, Graz, Austria). Five biological replicates were performed for each sample in all experiments.

Statistics of root system parameters

The root system of rice plant was put into a container filled with distilled water. In order to minimize the intercross among the roots during image scanning, the whole root system was carefully separated into 5 portions, and each portion was transferred to individual container for scanning. The number of lateral roots was obtained by scanning. The average lengths of root hairs (3 mm from the root tips) were measured by a LEICA MZ95 stereomicroscope with camera scale (Leica Instrument, Nusslosh, Germany).

Qualitative analysis of root-associated APase activity

Root APase staining was performed according to Bozzo's report [32]. The roots were excised from 7 days Pi-deprived and Pi-supplied seedlings and incubated with a 5-bromo-4-chloro-3-indolyl-phosphate (BCIP) content agar solution containing 50 mM sodium acetate (pH 5.5) with 10 mM $MgCl_2$, 0.6% agar and 0.08% BCIP at room temperature for 20 min. The blue color on the root surface, formed by hydrolysis of BCIP, was photographed using an EOS 40D camera (Canon Corporation, Tokyo, Japan). Protein (1 μg) was used for APase activity assay, and the protein was added to 620 μl of reaction buffer (50 mM NaAc pH 5.5 and 10 mM $MgCl_2$), and 10 μl of p-nitrophenol phosphate (10 mg ml^{-1} pNPP; Sigma). After incubation at 37°C for 10 min, the reaction was stopped by 1.2 ml of 1M NaOH, and then absorbance was measured at 412 nm wavelength. Phosphatase activity was counted as ng of pNPP accumulated μg^{-1} soluble protein min^{-1}. These experiments were repeated three times.

Results

Physiological and morphological evidence that OsARF16 is involved in cytokinin responses in rice seedlings

In previous studies, we have reported that the OsARF16 is required for auxin signaling and the phosphate starvation response in rice. The phenotype of *osarf16* mutant under Pi deficiency condition was well studied in our last publication [29]. In the present study, the mutant *osarf16* was used to investigate how auxin response factor is involved in cytokinin signaling. To explore the possible effects of OsARF16 on morphological and physiological responses to cytokinin (6-BA) treatment, wild-type (NIP) and mutant *osarf16* seeds were grown in nutrient solutions containing different concentrations of 6-BA. The concentrations of 6-BA used in this experiment were 0, 0.01, 0.1, 1.0 and 10 μM. After 7 days of treatment, the root growth was considerably inhibited by 6-BA in the NIP seedlings (Fig. 1A–E). The root lengths of NIP declined significantly from 9.59 cm to 0.72 cm when the 6-BA concentration increased from 0 to 10 μM. Under the same conditions, the root growth inhibition was less significant in the *osarf16* mutant. As the 6-BA concentration increased, the *osarf16* mutant root lengths fell from 10.44 cm to 2.21 cm (Fig. 1F). In addition to root growth inhibition, shoot elongation was also affected by 6-BA application in NIP. The shoot lengths of NIP declined significantly from 13.32 cm to 4.87 cm when the 6-BA concentration increased from 0 to 10 μM. Under the same conditions, the root growth inhibition was less significant in the *osarf16* mutant. Specially, the shoot length in NIP was only 68% of that in the *osarf16* mutant under the 0.1 μM 6-BA treatment (Fig. 1C, G).

Cytokinin regulates root system architecture (RSA) by disturbing cell division activity and increasing the length of the root hairs at the root tips [33,34]. In the present study, we measured the number of lateral roots on the main roots and the average lengths of the root hairs on the root tips of both NIP and the *osarf16* mutant under the 0.1 μM 6-BA treatment. The results showed that the number of lateral roots on *osarf16* was much greater than on NIP under 0.1 μM 6-BA treatment (Fig. 2A). Without 6-BA application, the number of lateral roots on NIP was similar to the *osarf16* mutant. The 0.1 μM 6-BA treatment reduced the number of lateral roots by 80% in the NIP seedlings. However, the reduction was only 45% in the *osarf16* mutant seedlings (Fig. 2B). The number of lateral roots in *osarf16* primary root was almost twice as much as NIP under 0.1 μM 6-BA treatment. When 6-BA was not applied, there were no root hairs on the root tips of both NIP and the *osarf16* mutant. On average, the lengths of the root hairs (3 mm from the root tips) could be increased to almost 500 μm on the tips of NIP, but only to 300 μm on the tips of the *osarf16* mutant under the 0.1 μM 6-BA treatment (Fig. 2C). The results revealed that the *osarf16* mutant was more insensitive to cytokinin treatment than NIP.

To confirm that the phenotypes described above were caused by the loss-of-function of *OsARF16*, we transformed the *OsARF16* gene back into the *osarf16* mutant and three independent complementation lines (*osarf16/C*) were chosen for analysis. The growth parameters for NIP and *osarf16/C* were approximately the same under the different 6-BA treatments (0, 0.01, 0.1, 1.0 and 10 μm) (Figure S1). Again, we used various cytokinins (kinetin, zeatin) to confirm the involvement of OsARF16 in the cytokinin

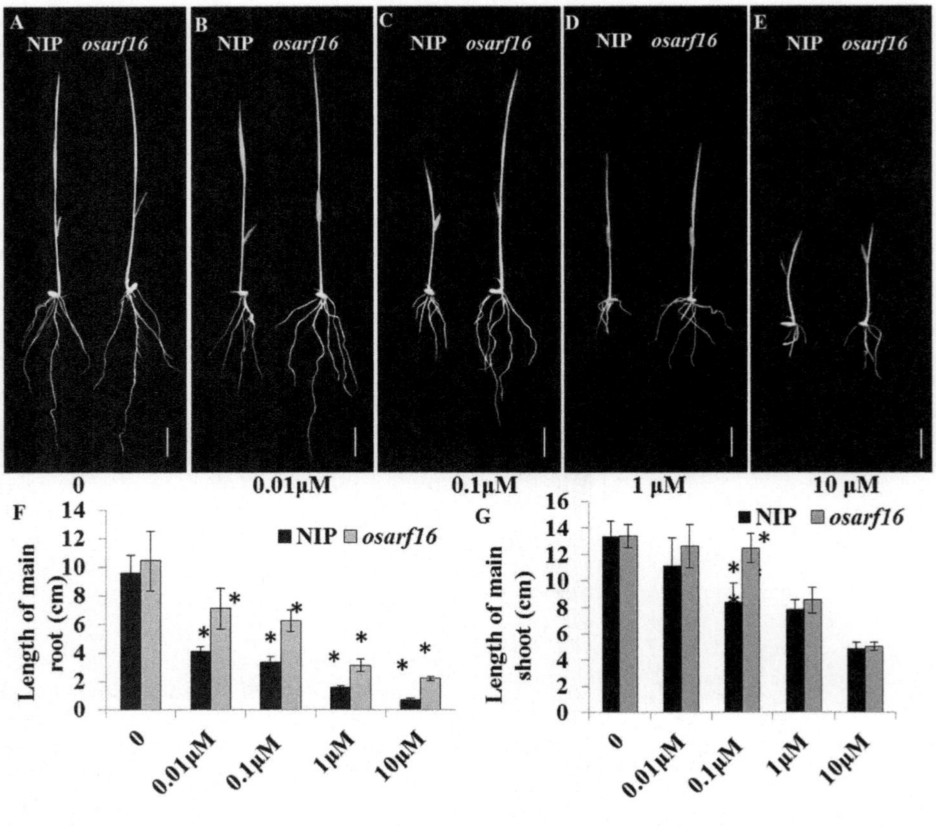

Figure 1. Physiological evidences for involvement of OsARF16 in cytokinin (6-BA) Responses. (A–E) Phenotype of NIP and *osarf16* mutant under different concentration of 6-BA treatments (0/0.01/0.1/1/10 μM) using 7-day-old seedlings (Bar represents 2 cm). The graphs represent statistics data of main root length (F) and main shoot length (G). Data are shown as the mean ± SD (n = 5). Significant ($P<0.05$) differences in length of main root and main shoot between *osarf16* and NIP are indicated by an asterisk.

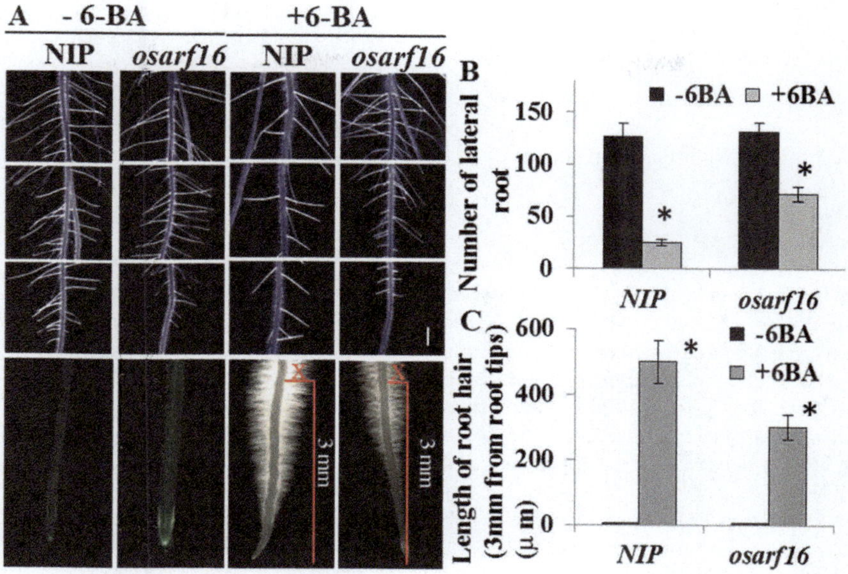

Figure 2. Effects of cytokinin on regulation of root system architecture (RSA). (A) 0.1 μM 6-BA treatment reduced the number of lateral root in primary root and induced root hair of NIP and *osarf16* mutant in primary root tips (The bar represents 5 mm). The graphs represent statistics data of lateral root number (B) and average root hairs length (C). Data are shown as the mean ± SD (n = 5). Significant ($P<0.05$) differences in number of lateral root in primary root and average length of root hairs in primary root tips between *osarf16* and NIP are indicated by an asterisk.

responses. Kinetin and zeatin inhibited root system elongation and shoot growth and *osarf16* was, again, less sensitive to both cytokinins than *NIP* (Figure S2). The results confirmed that OsARF16 was involved in plant responses to cytokinin.

OsARF16 affects cytokinin signaling

To further determine the cytokinin signaling changes in the *osarf16* mutant, we analyzed cell division activity in the primary root tips, the lateral roots in primary root and the lateral root primordia under 6-BA treatment using the *pCYCB1;1:Dbox-GUS* reporter [30]. Five positive transgenic lines were used for this analysis. Under the control treatment, the cell division activities, as indicated by the expressions of the *pCYCB1;1:Dbox-GUS* reporter in NIP and the *osarf16* mutant, were almost the same. After 6-BA treatment (3 hr), the cell division activities in the primary root tips, the lateral roots in primary root and the lateral root primordia considerably increased in NIP, whereas the induction of cell division activity in *osarf16* was much lower than in NIP (Fig. 3A, B). The GUS activity data is shown in Fig. 3C. The GUS activity under −6-BA treatment in NIP was almost the same as that in *osarf16* mutant. However, the GUS activity in NIP was 69.0% higher than that in *osarf16* mutant. The data was essentially in agreement with the GUS staining results showed in Fig. 3 A and B.

Many studies have investigated cytokinin signaling in rice [35,36]. We analyzed the expression of 37 cytokinin response genes in both NIP and the *osarf16* mutant in order to improve our understanding of the mechanism behind OsARF16 involvement in the cytokinin signaling system. These genes are all part of two-component regulatory systems. Five belong to HKs (cytokinin-response histidine protein kinase) (OsHK1–4, OsHKL1), five to HPs (histidine phosphotransfer proteins) (OsHP1–5), fifteen to type-A RRs (response regulators) (OsRR1–15), seven to type-B RR genes (OsRR16–22), and five are predicted pseudo-response regulators (OsPRR1–5) [37]. A qRT-PCR analysis was performed on the roots of 7-day-old NIP and *osarf16* seedlings grown in culture solutions with or without 6-BA (the primer sequences information is provided in Table S1). The results showed that all 37 cytokinin responsive genes were sharply induced by 6-BA treatment in the roots of the NIP seedlings. However, in the *osarf16* mutant seedlings, the induction of these cytokinin responsive genes was reduced under the 6-BA treatment compared to the NIP seedlings (Fig. 3D). This suggested that cytokinin signaling in *osarf16* was compromised compared to NIP.

The expression pattern of *OsARF16* was dramatically changed by cytokinin treatment

We used the GUS reporter gene to evaluate the *OsARF16* expression patterns in the roots and shoots under −6-BA treatment and +6-BA treatment conditions. Five positive transgenic lines were used for this analysis. Under the control treatment (−6-BA treatment), *OsARF16* was expressed in root tips and stele of both primary roots (Fig. 4 A, C) and lateral roots in primary root (Fig. 4 E). Under +6-BA treatment (3 hr), the *OsARF16* expression level increased several times compared to the control treatment, especially in the root hairs (Fig. 4A–F). The *OsARF16* expression level was almost undetectable in root hairs under the control treatment, but was strongly induced by the +6-BA treatment. Meanwhile, the expression level of *OsARF16* was also induced by 6-BA treatment in shoots (Fig. 4G, H). We used qRT-PCR to confirm the *OsARF16* expression pattern changes in roots under the control and +6-BA treatment conditions and the results were consistent with the GUS staining analysis results (Fig. 4I). The expression level was 7.5 folds induced by 6-BA treatment in the whole root organ.

Analysis of total phosphorus (P) content over a time-course in NIP and the *osarf16* mutant

The total P contents over a time-course were measured in NIP and the *osarf16* mutant seedlings in order to clarify the function of OsARF16 in the cytokinin-mediated inhibition of phosphate transport between roots and shoots. The Pi-deprived rice seedlings, which were grown in a Pi free nutrient solution for 7

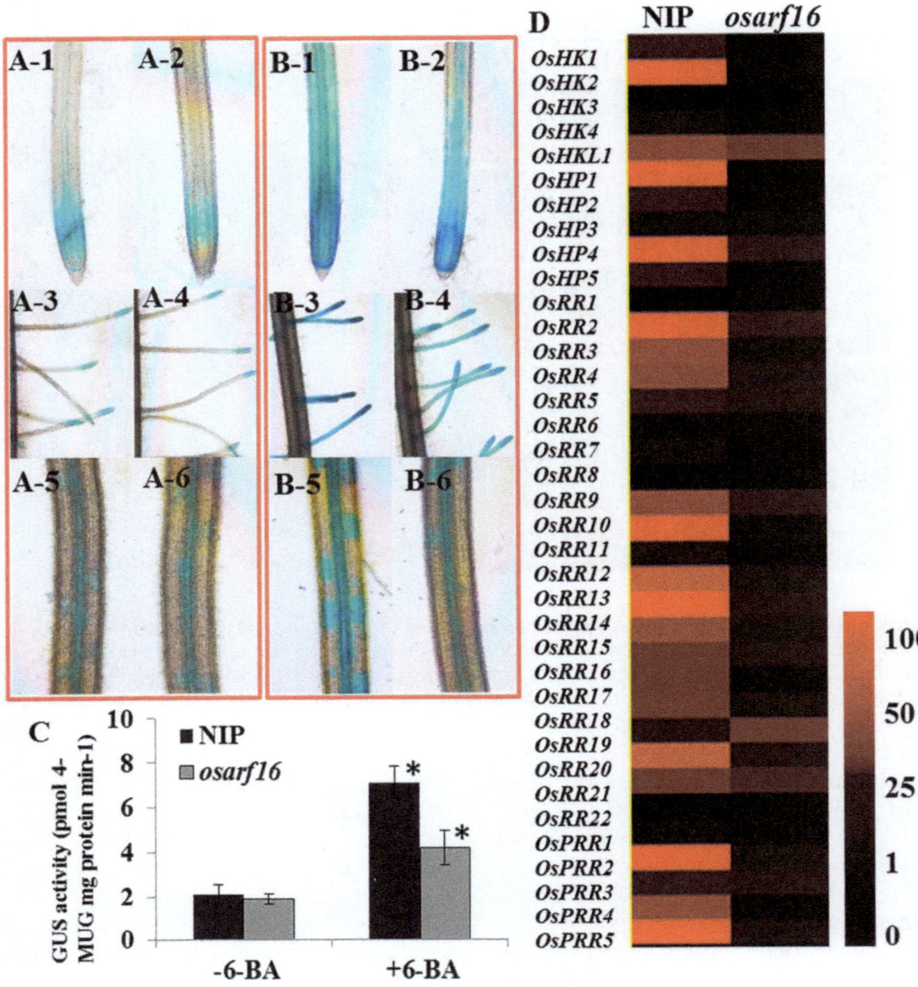

Figure 3. OsARF16 affects cytokinin signaling. (A, B) *pCYCB1;1:Dbox-GUS* staining was performed in primary root tips, lateral roots and lateral root primordia of NIP and *osarf16* mutant under −6-BA/+6-BA treatment. Concentration of 6-BA in the treatments was 0.1 μM. *pCYCB1;1:Dbox-GUS* staining in root tips of NIP (A-1) and *osarf16* mutant (A-2) under −6-BA treatment; *pCYCB1;1:Dbox-GUS* staining in lateral roots of NIP (A-3) and *osarf16* mutant (A-4) under −6-BA treatment; *pCYCB1;1:Dbox-GUS* staining in lateral root primordia of NIP (A-5) and *osarf16* mutant (A-6) under −6-BA treatment; *pCYCB1;1:Dbox-GUS* staining in primary root tips of NIP (B-1) and *osarf16* mutant (B-2) under +6-BA treatment; *pCYCB1;1:Dbox-GUS* staining in lateral roots of NIP (B-3) and *osarf16* mutant(B-4) under +6-BA treatment; *pCYCB1;1:Dbox-GUS* staining in lateral root primordia of NIP (B-5) and *osarf16* mutant (B-6) under +6-BA treatment. (C) Analysis of GUS activity. The entire root of the seedling was used for analysis. The graph represents statistics data of GUS activity in NIP and *osarf16* mutant roots. Data are shown as the mean ± SD (n = 5). Significant ($P<0.05$) differences in GUS activity between *osarf16* and NIP are indicated by an asterisk. (D) Analysis of the expression levels of cytokinin signaling related genes in *osarf16* mutant and NIP. The data of $2^{-\Delta\Delta Ct}$ (qRT data) refers to the fold difference in expression of cytokinin-related genes under cytokinin treatment (3 hr) compared with the untreated seedlings. Heat map representation was performed using the normalized $2^{-\Delta\Delta Ct}$ values with Treeview 1.6 to visualize the analysis data. The different colors correspond to the values of the gene change-fold ratio shown in the bar. The data were analyzed by three independent repeats.

days, were transferred to a nutrient solution containing 1 mM Pi in order to monitor P contents in the shoots over a 24 h period and in roots over a 12 h period after Pi application. After 7 days of Pi starvation, the total P contents were only 4.7 mg g^{-1} in the NIP roots and 6.2 mg g^{-1} in the *osarf16* mutant roots. When resupplied with 1 mM Pi, the P contents in the roots rapidly increased in both NIP (16.55 mg g^{-1}) and *osarf16* (18.35 mg g^{-1}) after 12 h of Pi resupply. Under the same conditions, the P contents in the shoots were 4.2 mg g^{-1} in NIP and 5.6 mg g^{-1} in the *osarf16* mutant. When resupplied with 1 mM Pi, the P content only increased to 12.96 mg g^{-1} in NIP roots with 6-BA application (78.3% of the 6-BA free treatment) after 24 h of Pi resupply and increased to 17.23 mg g^{-1} in the *osarf16* shoots (93.4% of the 6-BA free treatment). On the other hand, the P contents in the shoots also rose in both NIP (15.56 mg g^{-1}) and *osarf16* (17.65 mg g^{-1}) after 24 h of Pi resupply. With 6-BA application, the P content only increased to 8.2 mg g^{-1} in NIP shoots (52.9% of the 6-BA free treatment) after 24 h of Pi resupply and increased to 14.5 mg g^{-1} in the *osarf16* shoots (82.9% of the 6-BA free treatment). Cytokinin (6-BA) clearly inhibited Pi transport in NIP, but the inhibition of Pi transport by cytokinin was less obvious in the *osarf16* mutant seedlings compared to NIP (Fig. 5A, B).

Furthermore, we analyze shoot/root ratios of P contents in both NIP and *osarf16* mutant under different conditions. After 7 days Pi starvation, the shoot/root ratios of P contents in both NIP and

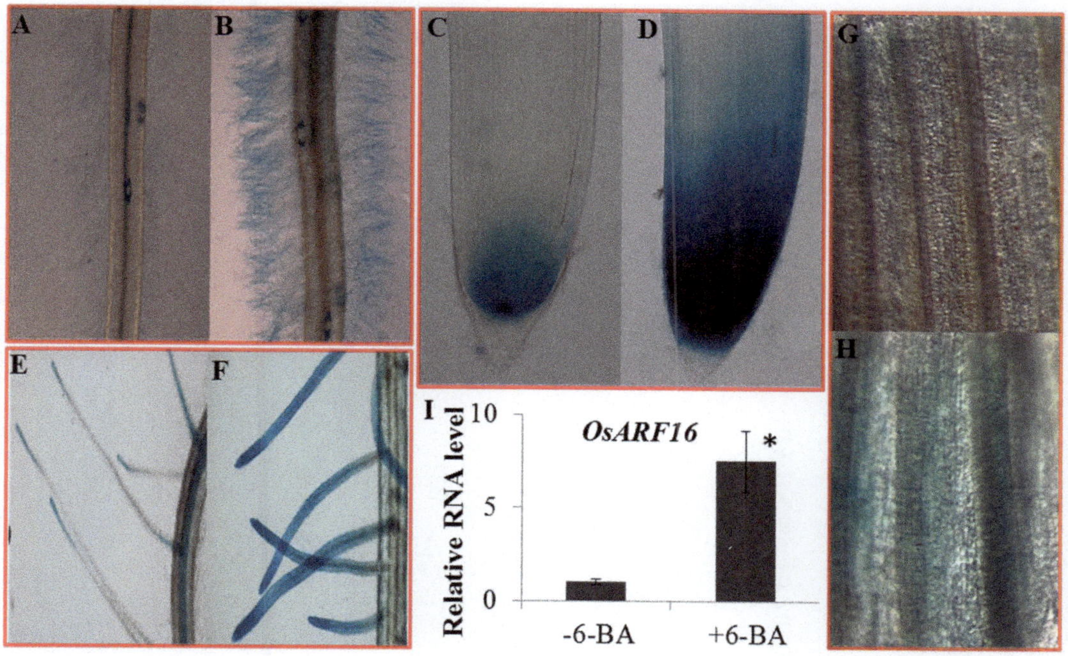

Figure 4. Analysis of *OsARF16* expression pattern under cytokinin treatment using *OsARF16*-Promoter: GUS lines. The expression pattern of *OsARF16* in primary root tips under (A) control and (B) 0.1 μM 6-BA treatment for 3 hr; The expression pattern of *OsARF16* in root hairs of primary root under (C) control and (D) 0.1 μM 6-BA treatment for 3 hr; The expression pattern of *OsARF16* in lateral roots of primary root under (E) control and (F) 0.1 μM 6-BA treatment for 3 hr. The expression pattern of *OsARF16* in leaves under (G) control and (H) 0.1 μM 6-BA treatment for 3 hr. (I) qRT-PCR method was used to confirm the *OsARF16* expression pattern changes in roots under 0.1 μM 6-BA treatment for 3 hr. Significant ($P<0.05$) differences in the expression level of *OsARF16* between 6-BA treatment and mock treatment is indicated by an asterisk.

osarf16 mutant were nearly 50%. When resupplied with 1 mM Pi, the shoot/root ratios of P contents were increased quickly after 3 hours Pi resupply and decreased to 50% gradually thereafter in both NIP and *osarf16* mutant. With 6-BA application, the shoot/root ratios of P contents increased to 62% after 3 hours Pi resupply and did not decreased obviously during 12 hour Pi resupply period in NIP. With 6-BA application, the shoot/root ratios of P contents increased to 65% after 3 hours Pi resupply and decreased obviously during 12 hour Pi resupply period in *osarf16* mutant (Figure S3). The 6-BA treatment affects the shoot/root ratios of P contents in NIP seedlings, but it was less obvious in the *osarf16* mutant.

Involvement of OsARF16 in cytokinin-mediated inhibition of Pi transport

The *PHOSPHATE TRANSPORTER1* (*PHT1*) gene family plays a number of critical roles in Pi uptake, translocation and homeostasis in *Arabidopsis*. Some of these genes have also been identified in rice [38,39]. In this study, we used qRT-PCR to analyze the expression levels of several members of the rice *PHT1* gene family (including *OsPT1/2/3/6/8/9* and *10*). The primer sequences information is provided in Table S2. Most of these phosphate transporter coding genes were up-regulated by Pi deficiency and their expression levels were inhibited in NIP roots and shoots by applications of 0.1 μM 6-BA (Fig. 5C, D). The expression levels of these *OsPTs* were higher in the *osarf16* mutant than in NIP under −Pi/+6-BA treatments.

Expression analysis of *phosphate (Pi) starvation-induced* (*PSI*) genes in NIP and the *osarf16* mutant

To test whether the *phosphate (Pi) starvation-induced* (*PSI*) genes were also affected by cytokinin treatment, we chose four genes: *OsIPS1*/*OsIPS2*, *OsSPX1* [40] and *OsSQD2* [41]. Then we analyzed the transcript abundance in NIP and the *osarf16* mutant and under +Pi/−6-BA, −Pi/−6-BA, +Pi (resupply)/−6-BA, +Pi/+6-BA, −Pi/+6-BA and +Pi (resupply)/+6-BA treatments (The primer sequences information are provided in Table S3). The expression level analysis revealed that these four genes were sharply up-regulated under the −Pi/−6-BA treatment and quickly fell down to background levels during the 24 h period after Pi resupply (Fig. 6). All four *PSI* genes were repressed by 6-BA application under the −Pi treatment in NIP. For example, *OsIPS1* was 230 times higher under the −Pi/−6-BA treatment compared to the +Pi/−6-BA treatment, but was only 25 times higher under −Pi/+6-BA treatment. In the *osarf16* mutant, the expression levels of these *PSI* genes were much higher than that in NIP under the −Pi/+6-BA treatment.

Analysis of acid phosphatase (APase) activity and expression of six *PAPase* genes

The induction of APase activity under Pi deficiency conditions helps to catalyze inorganic phosphate hydrolysis from organophosphates and is an important strategy that enables plants to deal with Pi deficiency [42]. The expressions of *PAPase* genes and APase activity were measured in NIP and the *osarf16* mutant in order to understand how OsARF16 is involved in Pi homeostasis through its effects on APase activity. Rice seedlings were grown under +Pi/−6BA, +Pi/+6BA, −Pi/−6BA and −Pi/+6BA conditions. In contrast, 6-BA application significantly suppressed the APase activity in NIP under Pi deficiency conditions (−Pi/+6-BA treatment). However, the APase activity was 50% higher in the *osarf16* mutant than in NIP the under −Pi/+6-BA treatment (Fig. 7A). The data were gained from the analysis of root intracellular APase activity on substrate pNPP (Fig. 7B). Many

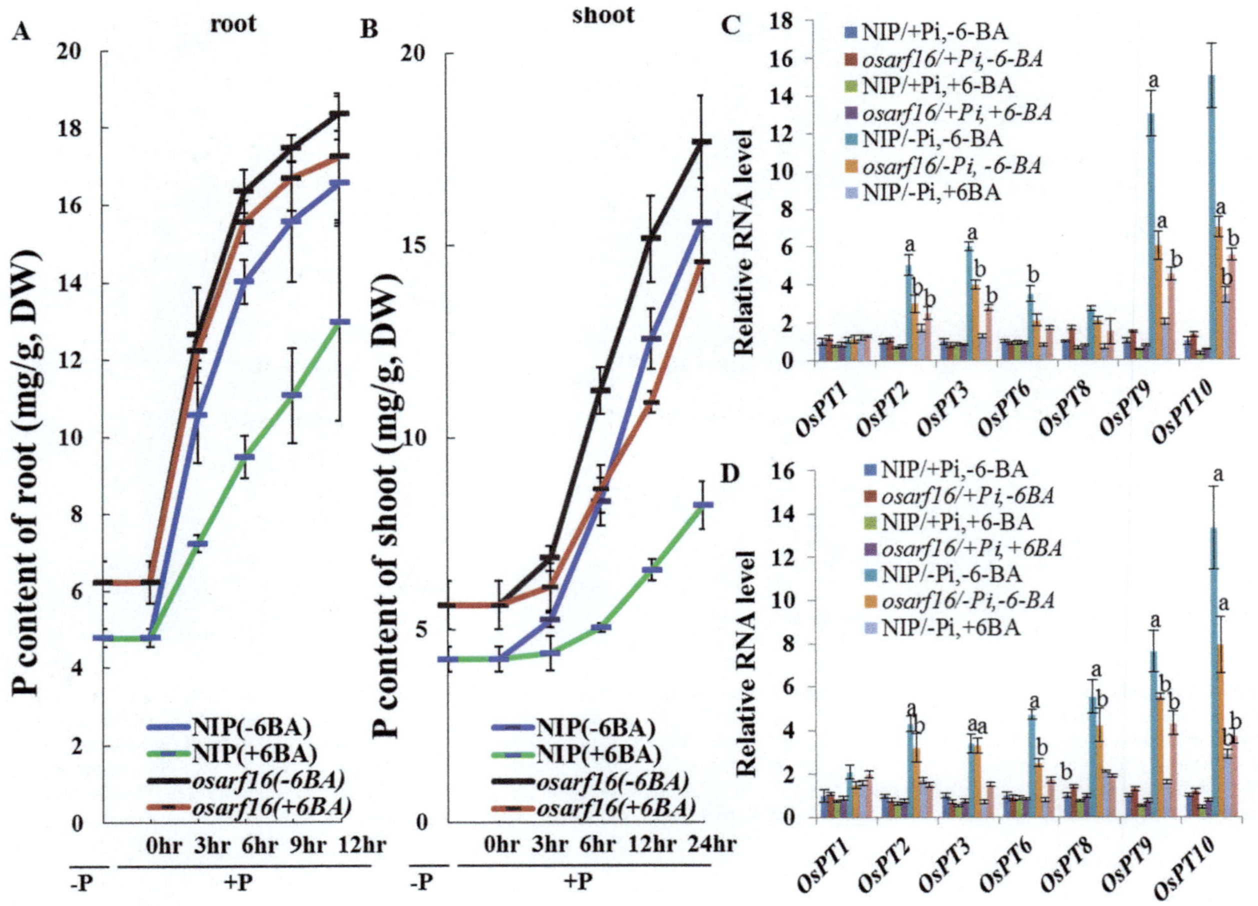

Figure 5. Analysis of total phosphorus (P) contents over a time-course in both NIP and *osarf16* mutant. (A,B) Total P contents in NIP and *osarf16* mutant after Pi resupply using 7-day-old Pi-deprived seedlings: (A) in roots and (B) in shoots. The statistics of P contents were calculated for five independent biological replications. (C, D) Expression of Pi transporter coding genes in Pi-sufficient (+Pi) and Pi-deficient (−Pi) treatments with or without 6-BA (C) in roots and (D) in shoots. Concentration of 6-BA in the treatments was 0.1 μM. The *ACTIN* gene of rice was used as the reference gene for qRT-PCR. Value ± SD of five independent replicates. "a" indicated significant difference in expression levels of *OsARF16* from treatments to mock at 1% by student's *t* test. "b" indicated significant difference in expression levels of *OsARF16* from treatments to mock at 5% by student's *t* test.

purple APase (PAPase) genes, which are involved in Pi uptake and Pi starvation responses, have been reported by Wang *et al.* (2009). We chose six of the PAPase genes for this experiment (*OsPAP9b*, *OsPAP10a*, *OsPAP10c*, *OsPAP20b*, *OsPAP23* and *OsPAP27a*) because they were induced by −Pi deficiency by more than two-fold (The primer sequences information is provided in Table S4). Notably, all six genes were down-regulated by 6-BA treatment under the −Pi deficiency treatment in NIP. The expression levels of these genes were higher in the *osarf16* mutant than in NIP under the −Pi/+6-BA treatment (Fig. 7C). This suggested that cytokinin-mediated inhibition of *APase* expression is dependent on OsARF16.

Discussion

OsARF16 is a key regulator of the cytokinin response in rice

ARFs are involved in many important biological functions, including plant development and nutritional homeostasis [8,43]. OsARF16 as a transcriptional activator that is highly homologous to AtARF7/19 in *Arabidopsis* and is implicated in Pi deficiency response related root system architecture changes [29,44]. In *Arabidopsis* meristems, cytokinin controls the differentiation rate of transit-amplifying cells by antagonizing the plant hormone auxin, which is needed to sustain cell proliferation [15]. In our previous studies, we identified the *osarf16* mutant and characterized its biological effects on auxin signaling and auxin distribution [29]. However, how ARFs interact with cytokinin signaling in rice is largely unknown.

Cytokinin suppresses the growth of rice seedlings and *OsARF16* loss-of-function attenuates sensitivity to exogenous cytokinin treatment (Fig. 1). High concentration 6-BA treatments cause severe growth defects in wild-type seedlings, including root system development and shoot elongation. In particular, there was an induction in lateral root numbers and main root growth in the *osarf16* mutant compared to NIP under 6-BA treatment. This suggested that OsARF16 was required for cytokinin responses in rice.

In rice plants, *PINs*, and *LAXs* are the two major gene families responsible for auxin transport *in vivo* [29]. Exogenous cytokinin treatment also inhibited the expression of auxin transport genes and disrupted auxin distribution, which is required for many of the physiological responses to cytokinin [45]. Our data showed that 6-BA treatment clearly down-regulated the expression levels of

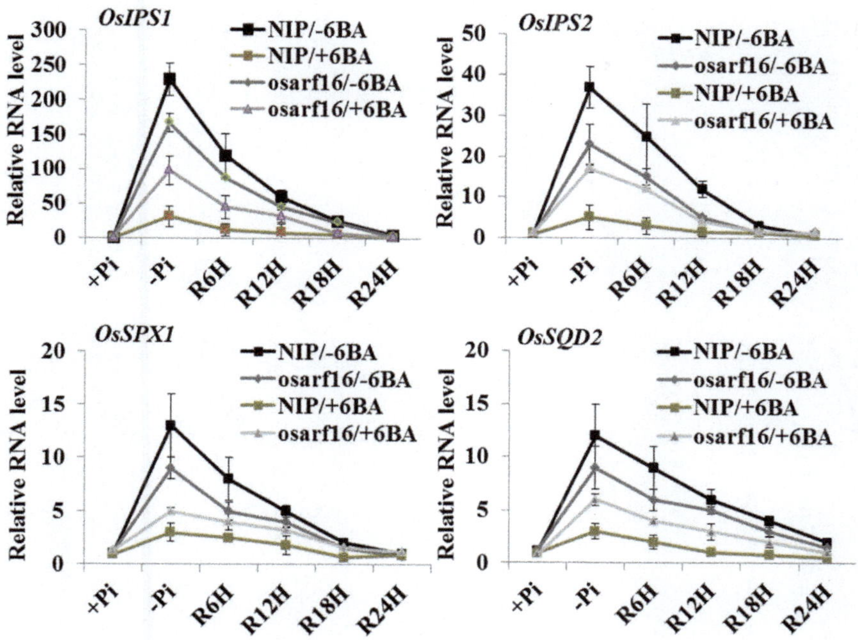

Figure 6. Expression analysis of *phosphate (Pi) starvation-induced* (*PSI*) genes, *OsIPS1, OsIPS2, OsSPX1, OsSQD2* in NIP and *osarf16* mutant. Analysis of expressions were performed under phosphate-sufficient (+Pi), phosphate-deficient (−Pi) and Pi resupply conditions (R6h, R12h, R18h and R24h refer to resupply times of 6, 12, 18 and 24 h, respectively) with or without 6-BA treatments. Concentration of 6-BA in the treatments was 0.1 μM. 7-day-old seedlings of NIP and *osarf16* mutant were used for quantitative reverse transcription polymerase chain reaction (qRT-PCR) analysis. Value ± SD of five independent replicates.

OsPINs and *OsLAXs* in NIP. However, the expression levels of the *OsPIN* and *OsLAX* family genes were much higher in the *osarf16* mutant than in NIP under the 6-BA treatment (Figure S4). OsARF16 may be involved in cytokinin responses by regulating the expression of auxin transporter coding genes.

OsARF16 is involved in cytokinin-mediated negative regulation of phosphate uptake and phosphate signaling in rice (*Oryza sativa L.*)

Cytokinin inhibits phosphate transport by regulating the expression of the phosphate transporter coding genes [22,23]. In this study we used 7-day-old Pi-deprived seedlings to investigate how OsARF16 is involved in the negative regulation of Pi uptake and transport by cytokinin. When transferred to the Pi supply nutrient solution, the phosphate contents in NIP seedlings rapidly rose (Fig. 5A) and cytokinin application could significantly block this process. However, in the *osarf16* mutant, the effect of cytokinin on preventing Pi transport from the roots to the shoots was weaker than in NIP (Fig. 5B). The expression levels of *OsPTs* were much higher in the *osarf16* mutant than that in NIP under the 6-BA treatment, which was consistent with the results of the Pi content measurements. These results indicated that OsARF16 was involved in cytokinin-mediated inhibition of Pi transport from the roots to the shoots by regulating *PT* expressions in rice.

OsARF16 is an important transcription factor involved in the crosstalk between cytokinin and Pi starvation signaling

It has been already reported that Pi deficiency signaling could be inhibited by cytokinin and that there was a crosstalk between cytokinin and Pi deficiency signaling [25]. Cytokinin may repress the Pi starvation response by increasing intracellular phosphate content. Under Pi deficiency condition, more organic Pi is released into cellular Pi pool (inorganic Pi) driven by exogenous 6-BA [26]. In our experiment, total Pi (organic Pi and inorganic Pi) have been measured in NIP and *osarf16* mutant under different conditions. After 7 days Pi starvation, the Pi level in rice seedlings was greatly depressed and the increased of total Pi content after Pi resupply was inhibited by 6-BA treatment (Fig. 5). 6-BA treatment elevates inorganic Pi content by reducing organic Pi concentration (major component of total Pi). However, the full physiological and molecular relevance of how cytokinin is involved in Pi signaling remains unclear. In this study, we discovered that OsARF16, an auxin response factor, was an important regulator involved in the crosstalk between cytokinin and Pi deficiency signaling. Pi signaling was repressed by cytokinin treatment and it depended on a functioning OsARF16 gene. Under Pi deficiency conditions, the expressions of *OsIPS1/2*, *OsSPX1* and *OsSQD2* were inhibited by 6-BA treatment in NIP. However, inhibition of *PSI* gene expressions by 6-BA treatment in the *osarf16* mutant was compromised compared to the NIP seedlings.

Acid phosphatase activity was also reduced by cytokinin treatment. An increase in APase activity in the roots is a typical Pi starvation response when plants are subjected to Pi deficiency. Induction and secretion of acid phosphatases (APases) is a highly efficient mechanism that helps plants to survive and grow under Pi deficient conditions [28,46,47]. According to the results (Fig. 7A, B), the APase activity in NIP roots was reduced by 6-BA treatment under Pi deficiency conditions. The APase activity was much stronger in the *osarf16* mutant roots compared to the NIP roots under the same conditions. The expression level of some purple APase coding genes, such as *OsPAP20b*, *23* and *27a* in the *osarf16* mutant under the −Pi/+6-BA treatment, were almost twice than in NIP. These results strongly demonstrated that

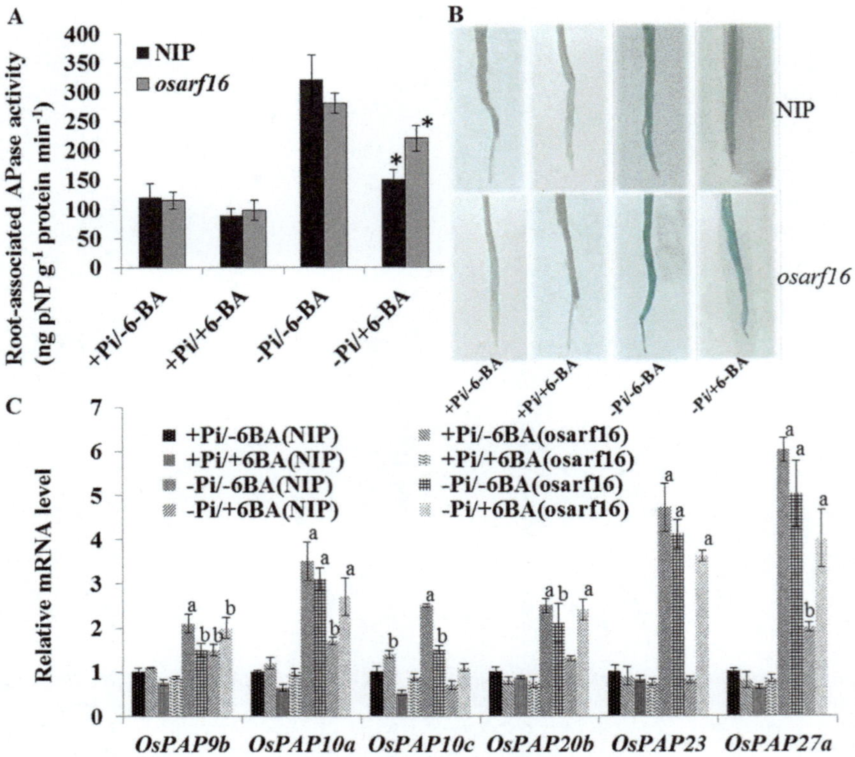

Figure 7. Analysis of acid phosphatase (APase) activity and expression of six purple APase coding genes. (A) Acid phosphatase (APase) activity in NIP and *osarf16* mutant under +Pi/−6BA, +Pi/+6BA, −Pi/−6BA and −Pi/+6BA conditions respectively. Concentration of 6-BA in the treatments was 0.1 μM. The graph represents statistics data of root-associated APase activity. (B) Acid phosphatase (APase) activity in the root surface of NIP and *osarf16* mutant under +Pi/−6BA, +Pi/+6BA, −Pi/−6BA and −Pi/+6BA conditions respectively. (C) Expression levels of six purple APase coding genes under +Pi/−6BA, +Pi/+6BA, −Pi/−6BA and −Pi/+6BA conditions respectively. Data are shown as the mean ± SD (n = 5). Significant ($P < 0.05$) differences in APase activity between *osarf16* mutant and NIP are indicated by an asterisk. "a" indicated significant difference in expression levels of *OsPAPs* from treatments to mock at 1% by student's *t* test. "b" indicated significant difference in expression levels of *OsPAPs* from treatments to mock at 5% by student's *t* test.

OsARF16 is involved in the crosstalk between cytokinin and Pi starvation signaling.

Conclusion

In summary, we found that OsARF16 was involved in cytokinin-mediated inhibition of phosphate transport and phosphate deficiency signaling in rice. Our data show that OsARF16 plays a critical role in auxin and Pi starvation responses and is involved in the crosstalk between cytokinin and Pi deficiency signaling. Our next study will focus on the direct downstream targets of OsARF16 in order to reveal how cytokinin regulates Pi transport and Pi signaling in rice.

Supporting Information

Figure S1 Physiological results for *osarf16/C* under 6-BA treatment.

Figure S2 Physiological evidence for OsARF16 was involved in various cytokinins (kinetin, zeatin) Responses.

Figure S3 Analysis of shoot/root ratios of P contents in both NIP and *osarf16* mutant.

Figure S4 The expression levels of *OsPIN* and *OsLAX* family genes.

Table S1 The RT primer sequences 37 cytokinin response genes.

Table S2 Primer sequences for phosphorus transporters.

Table S3 Primer sequences for the Phosphate Starvation Induced genes (*PSIs*).

Table S4 Primer sequences for purple acid phosphatase genes.

Acknowledgments

We are grateful to Dr. Yanhua Qi (College of Life Sciences, Zhejiang University) for the kind gifts of *osarf16* mutant and *OsARF16* promoter: GUS transgenic lines, to Dr. Dean Jiang for technical support.

Author Contributions

Conceived and designed the experiments: CS HW ST YY TS. Performed the experiments: RY YY ST. Analyzed the data: YY ST LZ. Contributed reagents/materials/analysis tools: CS YY. Wrote the paper: CS HW ST.

References

1. Mockaitis K, Estelle M (2008) Auxin receptors and plant development: a new signaling paradigm. Annu Rev Cell Dev Biol 24: 55–80.
2. Guilfoyle TJ, Hagen G (2007) Auxin response factors. Curr Opin Plant Biol 10: 453–460.
3. Schruff MC, Spielman M, Tiwari S, Adams S, Fenby N, et al. (2006) The AUXIN RESPONSE FACTOR 2 gene of Arabidopsis links auxin signalling, cell division, and the size of seeds and other organs. Development 133: 251–261.
4. Sessions A, Nemhauser JL, McColl A, Roe JL, Feldmann KA, et al. (1997) ETTIN patterns the Arabidopsis floral meristem and reproductive organs. Development 124: 4481–4491.
5. Hardtke CS, Berleth T (1998) The Arabidopsis gene MONOPTEROS encodes a transcription factor mediating embryo axis formation and vascular development. EMBO J 17: 1405–1411.
6. Li J, Dai X, Zhao Y (2006) A role for auxin response factor 19 in auxin and ethylene signaling in Arabidopsis. Plant Physiol 140: 899–908.
7. Goetz M, Vivian-Smith A, Johnson SD, Koltunow AM (2006) AUXIN RESPONSE FACTOR8 is a negative regulator of fruit initiation in Arabidopsis. Plant Cell 18: 1873–1886.
8. De Smet I, Lau S, Voss U, Vanneste S, Benjamins R, et al. (2010) Bimodular auxin response controls organogenesis in Arabidopsis. Proc Natl Acad Sci U S A 107: 2705–2710.
9. Wu B, Li YH, Wu JY, Chen QZ, Huang X, et al. (2011) Over-expression of mango (Mangifera indica L.) MiARF2 inhibits root and hypocotyl growth of Arabidopsis. Mol Biol Rep 38: 3189–3194.
10. Agusti J, Herold S, Schwarz M, Sanchez P, Ljung K, et al. (2011) Strigolactone signaling is required for auxin-dependent stimulation of secondary growth in plants. Proc Natl Acad Sci U S A 108: 20242–20247.
11. Sakamoto T, Morinaka Y, Inukai Y, Kitano H, Fujioka S (2013) Auxin signal transcription factor regulates expression of the brassinosteroid receptor gene in rice. Plant J 73: 676–688.
12. Vandenbussche F, Callebert P, Zadnikova P, Benkova E, Van Der Straeten D (2013) Brassinosteroid control of shoot gravitropism interacts with ethylene and depends on auxin signaling components. Am J Bot 100: 215–225.
13. Yang DL, Yang Y, He Z (2013) Roles of plant hormones and their interplay in rice immunity. Mol Plant 6: 675–685.
14. Bielach A, Duclercq J, Marhavy P, Benkova E (2012) Genetic approach towards the identification of auxin-cytokinin crosstalk components involved in root development. Philos Trans R Soc Lond B Biol Sci 367: 1469–1478.
15. Dello Ioio R, Nakamura K, Moubayidin L, Perilli S, Taniguchi M, et al. (2008) A genetic framework for the control of cell division and differentiation in the root meristem. Science 322: 1380–1384.
16. Marhavy P, Duclercq J, Weller B, Feraru E, Bielach A, et al. (2014) Cytokinin controls polarity of PIN1-dependent auxin transport during lateral root organogenesis. Curr Biol 24: 1031–1037.
17. Laplaze L, Benkova E, Casimiro I, Maes L, Vanneste S, et al. (2007) Cytokinins act directly on lateral root founder cells to inhibit root initiation. Plant Cell 19: 3889–3900.
18. Marhavy P, Bielach A, Abas L, Abuzeineh A, Duclercq J, et al. (2011) Cytokinin modulates endocytic trafficking of PIN1 auxin efflux carrier to control plant organogenesis. Dev Cell 21: 796–804.
19. Kitomi Y, Ito H, Hobo T, Aya K, Kitano H, et al. (2011) The auxin responsive AP2/ERF transcription factor CROWN ROOTLESS5 is involved in crown root initiation in rice through the induction of OsRR1, a type-A response regulator of cytokinin signaling. Plant J 67: 472–484.
20. Perilli S, Perez-Perez JM, Di Mambro R, Peris CL, Diaz-Trivino S, et al. (2013) RETINOBLASTOMA-RELATED protein stimulates cell differentiation in the Arabidopsis root meristem by interacting with cytokinin signaling. Plant Cell 25: 4469–4478.
21. Muller B, Sheen J (2008) Cytokinin and auxin interaction in root stem-cell specification during early embryogenesis. Nature 453: 1094–1097.
22. Karthikeyan AS, Varadarajan DK, Mukatira UT, D'Urzo MP, Damsz B, et al. (2002) Regulated expression of Arabidopsis phosphate transporters. Plant Physiol 130: 221–233.
23. Ribot C, Wang Y, Poirier Y (2008) Expression analyses of three members of the AtPHO1 family reveal differential interactions between signaling pathways involved in phosphate deficiency and the responses to auxin, cytokinin, and abscisic acid. Planta 227: 1025–1036.
24. Shin H, Shin HS, Chen R, Harrison MJ (2006) Loss of At4 function impacts phosphate distribution between the roots and the shoots during phosphate starvation. Plant J 45: 712–726.
25. Martin AC, del Pozo JC, Iglesias J, Rubio V, Solano R, et al. (2000) Influence of cytokinins on the expression of phosphate starvation responsive genes in Arabidopsis. Plant J 24: 559–567.
26. Wang X, Yi K, Tao Y, Wang F, Wu Z, et al. (2006) Cytokinin represses phosphate-starvation response through increasing of intracellular phosphate level. Plant Cell Environ 29: 1924–1935.
27. Franco-Zorrilla JM, Martin AC, Solano R, Rubio V, Leyva A, et al. (2002) Mutations at CRE1 impair cytokinin-induced repression of phosphate starvation responses in Arabidopsis. Plant J 32: 353–360.
28. Wang S, Zhang S, Sun C, Xu Y, Chen Y, et al. (2014) Auxin response factor (OsARF12), a novel regulator for phosphate homeostasis in rice (Oryza sativa). New Phytol 201: 91–103.
29. Shen C, Wang S, Zhang S, Xu Y, Qian Q, et al. (2013) OsARF16, a transcription factor, is required for auxin and phosphate starvation response in rice (Oryza sativa L.). Plant Cell Environ 36: 607–620.
30. Wang X, Du G, Wang X, Meng Y, Li Y, et al. (2010) The function of LPR1 is controlled by an element in the promoter and is independent of SUMO E3 Ligase SIZ1 in response to low Pi stress in Arabidopsis thaliana. Plant Cell Physiol 51: 380–394.
31. Jefferson RA, Kavanagh TA, Bevan MW (1987) GUS fusions: beta-glucuronidase as a sensitive and versatile gene fusion marker in higher plants. EMBO J 6: 3901–3907.
32. Bozzo GG, Dunn EL, Plaxton WC (2006) Differential synthesis of phosphate-starvation inducible purple acid phosphatase isozymes in tomato (Lycopersicon esculentum) suspension cells and seedlings. Plant Cell Environ 29: 303–313.
33. Chang L, Ramireddy E, Schmulling T (2013) Lateral root formation and growth of Arabidopsis is redundantly regulated by cytokinin metabolism and signalling genes. J Exp Bot 64: 5021–5032.
34. An L, Zhou Z, Sun L, Yan A, Xi W, et al. (2012) A zinc finger protein gene ZFP5 integrates phytohormone signaling to control root hair development in Arabidopsis. Plant J 72: 474–490.
35. Ito Y, Kurata N (2006) Identification and characterization of cytokinin-signalling gene families in rice. Gene 382: 57–65.
36. Hirose N, Makita N, Kojima M, Kamada-Nobusada T, Sakakibara H (2007) Overexpression of a type-A response regulator alters rice morphology and cytokinin metabolism. Plant Cell Physiol 48: 523–539.
37. Du L, Jiao F, Chu J, Jin G, Chen M, et al. (2007) The two-component signal system in rice (Oryza sativa L.): a genome-wide study of cytokinin signal perception and transduction. Genomics 89: 697–707.
38. Ai P, Sun S, Zhao J, Fan X, Xin W, et al. (2009) Two rice phosphate transporters, OsPht1;2 and OsPht1;6, have different functions and kinetic properties in uptake and translocation. Plant J 57: 798–809.
39. Sun S, Gu M, Cao Y, Huang X, Zhang X, et al. (2012) A constitutive expressed phosphate transporter, OsPht1;1, modulates phosphate uptake and translocation in phosphate-replete rice. Plant Physiol 159: 1571–1581.
40. Wang C, Ying S, Huang H, Li K, Wu P, et al. (2009) Involvement of OsSPX1 in phosphate homeostasis in rice. Plant J 57: 895–904.
41. Essigmann B, Guler S, Narang RA, Linke D, Benning C (1998) Phosphate availability affects the thylakoid lipid composition and the expression of SQD1, a gene required for sulfolipid biosynthesis in Arabidopsis thaliana. Proc Natl Acad Sci U S A 95: 1950–1955.
42. Trull MC, Deikman J (1998) An Arabidopsis mutant missing one acid phosphatase isoform. Planta 206: 544–550.
43. Inukai Y, Sakamoto T, Ueguchi-Tanaka M, Shibata Y, Gomi K, et al. (2005) Crown rootless1, which is essential for crown root formation in rice, is a target of an AUXIN RESPONSE FACTOR in auxin signaling. Plant Cell 17: 1387–1396.
44. Perez-Torres CA, Lopez-Bucio J, Cruz-Ramirez A, Ibarra-Laclette E, Dharmasiri S, et al. (2008) Phosphate availability alters lateral root development in Arabidopsis by modulating auxin sensitivity via a mechanism involving the TIR1 auxin receptor. Plant Cell 20: 3258–3272.
45. Zhang W, Swarup R, Bennett M, Schaller GE, Kieber JJ (2013) Cytokinin induces cell division in the quiescent center of the Arabidopsis root apical meristem. Curr Biol 23: 1979–1989.
46. Robinson WD, Park J, Tran HT, Del Vecchio HA, Ying S, et al. (2012) The secreted purple acid phosphatase isozymes AtPAP12 and AtPAP26 play a pivotal role in extracellular phosphate-scavenging by Arabidopsis thaliana. J Exp Bot 63: 6531–6542.
47. Tian J, Wang C, Zhang Q, He X, Whelan J, et al. (2012) Overexpression of OsPAP10a, a root-associated acid phosphatase, increased extracellular organic phosphorus utilization in rice. J Integr Plant Biol 54: 631–639.

20

Analysis of Genetic Diversity and Population Structure of Rice Germplasm from North-Eastern Region of India and Development of a Core Germplasm Set

Debjani Roy Choudhury[1,9], Nivedita Singh[1,9], Amit Kumar Singh[1], Sundeep Kumar[1], Kalyani Srinivasan[2], R. K. Tyagi[2], Altaf Ahmad[3], N. K. Singh[4], Rakesh Singh[1]*

1 Division of Genomic Resources, National Bureau of Plant Genetic Resources, New Delhi, 110 012, India, **2** Division of Germplasm Conservation, National Bureau of Plant Genetic Resources, New Delhi, 110 012, India, **3** Department of Botany, Faculty of Science, Jamia Hamdard (Hamdard University), New Delhi, 110062, India, **4** National Research Centre on Plant Biotechnology, Indian Agricultural Research Institute, New Delhi, 110 012, India

Abstract

The North-Eastern region (NER) of India, comprising of Arunachal Pradesh, Assam, Manipur, Meghalaya, Mizoram, Nagaland and Tripura, is a hot spot for genetic diversity and the most probable origin of rice. North-east rice collections are known to possess various agronomically important traits like biotic and abiotic stress tolerance, unique grain and cooking quality. The genetic diversity and associated population structure of 6,984 rice accessions, originating from NER, were assessed using 36 genome wide unlinked single nucleotide polymorphism (SNP) markers distributed across the 12 rice chromosomes. All of the 36 SNP loci were polymorphic and bi-allelic, contained five types of base substitutions and together produced nine types of alleles. The polymorphic information content (PIC) ranged from 0.004 for Tripura to 0.375 for Manipur and major allele frequency ranged from 0.50 for Assam to 0.99 for Tripura. Heterozygosity ranged from 0.002 in Nagaland to 0.42 in Mizoram and gene diversity ranged from 0.006 in Arunachal Pradesh to 0.50 in Manipur. The genetic relatedness among the rice accessions was evaluated using an unrooted phylogenetic tree analysis, which grouped all accessions into three major clusters. For determining population structure, populations K = 1 to K = 20 were tested and population K = 3 was present in all the states, with the exception of Meghalaya and Manipur where, K = 5 and K = 4 populations were present, respectively. Principal Coordinate Analysis (PCoA) showed that accessions were distributed according to their population structure. AMOVA analysis showed that, maximum diversity was partitioned at the individual accession level (73% for Nagaland, 58% for Arunachal Pradesh and 57% for Tripura). Using POWERCORE software, a core set of 701 accessions was obtained, which accounted for approximately 10% of the total NE India collections, representing 99.9% of the allelic diversity. The rice core set developed will be a valuable resource for future genomic studies and crop improvement strategies.

Editor: Swarup Kumar Parida, National Institute of Plant Genome Research (NIPGR), India

Funding: Financial support granted by Indian Council of Agricultural Research, New Delhi, India, under NPTC project, is gratefully acknowledged. The funders had no role in study design, data collection and analysis, decision to publish, or preparation of the manuscript.

Competing Interests: The authors have declared that no competing interests exist.

* Email: rsingh@nbpgr.ernet.in

9 These authors contributed equally to this work.

Introduction

Plant genetic resources are of paramount importance for the future and to ensure the food and nutritional security of an increasing population. A large number of genetic materials have been conserved in Genebank, but their use is being limited due to an unmanageable number of accessions and the continuous expansion of accessions numbers. Core germplasm development has been proposed for better management and use of collections available in Genebank [1]. This requires the development of a core set of accession to more precisely characterize, explore, and conserve Genebank resources, monitor the genetic drift during preservation, and identify gaps in genetic diversity [2,3]. The idea of the core set was proposed by Frankel and Brown in 1984 [2]. As proposed, a core set is a small set of accessions (usually 10 % of the population) chosen to represent the genetic spectrum of an entire collection [4].

The sampling percentage of a core set has long been under debate [5]; 20%–30% of the sampling percentage was suggested by Yonezawa *et al.* (1995) to adequately sample a given collection [6]. Mini core sets representing ~1% of total collections and have also been used to characterize very large collections [7–10]. In fact, a mini core subset from USDA rice gene bank has been developed using 26 phenotypic trait and 70 molecular markers [11]. Still the perfect ratio and fixed size for all core set does not exist, since different crops or different constructing goal needs different sampling percentages [5]. Several methods have been used to develop core set. The stratified random sampling method, in contrast to the simple random sampling method, has been successfully applied to the development of numerous core set, e.g., fodder crops, potatoes, etc. [12–16]. One common approach for constructing a core set is grouping whole collections into major ecotypes and then selecting representatives from each ecotype [5,17–18]. However, a core set formed for the purpose of capturing accessions with rare or extreme values of the desired trait(s) (e.g. high resistance to pest or high yield and yield contributing traits) should be evaluated separately with the intention of representing the (pattern of) genetic diversity in the collection. Indeed, the pattern of genetic diversity will have genetic variations among all accessions that have been accumulated as a result of natural processes, species' characteristics and historical events [19]. Until now, there is no universally accepted method for constructing a core set as many factors affect representativeness of core set, such as sampling percentage, data type, number of traits observed, genetic diversity of germplasm, grouping method and sampling method [18,20–21].

The rice genetic resource is an important source for rice breeding and makes a valuable contribution to global wealth and food security [22]. India is considered to be the origin of rice. Based on the phylo-geographical and archeological evidence it is suggested that rice was domesticated about 10,000 years ago from its wild ancestor *O. rufipogon* in the region South of Himalayan mountain range, likely in the present day eastern and North-Eastern (NE) India, extending eastward to Nepal, Myanmar and Thailand to Southern China [23–25]. The NE India is a large geographical area comprising of seven states namely Arunachal Pradesh, Assam, Manipur, Meghalaya, Mizoram, Nagaland and Tripura states, is home to a large number of indigenous rice varieties [26]. The geography of the NE states of India is unique, having snow capped peaks of the Himalayas, the ecological hot spots of the NE foothills and the Brahmaputra valley. High rainfall, humidity, varied topography and altitude, heavy natural selection pressures, environmental stresses have made the region rich both in floristic and crop diversities [27].

A large collection of the rice genetic resource from NE India is currently conserved at National Genebank (NGB), National Bureau of Plant Genetic Resources (NBPGR), New Delhi. Since these germplasm contains unique traits which can be exploited for future crop improvement programme but has not been properly characterized at molecular level. Therefore, in the present study this collection was analyzed (i) to study the genetic diversity and population structure of all 6,984 rice germplasm accessions using 36 SNP markers, which were developed and used earlier for diversity and population structure analysis of rice varieties [28] and (ii) to develop a core set representing the maximum diversity of unique NE India rice genetic resources. Taken together, the described core set can be used to effectively for genomic studies, in rice crop improvement and conservation programs in genebank.

Materials and Methods

Plant materials

A total of 6,984 accessions of NE India (Arunachal Pradesh, Assam, Manipur, Meghalaya, Mizoram, Nagaland and Tripura,) were drawn from NGB, NBPGR, New Delhi. The NE India germplasm collection also included 24 accessions with EC ID from IRRI, Philippines and two accessions from Ghana which were growing in farmer's field since their introduction in 1991. The purpose of their inclusion was to analyse their clustering in relation to Indian germplasm collection. The details of each accession along with passport information were obtained from the online database of NGB, (www.nbpgr.ernet.in) and presented in Table S1.

DNA extraction from rice seed

Seeds of each genotype (10–12 seeds) were dehusked and used for DNA isolation using QIAGEN DNeasy plant mini kit. Kernels were ground into fine powder using tissue lyser (Tissue lyser II Retsch, Germany) with a tissue lyser adapter set (QIAGENq). DNA was extracted following the procedures described by manufacturers.

Genotyping of rice accessions using SNP markers

Genomic DNA of all the 6,984 accessions was diluted to prepare working stocks of 10 ng/µl. Sequenom Mass ARRAY multiplex assays were designed for 36 SNPs (iPLEX gold chemistry), representing conserved single-copy rice genes [29], taking three genes per rice chromosome [28]. These SNP markers are located on the short arm, centromeric region and long arm of the twelve rice chromosome. The 36-plex assay(s) were designed and validated by Sequenom Corporation (San Diego). The 30-mer pre-amplification primers and variable length genotyping primers generated by the Assay Design 3.1 software were procured and used for the validation of SNPs according to the Sequenom user manual. Mass ARRAY Typer 3.4 Software was used for the visualization of SNPs and allele calling. Chromosome number, primer ID and physical position of 36 SNPs used in this study are given in Table S2.

Table 1. Estimation of gene diversity, heterozygosity, PIC and major allele frequency in NE Indian rice collection (6984)

S. No.	States	No. of accessions	Gene diversity	Heterozygosity	PIC	Major allele frequency
1	Arunachal Pradesh	663	0.006–0.49 (0.29)	0.003–0.16 (0.07)	0.006–0.37 (0.23)	0.51–0.99 (0.78)
2	Assam	2635	0.01–0.5 (0.31)	0.01–0.32 (0.10)	0.01–0.37 (0.25)	0.50–0.99 (0.76)
3	Manipur	549	0.02–0.50 (0.26)	0.004–0.26 (0.09)	0.02–0.37 (0.21)	0.50–0.98 (0.80)
4	Meghalaya	2427	0.03–0.49 (0.34)	0.01–0.29 (0.12)	0.03–0.37 (0.27)	0.50–0.98 (0.74)
5	Mizoram	107	0.01– 0.49 (0.28)	0.01–0.42 (0.16)	0.01–0.37 (0.23)	0.52–0.99 (0.79)
6	Nagaland	377	0.008–0.49 (0.31)	0.002–0.16 (0.07)	0.008–0.37 (0.25)	0.52–0.99 (0.76)
7	Tripura	226	0.004–0.49 (0.28)	0.004–0.38 (0.08)	0.004–0.37 (0.22)	0.51–0.99 (0.79)
	Total	6984				

Mean values are given in parentheses.

Table 2. State-wise Analysis of Molecular Variance (AMOVA) of NE Indian rice collections.

States	Source	df	SS	MS	Est. Var.	%
Arunachal Pradesh	Among Pops	2	1160.573	580.286	1.523	21
	Among Indiv	660	6728.744	10.195	4.305	58
	Within Indiv	663	1051.000	1.585	1.585	21
	Total	1325	8940.317		7.413	100
Assam	Among Pops	2	4963.974	2481.987	1.751	24
	Among Indiv	2632	25094.849	9.535	3.836	51
	Within Indiv	2635	4908.500	1.863	1.863	25
	Total	5269	34967.323		7.449	100
Manipur	Among Pops	2	905.987	452.994	1.422	21
	Among Indiv	546	5785.991	10.597	4.509	58
	Within Indiv	549	866.500	1.578	1.578	21
	Total	1097	7558.478		7.510	100
Meghalaya	Among Pops	4	9032.582	2258.145	2.422	30
	Among Indiv	2422	23045.557	9.515	3.745	46
	Within Indiv	2427	4914.000	2.025	2.025	35
	Total	4853	36992.138		8.192	100
Mizoram	Among Pops	2	108.378	54.189	0.747	9
	Among Indiv	104	1247.813	11.998	4.569	56
	Within Indiv	107	306.000	2.860	2.860	35
	Total	213	1662.192		8.176	100
Nagaland	Among Pops	2	40.936	20.468	0.709	8
	Among Indiv	12	168.464	14.039	6.319	75
	Within Indiv	15	21.000	1.400	1.400	17
	Total	29	230.400		8.428	100
Tripura	Among Pops	2	400.316	200.158	1.564	22
	Among Indiv	223	2084.863	9.349	3.946	57
	Within Indiv	226	329.500	1.458	1.458	21
	Total	451	2814.679		6.967	100

Statistical analyses

The major allele frequency, gene diversity, heterozygosity and Polymorphic Information Content (PIC) for each locus were calculated for SNP markers using Power Marker 3.5 [30]. In addition, Principal Component Analyses (PCA), genetic distances [31] across the genotypes and neighbor-joining (NJ) tree were calculated using Power Marker 3.5 [30]. Phylogenetic trees for all states were constructed using MEGA software version 6.0 [32]. Principle Coordinate Analysis (PCoA) and Analysis of Molecular Variance (AMOVA) were performed using software GenAlEx V6.5 [33]. SNP data were numerically coded as follows: A = 1, C = 2, G = 3, T = 4 and missing data was coded as 0 as suggested in GenAlEx V6.5 user manual [33]. The model-based program, STRUCTURE 2.3.3 [34] was used to infer the population structure. For each K, three replications were run. Each run was implemented with a burn-in period of 100,000 steps followed by 100,000 Monte Carlo Markov Chain replicates [34] derived for each K and then plotted to find the plateau of the ΔK values [35]. The "Structure harvester" program was used (http://taylor0.biology.ucla.edu) to determine the final population. POWERCORE software [36] was used for the core development. Since numbers of rice accessions from each NE states were not equal, ranging from 107 accessions from Mizoram to 2,635 accessions from Assam. Therefore, to avoid the dominance of collections from individual states within the core set, we first developed core sub-sets using the rice collections from each state. Next, all accessions of the core sub-sets were pooled together to create the final core set of NE India rice germplasm.

Results

Genetic diversity

Genetic diversity of NE India rice collections (6,984 accessions), available at NGB, was estimated using 36 SNP markers. Alleles generated with all 36 SNP markers were scored to study the genetic diversity. Since these accessions cover seven states of NE India, they were analyzed state-wise for precise estimation of the level of their genetic diversity. The accessions collected from Arunachal Pradesh (663), Assam (2635), Manipur (549), Meghalaya (2427), Mizoram (107), Nagaland (377) and Tripura (226) were analyzed for major allele frequency, heterozygosity, gene diversity and PIC (Table 1). The highest mean major allele frequency was present in Manipur accessions (0.80) and lowest was found in Meghalaya accessions (0.74). The comparison of mean heterozygosity of alleles across the seven states revealed that Mizoram accessions were highly heterozygous (0.16); whereas

Table 3. State-wise Principal Coordinate Analyses of NE Indian rice collections, with percentage of variation explained by the first 3 axes.

States	Per cent variation	1-axis	2-axis	3-axis
Arunachal Pradesh	%	26.74	9.23	7.30
	Cum %	26.74	35.97	43.27
Assam	%	20.79	8.30	6.50
	Cum %	20.79	29.09	35.58
Manipur	%	21.54	7.23	5.73
	Cum %	21.54	28.77	34.50
Meghalaya	%	22.45	11.45	5.55
	Cum %	22.23	33.54	39.05
Mizoram	%	19.10	7.96	6.62
	Cum %	19.10	27.06	33.68
Nagaland	%	24.95	11.05	7.30
	Cum %	24.95	36.01	43.31
Tripura	%	14.96	10.15	7.57
	Cum %	14.96	25.12	32.69

Cum- Cumulative.

heterozygosity in the accessions belonging to Arunachal Pradesh (0.07) and Nagaland (0.07) were recorded minimum (Table 1). The mean gene diversity across the seven states varied from 0.34 (in the collections from Meghalaya) to 0.26 in those from (Manipur) and PIC varied from 0.21 (in the collections from Manipur) to 0.27 (from Meghalaya). Out of 36 primers used for characterization of total rice collections, five SNP primers (01-608-4_C_375, 02-3029-1_C_474, 03-3478-1_C_206, 05-48-1_C_279 and 08-4218-5_C_129) exhibited the maximum PIC and gene diversity value 0.37 and 0.49, respectively (Table S2). The primer 04-19-4_C_240 was found to be the least informative with PIC and gene diversity values equaling 0.02 and 0.02, respectively, and major allele frequency of 0.99 (Table S2), SNP markers which showed low PIC values in the present study may not be considered for future studies on genetic diversity and population structure in rice. The maximum heterozygosity was observed with primer 04-1801-20_C_428 was 0.28, while the minimum heterozygosity was observed with primer 04-19-4_C_240 (0.01).

Cluster analysis was performed using the NJ method for all the collections from seven states and unrooted phylogenetic trees were constructed (Fig. S1). Rice collections of each state were grouped in to three major clusters. Genetic distance for collections of Arunachal Pradesh varied from 0.00 to 0.75, Assam from 0.00 to 0.66, Manipur from 0.00 to 0.74, Meghalaya from 0.00 to 0.77, Mizoram from 0.00 to 0.68, Nagaland from 0.00 to 0.66, and Tripura from 0.00 to 0.67.

Population structure

A model-based program, STRUCTURE, was used to determine the genetic relationship among individual rice accessions. The membership of each accession was run from K = 1 to K = 20 for all the collections from all seven states to estimate the number of populations. Structure Harvester (http://taylor0. biology.ucla.edu) was used to determine final number of populations. The number of populations in Meghalaya collections was estimated as five, in Manipur collections as four and three for collections belonging to remaining states, i.e. Arunachal Pradesh, Assam, Mizoram, Nagaland and Tripura (Fig. S2, Table S3). The collection from Arunachal Pradesh contained 265 pure and 114 admix individuals in population1, 65 pure and 16 admix individuals in population2 and 131 pure and 72 admix individuals were present in population3. The collection from Assam showed 300 pure and 329 admix in population1, 1073 pure and 568 admix in population2, 288 pure and 77 admix in population3. The Nagaland collections 61 pure and 10 admix were present in population1, 164 pure and 91 admix in population2 and 27 pure and 24 admix were present in population3. Tripura collections showed 14 pure and 1 admix in population1, 107 pure and 21 admix in population2, 54 pure and 29 admix in population3. The collections from Mizoram had 39 pure and 19 admix in population1, 4 pure and no admix in population2 and 31 pure and 14 admix in population3. Similarly, Manipur collections showed 100 pure and 63 admix in population1, 53 pure and 55 admix in population2, 22 pure and 29 admix in population3 and 100 pure and 128 admix in population4. The accessions included in the core are marked with boxes in the bar plot (Fig. S3).

Analysis of molecular variance (AMOVA)

An AMOVA study of all rice accessions, belonging to seven different NE states, was performed to analyze the distribution of genetic diversity between and within the populations. The number of populations for each state was considered the same, as revealed by the population structure. AMOVA analyses showed that 21% diversity exists among populations in the collection from Arunachal Pradesh, while the collections from Assam, Manipur, Meghalaya, Mizoram, Nagaland and Tripura showed 24, 19, 29, 9, 8 and 22% diversity, respectively (Table 2). Maximum diversity of 29% among population was observed in Meghalaya collections, while minimum diversity of 8% was observed in Nagaland collections (Fig. S4). The highest level of diversity among individuals was present in Nagaland collections (75%) and the least diversity was observed in Meghalaya collections (46%).

Principal coordinate analyses (PCoA)

PCoA revealed significant diversity in the NE India rice germplasm collections. Meghalaya collections, which exhibited

Table 4. A comparison of percent of alleles generated in total collection vs core set.

Alleles	Arunachal Pradesh: Total collection (663 acc.)	Arunachal Pradesh: Core set (60 acc.)	Assam: Total collection (2635 acc.)	Assam: Core set (259 acc.)	Manipur: Total collection (549 acc.)	Manipur: Core set (60 acc.)	Meghalaya: Total collection (2427 acc.)	Meghalaya: Core set (242 acc.)	Mizoram: Total collection (107 acc.)	Mizoram: Core set (14 acc.)	Nagaland: Total collection (377 acc.)	Nagaland: Core set (35 acc.)	Tripura: Total collection (226 acc.)	Tripura: Core set (31 acc.)
AA	20.08	21.36	19.43	20.63	17.73	18.02	20.66	20.93	18.27	19.07	21.23	20.98	19.33	20.31
CC	21.16	17.68	18.23	18.44	13.51	15.91	16.98	17.22	15.82	16.92	16.83	17.44	18.43	16.96
GG	16.72	18.44	21.10	18.93	24.67	21.07	20.07	18.90	19.55	19.07	20.53	19.86	22.19	20.01
TT	37.83	32.12	34.94	30.20	38.62	34.43	35.17	30.64	36.55	31.70	37.11	31.33	35.32	33.82
CT/TC	1.57	3.70	2.31	3.87	2.14	3.71	2.67	4.45	3.66	4.29	1.59	3.78	2.21	3.81
GA/AG	1.53	3.65	2.32	4.36	2.00	3.94	2.42	4.25	3.76	5.48	1.49	3.17	1.46	3.05
TA/AT	0.33	0.78	0.46	0.96	0.49	0.86	0.56	1.08	0.83	1.31	0.26	0.75	0.37	0.61
GT/TG	0.70	2.00	1.07	2.20	0.80	1.83	1.29	2.10	1.46	1.91	0.87	2.24	0.67	1.22
CA/AC	0.07	0.27	0.14	0.41	0.04	0.22	0.17	0.43	0.10	0.24	0.09	0.47	0.03	0.20

the maximum number of population ($k = 5$) showed a very distinct PCoA plot, population1 in Meghalaya collections concentrated only in quadrant 3 and 4, whereas population4 concentrated in 1 and 2 quadrant and unlike population3, population4 and population5 did not show intermixing with each other (Fig. S5d). PCoA of collections from Arunachal Pradesh showed that population2 was very distinct, forming a separate group (Fig. S5a). PCoA plot of Assam, Manipur and Nagaland collections showed that all three populations were intermixed (Fig. S5b, Fig. S5c & Fig. S5f). In Mizoram collection population2 and Tripura collection population1 were highly distinct and formed a tight group in quadrant 1 and 4, respectively (Fig. S5e, Fig. S5g). The maximum cumulative variation (%) was observed for Nagaland rice collections with a value of 43.31% (Table 3); whereas, the minimum variation was found in Tripura accessions (32.53%). Overall, <44% variation were observed by first three components of principal coordinates, which indicates that accessions from all seven states were very diverse from one another.

Development of Core Set

Of the 6,984 rice accessions studied, a core set of 701 accessions (i.e., 60 from Arunachal Pradesh, 259 from Assam, 60 from Manipur, 242 from Meghalaya, 14 from Mizoram, 35 from Nagaland, and 31 from Tripura) were selected using POWERCORE (Table S4). Nine allele types were produced by 36 SNP markers, of which four alleles were homozygous and five were heterozygous. Out of six types of substitutions possible (2 transitions and 4 transversion), five types of substitutions were observed and C/G or G/C type substitutions were not observed in our study. Allele frequency was calculated for all collections from seven states and core set (Table 4). A comparison of allele frequency showed that there was no loss of alleles in the resulting core set and they remained 99.9% identical.

Genetic diversity of Core Set

The genetic diversity of the core set was estimated to know the extent of diversity captured from total collection. Major allele frequency; gene diversity, heterozygosity and PIC were separately estimated for core-sets of individual states (Table 4). Comparisons of all genetic parameters (major allele frequency, gene diversity, heterozygosity and PIC) revealed that values for core set (Table 5) were greater than total collections (Table 1). For example, the Arunachal Pradesh gene diversity was 0.29, which increased to 0.39 in the core set. Similarly, the PIC for the Arunachal Pradesh collection was 0.23, while its core set PIC was 0.31. Cluster analysis showed that, core set accessions were getting distributed in to four clusters. The nine accessions with EC ID which were also part of core set were represented in three out of four clusters. The genetic diversity for the core set varied from 00.00 to 0.68 (Fig. 1).

Population structure of Core Sets

STRUCTURE analysis grouped core set of 701 rice accessions into three populations (Fig. 2). Populations were studied for the number of pure and admix individuals. Population1 had 62 pure and 74 admix, population2 had 111 pure and 53 admix and population3 had 250 pure and 151 admix (Fig. 3).

AMOVA and PCoA of Core Set

AMOVA analyses for core set showed 20% variance among populations (Table 6, Fig. 4); whereas, 48% variance was present among individuals. The partitioning of molecular variance was similar to the total collection and maximum diversity among individuals. PCoA could not be useful to identify significant

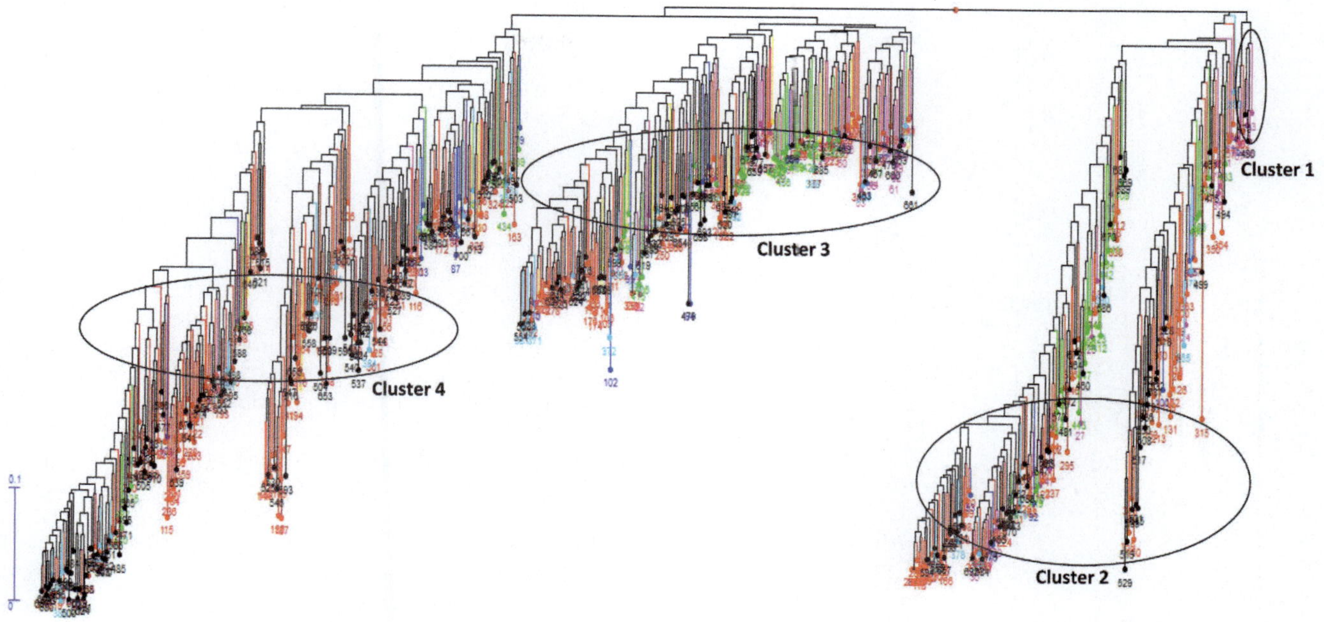

Figure 1. Phylogenetic tree of NE India rice core set (701 accessions) constructed based on SNP data. Germplasm accessions collected from the NE states are depicted with different colours Arunachal Pradesh - green, Assam - red, Manipur - purple, Meghalaya - black, Mizoram -yellow, Nagaland - light blue, Tripura - dark blue.

isolation in populations except, for population2, which was tightly clustered in quadrant 3 and 4 (Fig. 5). Maximum cumulative percent variation was explained by the first three coordinates with 43.08% variance (Table 7). A plot of percent allele frequency of total collection *versus* core set showed that nearly all alleles with similar frequency were represented in the core collection (Fig. 6) and 99.9% alleles have been retained in the collections designated as core set.

Discussion

Genome level profiling of huge germplasm collections in crop species is essential to identify small and diverse sets of accessions for their efficient use in crop improvement programmes. Despite the advancement in genomics, the Indian rice collection remained uncharacterized at the molecular level, with respect to parameters such as genetic diversity and population structure. This has been the major limiting factor in their utilization and development of improved cultivars. The present study is the first major effort to

Table 5. Estimation of gene diversity, heterozygosity, PIC and major allele frequency in NE India rice core set (701 accessions).

S. No.	States	core sample	Gene diversity	Heterozygosity	PIC	Major allele frequency
1	Arunachal Pradesh	60	0.0726–0.5000 (0.3962)	0.0377–0.2778 (0.1713)	0.0700–0.3750 (0.3124)	0.5000–0.9623 (0.6940)
2	Assam	259	0.1670–0.5000 (0.4018)	0.0843–0.3618 (0.1908)	0.1531–0.3740 (0.3172)	0.5000–0.9080 (0.69271)
3	Manipur	60	0.1372–0.5000 (0.3749)	0.0455–0.3455 (0.1745)	0.1278–0.3750 (0.2982)	0.5000–0.9259 (0.7142)
4	Meghalaya	242	0.2390–0.4995 (0.4100)	0.1013–0.3493 (0.2001)	0.2105–0.3747 (0.3231)	0.5161–0.8612 (0.6873)
5	Mizoram	14	0.0689–0.4970 (0.3506)	0.0714–0.4615 (0.22[illegible]3)	0.0665–0.3735 (0.2813)	0.5385–0.9643 (0.7416)
6	Nagaland	35	0.1087–0.5000 (0.3854)	0.0385–0.3846 (0.1682)	0.1028–0.3846 (0.3049)	0.5000–0.9423 (0.7033)
7	Tripura	31	0.0351–0.5000 (0.3596)	0.0323–0.3667 (0.1524)	0.0345–0.3750 (0.2875)	0.5000–0.9821 (0.7344)

Mean values are given in parentheses.

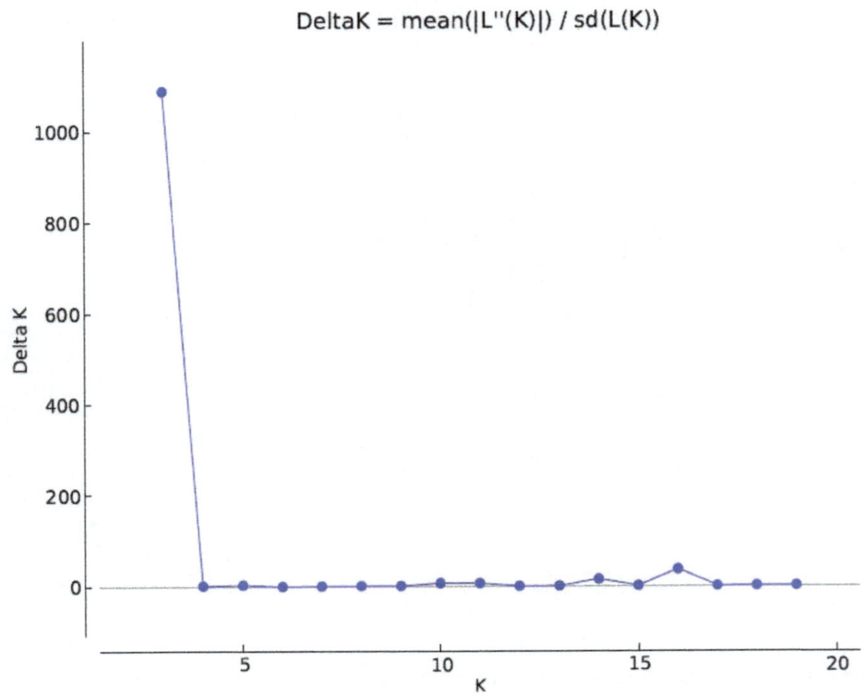

Figure 2. Estimation of populations in NE India rice core set using LnP(D) derived Δk for k from 1 to 20.

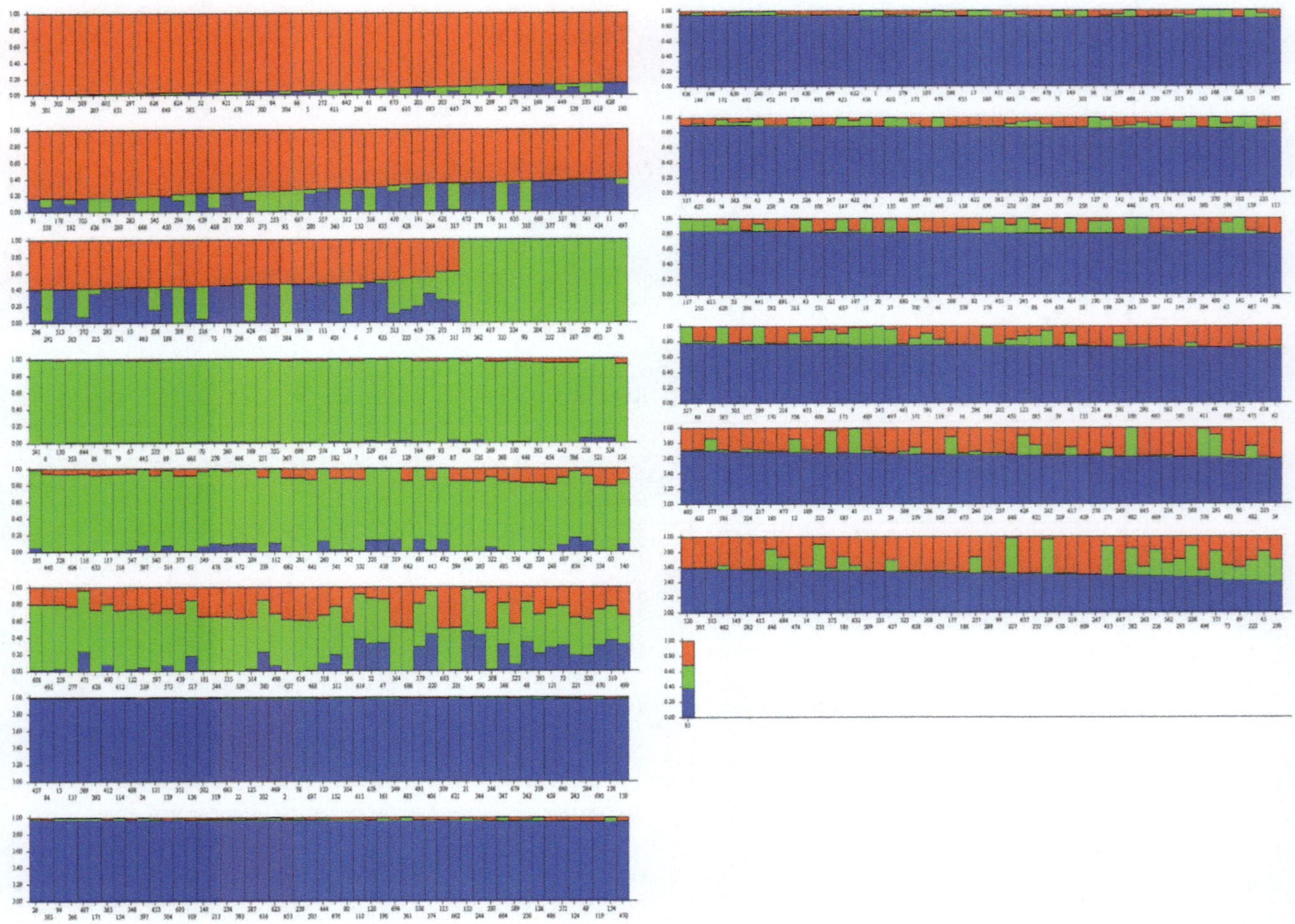

Figure 3. Model based clustering of the NE India rice core set (701 accessions).

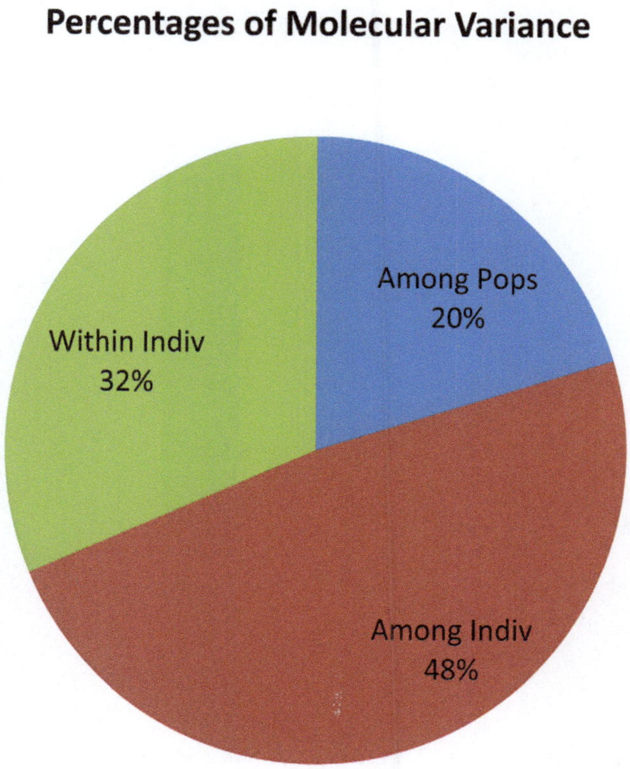

Figure 4. Analysis of Molecular Variance (AMOVA) analyses of NE India rice core set.

characterize the NE India rice collections. This study shows that highly efficient and selective SNP markers can be used to enhance genome based analysis, e.g., genome wide association mapping.

State wise analysis of the genetic diversity of the NE India collection showed interesting results. The average PIC values for these SNP markers did not vary much across the states and ranged from 0.23 (Manipur) to 0.27 (Meghalaya). The average PIC value recorded in the present study was in the range reported by researchers in previous studies. Singh *et al.*, (2013) [28] reported average PIC value 0.25 for the SNP markers in a set of 375 Indian rice varieties. The highest values for the parameters like PIC and gene diversity were found in collections from Meghalaya, which indicates the presence of a large rice genetic diversity in Meghalaya collections as compared to those from other six states. The same was also confirmed from PCoA, which displayed cluttered plot for the Meghalaya collection. The highest heterozygosity was observed in the collections from Mizoram, while Manipur had the highest major allele frequency.

Analysis of statistical genetic parameters across the collections from different states indicated a high level of variability in the NE India collection. There is relatively a high level of genetic diversity among the NE India rice collections, as reported by others using morphological [37] and molecular markers such as RAPD [38], ISSR [39] and SSR [26,40]. Comparison of our results with previous findings revealed that PIC and genetic diversity were nearly half of the SSR-based studies which were on the expected line [26,28,40]. Similar results have been reported in other species that have used SSR and SNP markers [41,42]. In the past, analysis of rice accessions from all seven states revealed three major clusters. Vairavan *et al.*, (1973) [37] reported three clusters among a set 400 accessions of NE India rice collection based on morphological analysis. Das *et al.*, (2013) [40] showed four groups were present in NE India rice varieties, while Choudhury *et al.*, (2013) [26] reported only two groups, based on SSR marker analysis.

Population structure study has been reported for the first time in NE rice collections and the number of populations estimated five for Meghalaya, four for Manipur and three for the remaining states. Admixture has been reported in all the states besides pure lines and this may be due to germplasm taken in this study which is more heterogeneous in comparison to varieties.

There have been no other studies for characterization of a large number of NE India rice accessions at the molecular level that

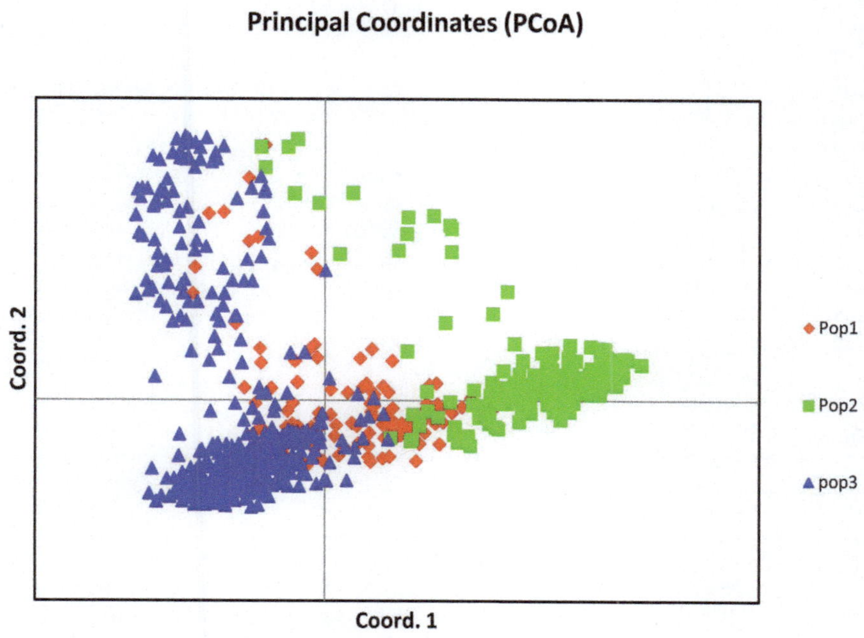

Figure 5. Principal Coordinate Analysis (PCoA) analyses of NE India rice core set.

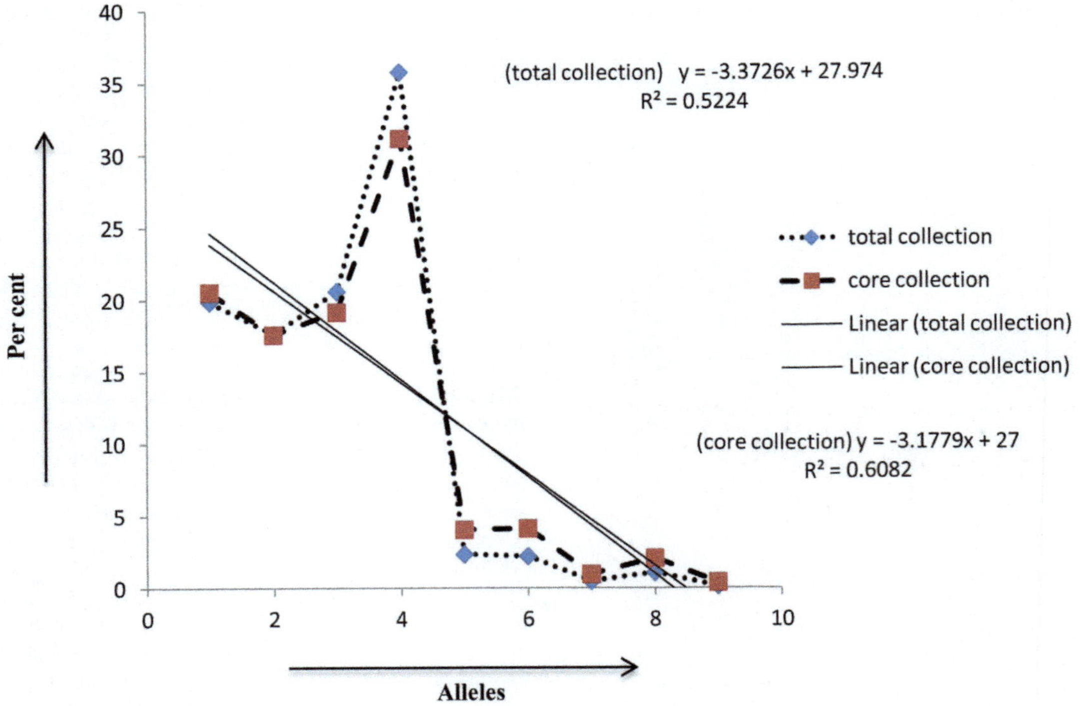

Figure 6. Comparison of allele frequency occurrence in total NE rice collections vs core set identified in the present study.

could be used for comparison of our results. AMOVA analysis showed maximum variation among individuals for accessions belonging to Nagaland (75%) and minimum for Meghalaya collection (37%). Choudhury *et al.*, (2013) [26] conducted AMOVA analysis on NE rice, which revealed 25% variation among individuals, 66% among varieties and 9% among cultivated types. The variation among individuals found in our study was higher than that reported previously, which may be attributed to the large sample size used in present study. Also partitioning of variability at population and individual levels in AMOVA among states were different because sample size taken from different states varied from 107 accessions from Mizoram to 2,635 accessions from Assam.

The genomic characterization revealed huge variability among the NE India rice collection; therefore, we decided to identify a core set for the NE India accessions to facilitate the rice breeders to effectively use the same in their crop improvement programmes and efficient conservation and management of germplasm in the genebank. In recent past, various types of molecular markers have been used to develop core sets for different annual crops. These markers include SNPs for Arabidopsis [43], SSRs for rice, wheat and common bean [44–46], AFLPs for barley [47], RFLPs for wild wheat [48] and RAPD for common bean [49]. The present study used advanced SNP markers due to their biallelic nature, co-dominance inheritance and amenability to high throughput analysis. Another crucial step for development of core set is to formulate a suitable strategy that can identify either major and minor variances or diversity, using a fully representative accession from the large collections. One of the most commonly followed approach for constructing a core set from a collection is to rely on grouping an entire collection based on growing regions and then selecting core accessions from each group followed by combining accessions from each group to form final core set [5].

In the current study, we divided the NE India rice collection into seven (on the basis of source states) subsets and later all the accessions of subsets were pooled to form a core set of NE India rice germplasm. Statistical parameters, genetic distance and cluster analysis were taken into consideration to develop this core set; these methods were followed in accordance with other studies [4,50,51,52,53,54]. We developed the NE India rice core set collection using SNP markers employing POWERCORE software (using advanced M strategy implemented through a modified heuristic algorithm). We relied on molecular data more than the passport and phenotypic data, because molecular markers can

Table 6. AMOVA of core set of NE Indian rice (701 accessions).

Source	df	SS	MS	Est. Var.	%
Among Pops	2	1649.152	824.576	1.996	20
Among Indiv.	698	8774.397	12.571	4.740	48
Within Indiv.	701	2166.500	3.091	3.091	32
Total	**1401**	**12590.049**		**9.826**	**100**

Table 7. Percentage of variation explained by the first 3 axes in Principal Coordinate Analyses of NE Indian rice core set (701 accessions).

Per cent variation	1-axis	2-axis	3-axis
%	18.41	18.04	6.63
Cum %	18.41	36.45	43.08

accurately represent the genetic diversity of the original collection and free of problems related to missing data or environmental interactions that are typically found in passport and phenotypic data. POWERCORE was used to develop a core set of 701 rice accessions (10% of total NE India rice collection). Nearly all alleles were represented in the core set; hence, there was no loss of alleles during the core set development.

In present study, we developed a core set of 10% of total collection; while there is no perfect core set constructing theory or universal methods to fit all crops, which might be due to large number of germplasm resources, absence of data for accessions, or complication and diversification of data type, and so on [50]. The core set was statistically supported with various other statistical parameters including genetic distance and cluster analysis that were taken into consideration in the final development of core set which is in accordance with previous studies [50]. The gene diversity value 0.54 and PIC value 0.48 estimated by SSR markers on rice core [22], is a higher range than the values found in our study (gene diversity value 0.39 and PIC value 0.31). This variation is obvious due to nature of marker used for development of core set. In present study, we have used SNP marker which is biallelic in nature whereas, Zhang et al., 2011, [22] have used SSR marker which is multi-allelic. A study conducted with SNP markers in tropical maize [51] reported a gene diversity value 0.29 and heterozygosity value 0.25 on core sets, and similar gene diversity was observed in present study. The gene diversity, heterozygosity, PIC values of core set were higher than those of total collection which was expected because the diversity increases with elimination of genetically similar accessions during core set development [11].

In conclusion, SNP marker based molecular characterization of NE India rice collection revealed that large variation exists among the accessions and the pattern of the genetic diversity and population structure varied across the seven states. Collections from Meghalaya were found to be the most diverse and possessed maximum number of populations. Core sets have been developed state wise as well as for the entire NE collection that can be potentially exploited by rice breeders in conventional breeding programmes and molecular breeders/researchers for genomic studies such as association mapping for gene discovery for various desirable traits and identification of suitable parents for their use in rice improvement. Further development of core set and sub-sets for NE India rice is likely to significantly reduce the task of the genebank curators as they will need to multiply and maintain small number accessions in the medium term storage as active collections to meet the requirement of the rice breeders.

Supporting Information

Figure S1 Circular phylogenetic trees of NE rice collection constructed based on SNP data using MEGA software (a) Arunachal Pradesh, (b) Assam, (c) Manipur, (d) Meghalaya, (e) Mizoram, (f) Nagaland and (g) Tripura. (Red marked accessions was selected for core by Power Core).

Figure S2 Estimation of populations in NE rice collections using LnP(D) derived Δk for k from 1 to 20 in (a) Arunachal Pradesh, (b) Assam, (c) Manipur, (d) Meghalaya, (e) Mizoram, (f) Nagaland and (g) Tripura.

Figure S3 Population structure of NE rice collection based on SNP data (a) Arunachal Pradesh, (b) Assam, (c) Manipur, (d) Meghalaya, (e) Mizoram, (f) Nagaland and (g) Tripura.

Figure S4 Analysis of Molecular Variance in NE rice collection based on SNP data (a) Arunachal Pradesh, (b) Assam, (c) Manipur, (d) Meghalaya, (e) Mizoram, (f) Nagaland and (g) Tripura.

Figure S5 Principal Coordinate Analysis in NE rice collection based on SNP data (a) Arunachal Pradesh, (b) Assam, (c) Manipur, (d) Meghalaya, (e) Mizoram, (f) Nagaland and (g) Tripura.

Table S1 List of NE rice collections studied using SNP markers.

Table S2 List of SNP primers used for genotyping of NE rice collection (6984 rice accessions) alongwith gene diversity, PIC and Major allele frequency.

Table S3 Output Table generated from Structure Harvester software for Evanno method. Yellow highlight shows the largest value in the Delta K column.

Table S4 List of rice accessions identified as NE core set based on SNP markers.

Acknowledgments

We are grateful to Dr James W. Lillard, JR., Professor and Associate Dean for Research, Morehouse School of Medicine, Atlanta, GA 30310, for manuscript editing. Thanks are due to the Director, NBPGR, New Delhi, for providing the facilities to carry out this study.

Author Contributions

Conceived and designed the experiments: RS NKS. Performed the experiments: DRC NS. Analyzed the data: RS AKS SK. Contributed reagents/materials/analysis tools: KS RKT AA. Wrote the paper: RS AKS RKT.

References

1. Frankel OH (1984) Genetic perspective of germplasm conservation. In: Arber WK. Llimensee, Peacock WL, Starlinger P, editors. Genetic manipulation: Impact on man and society. Cambridge University Press, Cambridge, England, pp. 161–170.
2. Frankel O H, Brown AHD (1984) Plant genetic resources today: A critical appraisal. In Holden JHW, Williams JT. Crop Genetic Resources: Conservation and Evaluation. London, UK: George Allen and Unwin Ltd. pp. 249–257.
3. Brown AHD (1989) The case for core collections. In: Brown AHD, Frankel OH, Marshall DR, Williams JT, editors. The Use of Plant Genetic Resources. Cambridge University Press, Cambridge, UK. pp. 136–155.
4. Ebana K, Kojima Y, Fukuoka S, Nagamine T, Kawase M (2008) Development of mini core collection of Japanese rice landrace. Breed Sci 58: 281–291.
5. Wang J, Guan Y, Wang Y, Zhu L, Wang Q (2014) A Strategy for finding the optimal scale of plant core collection based on Monte Carlo simulation. Sci. World Journal doi: 10.1155/2014/503473.
6. Yonezawa K, Nomura T, Morishima H (1995) Sampling strategies for use in stratified germplasm collections. In: Hodgkin T, Brown AHD, Hintum van Th JL, Morales EAV, editors. Core collections of plant genetic resources John Wiley and sons Chichester UK pp. 35–53.
7. Upadhyaya HD, Yadav D, Dronavalli N, Gowda CLL, Singh S (2010) Mini core germplasm collections for infusing genetic diversity in plant breeding programs. Electronic J Plant Breed 1: 1294–1309.
8. Jiang Q, Hou J, Hao C, Wang L, Ge H, et al. (2011) The wheat (*T. aestivum*) sucrose synthase 2 gene (*TaSus2*) active in endosperm development is associated with yield traits. Funct Integr Genomics 11: 49–61.
9. Upadhaya HD, Reddy KN, Sharma S, Varshney RK, Bhattacharjee R, et al. (2011) Pigeonpea composite collection and identification of germplasm for use in crop improvement programmes. Plant Genet Resour 9: 97–108.
10. Sharma M, Rathore A, Naga Mangala U, Ghosh R, Sharma S, et al. (2012) New sources of resistance to Fusarium wilt and sterility mosaic disease in a mini-core collection of pigeonpea germplasm. Eur J Plant Pathol 133: 707–714.
11. Agrama HA, Yan WG, Lee F, Fjellstrom R, Chen MH, et al. (2009) Genetic assessment of a mini-core subset developed from USDA Rice Genebank. Crop Science 49: 1336–1346.
12. Balfourier F, Charmet G, Prosperi JM, Goulard M, Monestiez P (1998) Comparison of different spatial strategies for sampling *a* core collection of natural populations *of* fodder crops. Genet Sel Evol 30:215–235.
13. Chandra S, Huaman Z, Hari Krishna S, Ortiz R (2002) Optimal sampling strategy and core collection size of Andean tetraploid potato based on isozyme data: A simulation study. Theor Appl Genet 104: 1325–1334.
14. Hu J, Zhu J, Xu HM (2000) Methods of constructing core collections by stepwise clustering with three sampling strategies based on the genotypic values of crops. Theor Appl Genet 101: 264–268.
15. Peeters JP, Martinelli JA (1989) Hierarchical cluster analysis as a tool to manage variation in germplasm collections. Theor Appl Genet 78: 42–48.
16. Franco J, Crossa J, Warburton ML, Taba S (2006) Sampling strategies for conserving maize diversity when forming core subsets using genetic markers. Crop Sci46: 854–864.
17. Reeves PA, Panella LW, Richards CM (2012) Retention of agronomically important variation in germplasm core collections: implications for allele mining. Theor Appl Genet 124: 1155–1171.
18. Upadhaya HD, Gowda CLL, Pundir RPS, Reddy VG, Singh S (2006) Development of a core subset of finger millet germplasm using geographical origin and data on 14quantitative traits. Genet Resour Crop Evol 53: 679–685.
19. Odong TL, Jansen J, van Eeuwijk FA, van Hintum TJL (2013) Quality of core collections for effective utilization of genetic resources review, discussion and interpretation. Theor Appl Genet 126: 289–305.
20. Díez CM, Imperato A, Rallo L, Barranco D, Trujillo I (2012) Worldwide core collection of olive cultivars based on simple sequence repeat and morphological markers. Crop Sci 52:211–221.
21. Rao ES, Kadirvel P, Symonds RC, Geethanjali S, Ebert AW (2012) Using SSR markers to map genetic diversity and population structure of *Solanum pimpinellifolium* for development of a core collection. Plant Genet Resour 10:38–48.
22. Zhang P, Li J, Li X, Liu X, Zhao X (2011) Population Structure and Genetic Diversity in a Rice Core Collection (*Oryza sativa L.*) investigated with SSR markers. PloS ONE 6 (12): e27565 doi: 10. 1371/journal.pone.0027565.
23. Chang TT (1976) The origin, evolution, cultivation, dissemination, and diversification of Asian and African rice. Euphytica 25: 425–441.
24. Khush GS (1997) Origin, dispersal, cultivation and variation of rice. Plant Mol Biol 35: 25–34.
25. Londo JP, Chiang YC, Hung KH, Chiang TY, Schaal BA (2006) Phylogeography of Asian wild rice, *Oryza rufipogon*, reveals multiple independent domestications of cultivated rice, *Oryza sativa*. Proc Natl Acad Sci USA 103: 9578–9583.
26. Choudhury B, Khan ML, Dayanandan S (2013) Genetic structure and diversity of Indigenous Rice varieties (*Oryza sativa*) in Eastern Himalayan region of Northeast India. Springer Plus 2: 228–237.
27. Hore DK (2005) Rice diversity collection, conservation and management in northeastern India. Genet Resour Crop Evol 52:1129–1140.
28. Singh N, Choudhury DR, Singh AK, Kumar S, Srinivasan K, et al. (2013) Comparison of SSR and SNP markers in estimation of genetic diversity and population structure of Indian rice varieties. PLoS ONE 8(12): e84136.
29. Singh NK, Dalal V, Batra K, Singh BK, Chitra G, et al. (2007) Single copy genes define a conserved order between rice and wheat for understanding differences caused by duplication, deletion, and transposition of genes. Funct Integr Genomics 7: 17–35.
30. Liu K, Muse SV (2005) Power Marker: an integrated analysis environment for genetic marker analysis. Bioinformatics 21: 2128–2129. doi:10.1093/bioinformatics/bti282.
31. Nei M, Tajima F, Tateno Y (1983) Accuracy of estimated phylogenetic trees from molecular data. II. Gene frequency data. J Mol Evol 19: 153–170.
32. Tamura K, Stecher G, Peterson D, Fillipski A, Kumar S (2013) MEGA6 Molecular Evolutionary genetics Analyses version 6.0. Mol Biol Evol 30: 2725–2729.
33. Peakall R, Smouse P (2012) GenAlEx 6.5: Genetic analysis in Excel. Population genetic software for teaching and research - an update. Bioinformatics 28: 2537–2539.
34. Pritchard JK, Stephens M, Donnelly P (2000) Inference of population structure using multilocus genotype data. Genetics 155: 945–959.
35. Evanno G, Regnaut S, Goudet J (2005) Detecting the number of clusters of individuals using the software STRUCTURE: a simulation study. Mol Ecol 14: 2611–2620.
36. Kim KW, Chung HN, Cho GT, Ma KH1, Chandrabalan D, et al (2007) PowerCore: a program applying the advanced M strategy with a heuristic search for establishing core sets. Bioinformatics 23: 2155–2162.
37. Vairavan S, Siddiq EA, Arunachalam V, Swaminathan MS (1973) A study on the nature of genetic divergence in rice from Assam and Northeast Himalayas. Theor Appl Genet 43:213–221.
38. Sarma RN, Bahar B (2005) Genetic variation of bora rice (glutinous rice) of Assam as revealed by RAPDs. Plant Genetic Resources Newsletter 144:34–38, FAO. IPGRI
39. Bhuyan N, Borah BK, Sarma RN (2007) Genetic diversity analysis in traditional lowland rice (Oryza sativa L.) of Assam using RAPD and ISSR markers. Curr Sci 93:967–972
40. Das B, Sengupta S, Parida SK, Roy B, Ghosh M, et al. (2013) Genetic diversity and population structure of rice landraces from Eastern and North Eastern States of India. BMC Genetics 14:71–85.
41. Van Inghelandt D, Melchinger AE, Lebreton C, Stich B (2010) Population structure and genetic diversity in a commercial maize breeding program assessed with SSR and SNP markers. Theor Appl Genet 120: 1289−1299.
42. Varshney RK, Baum M, Guo P, Grando S, Ceccarelli S, et al. (2010) Features of SNP and SSR diversity in a set of ICARDA barley germplasm collection. Mol Breed 26: 229−242.
43. McKhann HI, Camillery C, Berard A, Bataillon T, David JL, et al. (2004) Nested core collections maximizing genetic diversity in *Arabidopsis thaliana*. Plant J 38: 193– 202.
44. Yan WG, Agrama H, Jia M, Fjelistrom R, McClung A (2010) Geographic description of genetic diversity and relationships in the USDA rice world collection. Crop Sci 50: 2406–2417.
45. Balfourier F, Roussel V, Strelchenko P, Exbrayat-Vinson F, Sourdille P, et al. (2007) A Worldwide bread wheat core collection arrayed in a 384-well plate. Theor Appl Genet 114: 1265–1275.
46. Blair MW, Diaz LM, Buendia HF, Duque MC (2009) Genetic diversity, seed size association and population structure of a core collection of common beans (*Phaseolus vulgaris* L.). Theor Appl Genet 119: 955–972.
47. van Treuren R, Tchoudinova I, van Soest LJM, van Hintum TJL (2006) Marker- assisted acquisition and core collection formation: A case study in barley using AFLPs and pedigree data. Genet Res Crop Evol 53: 43–52.
48. Zaharieva M, Santoni S, David J (2001) Use of RFLP markers to study genetic diversity and to build a core collection of the wild wheat relative Ae-geniculata Roth (= Ae-ovarta L.). Genet Sel Evol 33S: 269–288.
49. Skroch PW, Neinhuis J, Beebe S, Tohne J, Pedraza F (1998) Comparison of Mexican common bean (*Phaseolus vulgaris* L.) core and reserve germplasm collections. Crop Sci 38: 488–496.
50. Xiao-ling L, Yong-gen L, Jin-quan L, Hai-ming X, Shahid MQ (2011) Strategies on sample size determination and qualitative and quantitative traits integration to construct core collection of rice (*Oryza sativa*). Rice Sci 18: 46–55.
51. Wen W, Franco J, Chavez-Tovar VH, Yan J, Taba S (2012) Genetic Characterization of a Core Set of a Tropical Maize Race Tuxpeño for Further Use in Maize Improvement. PLoS ONE 7(3): e32626.
52. Kuroda Y, Tomooka N, Kaga A, Vaughan D A (2009) Genetic diversity of wild soybean (*Glycine soja* Sieb. et Zucc.) and Japanese cultivated soybeans [*G. max* (L.) Merr.] based on microsatellite (SSR) analysis and the selection of a core collection. Genet Resour Crop Evol 56:1045–1055.

53. Jansen J, van Hintum TJL (2007) Genetic distance sampling: a novel sampling method for obtaining core collections using genetic distances with an application to cultivated lettuce. Theor Appl Genet 114:421–428

54. Soto-Cerda B J, Diederichsen A, Ragupathy R, Cloutier S (2013) Genetic characterization of a core collection of flax (*Linum usitatissimum* L.) suitable for association mapping studies and evidence of divergent selection between fiber and linseed types. BMC Plant Biol 13: 78–91.

Permissions

All chapters in this book were first published in PLOS ONE, by The Public Library of Science; hereby published with permission under the Creative Commons Attribution License or equivalent. Every chapter published in this book has been scrutinized by our experts. Their significance has been extensively debated. The topics covered herein carry significant findings which will fuel the growth of the discipline. They may even be implemented as practical applications or may be referred to as a beginning point for another development.

The contributors of this book come from diverse backgrounds, making this book a truly international effort. This book will bring forth new frontiers with its revolutionizing research information and detailed analysis of the nascent developments around the world.

We would like to thank all the contributing authors for lending their expertise to make the book truly unique. They have played a crucial role in the development of this book. Without their invaluable contributions this book wouldn't have been possible. They have made vital efforts to compile up to date information on the varied aspects of this subject to make this book a valuable addition to the collection of many professionals and students.

This book was conceptualized with the vision of imparting up-to-date information and advanced data in this field. To ensure the same, a matchless editorial board was set up. Every individual on the board went through rigorous rounds of assessment to prove their worth. After which they invested a large part of their time researching and compiling the most relevant data for our readers.

The editorial board has been involved in producing this book since its inception. They have spent rigorous hours researching and exploring the diverse topics which have resulted in the successful publishing of this book. They have passed on their knowledge of decades through this book. To expedite this challenging task, the publisher supported the team at every step. A small team of assistant editors was also appointed to further simplify the editing procedure and attain best results for the readers.

Apart from the editorial board, the designing team has also invested a significant amount of their time in understanding the subject and creating the most relevant covers. They scrutinized every image to scout for the most suitable representation of the subject and create an appropriate cover for the book.

The publishing team has been an ardent support to the editorial, designing and production team. Their endless efforts to recruit the best for this project, has resulted in the accomplishment of this book. They are a veteran in the field of academics and their pool of knowledge is as vast as their experience in printing. Their expertise and guidance has proved useful at every step. Their uncompromising quality standards have made this book an exceptional effort. Their encouragement from time to time has been an inspiration for everyone.

The publisher and the editorial board hope that this book will prove to be a valuable piece of knowledge for researchers, students, practitioners and scholars across the globe.

List of Contributors

Yilin Li and Weiming Shi
State Key Laboratory of Soil and Sustainable Agriculture, Institute of Soil Science, Chinese Academy of Sciences, Nanjing, China

Xingxiang Wang
State Key Laboratory of Soil and Sustainable Agriculture, Institute of Soil Science, Chinese Academy of Sciences, Nanjing, China
Jiangxi Key Laboratory of Ecological Research of Red Soil, Yingtan, Jiangxi, China

Peng Zhang
State Key Laboratory for Conservation and Utilization of Subtropical Agro-bioresources, South China Agricultural University, Guangzhou, China
State Key Laboratory of Rice Biology, China National Rice Research Institute, Hangzhou, China

Xiangdong Liu and Yonggen Lu
State Key Laboratory for Conservation and Utilization of Subtropical Agro-bioresources, South China Agricultural University, Guangzhou, China

Hanhua Tong
State Key Laboratory of Rice Biology, China National Rice Research Institute, Hangzhou, China

Jinquan Li
State Key Laboratory for Conservation and Utilization of Subtropical Agro-bioresources, South China Agricultural University, Guangzhou, China
Department of Plant Breeding and Genetics, Max Planck Institute for Plant Breeding Research, Cologne, Germany

Qichao Zhao, Minghong Liu, Miaomiao Tan and Zhicheng Shen
State Key Laboratory of Rice Biology, Institute of Insect Sciences, Zhejiang University, Hangzhou, China

Jianhua Gao
College of Life Science, Shanxi Agricultural University, Taigu, China

Marta Brozynska, Agnelo Furtado and Robert James Henry
Queensland Alliance for Agriculture and Food Innovation, The University of Queensland, Brisbane, Queensland, Australia

Jing Zhang and Qifa Zhou
College of Life Sciences, Zhejiang University, Hangzhou, China

Wenjiang Huang
Key Laboratory of Digital Earth Science, Institute of Remote Sensing and Digital Earth, Chinese Academy of Sciences, Beijing, China

Shalabh Dixit, Ma. Teresa Sta Cruz, Paul T. Maturan, Jhon Christian E. Ontoy and Arvind Kumar
International Rice Research Institute (IRRI), Los Baños, Laguna, Philippines

B. Emma Huang
Computational Informatics, CSIRO, Dutton Park, Queensland, Australia

Gia Khuong Hoang Hua, Lien Bertier, Saman Soltaninejad and Monica Höfte
Laboratory of Phytopathology, Department of Crop Protection, Faculty of Bioscience Engineering, Ghent University, Gent, Belgium

Helga Fernández, Nicolás Prandoni, Eduardo Díaz and Manuel Carmona
Department of Environmental Biology, Centro de Investigaciones Biológicas-CSIC, Madrid, Spain

Mercedes Fernández-Pascual, Susana Fajardo and César Morcillo
Plant Protection Department, Instituto de Ciencias Agrarias-CSIC, Madrid, Spain

Arijit Ganguly, Ranita Chakravorty, Dipak K. Mandal and Parimalendu Haldar
Department of Zoology, Visva-Bharati University, Santiniketan, West Bengal, India

Angshuman Sarkar
Department of Statistics, Visva-Bharati University, Santiniketan, West Bengal, India

Julieta Ramos-Elorduy and Jose Manuel Pino Moreno
Departamento de Zoología, Universidad Nacional Autonoma de Mexico, Mexico City, Distrito Federal, México

Jun Wang, Nicholas C. Marowsky and Chuanzhu Fan
Department of Biological Sciences, Wayne State University, Detroit, Michigan, United States of America

Danying Wang, Song Chen, Chenglin Ji, Chunmei Xu and Xiufu Zhang
State Key Laboratory of Rice Biology, China National Rice Research Institute, Hangzhou, Zhejiang, China

Zaiman Wang
Key Laboratory of Key Technology on Agricultural Machine and Equipment, South China Agricultural University, Guangzhou, Guangdong, China

Bhagirath Singh Chauhan
Queensland Alliance for Agriculture and Food Innovation, University of Queensland, Queensland, Australia

Yu Han, Jiarong Meng, Jie Chen, Wanlun Cai, Yu Wang, Jing Zhao, Yueping He and Hongxia Hua
Hubei Insect Resources Utilization and Sustainable Pest Management Key Laboratory, College of Plant Science and Technology, Huazhong Agricultural University, Wuhan, P.R. China

Yanni Feng
College of Life Science and Technology, Huazhong Agricultural University, P.R. China

Hiroshi Wada, Chisato Masumoto-Kubo, Satoshi Morita and Koichi Tsutsumi
Kyushu Okinawa Agricultural Research Center, National Agriculture and Food Research Organization, Chikugo, Fukuoka, Japan

Yousef Gholipour and Hiroshi Nonami
Department of Biomechanical Systems, Faculty of Agriculture, Ehime University, Matsuyama, Ehime, Japan

Fukuyo Tanaka
Agricultural Research Center, National Agriculture and Food Research Organization, Tsukuba, Ibaraki, Japan

Rosa Erra-Balsells
Department of Organic Chemistry-CIHIDECAR, Faculty of Exact and Natural Sciences, University of Buenos Aires, Buenos Aires, Argentina,

Kenzo Hiraoka
Clean Energy Research Center, The University of Yamanashi, Kofu, Yamanashi, Japan

Mehanathan Muthamilarasan, Rohit Khandelwal, Chandra Bhan Yadav, Venkata Suresh Bonthala, Yusuf Khan and Manoj Prasad
National Institute of Plant Genome Research, New Delhi, India

Muhammad Tariq Rafiq
Ministry of Education Key Laboratory of Environmental Remediation and Ecological Health, College of Environmental and Resource Sciences, Zhejiang University, Hangzhou, China
Department of Environmental Science, International Islamic University, Islamabad, Pakistan

Rukhsanda Aziz, Xiaoe Yang, Wendan Xiao and Tingqiang Li
Ministry of Education Key Laboratory of Environmental Remediation and Ecological Health, College of Environmental and Resource Sciences, Zhejiang University, Hangzhou, China

Peter J. Stoffella
Indian River Research and Education Center, Institute of Food and Agricultural Sciences, University of Florida, Fort Pierce, Florida, United States of America

Aamir Saghir
Institute of Statistics, Zhejiang University, Hangzhou, China

Muhammad Azam
College of Agriculture and Biotechnology, Zhejiang University, Hangzhou, China

Nuri Company, Anna Nadal, Cristina Ruiz and Maria Pla
Institute for Food and Agricultural Technology, University of Girona, Girona, Spain

Michio Murakami and Taikan Oki
Institute of Industrial Science, The University of Tokyo, Meguro, Tokyo, Japan
Japan Science and Technology Agency, Core Research for Evolutionary Science and Technology (CREST), Chiyoda, Tokyo, Japan

Christine D. Zanke, Jie Ling, Marion S. Röder and Kerstin Neumann
Leibniz Institute of Plant Genetics and Crop Plant Research (IPK), Gatersleben, Germany

Jörg Plieske and Martin W. Ganal
TraitGenetics GmbH, Gatersleben, Germany

Sonja Kollers, Erhard Ebmeyer and Viktor Korzun
KWS LOCHOW GMBH, Bergen, Germany

Odile Argillier
Syngenta France S.A.S., Orgerus, France

Gunther Stiewe and Maike Hinze
Syngenta Seeds GmbH, Bad Salzuflen, Germany

Chenjia Shen, Yanjun Yang, Huizhong Wang and Tao Sun
College of Life and Environmental Sciences, Hangzhou Normal University, Hangzhou 310036, China

Runqing Yue and Shuanggui Tie
Henan Academy of Agricultural Sciences, Zhengzhou 450002, China,

Lei Zhang
Department of Plant Pathology, Washington State University, Pullman, WA 99164-6430, United States of America

Debjani Roy Choudhury, Nivedita Singh, Amit Kumar Singh, Sundeep Kumar and Rakesh Singh
Division of Genomic Resources, National Bureau of Plant Genetic Resources, New Delhi, 110 012, India

Kalyani Srinivasan and R. K. Tyagi
Division of Germplasm Conservation, National Bureau of Plant Genetic Resources, New Delhi, 110 012, India

Altaf Ahmad
Department of Botany, Faculty of Science, Jamia Hamdard (Hamdard University), New Delhi, 110062, India

N. K. Singh
National Research Centre on Plant Biotechnology, Indian Agricultural Research Institute, New Delhi, 110 012, India

Index